Advances in
Carbohydrate Chemistry and Biochemistry

Volume 27

Advances in Carbohydrate Chemistry and Biochemistry

Editors

R. STUART TIPSON

DEREK HORTON

Board of Advisors

L. ANDERSON   D. FRENCH   W. W. PIGMAN   W. J. WHELAN
ROY L. WHISTLER

Board of Advisors for the British Commonwealth

A. B. FOSTER   SIR EDMUND HIRST   J. K. N. JONES   MAURICE STACEY

Volume 27

ACADEMIC PRESS   New York and London   **1972**

COPYRIGHT © 1972, BY ACADEMIC PRESS, INC.
ALL RIGHTS RESERVED.
NO PART OF THIS PUBLICATION MAY BE REPRODUCED OR
TRANSMITTED IN ANY FORM OR BY ANY MEANS, ELECTRONIC
OR MECHANICAL, INCLUDING PHOTOCOPY, RECORDING, OR ANY
INFORMATION STORAGE AND RETRIEVAL SYSTEM, WITHOUT
PERMISSION IN WRITING FROM THE PUBLISHER.

ACADEMIC PRESS, INC.
111 Fifth Avenue, New York, New York 10003

United Kingdom Edition published by
ACADEMIC PRESS, INC. (LONDON) LTD.
24/28 Oval Road, London NW1

LIBRARY OF CONGRESS CATALOG CARD NUMBER: 45-11351

PRINTED IN THE UNITED STATES OF AMERICA

CONTENTS

LIST OF CONTRIBUTORS vii
PREFACE ix

William Werner Zorbach (1916–1970)

G. A. JAMIESON

Text 1

Proton Magnetic Resonance Spectroscopy: Part I

BRUCE COXON

I. Introduction 7
II. Instrumentation 11
III. Applications of Superconducting Solenoids 18
IV. Fourier-transform, Nuclear Magnetic Resonance Spectroscopy 43
V. Analysis of Spectra 62

Non-aqueous Solvents for Carbohydrates

C. J. MOYE

I. Introduction 85
II. Solubility and Solvation 86
III. Inorganic Solvents 90
IV. Organic Solvents 96
V. Special Applications 123

Sugars Specifically Labeled with Isotopes of Hydrogen

J. E. G. BARNETT AND D. L. CORINA

I. Introduction 128
II. Preparation 128
III. Radiochemical and Chemical Stability 138
IV. Localization 140
V. Physical Properties and Their Applications 147
VI. Applications in Mechanistic Chemistry 151
VII. Applications in Mechanistic Biochemistry 155
VIII. General Applications in Biochemistry 176
IX. Tables of Known Sugars Specifically Labeled with Isotopes of Hydrogen . . . 181

The Use of Carbohydrates in the Synthesis and Configurational Assignments of Optically Active, Non-carbohydrate Compounds

T. D. INCH

I. Introduction 191
II. Use of Carbohydrates in Asymmetric Synthesis 192
III. Carbohydrates as Sources of Asymmetric Carbon Atoms for the Synthesis and Proof of Configuration of Biologically Important, Non-carbohydrate Compounds 205
IV. Stereoselective Synthesis of Asymmetric Sulfoxides 222

The Wittig Reaction in Carbohydrate Chemistry

YU. A. ZHDANOV, YU. E. ALEXEEV, AND V. G. ALEXEEVA

I. Introduction 227
II. General Considerations 228
III. Types of Final Products and Their Utilization in Synthesis 232
IV. Anomalous Wittig Reaction with Free and with Partially Protected Sugars . . 284
V. Conclusion 292
VI. Further Developments 292

Glycoenzymes: Enzymes of Glycoprotein Structure

JOHN H. PAZUR AND N. N. ARONSON, JR.

I. Introduction 301
II. Types of Glycoenzymes 305
III. Chemical and Physical Structure 309
IV. Biosynthetic Pathways 328
V. Biological and Structural Significance 337

AUTHOR INDEX FOR VOLUME 27 343
SUBJECT INDEX FOR VOLUME 27 362
CUMULATIVE AUTHOR INDEX FOR VOLUMES 1–27 384
CUMULATIVE SUBJECT INDEX FOR VOLUMES 1–27 393
ERRATA 405

LIST OF CONTRIBUTORS

Numbers in parentheses indicate the pages on which the authors' contributions begin.

YU. E. ALEXEEV, *Department of Chemistry, The Rostov State University, Rostov-on-Don, U.S.S.R.* (227)

V. G. ALEXEEVA, *Department of Chemistry, The Rostov State University, Rostov-on-Don, U.S.S.R.* (227)

N. N. ARONSON, JR., *Department of Biochemistry, The Pennsylvania State University, University Park, Pennsylvania 16802* (301)

J. E. G. BARNETT, *Department of Physiology and Biochemistry, University of Southampton, Southampton S09 5NH, England* (127)

D. L. CORINA, *Department of Physiology and Biochemistry, University of Southampton, Southampton SO9 5NH, England* (127)

BRUCE COXON, *National Bureau of Standards, Washington, D.C. 20234* (7)

T. D. INCH, *Chemical Defence Establishment, Porton Down, Salisbury, Wiltshire, England* (191)

G. A. JAMIESON, *American National Red Cross, Blood Research Laboratory, Bethesda, Maryland 20014* (1)

C. J. MOYE, *Cottee's General Foods Ltd., P.O. 520, Crow's Nest, N.S.W. 2065, Sydney, Australia* (85)

JOHN H. PAZUR, *Department of Biochemistry, The Pennsylvania State University, University Park, Pennsylvania 16802* (301)

YU. A. ZHDANOV, *Department of Chemistry, The Rostov State University, Rostov-on-Don, U.S.S.R.* (227)

PREFACE

The twenty-seventh volume in this serial publication presents the first part of a two-part article by Coxon (Washington) on the proton magnetic resonance spectroscopy of carbohydrates that emphasizes important recent advances and updates the article by Hall in Volume 19. Moye (Sydney) provides a useful account of non-aqueous solvents for carbohydrates that brings together in one place a great deal of widely scattered information; his contribution should prove especially useful for the technologist, but the fundamental investigator will also find much valuable information from a seldom-stressed viewpoint. The preparation, properties, and uses of sugars specifically labeled with isotopes of hydrogen are capably discussed by Barnett and Corina (Southampton); such compounds provide a most important tool for understanding organic and biochemical reaction-mechanisms of the sugars. Inch (Salisbury) delineates the use of carbohydrates in the synthesis and configurational assignments of optically active, non-carbohydrate compounds, some of which have powerful physiological effects as cholinergic and other drugs. Zhdanov, Alexeev, and Alexeeva (Rostov-on-Don) contribute a comprehensive account of applications of the Wittig reaction in carbohydrate chemistry; this exceptionally versatile reaction will undoubtedly play a key role in much future synthetic work in the field.

It is, perhaps, not widely recognized that many enzymes, especially the hydrolases, possess a covalently-linked carbohydrate moiety; in this Volume, Pazur and Aronson (University Park, Pennsylvania) bring into focus this aspect of enzyme structure by presenting a discussion of such glycoenzymes, that is, enzymes having a glycoprotein structure. The obituary in this Volume, by Jamieson (Washington), pays tribute to William Werner Zorbach, a scientist remembered as an outstanding teacher and a gifted experimentalist in the chemistry of carbohydrates.

The Subject Index was prepared by Dr. L. T. Capell.

Kensington, Maryland R. STUART TIPSON
Columbus, Ohio DEREK HORTON
October, 1972

1916–1970

WILLIAM WERNER ZORBACH

1916–1970

Bill Zorbach combined the qualities of an outstanding teacher, a gifted experimentalist, and a warm and cheerful friend. He could transmit to his students his enthusiasm, to his research associates his experimental insight, and to his friends his sincerity. He was a man without pretence, to himself or to others.

William Werner Zorbach was born in Sandusky, Ohio, on June 15, 1916, and died in Houghton, Michigan, of a stroke, on June 28, 1970. Although his active research life was limited to only twenty years, he made significant contributions to the chemistry of such important pharmacological agents as cardiac glycosides, antimalarial drugs, and carcinostatic agents, which greatly expanded our knowledge of structure–activity relationships.

His family was of Swiss origin and the name was supposedly a contraction of a form "Zu und vor Bach" (to-and-from the brook), adopted by an ancestor when surnames became necessary. Zorbach's father was a master carpenter, but did not want his son to become one; this resulted in his life-long ineptitude in the domestic aspects of carpentry. However, in the light of his early interests, his father built him, at the age of twelve or thirteen, a laboratory in the basement of their home. The occasional explosions not only blew a few holes in the ceiling, but certainly affected the equanimity of the music students to whom his mother gave piano lessons.

One of the explosions had serious consequences. At the age of fourteen, Zorbach was discharging dynamite caps by means of a relay system, and, when he went to investigate one that had not gone off, the resulting explosion necessitated the amputation of the thumb and the first finger to the first joint of his left hand. Despite this handicap, Zorbach decided that he wished to learn to play the piano, and he did so exceptionally well. In his later life, he usually played at least an hour a day, being particularly fond of the Bach Preludes and Fugues. At this time, he also started a permanent record collection that finally included all the recorded works of Mozart and all the recorded piano and organ works of Bach.

It was during his high-school period that Zorbach became acquainted with Dr. Norbert Lange, the author of Lange's *Handbook of Chemistry* which was published in Sandusky; and it was Lange who helped to direct his interests toward science. This friendship was maintained over the years ending with Zorbach's death, which was followed about a month later by that of Lange himself.

Following graduation from high school in the class of 1934, at the height of the Depression, he enrolled for courses in Mechanical Engineering at Bowling Green State University, but soon decided that this was not his field, and left after one academic year to work as an assistant traffic engineer (freight) at the Hinde and Dauche Paper Company until 1941, and then in the water purification section of the TNT division of the Trojan Power Company.

Zorbach served as a Corporal in the U. S. Army Engineers from 1943 to 1945, spending a year on Guam. It was during this period that his first wife, Ruth Hastings Zorbach, whom he had married in 1942, died.

Following his discharge from the Army in 1945, he returned to Bowling Green State University under the GI Bill, and graduated in Chemistry in two years, having attended summer school, as well as regular sessions, throughout this period. His first scientific paper, on the isolation of essential oils, arose from the modification of an experiment performed in his first organic chemistry course as an undergraduate. He was active in numerous scholastic honor societies, and was President of the Chemistry Club. At the end of his junior year, he married a fellow chemistry student, Miss Betty Canfield, who delayed her own graduation from June of 1947 to August so that they could graduate together. Their daughter, Judy, was born in 1954.

Following their graduation from Bowling Green, the Zorbachs went to Montreal, where he enrolled in Chemistry at McGill University, but, following his qualifying year, transferred to Biochemistry, where he received his Ph. D. in 1951 under the mentorship of Dr. R. D. H. Heard with a thesis on "Cholesterol-4-C^{14}—Synthesis and Studies in the Cholestane Series." This dual allegiance to chemistry and biochemistry was to be characteristic of Dr. Zorbach's work and teaching. At that time, Heard's health was failing, and the graduate students were forced into a close-knit group in which each student bolstered the knowledge and enthusiasm of the others. This was particularly necessary, because they were embarking on what was then a novel field in biochemistry, the tagging of steroid hormones with radiocarbon. Zorbach's project was particularly difficult, in that he was to open ring B, include ^{14}C in appropriate intermediate structures, and then cyclize to the labeled analogs. Despite the difficulties and disappointments, his own enthusiasm did not flag, nor did he allow others to become dispirited over their work, so that, by their mutual help, almost the whole group of students received their doctorates at the same convocation.

He returned briefly to Bowling Green State University in 1951–52, but was appointed Assistant Professor of Chemistry at Georgetown University, Washington, D. C., in 1952. Following a period as Research Chemist at the National Institute of Arthritis and Metabolic Diseases from 1956–58, Zorbach returned to Georgetown as Associate Professor in 1958, was Acting Chairman in 1959, and was promoted to full Professor in 1964. Dr. Zorbach became increasingly disenchanted with certain aspects of the academic life, and, in 1967, he accepted the position of Director of the Division of Bio-Organic Chemistry in the newly established Gulf South Research Institute in Louisiana. However, in this purely research capacity, he missed the contact with students and the stimulation of teaching, and, in 1969, accepted an appointment as Professor in the Michigan Technological University.

Professor Zorbach's work in carbohydrate chemistry was characterized by meticulous attention to detail, and emphasis on careful characterization and optimal yields at every step of multistep procedures. The importance he placed on the characterization of crystalline intermediates reflects the earlier periods of carbohydrate chemistry, before chromatographic techniques had simplified this problem. One of the cornerstones of Zorbach's work was his realization that p-nitrobenzoic esters, which were known to be excellent derivatives for the characterization of simple alcohols, might be equally suitable for the preparation of crystalline O-acylglycosyl halides. Thus, 2,6-dideoxy-D-*ribo*-hexose (D-digitoxose) was converted in excellent yield into its 1,3,4-tris-(p-nitrobenzoate) and thence, by means of hydrogen bromide, into the 1-bromide; this was found to be a highly reactive, crystalline solid which was, however, fairly stable under anhydrous conditions. This bromo derivative was then coupled directly to digitoxigenin, to yield a monodigitoxoside that showed a definite digitalis-like action despite the fact that it possessed an unnatural, α-D-glycosidic linkage, whereas the natural cardenolides are invariably β anomers. However, the synthesis was subsequently modified to provide both of the anomers, and the use of acylated 1-O-(p-nitrobenzoyl)aldoses for the preparation of acylated glycosyl halides was rapidly adopted by many other workers.

Because a major aim of Zorbach's work was to ascertain the effect of subtle structural changes in the carbohydrate components of cardenolides on their cardiotonic activity, he and his associates devised methods for the synthesis of 2,6-dideoxy-D-*arabino*-hexose, the C-3 epimer of digitoxose, and improved the methods then used for the synthesis of "2-deoxy-D-allose" and L-rhamnose. This last sugar was then used for the preparation of evomonoside (digitoxigenin α-L-rhamnopyranoside), thus confirming by synthesis the structure that had been assigned on the basis of degradative studies. This work was later extended to the synthesis of α- and β-D-rhamnosides of digitoxigenin and strophanthidin, the latter being the "unnatural" anomer. The β-D-rhamnoside did prove to have enhanced cardiotonic activity, but it was found to be less than that of the naturally occurring α-L-rhamnosides. In this vein, Zorbach and his colleagues synthesized 3-β-(2-deoxy-β-D-*lyxo*-hexopyranosyl)oxy-14β-hydroxy-5β-card-20(22)-enolide, because this has an axially attached 4′-hydroxyl group, unlike all other previously tested cardenolides, in which all of the hydroxyl groups were equatorially attached. At the same time, they synthesized 2-deoxy-3,4,6-tri-O-(*p*-nitrobenzoyl)-α-D-*lyxo*-hexosyl bromide, but this proved not to be useful in the synthesis of nucleosides, since, for unknown reasons, it failed to condense with dialkoxypyrimidines.

As evidence had been accumulating that suggested that deoxygenation at a carbon atom in the carbohydrate component of the glycoside leads to a loss in cardiotonic potency, Zorbach and his group set out to change the ω-methyl group to a hydroxymethyl group in the carbohydrate component of the potent convallatoxin (3-*O*-α-L-rhamnopyranosylstrophanthidin). The product "6-hydroxyconvallatoxin" (3-*O*-α-L-mannosylstrophanthidin) showed a molar potency 12% greater than that of convallatoxin, and is, therefore, the most potent of all known cardenolides.

Further work in the field of carbohydrate syntheses led Zorbach and his coworkers to the first crystalline methyl glycofuranosides and per-O-acylglycofuranosyl halides of 2-deoxyhexoses, to the first 2-deoxyhexofuranosyl nucleoside, and to the previously unknown 1,3-dideoxy-D-*erythro*-hexulose as a possible intermediate in the synthesis of 1′-methyl 2′-deoxynucleosides. In the course of this work, the use of *p*-nitrobenzoic esters was applied to the synthesis of crystalline 2-deoxy-3,5,6-tri-O-(*p*-nitrobenzoyl)-D-*ribo*-hexosyl bromide, and of the four methyl glycosides of 2-deoxy-D-*ribo*-hexose, as precursors of "unnatural" nucleosides containing the 2-deoxy-β-D-*ribo*-hexofuranosyl group. The interest in these modified nucleosides stemmed from the observation by Langen and Etzold that "2-deoxyglucosylthymine," which Zorbach had described earlier, is a powerful and specific inhibitor of pyrimidine nucleoside phosphorylase in Ehrlich ascites tumor cells, and enhances the incorporation of $2',5'$-dideoxy-$5'$-iodouridine into DNA. Zorbach and coworkers were able to show that this effect is also caused by the corresponding uracil derivatives that they prepared, although the *ribo* analogs are without effect.

Quite apart from these experimental advances, Zorbach made a major contribution to nucleoside chemistry in conceiving the idea of a series of volumes devoted to "Synthetic Procedures in Nucleic Acid Chemistry" and, together with R. S. Tipson as co-editor, bringing to final form a first volume that is indispensable to all workers in this area, as it assembles and summarizes the optimal methods that had been developed up to 1968. The manuscripts for the second volume, describing physical and physicochemical techniques used in the characterization and structural determination of nucleosides, nucleotides, and related compounds were sent to the publisher just a few months before Zorbach's death.

It is difficult to make an objective assessment of a teacher. In focusing on Zorbach's research achievements, it should be noted that his strong commitment to teaching caused him to return to Georgetown University in 1959, after an interim at the National Institutes of Health, and to the Michigan Technological University in 1968, after his period at the Gulf South Research Institute. As a teacher, he was clear and lucid, and extremely conscientious in the preparation of his notes and lectures. In a letter written to a friend a month before he died, Zorbach said "almost every evening has been taken up with class preparations, but I enjoy being back in academic life and I like the students."

For their part, his students liked him, because he conducted his classes at a fast but relaxed pace; informal and amusing, but at the same time demanding, he was able to impart to them a sense of his seriousness and his enthusiasm for his subject. He appreciated the give-and-take of teaching, his mastery of his topic, and the responsiveness of the students. Even in his examinations, he would try to lighten the tension by asking such true-or-false questions as "The cantomeric forms of D- and L-frugose can be resolved by transannulation."

His graduate students came to look on him more as a personal friend than as an academic mentor. Zorbach was, himself, an extremely hard worker, and he demanded from each a full day's work, but he was available at any time for advice and consultation.

He was an experimentalist in the true sense of the word, instrumental techniques and methods of analysis being merely a means to an end, rather than ends in themselves. To his graduate students, he was able to transmit his sense of the delights and satisfactions of structural determination and synthesis, and the experimental meticulousness necessary for their achievement; and his delight in the first crystallization of a new compound was as obvious and sincere as that of the student himself. In his memory, his friends and colleagues have established the Zorbach Memorial Prize, to be awarded annually for the best doctoral thesis in chemistry at Georgetown University.

If teaching and research were the twin enthusiasms of Zorbach's professional life, they were, nevertheless, not his whole life. His interest in music and his skill as a pianist dated, as already noted, from his high-school days. He was a skilled gardener, and an authority on the *Amyrillidaceae*; in fact, he published four papers on this family, one with his daughter in the "Amaryllis Year Book of 1968." He was a lover of good wine and good food, and could, himself, produce some memorable Indian meals with the curries and sauces prepared from recipes collected from his former students. He enjoyed the companionship of a succession of white Persian cats, but never forgot his love for his first one.

Bill Zorbach was a complex man with a great love of life that touched all who met him. His crew-cut hair and his mischievous blue eyes gave him a puckish, youthful appearance, with a ready smile that had nothing scholarly about it. To most, he will be remembered for his contributions in his chosen field; to a generation of students at Georgetown University he will be remembered as a gifted teacher; and by those of us who came to know him, he will be remembered as a friend.

G. A. JAMIESON

During the course of his scientific career, Professor Zorbach co-authored articles with the following scientists: J. Al-Kassir, Mrs. C. C. Bhat (C. C. de Kont), K. V. Bhat, W. Bühler, J. P. Ciaudelli, S. L. DeBernardo, G. Durr, W. H. Gilligan, N. Henderson, D. V. Kashelikar, H. R. Munson, A. P. Ollapally, T. A. Payne, G. Pietsch, S. Saeki, C. G. Salem, C. R. Tamorria, C. O. Tio, R. S. Tipson, G. Valiaveedan, J. E. Weber, and B. F. West. To several of these, and to Dr. Clair H. Yates, I am indebted for their reminiscences of their association with Dr. Zorbach, and to Mrs. Zorbach for her assistance and cooperation during the preparation of this tribute to her husband.

PROTON MAGNETIC RESONANCE SPECTROSCOPY: PART I

By Bruce Coxon

National Bureau of Standards, Washington, D. C.

I. Introduction 7
II. Instrumentation 11
   1. General 11
   2. Magnets and Probes.. 11
   3. Spectrometer Consoles 13
   4. Field–frequency Stabilization 14
   5. Signal-averaging Techniques 16
   6. Automatic Control and Data Acquisition........................... 17
III. Applications of Superconducting Solenoids........................... 18
   1. General Techniques and Advantages................................ 18
   2. Monosaccharides... 21
   3. Oligosaccharides.. 30
   4. Nucleosides and Nucleotides 33
   5. Polysaccharides 39
IV. Fourier-transform, Nuclear Magnetic Resonance Spectroscopy 43
   1. General Considerations.. 43
   2. Distinction of Continuous-wave and Pulsed Techniques 44
   3. Pulse Methods... 45
   4. Acquisition of the Free-induction, Decay Signal.................. 49
   5. Digital Processing of the Free-induction, Decay Signal........... 50
   6. Fast, Fourier Transformation 52
   7. Phase Correction.. 54
   8. Noise-stimulated Resonance....................................... 55
   9. Applications of Fourier-transform Techniques 56
V. Analysis of Spectra 62
   1. Non-equivalence of Nuclei in Carbohydrates....................... 62
   2. The Limitations of First-order Analysis 66
   3. Manual Analysis of Spectra 71
   4. Computerized, Iterative Analysis of Spectra 73

I. INTRODUCTION

Proton magnetic resonance (p.m.r.) spectroscopy is now firmly established as the most widely used technique for the structural, configurational, and conformational analysis of carbohydrates and their derivatives. Much of the earlier work on the application of p.m.r. spectroscopy to carbohydrates was inspired by the pioneering efforts of Lemieux and coworkers,[1] and, later, by Hall.[2] Lemieux's prediction[3] that an almost endless number of applications of high-resolution, nuclear magnetic resonance (n.m.r.) spectroscopy could be envisaged for gaining information on the immediate environment of atoms in organic molecules is steadily being fulfilled.

The first Chapter on nuclear magnetic resonance[2] in this Series was devoted principally to p.m.r. spectroscopy, because, up to 1964, virtually no magnetic resonance studies of other nuclei in carbohydrates and their derivatives had been made. The present Chapter is also concerned mainly with the p.m.r. technique, in the expectation that the broad subject of nuclei other than protons will be treated separately in this Series.

The progress of n.m.r. spectroscopy and of its applications to carbohydrates between 1964 and 1972 has been rapid and exciting. In each succeeding year, there have been major developments in instrumentation, applications, or technique. During this time, the magnetic field-strengths available have tripled, and there have been introduced such significant developments as: high-resolution, superconducting magnets; automatic, internal field–frequency and field–gradient stabilization; Fourier-transform techniques; determinations of relative signs of coupling constants; internuclear, double- and triple-resonance techniques; nuclear Overhauser effects; conformational equilibria determined by variable-temperature techniques; chemical-shift reagents; and many more. The full benefit of digital computers, both large and small, is becoming felt in the areas of spectral analysis and interpretation, the automatic control of spectrometers, and the acquisition and processing of data.

Owing to the great impact that p.m.r. spectroscopy has made on structural carbohydrate chemistry, it is not feasible to discuss here all of the reports in which this technique has been used merely to confirm assigned structures. In the present Chapter, some selectivity has been exercised in favor of novel structural and conformational applications, innovative techniques, information that could not be obtained by other methods, and a fundamental understanding of the magnetic parameters of carbohydrate molecules.

(1) R. U. Lemieux, R. K. Kullnig, H. J. Bernstein, and W. G. Schneider, *J. Amer. Chem. Soc.*, **80**, 6098 (1958).
(2) L. D. Hall, *Advan. Carbohyd. Chem.*, **19**, 51 (1964).
(3) R. U. Lemieux, *Trans. Roy. Soc. Can.*, **52**, 31 (1958).

Many texts[4-20] and annual[21,22] and biennial[23] reviews devoted exclusively to the general subject of magnetic resonance spectroscopy have appeared since 1964, as well as two new periodicals.[24,25] N.m.r. spectroscopy has also been reviewed regularly in the *Annual Review of Physical Chemistry*, *Annual Reports of the Chemical Society*, and, somewhat irregularly, in *Determination of Organic Structure by Physical Methods*.
The application of p.m.r. spectroscopy to deoxy sugars,[26] the con-

(4) J. W. Emsley, J. Feeney, and L. H. Sutcliffe, "High Resolution Nuclear Magnetic Resonance Spectroscopy," Pergamon Press, Oxford, 1965, Vol. 1; 1966, Vol. 2.
(5) R. H. Bible, Jr., "Interpretation of Nuclear Magnetic Resonance Spectra," Plenum Press, New York, 1965.
(6) "Nuclear Magnetic Resonance in Chemistry," B. Pesce, ed., Academic Press, New York, 1965.
(7) D. Chapman and P. D. Magnus, "Introduction to Practical High Resolution Nuclear Magnetic Resonance Spectroscopy," Academic Press, London, 1966.
(8) I. V. Aleksandrov, "The Theory of Nuclear Magnetic Resonance," translated by Scripta Technica, Inc., Academic Press, New York, 1966.
(9) P. L. Corio, "Structure of High-Resolution NMR Spectra," Academic Press, New York, 1966.
(10) G. Mavel, "Théories Moléculaires de la Résonance Magnétique Nucléaire, Applications à la Chimie Structurale," Dunod, Paris, 1966.
(11) "Nuclear Magnetic Resonance for Organic Chemists," D. W. Mathieson, ed., Academic Press, London, 1967.
(12) H. G. Hecht, "Magnetic Resonance Spectroscopy," Wiley, New York, 1967.
(13) "Magnetic Resonance in Biological Systems," A. Ehrenberg, B. G. Malmström, and T. Vängård, eds., Pergamon Press, Oxford, 1967.
(14) "Recent Developments of Magnetic Resonance in Biological Systems," S. Fujiwara and L. H. Piette, eds., Hirokawa, Tokyo, 1968.
(15) J. D. Memory, "Quantum Theory of Magnetic Resonance Parameters," McGraw-Hill, New York, 1968.
(16) F. A. Bovey, "Nuclear Magnetic Resonance Spectroscopy," Academic Press, New York, 1969.
(17) E. D. Becker, "High Resolution NMR," Academic Press, New York, 1969.
(18) R. M. Lynden-Bell and R. K. Harris, "Nuclear Magnetic Resonance Spectroscopy," Nelson, London, 1969.
(19) L. M. Jackman and S. Sternhell, "Application of Nuclear Magnetic Resonance Spectroscopy in Organic Chemistry," Academic Press, New York, 1969.
(20) B. I. Ionin and B. A. Ershov, "NMR Spectroscopy in Organic Chemistry," Plenum Press, New York, 1970.
(21) *Progr. Nuc. Magn. Resonance Spectrosc.*, **1-7** (1966-1971).
(22) *Annu. Rev. NMR Spectrosc.*, **1** (1968), **2** (1969), now *Annu. Rep. NMR Spectrosc.*, **3** (1970), **4** (1971).
(23) *Advan. Magn. Resonance*, **1-5** (1965, 1966, 1968, 1970, and 1971).
(24) *J. Magn. Resonance*, **1** (1969), *et seq.*
(25) *Org. Magn. Resonance*, **1** (1969), *et seq.*
(26) L. D. Hall and J. F. Manville, *Advan. Chem. Ser.*, **74**, 228 (1968).

figurational analysis[27] of carbohydrates, and the biochemistry of biopolymers[28] has been described, and Inch has provided an excellent general survey[29] of the p.m.r. spectroscopy of carbohydrates for the period 1965–1967. Other, less extensive, general reviews[30–32a] have appeared, as has also a modern text on the stereochemistry of carbohydrates.[33] Methods of using p.m.r. spectroscopy to determine carbohydrate conformations have been discussed,[34] as well as the conformational results obtained by these and other methods.[35]

The value of the n.m.r. technique lies to a large extent in the fact that the spectra obtained are usually readily amenable to direct, visual interpretation, in terms of specific structural or conformational features of the intact molecule, because the energy of the n.m.r. transitions is too low to permit molecular degradation or even conformational change. Depending on the needs of the problem at hand, or on the time and equipment available, further interpretations may then be made at one of many different levels of complexity, ranging from simple, first-order analysis to computerized, iterative analysis and the detailed assignment of energy-level diagrams. One drawback of n.m.r. spectroscopy is that it is a somewhat insensitive technique, owing to the very small difference in the spin populations of ground and excited states. Considerable progress in the enhancement of sensitivity has now been made, particularly by use of the pulse, Fourier-transform technique, which promises to revolutionize the acquisition of n.m.r. spectra, and which may also soon make the nuclear spin–lattice and spin–spin relaxation times (T_1 and T_2) as widely available for the purpose of molecular characterization as are chemical shifts and coupling constants. Accordingly, appreciable attention is given in this Chapter to the Fourier technique, because of its current

(27) R. J. Ferrier, *Progr. Stereochem.*, **4**, 43 (1969).
(28) J. J. M. Rowe, J. Hinton, and K. L. Rowe, *Chem. Rev.*, **70**, 42 (1970).
(29) T. D. Inch, *Annu. Rev. NMR Spectrosc.*, **2**, 35 (1969).
(30) R. D. Guthrie, R. J. Ferrier, and M. J. How, *Spec. Period. Repts. Carbohyd. Chem.* (London), **1**, 180 (1968).
(31) R. D. Guthrie, R. J. Ferrier, M. J. How, and P. J. Somers, *Spec. Period. Repts. Carbohyd. Chem.* (London), **2**, 189 (1969).
(32) R. D. Guthrie, R. J. Ferrier, T. D. Inch, and P. J. Somers, *Spec. Period. Repts. Carbohyd. Chem.* (London), **3**, 170 (1970).
(32a) J. S. Brimacombe, R. J. Ferrier, R. D. Guthrie, T. D. Inch, and J. F. Kennedy, *Spec. Period. Repts. Carbohyd. Chem.* (London), **4**, 149 (1971).
(33) J. F. Stoddart, "Stereochemistry of Carbohydrates," Wiley-Interscience, New York, 1971.
(34) B. Coxon, *Methods Carbohyd. Chem.*, **6**, 513 (1972).
(35) P. L. Durette and D. Horton, *Advan. Carbohyd. Chem. Biochem.*, **26**, 49 (1971).

importance, and because it presents some unique concepts and problems in the handling of data.

II. INSTRUMENTATION

1. General

Much of the credit for popularizing the n.m.r. method and diminishing its instrumental complexities to the point where it could be widely used and appreciated by chemists belongs undoubtedly to the staff of Varian Associates. The virtual monopoly once enjoyed by this company no longer exists, however, and important contributions to the design of spectrometers have been made by several companies around the world, and by many skilled individuals. As already indicated, sensitivity is often a limiting factor, even in p.m.r. spectroscopy, and this problem has been tackled from several directions other than by use of Fourier techniques.

2. Magnets and Probes

Most of the magnets in use up to 1972 have been of the electromagnet type, whose fields, although very susceptible to gross variations in the supply voltage (for example, those due to electrical storms), may readily be equipped with field-stabilization circuits by (a) coarse control of the a.c. supply voltage by means of a motor-driven, variable transformer, (b) corrections to the magnet current when variations of the internal or external magnetic flux cause a current to be induced in sensing coils wound about the pole-pieces of the magnet, and (c) nuclear stabilization brought about by corrections to the magnet current on the basis of changes in the resonance condition of a reference nucleus.

The requirement for a highly homogeneous magnetic field within a volume of ~ 1 ml is a stringent one, and the usable magnetic field-strengths in electromagnets have not been increased beyond the 2.3 T [1 T (tesla) = 10^4 gauss] that was already available in 1964. This field strength (corresponding to a resonance frequency of 100 MHz for protons) is now widely used, but is close to the saturation limit for cobalt–iron magnets. It appears, therefore, that any further increases in the field strengths of electromagnets must come from concentration of the flux density by specially designed polecaps. Nevertheless, the proton peak-width at half-height (resolution, δf) of ~ 0.3 Hz (which was available from high-field, twelve-inch electromagnets in 1964) has subsequently been decreased to ~0.05 Hz. This improvement is due partly to better design, and partly to a trend towards magnets having pole-pieces of fourteen, fifteen, or eighteen inches diameter, and a pole-gap larger than that used in twelve-inch magnets. The larger magnets afford a greater volume of highly homogeneous field, and even higher homogeneity at the center of this volume. One consequence of these developments is the possibility of using sample-tubes having diameters greater than 5 mm, for high-resolution, proton studies. For example, an increase of the sample-tube diameter from 5 to 10 mm increases the volume of sample sensed by the receiver coil of the spectrometer by a factor of $10^3/5^3 = 8$. The actual gain in sensitivity observed for protons is thought[36] to be nearer to a factor of 3, corresponding to a time-saving factor of 9 in signal-averaging processes. The larger electromagnets available in 1972 allow a resolution at least as good as 0.4–0.6 Hz for protons by use of 10-mm tubes and by the use of spinning sample-tubes up to a maximum diameter of 15 mm. This technique is especially useful for compounds of low solubility.

Solid-state power-supplies for electromagnets have been widely adopted since 1964, thus considerably diminishing the problem of dissipating large amounts of heat from hot vacuum-tubes. Closed-loop cooling-circuits for magnets, and specially designed ("dedicated") refrigeration systems have also come into general use.

Permanent magnets are used for some of the less expensive spectrometers, but suffer from the disadvantage that the stability of the field depends on extremely accurate control of the temperature of the magnet enclosure; this temperature may be affected by variable-temperature operations, unless the probe is thermally well-insulated.[37]

Considerable improvements in the design of probes for research spectrometers have been made, with emphasis on rapid changeover for operation with one nucleus to that for another by means of plug-in sample-inserts, solid-state preamplifiers, circuit boxes for different modes of field–frequency stabilization, and matching networks for tuning the probe to any measuring or decoupling frequency required. The use of field-effect transistors having low noise and high gain in the first stage of the preamplifier has afforded a significant improvement in sensitivity. The need to transmit several different frequencies into the probe for more sophisticated, multiple-resonance experiments has been met by multiple tuning of a single transmitter coil,[38]

(36) Technical literature, Bruker Scientific, Inc., 1969.
(37) R. H. Bible, Jr., *Appl. Spectrosc.*, **24**, 326 (1970).
(38) See, for example, R. Burton and L. D. Hall, *Can. J. Chem.*, **48**, 59 (1970).

or by providing extra coils,[38] in some instances wound directly onto a replaceable sample-insert. The range of variable-temperature operation has been extended to between $-150°$ and $+200°$ in most of the newer research spectrometers.

3. Spectrometer Consoles

The full impact of the advances in electronic technology has resulted in all-solid-state, spectrometer consoles, again with an emphasis on rapid conversion for observation of different nuclei, by switching of radiofrequencies (r.f.), instead of by exchange of bulky r.f. units. The use of semi-conductor components should result in a more reliable and constant performance of spectrometers. In the past, some operators have considered that, in order to maintain top performance of vacuum-tube spectrometers, annual replacement of all of the tubes was necessary.

The older methods for generating several similar radiofrequencies by audio-modulation of one parent radiofrequency have tended to be replaced by frequency-synthesizer methods that generate separate radiofrequencies, which are synchronized with a master oscillator by means of phase-comparator schemes. This arrangement permits use of independent attenuators, and has removed the undesirable changes of spectrum phase during sweeps that were a feature of some of the older 100-MHz instruments; it is also expected to result in better long-term stability of the frequency relationships. Spectrometers have become available that display a variety of sweep-modes, so that either the measuring frequency (f_1), the double-resonance frequency (f_2), or the static magnetic field (H_0) may be swept at will.[39] The frequency sweeps are now often generated by application of either an analog or a digitally stepped voltage to a voltage-to-frequency converter. Digital frequency-sweeps have also been obtained with older spectrometers by modifications in which a programmed, small computer is used to generate the digital sweep-ramp.

Facilities for a variety of multiple-resonance experiments, including broad-band, homo- or hetero-nuclear decoupling, are now generally built into the spectrometer, instead of being provided by auxiliary audio-oscillators. Oscilloscopes for optimization of magnetic-field homogeneity, precalibrated recorder-charts, and frequency counters for accurate calibration of charts and frequencies for multiple

(39) Although common in theoretical discussions, the use of the angular frequency ω is avoided in this Chapter, on the grounds that spectra are never calibrated in radians/sec.

resonance are now standard features of many spectrometers. There has been a welcome simplification in the design of control panels, which, on older instruments, often consisted of an illogical arrangement of miscellaneous electronic units.

4. Field–frequency Stabilization

A constant magnetic field-strength is an essential requirement for the use of precalibrated recorder-charts and for coherent accumulation both of continuous-wave spectra and most pulsed-spectra, by signal-averaging. However, in general, the required stability[36] of about one part in 10^{10} (~0.01 Hz) is not available from typical electromagnets unless a control circuit based on the nuclear resonance-condition is provided. Thus, small drifts in the magnetic field lead to large displacements of the spectra. In practice, it is convenient to stabilize the field-to-frequency ratio (H_0/f_0) by means of corrections to the magnet current based on positive or negative deviations from the zero-crossing point of a dispersion-mode signal[40,41] obtained by irradiating a reference nucleus with a constant frequency (f_0). This procedure maintains a constant condition of resonance for the reference nucleus, and, in addition to correcting any small variations of magnetic field due to the power supply or external magnetic objects, may also be designed[41] to compensate for any small variations of f_0. The latter variations are expected[36] to be less than one part in 10^8 if f_0 is generated from a quartz-crystal oscillator enclosed in a thermostatically controlled oven.

The reference nuclei may be part of the sample (internal, field–frequency stabilization, or *internal lock*) or contained in a separate ampule within the homogeneous volume of magnetic field (external, field–frequency stabilization, or *external lock*). Also, the reference nuclei may be of the same species as those whose spectra are being observed (*homonuclear lock*), or of a different species (*heteronuclear lock*). Each of these methods has advantages and disadvantages.

Internal lock has the advantages that (*a*) the magnetic field is stabilized in the region of the sample itself, thus giving maximum potential stability of its spectrum, and (*b*) only a limited volume of homogeneous field is required. It has the disadvantages of (*a*) a relatively complex operating-technique, and (*b*) instability if there are particles in the sample. Homonuclear, internal lock has been widely used since 1964, but has the disadvantage that the region of

(40) H. Primas, *Europ. Congr. Mol. Spectrosc.*, 5th, Amsterdam, 1961.
(41) R. Freeman and W. A. Anderson, *J. Chem. Phys.*, **37**, 2053 (1962).

the spectrum where f_0 is applied to the reference nucleus is obscured by a strong beat-frequency (and sidebands thereof) between f_0 and f_1. This had not been a significant disadvantage for carbohydrates, as few of them display signals in the region of the tetramethylsilane (Me$_4$Si) that is most generally used to provide the lock signal and the reference for proton chemical-shifts. However, the advent of *chemical-shift reagents* (see Part II, Section IX) that may move carbohydrate proton-resonances upfield into the region of the tetramethylsilane resonance or beyond it, poses problems in the use of this compound for internal lock. It may be desirable to use other compounds that have a resonance at *low field;* for example, sulfuric acid contained by a capillary, or benzene.

An additional disadvantage of homonuclear, internal lock in the p.m.r. analysis of small quantities of material is that the high amplification required exaggerates any slight errors in the setting of the spectrum phase-detector, with the result that the baseline of the spectrum slopes either up or down, towards the lock signal, particularly after the cumulative effect of signal-averaging. Indeed, it is usually not desirable, during signal-averaging, to sweep repetitively through the lock signal, because of the disruptive effect, which may cause the lock circuit to disengage or to disturb the operation of an automatic, field-gradient control-circuit based on the signal level of the lock. For these reasons, a heteronuclear, internal lock is preferable for some applications of p.m.r. spectroscopy to carbohydrates, but this type of lock has the disadvantages that more-complex circuitry is required in the probe and receiver, and that it also necessitates a greater number of tuning adjustments. Heteronuclear, internal lock is now widely used for observation of nuclei other than protons, but it is coming into use only slowly for proton studies. The existing heteronuclear lock-systems are based on irradiation of either fluorine or deuterium nuclei.

The volume of homogeneous field in the early twelve-inch magnets was insufficient to allow the inclusion of an external lock in 100-MHz spectrometers.[37] This failing has been remedied, and many commercial instruments now offer both external and internal lock. External lock (for example, on an ampule of water or hexafluorobenzene) does not afford quite such good field–frequency stability as an internal lock, but it has an advantage in routine operating convenience, in that sample tubes may be exchanged without any risk of disengagement of the lock. Some spectrometers have been designed to switch from internal lock to external lock when the sample is removed from the magnet, thus maintaining the value of the field until the next sample is inserted, at which point the internal lock operates once more. The performance characteristics of commercially available spectrometers have been compared in detail.[37]

5. Signal-averaging Techniques

As implied already, the improvement in signal-to-noise ratio gained by repetitive scanning of a spectrum and accumulation in a suitable memory-device is expected[42] to be proportional to the square root of the number of scans (the "$\sqrt{N_s}$ law"), because the signal-improvement is proportional to N_s, but the noise increase, to $\sqrt{N_s}$. The subject of sensitivity-enhancement in magnetic resonance has been reviewed in detail[42] from the point of view of communications theory, and it was shown that the $\sqrt{N_s}$ law holds for random noise if its power spectrum is a reasonably smooth function of frequency. This condition often appears to be satisfied by practical spectrometers that contribute a thermal-noise component which is independent of frequency, plus noise that increases towards low frequencies and which originates from drift, instabilities, or noise in semi-conductors.[42] The $\sqrt{N_s}$ law breaks down for coherent patterns of noise that may be associated with particular regions of the spectrometer sweep, or with small sub-groups of core memory in the signal-averaging computer. The latter source of coherent noise is compensated in some computers by the alternate addition and subtraction of this noise during successive scans.

In 1964, only one signal-averaging computer, having 1,024 words of memory, was available for averaging of n.m.r. spectra. This memory size is not sufficient for good resolution in proton spectra, and the better signal-averaging systems now have memory sizes of from 4,096 to 65,536 words, and range from small, variably programmed or fixed (wired) program computers to large processors and magnetic-disc systems. A word-length of at least 16 binary digits (16 bits) is generally considered desirable for proton spectra, so as to give a vertical resolution at least as good as one part in 65,536 (2^{16}).

The division of the total computer-memory into sections allows the contents of one section to be added to or subtracted from those of another. This technique may sometimes be used to obtain the spectrum of one component of a mixture by subtraction of the spectrum due to other components from that of the mixture. For example, a 100-MHz spectrum of α-D-ribopyranose tetrabenzoate (1) has been

(42) R. R. Ernst, *Advan. Magn. Resonance*, **2**, 1 (1966).

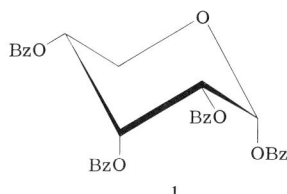

1

obtained by subtraction of the spectrum of the pure β-D anomer from that of the mixture of anomers.[43] In this example, some signal-averaging was performed to average out short-term variations in the gain of the spectrometer and to improve the precision of the peak positions.[43] Undesired solvent-peaks may be removed in a similar way if there are no differences of chemical shift between solvent and solution. In the latter regard, horizontal shifting of data within the memories of signal-averaging computers would be useful, but not all such computers have this feature.

Even the signal-averaging computers having wired programs now include a variety of options for performing mathematical manipulations of the spectra stored; for example, addition or subtraction of constants or ramps (baseline corrections), digital integration and differentiation, digital smoothing by means of a running average, scaling, and transfers of fractions of the contents of one section of memory into another.

The most sensitive technique for p.m.r. analysis of impurities in bulk materials, of small quantities of rare sugars, or of minor by-products of chemical reactions, is by the signal-averaging of pulsed spectra, followed by Fourier transformation (see Section IV, p. 43).

6. Automatic Control and Data Acquisition

The generation of frequency sweeps under computer control, described in Section II, 3 (p. 13) is one small aspect of a general tendency towards control of analytical instrumentation by means of digital computers, through appropriate interfacing devices. Several commercial and individually built systems have appeared in which all of the instrumental parameters for frequency sweeps or pulse-excitation (see Section IV, p. 45) are selected by teletype or numerical keyboard input to the computer, which then acquires the spectral data automatically and performs any further processing required. Automatic analysis of spectral peak positions and areas by a computer, and printing of the numerical results on a teletype or

(43) B. Coxon, *Tetrahedron*, **22**, 2281 (1966).

line-printer, promise to remove one of the major bottlenecks in the analysis of the n.m.r. spectra of carbohydrates, namely, the tedious, manual measurements of peaks on recorder charts.

In the absence of completely automated peak-analysis, a useful alternative is the use of a signal-averaging computer, with numeric readout (on an oscilloscope) of the contents of any computer address selected. In this method, an intensified spot of light (cursor) on the oscilloscope screen is moved to the top of a peak or integral step, and the numeric contents (counts) of that particular address are then displayed on the screen. The frequency of the peak may then be obtained by interpolation of its address number between the known frequencies of the first and last address numbers of the sweep and the integral from the difference in the number of counts in addresses before and after the step. These procedures have been applied to quantitative analysis[44] of solutions of α- and β-D-glucose in methyl sulfoxide-d_6 (see Part II, Section X), and to ^{13}C n.m.r. spectra of D-mannitol in water.[45] Dedicated, digital computers have also been used for automatic optimization of several field-gradients in the magnet, for enhancement of resolution, and for simulation of spectra.

Another control mechanism that is coming into increasing use is the use of alternate duty-cycles for transmitter and receiver (time-sharing modulation).[42,46] In this method, the transmitter and receiver are, in turn, switched on and off for short, non-overlapping intervals, at a rate of several kHz. The advantages claimed[42] are that (a) there results greater sensitivity, owing to the lack of detectable transmitter noise (and electrical leakage), and (b) field modulation is not necessary.

III. APPLICATIONS OF SUPERCONDUCTING SOLENOIDS

1. General Techniques and Advantages

An instrumental development of considerable importance to the p.m.r. spectroscopy of carbohydrates has been the introduction of high-resolution magnets based on superconducting solenoids.[47] As already mentioned, the (homogeneous) magnetic field-intensity in conventional electromagnets is restricted by the properties of the iron core and by the fact that the addition of auxiliary coils that are

(44) B. Coxon and R. Schaffer, *Anal. Chem.*, **43**, 1565 (1971).
(45) B. Coxon, unpublished results.
(46) E. Grunwald, C. F. Jumper, and S. Meiboom, *J. Amer. Chem. Soc.*, **84**, 4664 (1962).
(47) F. A. Nelson and H. E. Weaver, *Proc. Colloque Ampère, 14th, Ljubljana, 1966*, published by North-Holland, Amsterdam, 1967, p. 917.

resistive at room temperature would be attended[48] by a severe cooling problem for the high currents that are necessary for increasing the field-strength beyond ~2.3 T (see p. 11). This difficulty has been solved by removing the iron core and using coils (of a special alloy) that are non-resistive and, hence, infinitely conductive when cooled below a critical temperature. For the alloys of niobium that have been used,[47] this temperature is in the region of 10 K, achieved by immersing the solenoid in a Dewar vessel of liquid helium at 4 K.

The magnetic field inside a superconducting solenoid is not particularly homogeneous, and compensation of the field gradients is quite a complex problem that has been solved[49] by the provision of mathematically designed arrays of superconducting or normal, resistive shim-coils, or both.

Following four years of difficult development,[47,48] availability of a commercial, superconducting, p.m.r. spectrometer was announced in 1966. The instrument prototype was tested at a field strength of 5.87 T, corresponding to a resonance frequency of 250 MHz for protons, but, at this field value, abrupt changes of flux in the solenoid caused quenching within time periods ranging from a few hours to a few days.[47] Consequently, the field was decreased to 5.17 T, and the p.m.r. frequency to 220 MHz, with a resultant decrease in the number of quenches suffered by the solenoid during training, and better long-term stability.[47] Once established, the magnetic field in the solenoid has usually been sufficiently stable as not to require a field-frequency, control system for routine operation.[50] Indeed, a self-stabilizing phenomenon has been described[48] in which current increments induced in the solenoid by externally changing magnetic fields (caused, for example, by ferromagnetic objects being moved) so modify the gross current flowing in the solenoid as to balance out the effect of the external field.

The high cost of the first 220-MHz spectrometers and their restriction, at least initially, to observation of protons only (in 5-mm sample tubes) and to field-sweep operation, has encouraged a number of research groups[51] outside the spectrometer industry to construct

(48) J. K. Becconsall and M. C. McIvor, *Chem. Brit.*, **5**, 147 (1969).
(49) For example, see J. Dadok, *Abstr. Experimental NMR Conf., 10th, Pittsburgh,* 1969.
(50) L. F. Johnson, *Anal. Chem.*, **43**, No. 2, 28A; No. 3, 101A (1971).
(51) (a) F. A. L. Anet, V. J. Basus, C. H. Bradley, and A. Cheng; (b) R. Sprecher, J. Dadok, A. A. Bothner-By, and T. Link; (c) L. M. Huber and E. B. Baker, *Abstr. Experimental NMR Conf., 12th, Gainesville,* 1971. See also, abstracts of these conferences for 1969 and 1970. (d) F. A. L. Anet, C. H. Bradley, M. A. Brown, W. L. Mock, and J. H. McCausland, *J. Amer. Chem. Soc.*, **91**, 7782 (1969); F. A. L. Anet and J. J. Wagner, *ibid*, **93**, 5266 (1971).

individual high-resolution, superconducting systems based on commercially available, general purpose, superconducting solenoids and computer-controlled, frequency synthesizers. All of these systems[51] employ the time-sharing arrangement of transmitter and receiver mentioned in Section II (p. 18), and the field strengths achieved are in the range of 5.9–6.5 T, corresponding to p.m.r. frequencies of 250–260 MHz. The spectrometer constructed by F.A.L. Anet is a versatile example that has been used to observe ^{1}H, ^{2}H, ^{11}B, and ^{13}C nuclei.[51a,d] It is equipped with internal-lock, variable-temperature operation to below $-170°$, broad-band decoupling of protons at 251 MHz, accommodation for 5- or 10-mm sample tubes, and Fourier transformation (see Section IV, p. 52) of 8,192 datum points in approximately 30 seconds. For protons, a resolution of better than 0.15 Hz has been achieved.[51a]

The high cost of commercial superconducting spectrometers and of the continuing supply of liquid helium needed has limited the number of installations of these instruments, and, in some countries, such high-frequency p.m.r. spectra are available only from a national service center. However, some savings are realized because of the fact that, once established in the persistent mode, superconducting solenoids require neither a power supply nor cooling water.

Further improvements in superconducting materials[50] permitted the development in 1970 of a 300-MHz spectrometer that is equipped with a low-loss Dewar vessel, an improved sweep-system that allows both field sweeps and frequency sweeps up to a maximum width of 20 kHz, and accessories for a variety of nuclei and pulse–Fourier-transform techniques.

The principal advantages of superconducting, n.m.r. spectrometers are (a) greatly increased chemical shift, and (b) higher sensitivity. The magnified chemical shift is particularly useful in p.m.r. spectroscopy, because of the small range of chemical shifts of protons compared with those of other nuclei, and it is especially relevant to carbohydrates as these often contain a large number of protons in similar chemical environments; for example, methine protons attached to carbon atoms that bear hydroxyl or acetoxyl groups. Because chemical shifts (in Hz) are directly proportional to the magnetic field-strength, whereas true coupling-constants (not spacings) are independent of the field, proton multiplets that overlap at fields of 2.3 T or less are often separated at fields of 5 T or more. This procedure simplifies the analysis of the spectra in at least two ways; firstly, it affords resolution of more of the lines in the multiplets, and secondly, because of the strong trend towards first-order spectra at very high fields, the occurrence of extra splittings and broadenings in the spectra (second-order effects) is diminished, so that the coupling constants may be obtained more frequently by direct measurement of the spacings in the multiplets, instead of by computation.

In connection with the increased sensitivity at higher fields, it has been stated[42] that the magnetization of the sample and the voltage induced in the receiver coil are each proportional to the field strength H_0, so that the signal at the detector is expected to increase with H_0^2. However, owing to greater signal-losses in the receiver circuits at higher frequencies,[42] it appears that, in practice,[42,47,48] the improvement in sensitivity ranges between factors proportional to H_0 and $H_0^{3/2}$. This improvement is of definite utility in the analysis of dilute samples.

2. Monosaccharides

Spectra recorded at 220 MHz were soon used to confirm the assignments made from the 100-MHz p.m.r. spectra of poly-O-acyl-D-pentopyranosyl[52] and -hexopyranosyl[53,54] fluorides, although, for solutions of the latter derivatives in chloroform-d and acetone-d_6, the multiplets of H-5, H-6, and H-6' frequently overlap,[54] and, hence, are not completely first-order, even at 220 MHz. The magnitudes of the proton–proton coupling-constants of these fluorides have been used to determine their configuration, the shapes of their pyranoid rings, and, often, the approximate populations of rotamers derived by rotating O-6 about the C-5–C-6 bond.[54] Some of these assignments of rotameric populations have since been disputed, and the limitations of the method have been emphasized.[55] Definition of the ring shapes of the poly-O-acyl-aldohexopyranosyl fluorides permitted the first correlations of the chemical shifts and coupling constants of ^{19}F nuclei with the stereochemistry of these carbohydrate derivatives. The vicinal, $^{19}F-^1H$ coupling-constants were found[54] to display an even more marked dependence on dihedral angle and configuration than do vicinal $^1H-^1H$ coupling-constants,[2] so that $J_{F,H}$ (trans) = ~24 Hz, whereas J_{F_a,H_e} = 1.0–1.5 Hz, and J_{F_e,H_a} = 7.5–12.6 Hz.

Bhacca, Horton, and coworkers have made extensive applications of 220-MHz, p.m.r. spectra to the stereochemistry of a variety of peracetylated carbohydrate derivatives having different substituents at C-1. For example, the H-1–H-4 signals of 1-thio-α-L-arabinopyranose tetraacetate (**2**) in chloroform-d were found to overlap at 100

(52) L. D. Hall and J. F. Manville, *Carbohyd. Res.*, **4**, 512 (1967).
(53) L. D. Hall and J. F. Manville, *Can. J. Chem.*, **45**, 1299 (1967).
(54) L. D. Hall, J. F. Manville, and N. S. Bhacca, *Can. J. Chem.*, **47**, 1 (1969).
(55) R. U. Lemieux and J. C. Martin, *Carbohyd. Res.*, **13**, 139 (1970).

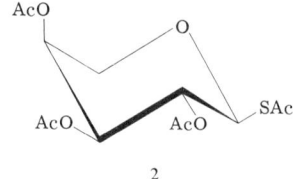

2

MHz, giving a complex multiplet, but were separated when recorded at 220 MHz, thereby allowing the assignment of the α-L configuration and $^4C_1$ (L) conformation to this 1-thio derivative.[56] The use of different solvents had failed to give readily interpretable spectra at 100 MHz. For other 1-thioaldopyranose derivatives,[56] the spacings within the multiplets of the 220-MHz spectra were often found to be slightly greater than those in the 100-MHz spectra, and the former spacings were presumed to be closer to the true values of the coupling constants.

An important benefit of the increased chemical-shift at higher field-strengths is the raising of the temperature (T_c) at which coalescence of the separate signals of interconverting structural or conformational species is observed. If T_c is very much below room temperature, as is commonly true for interconverting conformers, a higher T_c allows the separate signals of the conformers to be observed within the more accessible range of temperatures below T_c. This effect has been exploited in variable-temperature studies[57,58] of β-D-ribopyranose tetraacetate. At 20°, the 220-MHz spectrum of a solution of this compound in acetone-d_6 was completely first-order, but, as had been pointed out for 100-MHz spectra,[59] the value of $J_{1,2}$ observed (4.8 Hz) was intermediate between the values expected for the $^4C_1$ and $_1C^4$ conformations, thus suggesting a mixture of rapidly interconverting conformers. As the temperature of the solution was lowered to −60°, the signals of the ring protons at 220-MHz first broadened, and then, below −60°, gradually sharpened, until, at −84°, a separate singlet and doublet were clearly observed[57,58] for the anomeric protons of the individual chair conformers. Separate signals for these conformers could also be observed at 60 and 100 MHz, but at lower temperatures than at 220 MHz, and with somewhat less certainty, because of the smaller spectral dispersion.[58] A tentative identification of separate conformers (**3a** and **3b**) of the related tri-O-

(56) C. V. Holland, D. Horton, M. J. Miller, and N. S. Bhacca, *J. Org. Chem.*, **32**, 3077 (1967); N. S. Bhacca and D. Horton, *Chem. Commun.*, 867 (1967).
(57) N. S. Bhacca and D. Horton, *J. Amer. Chem. Soc.*, **89**, 5993 (1967).
(58) P. L. Durette, D. Horton, and N. S. Bhacca, *Carbohyd. Res.*, **10**, 565 (1969).
(59) R. U. Lemieux and J. D. Stevens, *Can. J. Chem.*, **43**, 2059 (1965).

benzoyl-β-D-ribopyranosyl cyanide (**3**) has been made[43] at 60 MHz.

The 220-MHz, p.m.r. spectrum of α-D-idopyranose pentaacetate (in acetone-d_6) is also first-order, in contrast to its 100-MHz spectrum, which shows strong distortions of the intensities of the lines in the H-2 and H-3 multiplets.[60] Methine-proton signals that overlap at 100 MHz are separated in the 220-MHz spectra of N-acetyl-α-D-glucofuranosylamine (**4**), its β-D-*galacto* isomer[61] (**5**), and 3,4,5,6-

tetra-O-acetyl-1,2-dideoxy-1-nitro-D-*ribo*-hex-1-enitol[62] (**6**), but not in

the spectrum of the D-*xylo* isomer of this hexene derivative, which has been analyzed by computation.[63]

(60) N. S. Bhacca, D. Horton, and H. Paulsen, *J. Org. Chem.*, **33**, 2484 (1968).
(61) A. B. Zanlungo, J. O. Deferrari, and R. A. Cadenas, *Carbohyd. Res.*, **10**, 403 (1969).
(62) J. M. Williams, *Carbohyd. Res.*, **11**, 437 (1969).
(63) C. W. Haigh and J. M. Williams, *J. Mol. Spectrosc.*, **32**, 398 (1969).

Early efforts to use p.m.r. spectroscopy to establish the configurations of diastereoisomeric quercitols were unsuccessful, with two exceptions,[64] because of complex spin–spin coupling, and overlapping of multiplets at 60 and 100 MHz. However, in the p.m.r. spectrum of (+)-*proto*-quercitol (**7**) in deuterium oxide at 220 MHz (see Fig. 1),

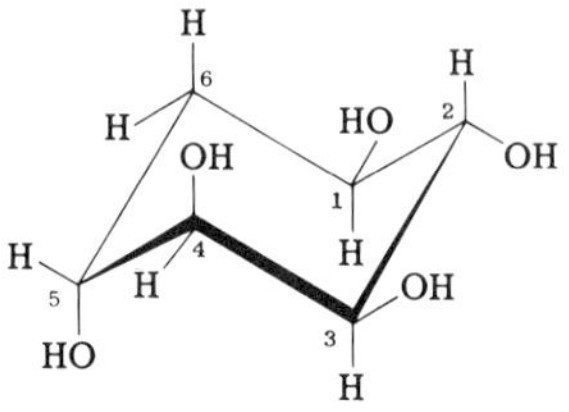

[Structure 7: cyclohexane ring with H and OH substituents, numbered 1-6, with HO groups]

7

the multiplets of most of the ring protons were found to be well separated, thus allowing complete verification[64,65] of the configuration determined much earlier from laborious chemical correlations. For *proto*-quercitol, H-1 and H-3 have very similar chemical environments, and, at 220 MHz, their signals overlap to give a complex multiplet composed of approximately nine lines.[50,64] At 300 MHz, these signals still overlap[50] (see Fig. 1), but, nevertheless, the octet of H-1 and the quartet for H-3 may be discerned, and analyzed in detail. Similarly, a 220-MHz spectrum permitted a clear, first-order analysis to be made of the symmetrical, A_2BX_2 system and structure of mytilitol hexaacetate (**8**), in contrast to its 60- and 100-MHz spectra.[66]

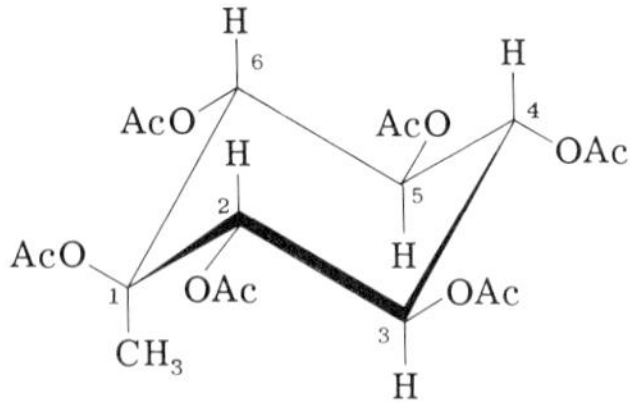

[Structure 8: cyclohexane ring with AcO, OAc, H, and CH₃ substituents, numbered 1-6]

8

(64) G. E. McCasland, M. O. Naumann, and L. J. Durham, *J. Org. Chem.*, **33**, 4220 (1968).
(65) G. E. McCasland, M. O. Naumann, and L. J. Durham, *Carbohyd. Res.*, **4**, 516 (1967).
(66) A. W. Sangster, S. Thomas, and L. F. Johnson, *Org. Magn. Resonance*, **3**, 255 (1971).

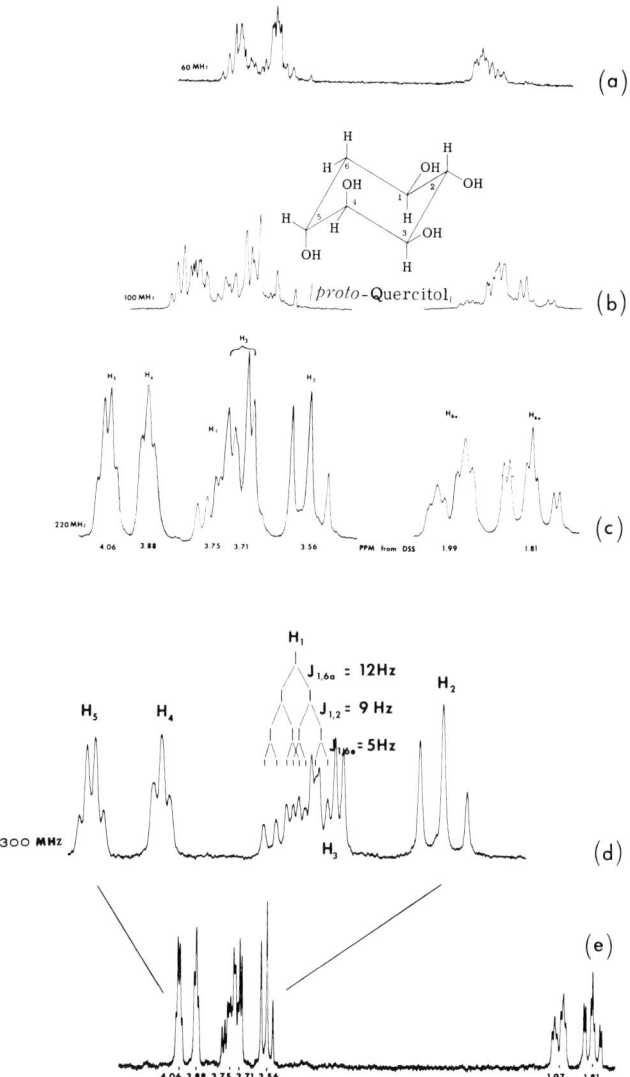

FIG. 1.—Proton Magnetic Resonance Spectra of *proto*-Quercitol (**7**) in Solution in Deuterium Oxide, at (a) 60, (b) 100, (c) 220, and (d) and (e) 300 MHz. [Chemical-shift values are downfield from a reference standard of sodium 4,4-dimethyl-4-silapentane-1-sulfonate (DSS).]

The p.m.r. spectra of simple derivatives of D-glucopyranose are often complex, because of the similar stereochemical environments of the ring protons. Thus, the 220-MHz spectra of the enamine derivatives **9** and **10**, formed by condensation of 2-amino-2-deoxy-D-

glucose and 2-amino-2-deoxy-D-mannose with 2,4-pentanedione, were not fully resolved, but allowed confirmation of their structure, α-D configuration, and $^4C_1$ conformation.[67] Measurements of large values of $J_{1,2}$ (7–8 Hz) from 220-MHz spectra have been used for establishing the β-D configuration for methyl (cholest-5-en-3β-yl 2,3,4-tri-O-acetyl-β-D-glucopyranosid)uronate[68] (**11**) and a number of aryl D-

glucopyranosides and D-glucopyranosiduronates (for example,[69] **12**).

---

(67) N. S. Bhacca and J. J. Ludowieg, *Carbohyd. Res.*, **11**, 432 (1969).
(68) J. J. Schneider and N. S. Bhacca, *J. Org. Chem.*, **34**, 1990 (1969).
(69) J. Kiss and F. Burkhardt, *Carbohyd. Res.*, **12**, 115 (1970).

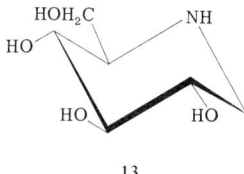

12

Such spectra have also assisted determination of the D-*gluco* stereochemistry of 5-amino-1,5-anhydro-5-deoxy-D-glucitol (**13**), a reduction

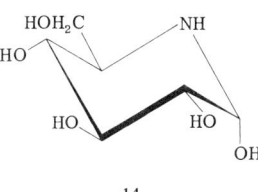

13

product of nojirimycin[70] (**14**), and the structure of diethyl 2,3,4,6-

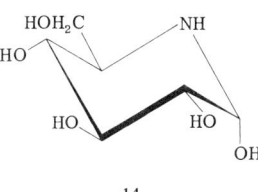

14

tetra-*O*-acetyl-β-D-glucopyranosylmalonate[71] (**15**).

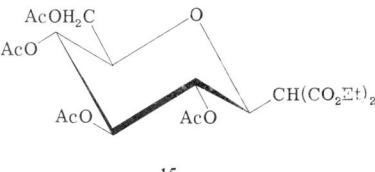

15

(70) S. Inouye, T. Tsuruoka, T. Ito, and T. Niida, *Tetrahedron*, **23**, 2125 (1968).
(71) S. Hanessian and A. G. Pernet, *Chem. Commun.*, 755 (1971).

A comparison[72] of the complex, 220-MHz spectrum of a mutarotated solution of α- and β-D-glucose in deuterium oxide at 60° with various decoupled and undecoupled 100-MHz spectra[73] of α- and α,β-D-glucose-5,6,6'-d_3 and α,β-D-glucose-5-d permitted measurement of all of the coupling constants between protons attached to the carbon chains of α- and β-D-glucose. The spacings measured at 220 MHz were in accord with $^4C_1$ conformations and were found[72] to differ by no more than 0.5 Hz from those determined at 100 MHz.

In contrast to spectra measured at 60 and 100 MHz, the 220-MHz spectrum of 5-S-acetyl-2,2¹-anhydro-3-deoxy-2-C-(hydroxymethyl)-5-thio-D-*erythro*-pentono-1,4-lactone was amenable to first-order analysis and supported the structure (**16**) expected from the mode of

<chemical structure 16>

preparation and from the results of mass-spectrometric studies.[74]

In conformational studies, 220-MHz spectra have been used to obtain complete proton chemical-shift and coupling-constant data for methyl tri-O-acetyl- and tri-O-benzoyl-α-D-lyxopyranosides,[75] and most of the proton coupling-constants for 1-thio-α-D-lyxopyranose tetraacetate,[76] dissolved in acetone-d_6. At 220 MHz, the H-5 and H-5' multiplets of the latter derivative[76] still showed virtual-coupling effects[77] due to strong coupling of H-3 with H-4 (large $J_{3,4}/\delta_{3,4}$ ratio).

Spectra recorded at 220 MHz have allowed determination of the chemical shifts and coupling constants for H-1 and H-2 of the various ring and anomeric forms[78] and cytosine derivatives (**17, 18, 19,** and **20**)[79] of 2-deoxy-2(S)- and 2(R)-deuterio-D-*erythro*-pentose (**21** and **22**), and have also simplified interpretations of the spectra of methyl 2-deoxy-β-D-*erythro*-pentofuranoside[80] (**23**) and various deoxyhalo-

(72) H. J. Koch and A. S. Perlin, *Carbohyd. Res.*, **15**, 403 (1970).
(73) W. Mackie and A. S. Perlin, *Can. J. Chem.*, **43**, 2921 (1965).
(74) D. R. Strobach, *Carbohyd. Res.*, **17**, 457 (1971).
(75) P. L. Durette and D. Horton, *Carbohyd. Res.*, **18**, 403 (1971).
(76) P. L. Durette and D. Horton, *Carbohyd. Res.*, **18**, 419 (1971).
(77) J. I. Musher and E. J. Corey, *Tetrahedron*, **18**, 791 (1962).
(78) B. Radatus, M. Yunker, and B. Fraser-Reid, *J. Amer. Chem. Soc.*, **93**, 3086 (1971).
(79) B. Fraser-Reid and B. Radatus, *J. Amer. Chem. Soc.*, **93**, 6342 (1971).
(80) R. U. Lemieux, L. Anderson, and A. H. Conner, *Carbohyd. Res.*, **20**, 59 (1971).

geno sugar derivatives, including 1,5-anhydro-4,6-O-benzylidene-2,3-dideoxy-3-(iodomethyl)-D-*ribo*- and D-*arabino*-hex-1-enitols[81] (**24, 25**), and several 5-iodo[82] (**26, 27**) and 5-fluoro[83] (**28**) derivatives of 5-deoxy-1,2-O-isopropylidene-α-D-xylofuranose. Spectra at 220 MHz have also proved useful for differentiation of the signals of various types of C-methyl groups.[84]

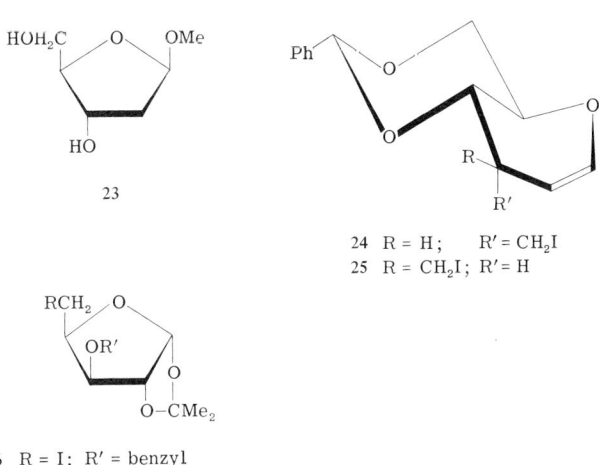

The 100-MHz spectra of solutions of 6-deoxy-1,2:3,5-di-O-isopropylidene-6-phthalimido-α-D-glucofuranose (**29**) and of its 6-^{15}N-labeled

(81) B. Fraser-Reid and B. Radatus, *J. Amer. Chem. Soc.*, **92**, 6661 (1970).
(82) R. C. Young, P. W. Kent, and R. A. Dwek, *Tetrahedron*, **26**, 3983 (1970).
(83) P. W. Kent and R. C. Young, *Tetrahedron*, **27**, 4057 (1971).
(84) P. M. Collins, *J. Chem. Soc. (C)*, 1960 (1971).

derivative (**30**) in a variety of solvents were found to be complex, and not amenable to a simple analysis.[85] However, the 220-MHz spectra of these compounds in benzene-d_6 solutions were first-order for H-1–H-5, but showed H-6 and H-6' as overlapping, A and B parts of an ABX sub-system (see Fig. 5, p. 76).

Despite the general improvement in spectral dispersion observed at high field-strengths, some examples of monosaccharide derivatives have been reported[86–88] for which 220-MHz spectra failed to give more readily interpretable results, or more information, than had been obtained at 100 MHz.

3. Oligosaccharides

The p.m.r. spectra of per-O-acetyl derivatives (**31**, **32**, and **33**) of

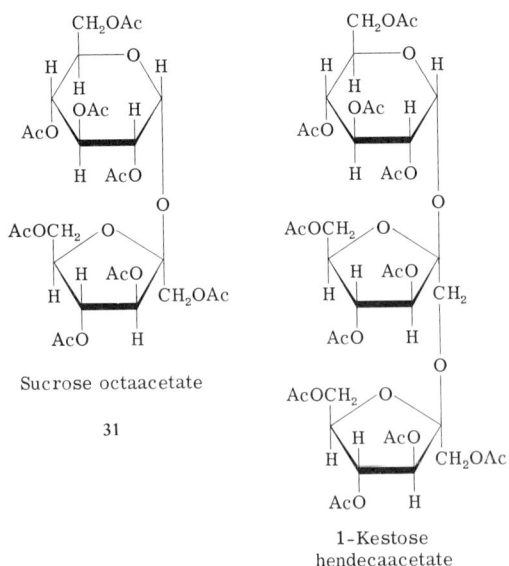

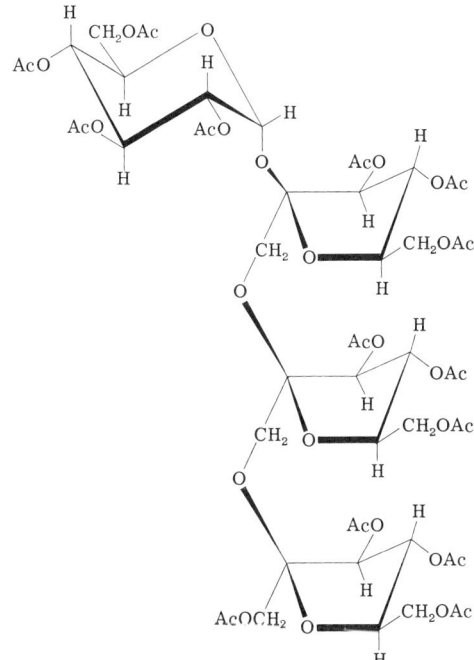

Nystose tetradecaacetate

33

sucrose and its homologous oligosaccharides, 1-kestose [β-D-Fruf-(2→1)-β-D-Fruf-(2↔1) α-D-Glcp] and nystose [β-D-Fruf-(2→1)-β-D-Fruf-(2→1)-β-D-Fruf-(2↔1) α-D-Glcp], have been obtained at 100 and at 220 or 230 MHz.[89] For each of these compounds, separate signals were observed for H-1–H-4 of the α-D-glucopyranosyl group and for H-3 and H-4 of the β-D-fructofuranosyl group(s). The high-frequency spectra were more useful for the tri- and tetra-saccharide peracetates than for sucrose octaacetate (31), which had previously been analyzed[90] at 100 MHz. The observation[89] of simple AB patterns in the intermediate-field region for 1-kestose hendecaacetate (32) and nystose tetradecaacetate (33) confirmed the presence of isolated pairs

(85) B. Coxon and L. F. Johnson, Carbohyd. Res., 20, 105 (1971).
(86) J. G. Buchanan, R. Fletcher, K. Parry, and W. A. Thomas, J. Chem. Soc. (B), 377 (1969).
(87) G. E. McCasland, M. O. Naumann, and L. J. Durham, J. Org. Chem., 34, 1382 (1969).
(88) G. R. Gray and R. Barker, Biochemistry, 9, 2454 (1970).
(89) W. W. Binkley, D. Horton, and N. S. Bhacca, Carbohyd. Res., 10, 245 (1969).
(90) R. U. Lemieux and R. Nagarajan, Can. J. Chem., 42, 1270 (1964).

of non-equivalent, bridge-methylene protons,[91] and, hence, of the (2 → 1) linkages of the D-fructofuranosyl groups. Analysis of the spectra obtained for these peracetates in each of three solvents was assisted by spin decoupling and by comparisons with similar spectra of methyl α-D-glucopyranoside tetraacetate.[91a]

Spin decoupling at 100 MHz, together with 220-MHz spectra, have also been used in confirming the spectral assignments for the inclusion complexes (for example, **34**) of cycloheptaamylose with a

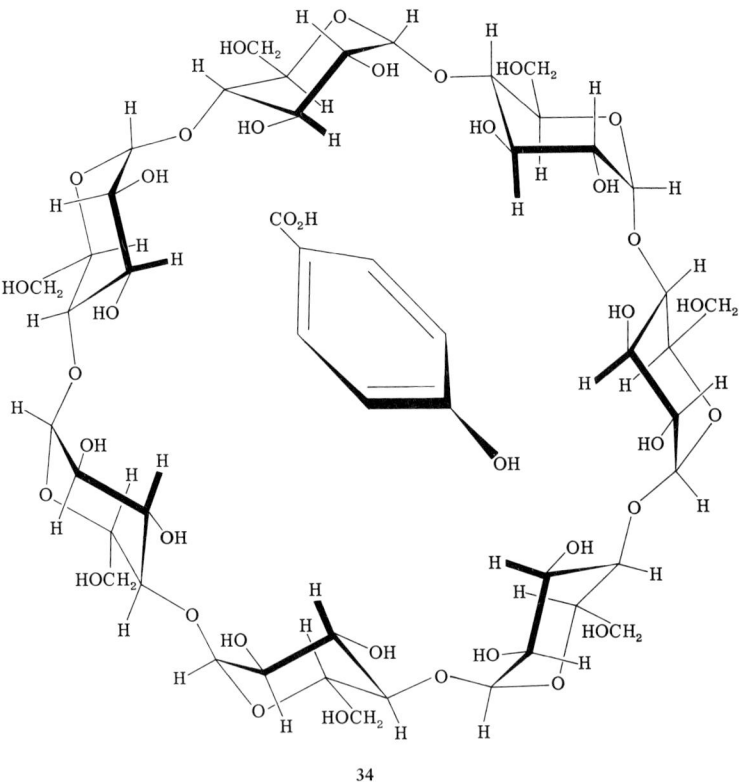

34

Static form of inclusion complex of cycloheptaamylose
and p-hydroxybenzoic acid

[On formation of the complex, the H-3 and H-5 resonances undergo a larger upfield shift (0.14 and 0.21 p.p.m., respectively) than those of H-1, H-2, H-4, and H$_2$-6 (0.04, 0.04, 0.04, and 0.06 p.p.m.).]

(91) *cf.*, F. Arcamone, W. Barbieri, G. Cassinelli, and C. Pol, *Carbohyd. Res.*, **14**, 65 (1970).
(91a) D. Horton and J. H. Lauterbach, *J. Org. Chem.*, **34**, 86 (1969).

variety of aromatic substrates[92] and for 3,6-anhydro-α-D-glucofuranosyl 3,6-anhydro-α-D-glucofuranoside tetrabenzoate[93] (**35**). The

$$\begin{bmatrix} \text{H} & \text{H} & \text{H} & \text{OBz} \\ & & \text{O} & \text{H} \\ \text{O} & \text{H} & \text{H} & \\ & \text{H} & \text{OBz} & \\ & 35 & & \end{bmatrix}_2 \text{O}$$

latter study provided further examples of the simplified spectra that are observed for symmetrical disaccharide derivatives in which the monosaccharide moieties are identical.[94-101]

4. Nucleosides and Nucleotides

High-frequency, p.m.r. spectroscopy has proved to be an extremely important tool in studies of the structures, conformations, and inter- and intra-molecular base-stacking interactions of nucleosides and nucleotides. The temperature dependence of the chemical shift of the base proton at position 6 (H-6) has been studied at 220 MHz for uridine (**36**), cytidine (**38**), and a number of their mono- (**37, 39**) and

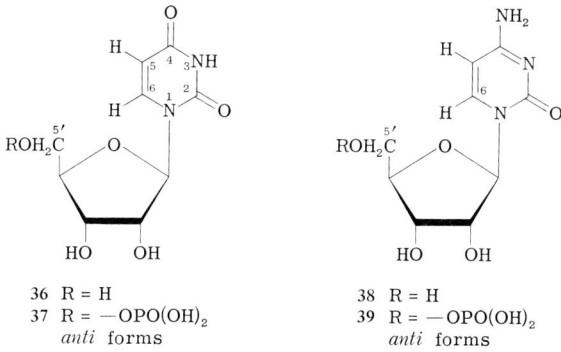

36 R = H
37 R = —OPO(OH)$_2$
anti forms

38 R = H
39 R = —OPO(OH)$_2$
anti forms

(92) P. V. Demarco and A. L. Thakkar, *Chem. Commun.*, 2 (1970).
(93) G. Birch, C. K. Lee, and A. C. Richardson, *Carbohyd. Res.*, **19**, 119 (1971).
(94) B. Coxon, H. J. Jennings, and K. A. McLauchlan, *Tetrahedron*, **23**, 2395 (1967).
(95) H. H. Baer and F. Rajabalee, *Can. J. Chem.*, **47**, 4086 (1969).
(96) G. Birch and A. C. Richardson, *Carbohyd. Res.*, **8**, 411 (1968).
(97) G. Birch and A. C. Richardson, *J. Chem. Soc.* (C), 749 (1970).
(98) Y. Ali, L. Hough, and A. C. Richardson, *Carbohyd. Res.*, **14**, 181 (1970).
(99) G. Birch, C. K. Lee, and A. C. Richardson, *Carbohyd. Res.*, **16**, 235 (1971).
(100) L. Hough, P. A. Munroe, and A. C. Richardson, *J. Chem. Soc.* (C), 1090 (1971).
(101) L. Hough, A. C. Richardson, and E. Tarelli, *J. Chem. Soc.* (C), 1732, 2122 (1971).

di-nucleotide derivatives.[102] The shift to higher field of each H-6 signal that was observed on increasing the temperature was attributed[102] to removal of the specific, deshielding effect of a 5'-hydroxyl or 5'-phosphate group by gradual conversion of the *anti* form (for example, **36, 37, 38,** and **39**),[103–106] in which H-6 is close to the 5'-groups of the D-ribofuranoid ring, to a mixture of conformers favoring the *syn* form,[106] in which H-6 is remote from these deshielding groups.

The proton signals of uridine (**36**) dissolved in deuterium oxide are well separated at 220 MHz, and a complete analysis and simulation of the spectrum by means of the LAOCOON II computer program (see Section V, p. 74) has been reported.[107] It was concluded from the temperature-independence of the coupling constants and chemical shifts of the D-ribofuranoid moieties of uridine, β-pseudouridine (**40**), and α-pseudouridine (**41**), that each of these compounds

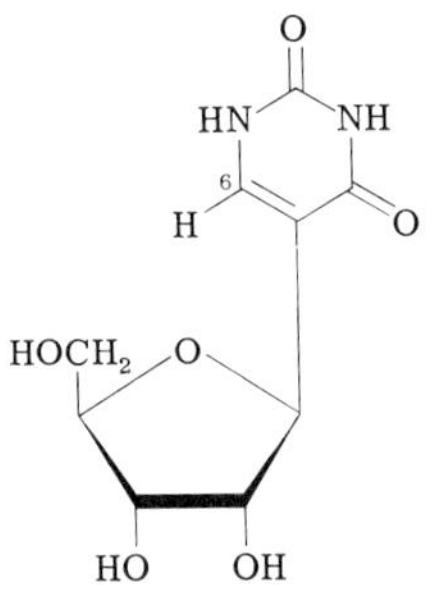

40
anti form

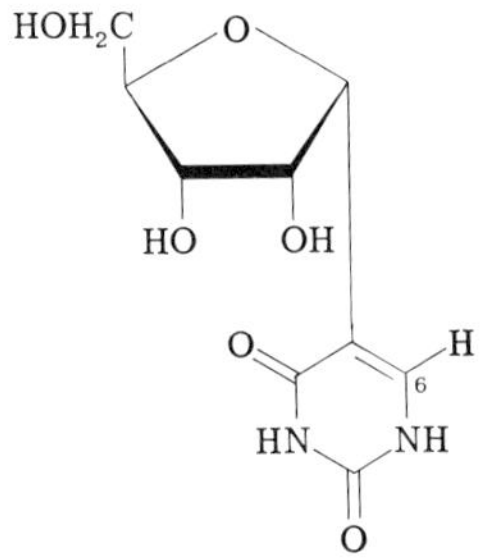

41
syn form

exists as a mixture of rapidly interconverting, classical, puckered conformations having similar energy content.[104,105,107] Small values (3.0 and 4.4 Hz, respectively) were found for the coupling constants $J_{4',5'}$ and $J_{4',5''}$ of uridine, and interpretation of these values in many

(102) I. C. P. Smith, B. J. Blackburn, and T. Yamane, *Can. J. Chem.,* **47,** 513 (1969).
(103) M. P. Schweizer, A. D. Broom, P. O. P. Ts'O, and D. P. Hollis, *J. Amer. Chem. Soc.,* **90,** 1042 (1968).
(104) A. A. Grey, I. C. P. Smith, and F. E. Hruska, *J. Amer. Chem. Soc.,* **93,** 1765 (1971).
(105) F. E. Hruska, A. A. Grey, and I. C. P. Smith, *J. Amer. Chem. Soc.,* **92,** 214, 4088 (1970).
(106) F. E. Hruska, *J. Amer. Chem. Soc.,* **93,** 1795 (1971).
(107) B. J. Blackburn, A. A. Grey, I. C. P. Smith, and F. E. Hruska, *Can. J. Chem.,* **48,** 2866 (1970).

different ways, including exchange of the values, the use of a range of values for J_0 in the Karplus equation, and consideration of classical rotamers staggered about the C-4′–C-5′ bond by 60°, or of rotamers staggered by 75° to take into account maximal oxygen–oxygen repulsion, invariably indicated[107] that rotamer **42** (*gauche-gauche*), in

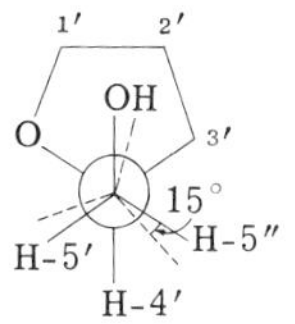

42

gauche - gauche rotamer
(The broken lines indicate a 15° distortion of the rotamer to allow for O-O repulsion.)

which the 5′-hydroxyl group is directed toward H-6 of the uracil moiety, is favored according to a mole fraction of 0.46–0.65. This rotamer is favored slightly for α- and β-pseudouridine (**41** and **40**).[104,105] Other studies of ribo- and deoxyribo-nucleosides at 220 MHz have indicated that the proportion of the *gauche–gauche* rotamer decreases as the proportion of conformers having C-2′ *endo* (or C-3′ *exo*, or both) increases,[108] or if there is a repulsion between the 5′-hydroxyl group and an α-keto group of the base, as in N-β-D-ribofuranosylcyanuric acid[109] (**43**).

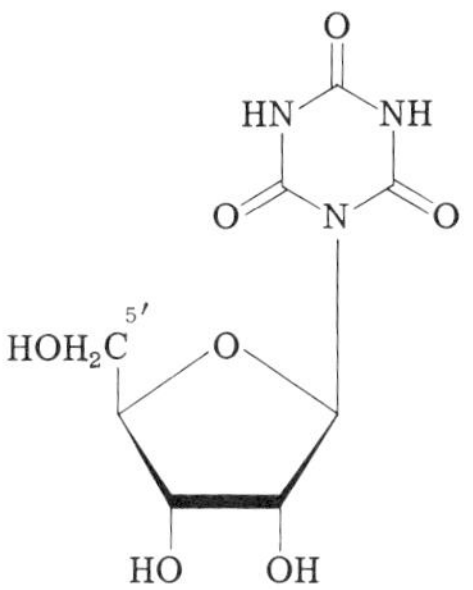

43

(108) F. E. Hruska, A. A. Smith, and J. G. Dalton, *J. Amer. Chem. Soc.*, **93**, 4334 (1971).
(109) H. Dugas, B. J. Blackburn, R. K. Robins, R. Deslauriers, and I. C. P. Smith, *J. Amer. Chem. Soc.*, **93**, 3468 (1971).

Many extensive studies[110-126] at 60, 100, or 220 MHz of dinucleotides and dinucleoside monophosphates have confirmed quite conclusively that the magnetic anisotropy of one base moiety influences the chemical shifts of the protons of the other base and, therefore, that these compounds exist in one or more folded conformations, in which the rings of the bases are stacked in parallel planes. Intra- and inter-molecular base-stacking both appear to be extremely common,[103,122-126] and the chemical shifts experienced by H-1′, H-2, and H-8 of adenylyl-(3′ → 5′)-adenosine (**44**) on inter-

(110) W. L. Meyer, H. P. Mahler, and R. H. Baker, Jr., *Biochim. Biophys. Acta*, **64**, 353 (1962).
(111) O. Jardetzky and N. G. Wade-Jardetzky, *J. Biol. Chem.*, **241**, 85 (1966).
(112) R. H. Sarma, V. Ross, and N. O. Kaplan, *Biochemistry*, **7**, 3052 (1968).
(113) D. P. Hollis, *Org. Magn. Resonance*, **1**, 305 (1969).
(114) R. U. Lemieux and J. W. Lown, *Can. J. Chem.*, **41**, 889 (1963).
(115) D. P. Hollis, *Biochemistry*, **6**, 2080 (1967).
(116) R. H. Sarma and N. O. Kaplan, *J. Biol. Chem.*, **244**, 771 (1969).
(117) D. J. Patel, *Nature*, **221**, 1241 (1969).
(118) R. H. Sarma and N. O. Kaplan, *Biochemistry*, **9**, 539 (1970).
(119) R. H. Sarma, M. C. Moore, and N. O. Kaplan, *Biochemistry*, **9**, 549 (1970).
(120) R. H. Sarma and N. O. Kaplan, *Biochemistry*, **9**, 557 (1970).
(121) W. A. Catterall, D. P. Hollis, and C. F. Walter, *Biochemistry*, **8**, 4032 (1969).
(122) S. I. Chan and J. H. Nelson, *J. Amer. Chem. Soc.*, **91**, 168 (1969).
(123) B. W. Bangerter and S. I. Chan, *J. Amer. Chem. Soc.*, **91**, 3910 (1969).
(124) K. N. Fang, N. S. Kondo, P. S. Miller, and P. O. P. Ts'O, *J. Amer. Chem. Soc.*, **93**, 6647 (1971).
(125) P. S. Miller, K. N. Fang, N. S. Kondo, and P. O. P. Ts'O, *J. Amer. Chem. Soc.*, **93**, 6657 (1971).
(126) I. Feldman and R. P. Agarwal, *J. Amer. Chem. Soc.*, **90**, 7329 (1968).

calation of added purine between the bases of the folded forms, or self-intercalation of these forms, have been found to provide a sensitive probe for the relative strength of the intramolecular-stacking interaction.[122]

It had been anticipated[112,113] that the methylene protons at C-4 of the dihydronicotinamide moiety of reduced nicotinamide adenine dinucleotide (**45**), a biochemically and clinically important coenzyme,

should be chemically non-equivalent, because, in a folded conformation, one of these protons is near to the anisotropic, adenine ring, whereas the other proton is remote. The earlier studies at[110–112,114] 60 and[115] 100 MHz failed to detect this non-equivalence, but later spectra, recorded[116–118] at 220 MHz, show the C-4 methylene resonances as an AB quartet having an apparent chemical-shift (δ_{AB}) of 15 Hz. There has been considerable debate[113,116–118] as to whether or not this non-equivalence, which was also barely detected[113] by 400 scans at 100 MHz, is related to the known stereospecificity of dehydrogenases towards the nicotinamide ring, a specificity manifested in the existence of two types of enzyme, one of which transfers the H-4 lying above the ring and the other, the H-4 below the ring.[113,118] The AB quartet observed for the H_2C-4 group of **45** in deuterium oxide at temperatures of 5–29° collapses at higher temperatures (59–74°) to a broad singlet that, simultaneously, moves *downfield* and appears near to the corresponding resonance of the reduced nicotinamide mononucleotide.[118] As an averaged chemical-shift of the A and B nuclei is not obtained, the disappearance of their nonequivalence[118] is almost certainly due to unfolding[118,121] of the conformations of **45** at higher temperatures, but there has been controversy as to exactly what is implied in terms of the relative proportions of folded and unfolded forms, and their rates of interconversion.[112,113,116-118] This situation has been clarified by detailed studies[118] at 220 MHz, which showed that the non-equivalence is quite common in dinucleotides, but not in mononucleotides, and, more importantly, that **45** which has been stereospecifically monodeuterated at C-4 of the nicotinamide ring by enzymic transfer of deuterium from ethanol-d_6 to the oxidized form (**46**) of **45** shows *two* peaks (of unequal intensity), for the *remaining* proton at C-4, that collapse to a singlet at 74°. The corresponding, monodeuterated nicotinamide mononucleotide (**47**), prepared from **45**-4-*d* by the

47

R configuration at C-4

action of snake-venom phosphate diesterase, showed,[118] at 5°, only a singlet for the remaining H-4 proton. These observations appear to be best interpreted in terms of unequal proportions of M and P (left- and right-handed) helical, folded forms of **45**-4-*d* that are interconverting sufficiently slowly (rate <90 Hz) at room temperature for a separate signal to be observed for the remaining H-4 in each of its two environments.[118] For **45**, this explanation implies that two AB quartets could be observed, in principle, for the methylene protons at C-4, but that, owing to accidental equivalence and low resolution in the 220-MHz spectra, these quartets overlap to give a single, broad, asymmetric, "AB" quartet.[118]

Even at 220 MHz, the spectral regions corresponding to the D-ribofuranosyl moieties of mono- and di-nucleotides are complex,[120] and complete analysis of their signals will clearly require even higher frequencies, or shifts produced by suitable paramagnetic ions in aqueous solution. Nevertheless, a selection of values of $J_{1',2'}$, $J_{2',3'}$, and

$J_{3',4'}$ has been obtained at 220 MHz for the D-ribofuranoid rings attached to either the adenine or pyridine residues of mono- and dinucleotides.[120] The increase (5.1 → 8.1 Hz) observed in the value of $J_{1',2'}$ for the D-ribofuranoid ring attached to the nicotinamide ring when **46** is reduced to **45** has been attributed to different shapes for the furanoid rings in these compounds, rather than to removal of the electronegative effect of the positively charged, nitrogen atom.[120] The value of $J_{1',2'}$ for the adenine–ribose moiety changes very little[120] when **46** is reduced to **45**. The spectral assignments made for the reduced and oxidized coenzymes at 220 MHz have been discussed in detail, and those for the H-1' signals of the adenine– and pyridine–ribofuranoid rings of the reduced coenzyme were shown to disagree[120] with those made earlier[111] at 60 MHz.

5. Polysaccharides

Well-resolved, 220-MHz spectra have been obtained from chloroform-d solutions of a number of fully acetylated, methylated, and benzoylated derivatives of polysaccharides.[127–129] For example, in the 220-MHz spectrum of a 2,3,6-tri-O-acetylcellulose (molecular weight ~60,000), three acetyl methyl singlets were observed, and also separate multiplets for each group of protons (H-1, H-2, . . . H-6') attached to the carbon chain of the polysaccharide.[127] The multiplets of H-1 to H-6' were assigned in detail, and measurement of their spacings yielded large values for the coupling constants $J_{1,2}$, $J_{2,3}$, $J_{3,4}$, and $J_{4,5}$ (8–9 Hz) that completely confirmed the generally accepted structure, configuration, and conformation of cellulose.[127,128] The 220-MHz spectrum of 2,3,6-tri-O-acetylamylose (**48**) was equally

48

informative, and yielded a small value for $J_{1,2}$ (~3 Hz) and large values for $J_{2,3}$, $J_{3,4}$, and $J_{4,5}$ (8–9 Hz) that were in accord with amylose's

(127) H. Friebolin, G. Keilich, and E. Siefert, *Angew. Chem. Int. Ed. Engl.*, **8**, 766 (1969).
(128) H. Friebolin, G. Keilich, and E. Siefert, *Org. Magn. Resonance*, **2**, 457 (1970).
(129) G. Keilich, E. Siefert, and H. Friebolin, *Org. Magn. Resonance*, **3**, 31 (1971).

being composed of α-D-linked D-glucose residues in the $^4C_1$ (D) conformation.[127,129]

Similar results were obtained for the corresponding tri-O-benzoyl derivatives of cellulose and amylose, but, for polysaccharide derivatives having a degree of polymerization (d.p.) in excess of 1,000 (molecular weight >500,000), the signals of the ring protons, although still separated, were so broadened that their coupling constants could not be measured.[127] The spectra of derivatives of dextran and mannan were also analyzed, and the spectrum of tri-O-benzoylpullulan was interpreted as a superposition of the spectra of the corresponding amylose [α-D-(1 → 4)] and dextran [α-D-(1 → 6)] derivatives.[127] The spectrum of 2,3,4-tri-O-benzoyldextran (**49**) (d.p. 50) yielded the

values $J_{1,2} < 4$, $J_{2,3} = J_{3,4} = J_{4,5} \sim 9$, and $J_{6,6'} \sim 10$ Hz, in agreement with the stereochemistry expected.[129] The best results were obtained from O-acetyl and O-methyl derivatives of d.p. <200 and from benzoyl derivatives having d.p. <100.

These studies[127-129] and others at 100 MHz suggested that reasonably well-resolved p.m.r. spectra can frequently be obtained from polysaccharides if these are examined as solutions of their per-O-acyl[130] or per-O-methyl[131,132] derivatives in *organic* solvents. Highly viscous, aqueous solutions of the free polysaccharides are less useful in this regard,[131,133-135] although better results have been obtained at 70–80° than at room temperature.[133,136]

(130) B. Casu, M. Reggiani, G. G. Gallo, and A. Vigevani, *Carbohyd. Res.*, **12**, 157 (1970).
(131) J. N. C. Whyte and J. R. Englar, *Can. J. Chem.*, **49**, 1302 (1971).
(132) J. N. C. Whyte, *Carbohyd. Res.*, **16**, 220 (1971).
(133) C. A. Glass, *Can. J. Chem.*, **43**, 2652 (1965).
(134) M. Falk, D. G. Smith, J. McLachlan, and A. G. McInnes, *Can. J. Chem.*, **44**, 2269 (1966).
(135) D. A. Rees and A. W. Wight, *J. Chem. Soc.* (B), 1366 (1971).
(136) A. S. Perlin, B. Casu, G. R. Sanderson, and L. F. Johnson, *Can. J. Chem.*, **48**, 2260 (1970).

Characteristic patterns for the H-1 signals in the 100-MHz spectra of many different polysaccharides have been observed that evidently arise from variations in side-chain structure.[137–139] Enzymolyses of such polysaccharides have been monitored by observation of the changes in the H-1 regions of the spectra.[137]

More-complex, 220-MHz spectra have been recorded[136] for each of the glycosaminoglycans heparin, heparitin, chondroitin 4- and 6-sulfate, dermatan sulfate, hyaluronic acid, and keratan sulfate, in solution in deuterium oxide at 50–70°. On the basis of the improved separation and integration of signals at 220 MHz (as compared with 100 MHz), assignments for the spectra of heparin were made in more detail, with some revision of the interpretations proposed previously.[140] The 220-MHz spectra of heparin (and chemical evidence[141]) suggest that major structural components are alternating, (1 → 4)-linked α-L-idopyranosyluronic acid 2-sulfate (**50**) and 2-deoxy-2-sulfoamino-α-D-glucopyranosyl 6-sulfate (**51**) groups.[136] Because strong signals for

D-glucopyranosyluronic acid were not detected, this moiety was presumed[136] to be a minor component of heparin. The assignments of anomeric configuration for the monosaccharide residues, and confirmation of the major repeating-unit, were based on the observation[136] of narrow signals at low field for H-1 of each residue in heparin, and on the isolation of the disaccharide 2-deoxy-4-O-(4-deoxy-α-L-*threo*-hex-4-enopyranosyluronic acid 2-sulfate)-2-sulfoamino-D-glucopyranose 6-sulfate (**52**) in ~75% yield by enzymolysis of heparin.[142]

---

(137) P. A. J. Gorin, J. F. T. Spencer, and D. E. Eveleigh, *Carbohyd. Res.*, **11**, 387 (1969).
(138) P. A. J. Gorin, J. F. T. Spencer, and S. S. Bhattacharjee, *Can. J. Chem.*, **47**, 1499 (1969).
(139) P. A. J. Gorin, J. F. T. Spencer, and M. J. Magus, *Can. J. Chem.*, **47**, 3569 (1969).
(140) A. S. Perlin, M. Mazurek, L. B. Jaques, and L. W. Kavanagh, *Carbohyd. Res.*, **7**, 369 (1968), and references cited therein.
(141) M. L. Wolfrom, S. Honda, and P. Y. Wang, *Chem. Commun.*, 505 (1968); *Carbohyd. Res.*, **10**, 259 (1969); A. S. Perlin and G. R. Sanderson, *Carbohyd. Res.*, **12**, 183 (1970).
(142) A. S. Perlin, D. M. Mackie, and C. P. Dietrich, *Carbohyd. Res.*, **18**, 185 (1971).

52

The α-L anomeric configuration of the uronic acid portion of this disaccharide was supported by the observation, at 100 MHz, of a long-range coupling-constant ($J_{2,4}$ 1.1 Hz) that was found to be characteristic of 4-deoxy-α-L-*threo*-hex-4-enopyranose derivatives, but not of their β-L anomers.[142]

A disaccharide of related structure, namely, 2-acetamido-2-deoxy-3-O-(4-deoxy-α-L-*threo*-hex-4-enopyranosyluronic acid)-α-D-galactopyranose, has been studied[143] at 60 and 100 MHz, and the conformation **53** proposed for it. The presence of N-acetyl signals of

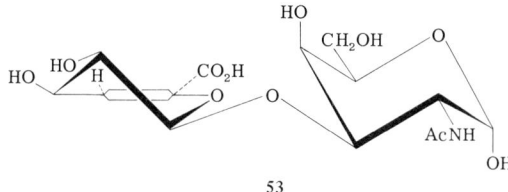

53

variable intensity is a puzzling feature of the spectra of various preparations of heparin, and suggests[136] either that 2-acetamido-2-deoxy-D-glucopyranosyl residues form a minor portion of the structure of heparin, or that some preparations are contaminated with another polysaccharide, for example, heparitin, to the extent of ~20%.

The presence of a number of signals, of differing relative intensities, in the 220-MHz spectra of heparitin suggests that it has a structure more heterogeneous than those of heparins, and that it contains approximately equal proportions of 2-acetamido-2-deoxy- and 2-

(143) S. Hirano, *Org. Magn. Resonance*, **2**, 577 (1970).

deoxy-2-sulfoamino-α-D-glucopyranosyl residues, and also D-glucuronic acid residues.[136]

The 220-MHz spectra of chondroitin 4- and 6-sulfate, and of dermatan sulfate, were consistent with structures that had previously been proposed for these compounds, in contrast to the 220-MHz spectrum of keratan sulfate.[136] The spectra of chondroitin 6-sulfate, keratan sulfate, and viscous solutions of hyaluronic acid showed broad bands, and, for hyaluronic acid, little structural information could be deduced from its spectrum.[136]

Clearly, the recording of p.m.r. spectra of complex molecules at high frequencies does not *guarantee* satisfactory interpretation, but the chances of success are considerably increased. General applications of the technique have been reviewed.[144]

IV. FOURIER-TRANSFORM, NUCLEAR MAGNETIC RESONANCE SPECTROSCOPY

1. General Considerations

Undoubtedly, the most significant advance in n.m.r. spectroscopy since 1964 has been the development of Fourier-transform techniques.[42,145] The conventional methods of recording a spectrum, either by sweeping the radiofrequency at a constant magnetic field or by sweeping the magnetic field while maintaining the radiofrequency constant, both suffer from the disadvantages that the resonances are scanned at different times and that much of the total observation time may be wasted in scanning empty regions (the baseline). Moreover, to prevent broadening and distortion of the resonance lines, it is necessary to use relatively low sweep-rates (of the order of the resolution desired) which, in turn, necessitate the use of low power-levels of the radiofrequency, so that saturation effects are avoided. The use of such a weak radiofrequency field leads to excitation of only a narrow region of the spectrum, perhaps 0.5 Hz in width.

Fortunately, radiofrequency pulses, or a noise-modulated, constant radiofrequency, provide a means for simultaneous excitation of all nuclei of a particular isotopic species; this technique offers an al-

(144) W. Naegele, in "Determination of Organic Structures by Physical Methods," F. C. Nachod and J. J. Zuckerman, eds., Academic Press, New York, 1971, Vol. 4, p. 1.
(145) R. R. Ernst and W. A. Anderson, *Rev. Sci. Instrum.*, **37**, 93 (1966).

ternative mode of operation that, for averaging of signals from dilute samples, or from nuclei of low natural abundance or sensitivity, is much more efficient than the conventional, continuous-wave (c.w.) mode.

2. Distinction of Continuous-wave and Pulsed Techniques

In order to appreciate the application of pulsed n.m.r. spectroscopy, it is instructive to consider the macroscopic features of an assembly of magnetic moments of the nuclei.

For the purposes of a discussion related to an experimental spectrometer, it will be assumed that the construction of its probe is based on the common, crossed-coil[146] arrangement, in which the static magnetic field (H_0) is applied along the z axis, a transmitter coil is oriented

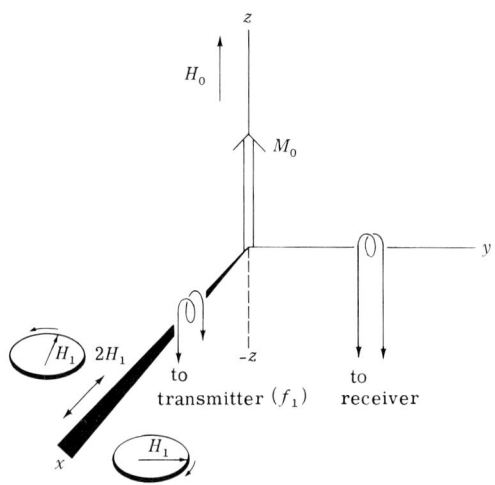

Crossed-coil arrangement in a n.m.r. probe, and resolution of the r.f. field ($2H_1$) into two counter-rotating components (H_1 and H_1).

with its axis along the x axis, and a receiver coil along the y axis. According to the Boltzmann distribution,[147] there exists a slight excess of nuclear moments precessing about, and having a uniform dis-

---

(146) This is not mandatory, and, in fact, a single-coil arrangement in which the pulse is applied to a receiver coil that is close to the sample gives good results also.
(147) $N_{+z}/N_{-z} = \exp(2\mu H_0/kT)$, where N_{+z} and N_{-z} are the numbers of nuclei aligned either with the field (+z) or against it (−z), and μ is the nuclear magnetic moment; J. D. Roberts, "Nuclear Magnetic Resonance," McGraw–Hill, New York, 1959, p. 9.

tribution about, the $+z$ axis, which gives a resultant, net, macroscopic magnetization (M_0) along this axis. As the receiver coil is orthogonal to M_0, no signal is detected thus far. However, any n.m.r. experiment in which the vector M_0 is tipped out of the z axis will give, in the x–y plane, a component of magnetization that will then induce a signal in the receiver coil.

In c.w.-n.m.r. spectroscopy, a relatively weak, but rapidly oscillating, magnetic field is produced on the x axis by the application of a continuous, low-powered radiofrequency (r.f.) to the transmitter coil(s). As this radiofrequency approaches the resonance frequency, the magnetization vector is very slightly tipped out of the z axis, and precesses about this axis. When this frequency of precession is matched by the r.f. applied (the resonance condition), some of the individual, nuclear moments undergo transitions to the less-stable energy-level represented by precession about the $-z$ direction, accompanied by absorption of energy from the transmitter coil.

For pulsed n.m.r., the quantum-theoretical concept of $2I + 1$ discrete, nuclear energy-levels (I is the spin-quantum number) produced by quantization of individual, nuclear, magnetic moments along the z axis is less useful than for c.w.-n.m.r., and therefore pulsed n.m.r. is usually described, both experimentally and theoretically, in terms of a continuum of extensive, spatial manipulations of the macroscopic magnetization-vector, in a classical, mechanical sense.

3. Pulse Methods

In most pulsed n.m.r. experiments, the r.f. power applied to the transmitter coil is much greater than that used for c.w.-n.m.r., and, typically, may be in the range of 50–35,000 W, or more. This power is applied in the form of a narrow, square pulse; that is, the r.f. is switched on very rapidly, and then a short time later switched off rapidly, typically after a period of the order of 1–200 μsec. As the receiver of the spectrometer is saturated while the r.f. pulse is on, no useful signal can be observed until this pulse is turned off. However, at this point, any persistent components of magnetization in the x–y plane that have been produced by the action of the pulse can then be measured by means of the signal induced in the receiver coil. Because this signal decays exponentially, due to relaxation effects,[145] and is observed in the absence of the exciting radiation, it is called the free-induction decay (f.i.d.) signal,[145] or free-precession signal.[148]

(148) H. D. W. Hill and R. Freeman, "Introduction to Fourier Transform NMR," Varian Associates, Palo Alto, 1970, p. 2.

Interpretation of the motion of the macroscopic, magnetization vector (M) is simpler, both visually and theoretically, when referred to a system of Cartesian coordinates that rotates about the z axis at an angular velocity corresponding to the resonance frequency of the nuclei under consideration.[145,149] In the laboratory coordinate-system, the r.f. applied produces a magnetic field that varies sinusoidally along the x axis and that may be regarded[150] as the resultant of two magnetic fields that are counter-rotating in the x–y plane. However, only the circulating component that rotates in the same sense as the precessing nuclei is effective in interacting with them. In the *rotating*

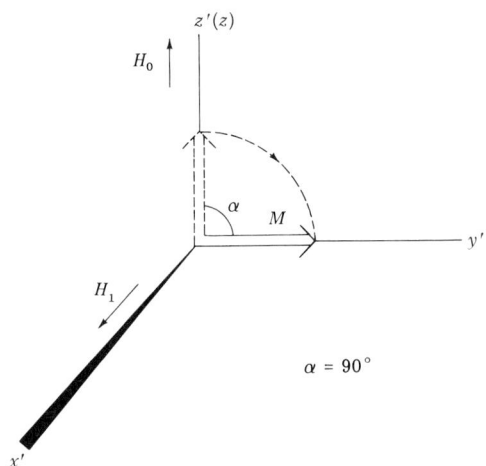

Effect of a "90° pulse" on the magnetization vector (M) in a frame of coordinates (x', y', z') rotating about z at a frequency f_1.

frame of reference defined by the new axes, x', y', and z', this component is rotating at the same velocity as the frame of reference, and, therefore, has a constant amplitude of magnetic field along x'. The effect of a sudden application of a very strong magnetic field (H_1) along x', is to start precession of M about this axis. However, this precession ceases when the r.f. pulse is turned off. It is the precession angle (α) described by M while the pulse is on that is of much concern in pulsed n.m.r. spectroscopy. A pulse that causes M to rotate about x' through an angle of 90° is described as a "90° pulse" and is an important type of single pulse, as it rotates M into the y'

(149) T. C. Farrar and E. D. Becker, "Pulse and Fourier Transform NMR," Academic Press, New York, 1971, p. 8.
(150) Ref. 4, p. 13.

axis, thus producing the maximum magnetization in the x–y plane of the laboratory frame of reference and, therefore, the maximum signal induced in the receiver coil as M (aligned along y') passes through the y axis.

In the laboratory frame of reference, the perturbed vector M describes a complex motion given by the resultant of a rapid precession about $z(z')$ (for example, at a frequency of 40–300 MHz) and a slower precession about the rotating axis x' (for example, at 1–250 kHz). The introduction of the concept of a rotating frame, therefore, simplifies the apparent motion of M by nullifying the rapid precession about the static magnetic field H_0.

In general, the "flip angle" (α) of the magnetization vector about x' is given[145] by $\alpha = 180\gamma H_1 t_p/\pi$ (degrees), where γ is the constant, gyromagnetic ratio of the nucleus, and t_p is the time during which the pulse is on (pulse-width). In order to obtain different values of α, it is customary to vary t_p instead of the power of the r.f. pulse ($2H_1$). If this power is sufficiently large, t_p is very much shorter than the relaxation times of the nucleus, and, therefore, relaxation during the pulse may be neglected.[145]

Pulsing of the radiofrequency has a further beneficial effect. It allows excitation of nuclei over a range of resonance frequencies. This is possible because square-wave modulation (pulsing) of a single r.f. carrier (f_1) generates[151] a spectrum of Fourier components over an approximate, effective, frequency range of $f_1 + t_p^{-1}$. For example, proton resonance over a range of $f_1 \pm 1,000$ Hz would require a pulse-width of 1 msec, or less. A nucleus such as ^{13}C, which has a chemical-shift range of ~400 p.p.m. (~10,000 Hz at $f_1 = 25$ MHz) requires a pulse-width of no more than 100 μsec in order to excite the whole range.

The real benefit of pulse techniques arises in the signal-averaging process, because, if the pulses are repeated rapidly enough, many more f.i.d. signals may be accumulated coherently in the memory of a computer, in a given time, than can be obtained from slowly swept c.w. spectra. Ernst and W. A. Anderson have proved[145] that repetitive pulsing of the r.f., with an interval of T sec *between* pulses, so modulates f_1 as to give a spectrum of sidebands having a spacing of T^{-1} Hz. For example, on this basis alone, a pulse interval of 10 sec would lead to an excitation interval of 0.1 Hz. However, because of several factors, it is not worth while to attempt the observation of a f.i.d. signal over periods greater than 10 sec. These factors are

(151) Ref. 149, pp. 16 and 68.

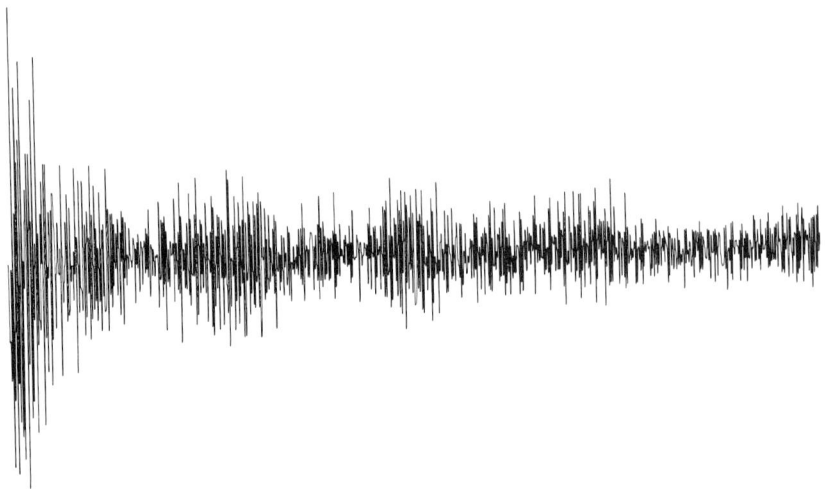

FIG. 2.—Proton, Free-induction Decay Signal[45] of a Solution of 6-Deoxy-1,2:3,4-di-O-isopropylidene-6-phthalimido-α-D-galactopyranose (54) (0.06 mg) in 4:1 (v/v) Chloroform-d (Commercial, "100%")–Hexafluorobenzene (0.06 ml) Contained in a Microcell. [The f.i.d. signal was acquired in 419 sec by using the crossed-coil mode with heteronuclear, internal, field–frequency stabilization on fluorine-19, with 1,024 pulses at 90 MHz, t_p 40 μsec, δt 400 μsec, and 1,024 datum points (for definitions, see text).]

(a) the need to maximize the signal:noise ratio of the f.i.d. signals over a given, total observation-time by collecting as many of them as possible, (b) limited computer-memory, and (c) the poor signal:noise ratio of the later part of the f.i.d. signal. Typically, pulse intervals in the range of 0.2–2.0 sec are employed. If relatively short pulse-intervals are used (compared with T_1), M will be unable to relax fully to the z axis before the next pulse is applied; this soon leads to a steady-state condition wherein the flip angle $α$ is just balanced by partial relaxation between pulses. The theoretical expression $\cos α_{\text{opt av}} = \exp(-T/T_1)$ for the average, optimum flip-angle has been developed,[145] but this angle is normally determined experimentally, merely by maximizing the amplitude of the f.i.d. signal. Presumably, this maximum is obtained in the steady state by relaxation of M from 90° to 90° $-α$ during the pulse interval, followed by excitation of M from 90° $-α$ to 90° while the pulse is on.

The use of a phase-sensitive detector to demodulate the f.i.d. signal emerging from the receiver coil is equivalent to providing a rotating frame of reference for the spectrometer itself, as the detector is referenced to the constant carrier r.f. (f_1). The f.i.d. signal detected, therefore, contains only phase and intensity information for the *difference* frequency between the carrier frequency applied and the resonance frequency of the nucleus in question. The differing frequency differences (sine waves, approximately) of chemically shifted nuclei interfere with one another to produce a f.i.d. signal that is a complex interferogram (see Fig. 2 for a f.i.d. signal of 0.06 mg of 6-deoxy-1,2:3,4-di-O-isopropylidene-6-phthalimido-α-D-galactopyranose, **54**) that cannot be interpreted directly if it is comprised of more

54

than about four different sine-waves. However, it has been known for many years that an exponential, f.i.d. spectrum in the time domain and a Lorentzian, c.w. spectrum in the frequency domain are Fourier transforms of one another.[145,152,153] Thus, with the assumption that a f.i.d. signal of sufficient signal:noise ratio can be collected and stored, Fourier analysis gives phase, frequency, and intensity information for the individual frequency-components of the complex f.i.d. signal. A single, phase-sensitive detector cannot distinguish between positive and negative difference frequencies and, therefore, it is customary to set the value of f_1 so that it lies just *outside* the range of nuclear resonance frequencies.

4. Acquisition of the Free-induction, Decay Signal

The high information-content of the f.i.d. signal (see Fig. 2) and the speed with which it must be acquired impose somewhat special requirements on the data-handling system. Successive analog signals from the phase-sensitive detector are usually digitized, and accumulated coherently within specific addresses of a digital-storage device, commonly either a small computer having fixed (wired) programs, a programmable, small or medium-sized computer, or a magnetic disk. From sampling theory, it is known that, in order to sample a sine wave unambiguously, at least two datum points per cycle must be taken.[154]

(152) I. J. Lowe and R. E. Norberg, *Phys. Rev.*, **107**, 46 (1957).
(153) Ref. 149, p. 29.
(154) Ref. 148, p. 9; Ref. 149, p. 70.

Thus, if the highest frequency present in the f.i.d. signal is F kHz, the analog-to-digital converter of the data-storage device must be capable of digitizing datum points at a rate of at least $2F$ kHz. It has been demonstrated[154] that, if a frequency is sampled at a rate of less than twice this frequency, it is folded back into the range defined by the sampling rate. For example, a frequency $F + \delta F$ then appears in the spectrum at $F - \delta F$ if it is sampled at a rate of $2F$ kHz. It is obvious that this process converts high-frequency noise into low-frequency noise, which is difficult to remove once it is present in the spectrum. This difficulty may be alleviated, or the sampled frequency-range may be restricted, by passing the f.i.d. signal through a low-pass, analog filter before digitization.

The instrumental parameter that determines the frequency range F (sweep width of the spectrum) is the sampling interval (or dwell time) δt, which is the time spent by a single computer-address in averaging or integrating one point in the f.i.d. signal. The maximum frequency F sampled (sometimes called the Nyquist frequency) is given[155] by $(2\delta t)^{-1}$. Thus, if δt is chosen on the basis of the frequency range required, the number (N) of addresses of computer memory available, or selected, then determines the acquisition time $(N\delta t)$ for the f.i.d. signal. This time is often less than that required for decay of the f.i.d. signal to an amplitude comparable with the noise, because of the limitations imposed on N by cost, or by the observation time chosen. Commonly, therefore, the later, weak part of the f.i.d. signal is truncated, and the next pulse is applied immediately, so as to maximize the number of f.i.d. signals accumulated. However, for nuclei having long relaxation times, a waiting period between acquisition of the last datum point of the f.i.d. signal and application of the next pulse may be beneficial. Short acquisition-times $(N\delta t)$ that result in poor resolution $(1/N\delta t)$ can be artificially lengthened by appending blocks of zeroes to the f.i.d. data before transformation ("zero filling"); for example, the transformation of 1,024 f.i.d. datum points followed by 3,072 zeroes (see Fig. 3) improves the resolution of the transformed spectrum by a factor of four, if field inhomogeneity can be ignored.

5. Digital Processing of the Free-induction, Decay Signal

Restriction of the frequency range of the f.i.d. signal by analog filtering has already been mentioned, but, in fact, some digital processing of the signal may be advantageous before Fourier transformation is performed.

(155) Technical literature, Nicolet Instrument Corporation, 1970.

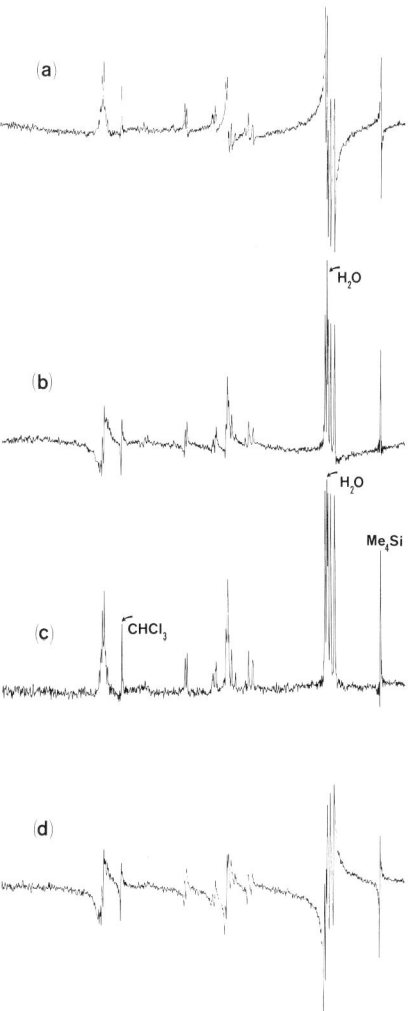

Fig. 3.—Fourier-transform, Proton Magnetic Resonance Spectra[45] of 6-Deoxy-1,2:3,4-di-O-isopropylidene-6-phthalimido-α-D-galactopyranose (**54**) (0.06 mg) at 90 MHz, Obtained by Transformation ($N = 4{,}096$) of the Free-induction Decay Signal (1,024 Datum Points, see Fig. 2), After the Appendation of 3,072 Zero, Datum Points ("Zero-filling," See Text). [(a) Spectrum associated with the real part of the transform, and (b) with the imaginary part; (c) absorption-mode spectrum computed by phase correction of the spectrum in (a); and (d) dispersion-mode spectrum computed by phase correction of the spectrum in (b). Parameters for phase correction, $A -255°$ and $B -215°$. Note that the phase of the tetramethylsilane and chloroform signals in (c) is slightly different from that of the carbohydrate derivative. By coincidence, the peak for residual water in spectrum (c) has almost the same intensity as the methyl signals, and could have been mistaken for one, had other spectra not been recorded.]

Thus, a further advantage of the Fourier-transform method is that, before transformation, either the sensitivity or the resolution of the transformed spectrum may be enhanced by digital filtering of the f.i.d. signal in the computer. The use of a "matched filter" for enhancement of sensitivity at the expense of resolution has been discussed in detail by Ernst,[42] and is based[42,145] on multiplication of the f.i.d. signal by the decreasing function $\exp(-n\delta t/T_2^*)$, where n is the index ($n = 0$ to $N - 1$) of the datum points of the signal, and T_2^* is an apparent relaxation time that is estimated from the decay time of the f.i.d. signal, and hence is matched with the experiment. The effect of this processing is to weight the points in the f.i.d. signal with their own signal:noise ratio, thus giving greater weight to the earlier points which already have a high signal:noise ratio, and, therefore, contribute most heavily to the sensitivity of the transformed spectrum.

Conversely, multiplication of the f.i.d. signal with an *increasing* exponential function $\exp(n\delta t/T_2^*)$ gives greater weight to the later points in the signal and enhances the resolution of the transformed spectrum, at the expense of sensitivity.[155a] In order to produce a strong improvement in resolution, it has been claimed[42] that the pulse interval T must be significantly longer than T_2. The Fourier transform of an infinitely long sine wave is an infinitely narrow peak. In practice, this condition is not realized and, in general, processes that shorten the observation time of the f.i.d. signal, such as a more rapid decay (for example, shorter T_2^* due to poorer field homogeneity), apodization (graduated truncation), multiplication of the f.i.d. signal with a decreasing exponential function, or a shorter pulse-interval, broaden the peaks in the transformed spectrum.

6. Fast, Fourier Transformation

Fourier-transform n.m.r. spectroscopy did not become a practical tool until the development of efficient algorithms for computerized, fast, Fourier transformation (f.f.t.) The first such algorithm to be widely appreciated was described by Cooley and Tukey[156] in 1965, although Cooley and coworkers[157] have pointed out that a somewhat

---

(155a) R. R. Ernst, R. Freeman, B. Gestblom, and T. R. Lusebrink, *Mol. Phys.*, **13**, 283 (1967).

(156) J. W. Cooley and J. W. Tukey, *Math. Comput.*, **19**, 297 (1965).

(157) J. W. Cooley, P. A. W. Lewis, and P. D. Welch, *Proc. I.E.E.E.*, **55**, 1675 (1967); *I.E.E.E. Trans. Audio Electroacoustics*, **AU-17**, 77 (1969).

similar method had been suggested in 1924 and published in 1942, but had been virtually ignored, except by one or two isolated groups working in the fields of mathematics or X-ray analysis.

The computation desired for a Fourier transform is[42]

$$F(f) = \int_{-\infty}^{\infty} A(t) \exp(-2\pi ift) dt,$$

where $F(f)$ is the function that defines the frequency (f) spectrum, $A(t)$ is the amplitude function of the f.i.d. signal in the time (t) domain, and $i = \sqrt{-1}$. For practical evaluation, this expression can be rewritten[155] as the discrete sum

$$F(j\delta f) = \sum_{n=0}^{N-1} A(n\delta t)[\cos(2\pi j\delta f.n\delta t) - i\sin(2\pi j\delta f.n\delta t)],$$

when n, the f.i.d. data-stepping index, takes values from 0 to $N-1$, and j, the frequency-stepping index, from 0 to $N/2$. The frequency-stepping increment δf $(=1/N\delta t)$ gives the resolution (line-width at half height) of the transformed spectrum. Cooley and coworkers strongly encouraged a trend toward efficient algorithms that perform pair-wise exchanges of initial and intermediate contributions from data points, and thereby accumulate progressively larger weighted sums until a complete set of Fourier coefficients is reached.[156-158] The number of operations required for these algorithms is of the order $2N\log_2 N$, as compared with N^2 separate multiplications and additions, so that the computation time is improved[156-158] by a factor of $N^2/2N\log_2 N = N/2\log_2 N$. For the transforms commonly performed on 4,096, 8,192, or 16,384 datum points, these factors are 171, 311, and 585, respectively.

An additional economical aspect is that, as the intermediate sums are generated, they replace the original data values (which are no longer required), thus restricting the computer-memory requirement to the N addresses needed for the data array plus approximately 4,000 other addresses for storage of the f.f.t. program.

Thus far, most f.f.t. computations for n.m.r. have been performed sequentially by general-purpose computers that typically require from 10–200 sec to compute a 4,096-point transform. The values of the trigonometric functions are usually obtained from a "look-up" table stored in the computer; alternatively, they may be calculated directly by use of about seven terms of a suitable, infinite series.

(158) J. W. Cooley, R. L. Garwin, C. M. Rader, B. P. Bogert, and T. G. Stockham, Jr., *I.E.E.E. Trans. Audio. Electroacoustics*, **AU-17**, 66 (1969).

From N datum points in the f.i.d. signal, the computation produces $N/2$ real (R) coefficients of amplitudes of points in the frequency spectrum (see Fig. 3a, p. 51) and a further $N/2$ coefficients associated with the imaginary (I) part of the transform (see Fig. 3b), so that[155]

$$F(j\delta f) = R(j\delta f) + iI(j\delta f).$$

One of these sets of amplitudes is redundant, in that, when correctly phased, the real set defines the absorption-mode spectrum (see Fig. 3c), and the "imaginary" set, the dispersion-mode spectrum (see Fig. 3d), which is 90° out of phase with the absorption mode.

7. Phase Correction

In general, the real and imaginary parts of the transform (see Figs. 3a and 3b) contain phase errors derived from several sources.[159] Incorrect adjustment of the phase detector in the receiver results in a constant phase-error. However, the finite duration of the pulse, the need to delay acquisition of the f.i.d. signal until after the receiver has recovered from the pulse, and any analog filters inserted between the detector and digitizer, each causes a frequency-dependent, phase-error that is usually assumed to vary linearly with frequency. The effect of this phase error is to mix the dispersion-mode spectrum with the absorption-mode spectrum (and *vice versa*) to extents that vary with the frequency. With the assumption of a linear dependence, the two modes can be unravelled by computing the linear transformations[155]

$$R'(j\delta f) = R(j\delta f)\cos\theta - I(j\delta f)\sin\theta$$

and

$$I'(j\delta f) = R(j\delta f)\sin\theta + I(j\delta f)\cos\theta,$$

where $R'(j\delta f)$ and $I'(j\delta f)$ are the new arrays of real and imaginary coefficients for amplitudes in the frequency spectrum, and θ is the variable phase error given by

$$\theta(j\delta f) = A + (B/F)j\delta f.$$

The constants A and B may be determined from a display of the spectrum on an oscilloscope, and entered into the computer either by teletypewriter,[155] or directly, by digitization of the analog voltages from several potentiometers.[160] Automatic phase-correction programs

(159) Ref. 149, p. 76.
(160) R. J. Cushley, D. R. Anderson, and S. R. Lipsky, *Anal. Chem.*, **43**, 1281 (1971).

have been described,[161] but are considered[160] to be less useful than manual corrections, because the automatic programs cannot deal so effectively with spectra having a very low signal:noise ratio.

8. Noise-stimulated Resonance

An alternative method for excitation of nuclei over a range of chemical shifts is by irradiation with a weak, noise-modulated radio-frequency, instead of with strong r.f. pulses. In one realization of this method, protons were irradiated with repetitive sequences of noise that was truly random,[162] and, in another,[163] fluorine nuclei were excited by "pseudo-random noise" generated by amplitude modulation of the r.f. with "maximum-length" sequences of pulses from a computer or shift register (a series of flip-flop devices connected by feedback loops). With the carrier wave suppressed, the latter process is equivalent to phase modulation of the r.f. by $+\pi/2$ radians when the pulse is turned on, and by $-\pi/2$ radians when it is turned off. This method is identical with that used in most broadband, heteronuclear, noise decouplers, except that greater power is required for decoupling.

The response detected by the spectrometer on stimulation of the nuclear spin-system by r.f. noise is noise that is coherent in some way with the noise input, plus incoherent noise added by the spectrometer.[163] In order to separate the latter two noise-signals, it is necessary to record both the input and output noise (for example, by digitization and storage in the memory of a computer or on magnetic tape), and then to compute a cross-correlation function of these signals by summing permuted products of amplitude values of the input and output noise at small intervals in the time domain.[162,163] Fourier transformation of the cross-correlation function then gives sets of real and imaginary coefficients in the frequency domain that are approximations to the amplitudes of the absorption- and dispersion-mode, c.w. spectra, respectively.

Each of the two noise-methods mentioned uses a periodic excitation by noise to produce a periodic response, apparently, so that signal-averaging techniques may be used to (a) diminish the total quantity of data stored, and (b) permit the computation of cross-correlation and Fourier-transform functions as discrete sums of a reasonably finite number of terms.[162,163]

(161) R. R. Ernst, *J. Magn. Resonance*, **1**, 7 (1969).
(162) R. Kaiser, *J. Magn. Resonance*, **3**, 28 (1970).
(163) R. R. Ernst, *J. Magn. Resonance*, **3**, 10 (1970).

9. Applications of Fourier-transform Techniques

a. Enhancement of Sensitivity, and General Applications.—Most applications of Fourier-transform n.m.r. spectroscopy have thus far been directed to enhancement of sensitivity. The signal:noise ratios of the pulse-Fourier-transform and c.w., slow-sweep techniques achievable in the same total observation time by detection of the centerband have been analyzed theoretically, with certain assumptions, and their ratio found[145] to be

$$\frac{(S/N) \text{ pulse}}{(S/N) \text{ c.w.}} = 0.799 (F/\delta f)^{1/2} G,$$

where $G = \left\{ \dfrac{2T_1[1 - \exp(-T/T_1)]^2}{T[1 - \exp(-2T/T_1)]} \right\}^{1/2}$

The function G has values of ~ 1 when the ratio (T/T_1) of pulse interval and longitudinal relaxation time is in the range 0.1–1. As T/T_1 increases from 1 to 100, G decreases slowly to[145] ~ 0.15. Thus, the enhancement of signal:noise ratio possible by the pulse technique is proportional to the square root of the ratio of total sweep-width F to the line width δf. For protons at 100 MHz, there results a theoretical improvement-factor of $0.8(F/\delta f)^{1/2} = 0.8(1,000/0.5)^{1/2} = 36$, and, for ^{13}C at 25 MHz, a factor of $0.8(10,000/1.0)^{1/2} = 80$.

In practice, slow-sweep conditions are rarely used for the c.w. technique, and, very often, better sensitivity can be obtained by sweeping faster and accepting some broadening and distortion of the lines. On this basis, enhancement factors have tended[160,164] to be ~ 10–20, corresponding to a time-saving factor of ~ 100–400. Ernst[42] has emphasized that the Fourier-transform technique is particularly useful for systems having low T_1/T_2 ratios (for example, protons), but less advantageous for systems having high T_1/T_2 ratios, for which rapid, multiple-scan, c.w. methods can give a significant improvement.[42]

A theoretical analysis has shown that the maximum sensitivities of pulse-Fourier and noise-Fourier methods are identical.[163] However, maximum sensitivity is only obtained from a simple pulse method if the pulse interval is very short, which is not consistent with good resolution. For the noise method, Ernst[163] has proposed that the sensitivity is determined by the amplitude of the noise input, up to a maximum of $(2/T_1)^{1/2}$, but that resolution is inversely proportional to the period (either total, or subdivided) for which the noise is applied.

(164) Ref. 148, p. 3.

This allows the possibility that optimum enhancement of sensitivity and resolution may be achieved simultaneously by the noise method.[163]

Many different *multiple*-pulse methods have been proposed; for example, driven-equilibrium, Fourier transform[165] (d.e.f.t.), spin-echo Fourier transform[166] (s.e.f.t.), and z-restoration spin-echo[167] (z.r.s.e.) These will not be described in detail, because (a) their application is restricted mainly to nuclei that do not readily display homonuclear spin-coupling (for example, ^{13}C), and (b) their utility for enhancement of sensitivity has been questioned.[168,169] It has been flatly stated that (a) even a simple pulse-sequence has the property of refocusing the components of magnetization that have been dispersed by field inhomogeneity, giving rise to a spin echo at the time of the next pulse, and (b) the refocusing of the magnetization that occurs in the steady-state response to a regular sequence of pulses enhances the sensitivity by an amount comparable with that achieved by the d.e.f.t. and s.e.f.t. techniques.[168]

By means of Fourier-transform, p.m.r. spectroscopy, spectra having a good signal:noise ratio have been rapidly obtained from 0.06 mg of 6-deoxy-1,2:3,4-di-*O*-isopropylidene-6-phthalimido-α-D-galactopyranose[45] (**54**) [see Fig. 3 (p. 51) and Fig. 4]. The 90-MHz spectra shown in Fig. 3 were obtained by Fourier transformation (in 100 sec, plus 50 sec for phase correction, if used) of 1,024 f.i.d. signals (see Fig. 2, p. 48) that had been acquired by signal averaging in a total time of 419 sec. Before transformation, 3,072 zero datum-points were inserted after the 1,024 points of the f.i.d. signal, so as to maximize the resolution in the transformed spectrum (zero-filling). This technique permitted the acquisition time for the f.i.d. signal to be diminished by a factor of four. An acquisition time of 112 min (16,384 pulses) gave a spectrum (see Fig. 4b) having an improved signal:noise ratio.

F.f.t., p.m.r. spectra at 100 MHz have been obtained[168a] from 0.36 mg of nicotinamide adenine dinucleotide (**46**, see p. 37). This study extended the range of concentration accessible for **46** to mM, thus permitting measurements of the variation of chemical shifts with concentration over a range much wider than had previously been

(165) E. D. Becker, J. A. Ferretti, and T. C. Farrar, *J. Amer. Chem. Soc.*, **91**, 7784 (1969).
(166) A. Allerhand and D. W. Cochran, *J. Amer. Chem. Soc.*, **92**, 4482 (1970).
(167) J. S. Waugh, *J. Mol. Spectrosc.*, **35**, 298 (1970).
(168) R. Freeman and H. D. W. Hill, *J. Magn. Resonance*, **4**, 366 (1971).
(168a) R. H. Sarma and R. J. Mynott, "N.m.r. Notes No. 8," Digilab Inc., Cambridge, Mass., 1971.

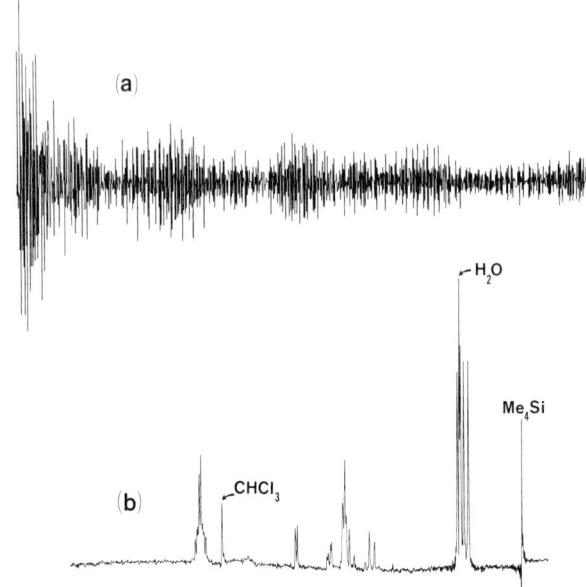

FIG. 4.—(a) Proton, Free-induction Decay Signal (1,024 datum points) of 6-Deoxy-1,2:3,4-di-O-isopropylidene-6-phthalimido-α-D-galactopyranose (**54**) (0.06 mg), Obtained[45] in the Same Way as the Signal Shown in Fig. 2, Except that 16,384 Pulses at 90 MHz Were Used, Together with Signal-averaging Over a Period of 112 min. (b) P.m.r. Spectrum[45] (Absorption-mode) Obtained by Fourier Transformation of the F.i.d. Signal (a) After Zero-filling (N = 4,096). [Although the signal:noise ratio is generally better than that of the spectrum in Fig. 3c, the strong methyl signals are flanked by side lobes (coherent noise in the baseline) due to the sudden truncation of the f.i.d. signal before it has essentially decayed to zero (see Fig. 4a). The inclusion of a suitable apodization function in the computer program would diminish this effect.]

feasible. From the results obtained, it was deduced that **46** undergoes intermolecular base-stacking in aqueous solution.[168a]

The rapidity with which a pulse-induced, f.i.d. signal may be acquired (for example, 0.1–1 sec), suggests other applications based on (a) the detection of transient chemical intermediates or tautomeric forms, and (b) other time-dependent phenomena, such as reaction kinetics, transfer of magnetization by chemical exchange, chemically induced, dynamic, nuclear polarization, and transient nuclear Overhauser effects. If a large number of unique f.i.d. signals are to be stored, a magnetic disc or tape unit is needed. F.f.t., p.m.r. spectroscopy has been used to monitor the kinetics of reactions of D-glucosyl phosphate and sucrose phosphorylase with arsenate, including the disappearance of this phosphate, the formation and subsequent hydrolysis of D-glucosyl arsenate, and the formation and subsequent mutarotation of α-D-glucose.[168b]

b. **Measurement of Nuclear Relaxation Times.**—Before the advent of Fourier transforms, the most important applications of pulsed n.m.r. spectroscopy were to the measurement of relaxation times. The relaxation times for different nuclei in a molecule could generally be measured only as averaged values, because the individual frequency components of the f.i.d. signal were not separated. Pulsed, Fourier-transform techniques have, however, been used to measure relaxation times of individual ^{13}C nuclei of a number of carbohydrates.

In the *partially relaxed, Fourier-transform* method[169] for measurement of spin–lattice relaxation (T_1), a 180° pulse is first applied in order to invert the magnetization vector M from the $+z$ direction to the $-z$ direction. As there is no component of M in the x–y plane, only longitudinal relaxation occurs, and the magnitude of M along $-z$ decreases, passes through zero, and then increases in the $+z$ direction until relaxation is complete. The magnitude of M remaining along $-z$, or growing into the $+z$ direction, is measured at various times t after the 180° pulse by applying a 90° pulse that rotates M into the $-y'$ axis if M remains along $-z$, or into the $+y'$ axis if M has grown into the $+z$ direction, thus permitting collection of an f.i.d. signal. These two situations give, respectively, negative and positive peaks in the Fourier-transform spectra, and the rate of recovery of the full positive intensity of the line is characterized by the value of T_1. If signal averaging is employed, the recycle time between successive 180°, t, 90° *pulse sequences* should be made at least three times longer than the longest T_1 of the sample.[169] The peak amplitudes A are given by

$$A = A_0[1 - 2\exp(-t/T_1)],$$

where A_0 is the amplitude at equilibrium, that is, the maximum, positive peak-height. The values of T_1 are obtained from the slopes of the straight lines obtained by plotting $\log(A_0 - A)$ against t, for each line in the spectrum.[169]

This technique has been used for measuring T_1 for each of the ^{13}C nuclei of sucrose and adenosine 5'-phosphate[169] (**55**). For sucrose, the

---

(168b) J. Grimaldi, S. L. Patt, and B. D. Sykes, *Abstr. Experimental NMR Conf., 13th*, Pacific Grove, 1972.
(169) A. Allerhand, D. Doddrell, and R. Komoroski, *J. Chem. Phys.*, **55**, 189 (1971).

^{13}C resonances were not assigned in detail, but the values of T_1 were used to infer that the motions of this molecule in aqueous solution are isotropic.[169] The T_1 values (0.19–0.23 sec) of the ^{13}C nuclei of the furanoid ring of **55** were found to be measurably longer than those (0.15–0.16 sec) of the protonated ^{13}C nuclei of the base, thus indicating that the rotational motion of the base is more restricted than that of the D-ribose moiety, possibly as a result of intermolecular base-stacking.[169]

In other ^{13}C studies of raffinose, stachyose,[170] various vitamin B_{12} derivatives,[171] and the anomeric equilibria of D-fructose and D-turanose in aqueous solution,[172] partially relaxed, Fourier-transform spectra, and the values of T_1 obtained from them, have been elegantly used to confirm the spectral assignments. The T_1 values of the ^{13}C nuclei of thermally denatured, yeast *t*RNA indicate that its D-ribose phosphate backbone undergoes rapid segmental motion at 81°. In contrast, the values of T_1 for folded *t*RNA yielded no evidence of segmental motion.[172a]

The spin–lattice relaxation-time T_1 may also be measured by use of a 90°, t, 90° pulse sequence that employs decreasing values of t to produce a progressive saturation effect.[173]

Values of the spin–spin relaxation-time (T_2) for individual spectral lines may be measured by Fourier transformation of the echoes produced by a Carr–Purcell–Meiboom–Gill type of pulse sequence,[174] but only in a simple manner, if there is no homonuclear spin-coupling present.[175] Refocusing of the dispersing magnetization-vector by

---

(170) A. Allerhand and D. Doddrell, *J. Amer. Chem. Soc.*, **93**, 2777 (1971).
(171) D. Doddrell and A. Allerhand, *Proc. Nat. Acad. Sci. U. S.*, **68**, 1083 (1971).
(172) D. Doddrell and A. Allerhand, *J. Amer. Chem. Soc.*, **93**, 2779 (1971).
(172a) A. Allerhand and R. Komoroski, *Abstr. Experimental NMR Conf., 13th, Pacific Grove*, 1972.
(173) R. Freeman and H. D. W. Hill, *J. Chem. Phys.*, **54**, 3367 (1971).
(174) Ref. 149, p. 32.
(175) R. Freeman and H. D. W. Hill, *J. Chem. Phys.*, **54**, 301 (1971).

application of 180° pulses removes the effect of field inhomogeneity, so that the rate of decay of the peaks obtained by Fourier transformation of each echo characterizes the true, or natural, value of T_2. If homonuclear coupling is present, this so modulates the spin echoes that refocusing is incomplete.[175] This so-called "J-modulation" has been turned to good advantage in a new type of high-resolution spectrum obtained by Fourier transformation of the *envelope* of spin echoes.[175] Normally, the fine structure of these spin echoes is a complex waveform that contains information about chemical shifts *and* coupling constants. However, it has been shown that, under certain limiting conditions, the amplitudes of the echo maxima are modulated only by the coupling constants, so that, if these amplitudes are collected as data points, the chemical-shift and coupling-constant information present in the fine structure of the echoes is effectively ignored.[175]

The general term "spin-echo spectrum" was suggested for the Fourier transformation of a spin echo, but the spectrum was termed a "J spectrum" if the following conditions were satisfied:[175] (a) the spin system is first-order, (b) the pulse interval T must be sufficiently long in comparison with the chemical-shift difference δ that $2\pi T\delta \gg 1$, (c) the 180° pulses that refocus the magnetization vector to give the echo must be exactly 180°, or compensated by the Meiboom–Gill technique,[174] and (d) other sources of modulation of spin echoes must be absent.[175] Under these conditions, the Fourier transform yields a novel type of spectrum in which the peaks have frequencies equal to half of the sums and differences of the coupling constants, thus giving peak spacings equal to the coupling constants. The most remarkable aspect of these J spectra is that their peak widths are equal to $1/\pi T_2$ instead of $1/\pi T_2^\circ$; this relationship evidently arises because the data points subjected to transformation have been taken from the maxima of the spin echoes. At these maxima, the magnetization vector has refocused and it, therefore, does not display the effect of the field inhomogeneity that often dominates[176] the con-

---

(176) The transverse relaxation rate (R°), which characterizes the rate at which the component of magnetization in the x–y plane disappears (and, hence, the rate of decay of the f.i.d. signal), is given by

$$R^\circ = \frac{1}{T_2^\circ} = \frac{1}{T_2} + \frac{1}{T_2^{inhomo}} = \frac{1}{T_2} + \frac{\gamma \delta H_0}{2}$$

where, as before, T_2 is the true or natural relaxation-time due to exchange of energy with neighboring spins, and T_2^{inhomo} characterizes the part of the experimental relaxation time T_2° that is due to the field inhomogeneity δH_0; see Ref. 149, p. 4.

tributions to T_2^*. This circumstance amounts to a resolution-enhancement method for measurement of very small coupling-constants, wherein one can "look inside" the inhomogeneity of the magnetic field.

For 3-bromothiophene-2-aldehyde (**56**), a long-range coupling-constant of 0.051 Hz was measured between H-4 and the aldehyde proton.[175] From the peak widths of the J spectra, the value of T_2 for the aldehyde proton was found to be 11.2 sec. Partial J spectra were obtained for this proton, and also for the ring protons of methyl furoate (**57**), by restricting the range of modulation frequencies

detected from the spin echo by means of a narrow-band filter. Partial J spectra allow the complete J-spectrum to be readily assigned.[175]

Thus far, it has been assumed that successive f.i.d. signals or spin echoes must be accumulated coherently by maintenance of a very stable field–frequency ratio. This requirement is not absolute, as has been shown by the development of a new *Fourier difference-spectroscopy* method,[177] in which an envelope detector is used to extract, from the f.i.d. signal, difference frequencies that are, to a good approximation, independent of magnetic-field variations, so that a lock system is not required. The difference frequencies are measured with respect to a reference compound that must be present in high concentration, but whose response is best removed before Fourier transformation, preferably by digital subtraction of a polynomial approximation of the response. For proton spectra, the method was shown[177] to give a sensitivity-enhancement factor of 9.5, which is close to that observed in conventional Fourier, p.m.r. spectroscopy.[145]

V. ANALYSIS OF SPECTRA

1. Non-equivalence of Nuclei in Carbohydrates

Experience with the p.m.r. spectra of carbohydrate derivatives has shown that most protons (or groups of protons) in similar chemical

(177) R. R. Ernst, *J. Magn. Resonance*, **4**, 280 (1971).

environments should, in principle, always be *assumed* to have different chemical shifts (or be anisochronous[179]), until they can or cannot be proved to be chemically equivalent (isochronous[178,179]) by analysis or by considerations of symmetry.[179] Typical examples of such nonequivalence of chemical shifts are the benzylic protons of benzyl glycosides[43] and ethers,[82] the chain-terminal or ring-methylene protons of many carbohydrates and their derivatives, the methyl groups of isopropylidene acetals, acetylated aldopentose dimethyl acetals,[180] 4-*C*-isopropyl-1,2-*O*-isopropylidene-α-D-*xylo*-tetrofuranose[181] (**58**), and the ethyl groups of bis(ethylsulfonyl)-(2,3,4-tri-*O*-acetyl-β-D-ribopyranosyl)methane[182] (**59**). These examples are described by

the general concept of *diastereotopism*;[179] that is, the nuclei or groups of nuclei in question are *diastereotopic* if they reside in diastereomeric environments and, hence, cannot be interchanged by any symmetry operation. The diastereotopic nuclei or groups of nuclei may, in principle, have different chemical shifts in chiral and achiral solvents.[179] In carbohydrate molecules, the diastereomeric environments are caused most frequently by dissymmetry in the remainder of the molecule, although, in a broad sense, this dissymmetry is neither a necessary nor a sufficient condition for the presence of diastereotopic groups.[179] Diastereotopic nuclei may, of course, accidentally be isochronous.

There are two types of molecular symmetry that cause chemical-shift equivalence. Nuclei or groups of nuclei that are interchangeable by a symmetry operation involving a simple *n*-fold axis of symmetry (C_n) have been termed *equivalent*, and are isochronous in chiral and

(178) A. Abragam, "The Principles of Nuclear Magnetism," Clarendon, Oxford, 1961, p. 480.
(179) K. Mislow and M. Raban, *Top. Stereochem.*, **1**, 1 (1967).
(180) J. Defaye, D. Gagnaire, D. Horton, and M. Muesser, *Carbohyd. Res.*, **21**, 407 (1972).
(181) D. Horton and C. G. Tindall, Jr., *Carbohyd. Res.*, **8**, 328 (1968).
(182) B. Coxon and L. Hough, *Carbohyd. Res.*, **8**, 379 (1968).

achiral solvents.[179] On the other hand, nuclei or groups of nuclei that may be interchanged by a rotation–reflection operation (S_n) reside in enantiomeric environments, have been termed *enantiotopic*, and are expected[179] to be isochronous in achiral solvents, but anisochronous in chiral solvents because of diastereomeric interactions. Symmetry rules for ascertaining sets of equivalent nuclei in rapidly interconverting conformations,[183] and general discussions[184] of equivalence and nonequivalence of nuclei in high-resolution, n.m.r. spectroscopy have been given.

Common usage of the term "magnetic nonequivalence" has been criticized[179] on the grounds that it is used to refer to nonequivalence either of chemical shifts or of coupling constants. However, from the point of view of simple logic, this usage is *not* unreasonable, as the negated form of magnetic equivalence, "not (chemical-shift equivalence and spin-coupling equivalence)" is the same as "not chemical-shift equivalence *or* not spin-coupling equivalence" (DeMorgan's theorem[185]). Some of the confusion and duplication of terms has arisen because the primary definitions of chemical-shift equivalence, symmetrical equivalence, and magnetic equivalence given in standard texts[186] refer only to the positive property of equivalence and not to its negated form.

A simple, two-fold axis of symmetry (C_2), and, hence, equivalence of the protons interchanged by this symmetry operation, is quite common in specific conformations of inositols and their derivatives, for example, the half-chair conformation (**60**) of *myo*-inosose-2 phenylosotriazole derivatives,[187] and in derivatives of alditols having an even number of chain carbon atoms,[188] for example, 2,3:4,5-dianhydro-D-iditol and its 1,6-diacetate and -benzoate[189] (**61**). The 2,5-*O*-meth-

60 R = H or Ac   61 R = H, Ac, or Bz

(183) F. S. Mortimer, *J. Magn. Resonance*, **1**, 1 (1969).
(184) M. van Gorkom and G. E. Hall, *Quart. Rev.* (London), **22**, 14 (1968); T. H. Siddall and W. E. Stewart, *Progr. Nuc. Magn. Resonance Spectrosc.*, **5**, 33 (1969).
(185) L. W. Bell, "Digital Concepts," Tektronix Inc., Beaverton, Oregon, 1968, p. 31.
(186) For example, see Ref. 4, Vol. 1, p. 284.
(187) A. J. Fatiadi, *Chem. Ind.* (London), 617 (1969).
(188) Ref. 33, pp. 21, 23, and 218.
(189) R. S. Tipson and A. Cohen, *Carbohyd. Res.*, **7**, 232 (1968).

ylene protons of 1,3:2,5:4,6-tri-*O*-methylene-D-mannitol (**62**) are isochronous, even at −82°, and this property has been cited as evidence that the 1,3-dioxepan ring of this acetal exists in skew forms having a C_2 axis of symmetry that passes through the 2,5-*O*-methylene carbon atom.[190] Conformational analysis favors the skew form, which is thought to be even more probable for 2,5-*O*-isopropylidene-1,3:4,6-di-*O*-methylene-D-mannitol (**63**), the methyl groups of which were also found to be isochronous.[190] The chemical shifts of the ethylidene protons of 1,3:2,5:4,6-tri-*O*-ethylidene-D-mannitol (**64**) and 2,5-*O*-ethylidene-1,3:4,6-di-*O*-methylene-D-mannitol (**65**) have been

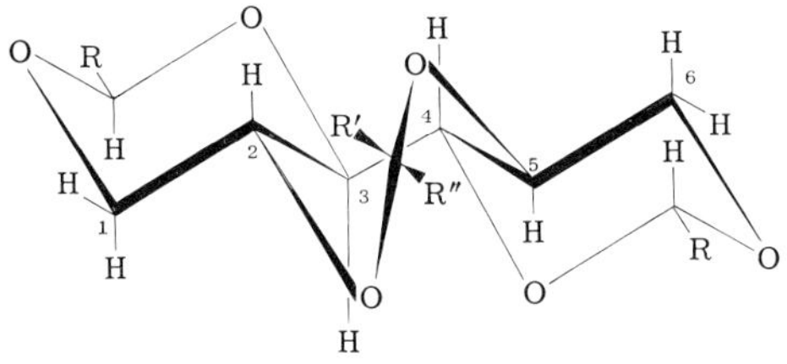

62 R = H; R′ = H; R″ = H
63 R = H; R′ = Me; R″ = Me
64 R = Me; R′ = Me; R″ = H
65 R = H; R′ = Me; R″ = H

interpreted in favor of the same skew form.[191] In contrast, the equivalence of the methylene protons of 1,6-dideoxy-2,5-*O*-methylene-D-mannitol (**66**) is believed[192] to be due to fast interconversion of skew

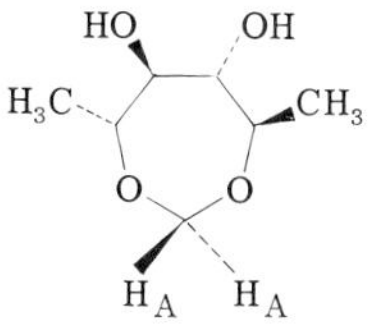

66

and chair forms, instead of to the exclusive existence of a form having C_2 symmetry. These arguments appear to be somewhat compromised by the discovery that the diastereotopic methylene protons of 3,4-*O*-benzylidene-2,5-*O*-methylene-D-mannitol (**67**) and three of its derivatives (**68, 69,** and **70**) are accidentally equivalent.[192] Accidental equivalence has also been observed[193] for the diastereotopic 2,3-*O*-methylene protons of methyl 4,6-*O*-benzylidene-2,3-*O*-methylene-α-

(190) T. B. Grindley, J. F. Stoddart, and W. A. Szarek, *J. Chem. Soc.* (*B*), 172 (1969).
(191) T. B. Grindley, J. F. Stoddart, and W. A. Szarek, *J. Chem. Soc.* (*B*), 623 (1969).
(192) J. F. Stoddart and W. A. Szarek, *J. Chem. Soc.* (*B*), 437 (1971).
(193) J. S. Brimacombe, A. B. Foster, B. D. Jones, and J. J. Willard, *J. Chem. Soc.* (*C*), 2404 (1967).

67 R = CH$_2$OH   69 R = CH$_2$OTs
68 R = CH$_2$OBz   70 R = CH$_3$

D-glucopyranoside, but not for those of its D-*galacto* isomer, for which it has been pointed out that supposition[193] of preferential shielding of one of the O-methylene protons is not necessary to account for their nonequivalence.[194]

2. The Limitations of First-order Analysis

First-order analysis[2] of p.m.r. spectra (in which the spacings within the multiplets are taken to be the coupling constants, and the nonweighted, mean positions of the peaks within the multiplets are reported as the chemical shifts) is very widely used by carbohydrate chemists, but not always justifiedly. In compiling values reported for geminal and vicinal proton coupling-constants of organic compounds, Bothner-By[195] commented in 1965 that it was necessary to regard with suspicion ~90% of values obtained in this way, because of their unreliability. This comment typifies the dissatisfaction of the spectroscopy specialist with the p.m.r. parameters reported by many organic chemists—parameters that have often been obtained by first-order analysis of spin systems that are not first-order; such parameters may furnish useful, descriptive "fingerprints" for characterization, but the extent to which the reported values may differ from the absolute values is often not stated explicitly. It is to be hoped that this situation has improved since 1965 because of the greater availability of high-field spectrometers that tend to produce spectra that are more nearly first-order spectra than do the low-field instruments, but, in some respects, this gain has been offset by the tremendous increase in the number of data published.

It has been emphasized[196] that the measurement of more accurate parameters is important for several reasons: (*a*) more-rigorous testing

(194) T. B. Grindley, J. F. Stoddart, and W. A. Szarek, *J. Amer. Chem. Soc.*, **91**, 4722 (1969).
(195) A. A. Bothner-By, *Advan. Magn. Resonance*, **1**, 195 (1965).
(196) B. Coxon, *Carbohyd. Res.*, **18**, 427 (1971).

is possible of the many theoretical and semi-empirical equations[34] that relate vicinal and long-range coupling-constants to interatomic dihedral angles and to atomic electronegativities; (b) determination of the populations of conformers at different temperatures may depend on the measurement of quite small differences in coupling constants or chemical shifts, and, hence, it is desirable to eliminate any large, systematic errors in these parameters; and (c) the availability of more accurate parameters should allow more precise correlations with molecular structure and conformation.

Comparison of first-order parameters with the parameters obtained by a more refined analysis gives an indication of the errors that may arise in first-order analysis. For a solution of 1,2:4,6-di-O-benzylidene-3-O-(methylsulfonyl)-α-D-glucopyranose in pyridine, computerized iterative analysis[197] of the signals of the H-1–H-6' nuclei (which are not even coupled particularly strongly) at 100 MHz indicated that the computed values of the coupling constants differed by up to 13% from the values derived by first-order analysis, with a mean deviation of 6%. Larger deviations may be anticipated for spin-systems that are coupled more strongly.

In conjunction with a vicinally located methine proton, the chain-terminal- or ring-methylene protons of carbohydrates frequently form a strongly coupled ABX sub-system, because of a geminal coupling-constant that is often large (10–14 Hz) and a chemical shift, between the methylene protons, that is often small. Many authors have merely pointed out the existence of ABX spin-systems (and others) in their molecules, thus lending to their work an aura of pseudo-respectability, without involving them in the labor of performing a detailed analysis. If the ABX signals are well separated from those of the other protons, a second-order treatment in the form of an ABX sub-spectral analysis is often justified, and such analyses have been made in a number of careful investigations of pentose and hexose derivatives.[1,43,56,59,62,63,75,76,80,82,94,198–207] The conditions

(197) B. Coxon, Carbohyd. Res., **14**, 9 (1970).
(198) B. Coxon, Tetrahedron, **21**, 3481 (1965).
(199) B. Coxon, Carbohyd. Res., **8**, 125 (1968).
(200) B. Coxon, Carbohyd. Res., **12**, 313 (1970).
(201) B. Coxon, Carbohyd. Res., **13**, 321 (1970).
(202) C. Cone and L. Hough, Carbohyd. Res., **1**, 1 (1965).
(203) D. Horton and J. D. Wander, Carbohyd. Res., **13**, 33 (1970).
(204) P. L. Durette and D. Horton, Carbohyd. Res., **18**, 57 (1971).
(205) P. L. Durette and D. Horton, Carbohyd. Res., **18**, 289 (1971).
(206) P. L. Durette and D. Horton, Carbohyd. Res., **18**, 389 (1971).
(207) A. A. Magnin, K. G. R. Pachler, and A. M. Stephen, Tetrahedron, **25**, 4543 (1969).

under which the first-order spacings (S_{AX} and S_{BX}) in the AB octet of an ABX system give good values of the coupling constants J_{AX} and J_{BX} have been investigated for a series of derivatives of D-ribopyranose.[43] Complex, but explicit, expressions connecting S_{AX} and S_{BX} with J_{AX}, J_{BX}, and δ_{AB} (the chemical shift between A and B) were derived from the transition energies for the ABX system, and, from these expressions, it is evident that S_{AX} and S_{BX} tend towards J_{AX} and J_{BX}, respectively, either for values of δ_{AB} that are large compared with J_{AB} and $J_{AX} - J_{BX}$, or for small values of $J_{AX} - J_{BX}$.

A striking example[43] where a first-order analysis would have been quite inappropriate is provided by the 60-MHz spectrum of a solution of 2,3,4-tri-O-benzoyl-β-D-ribopyranosyl cyanide (**3**, see p. 23) in pyridine-d_5, which yielded S_{AX} 4.0 and J_{AX} 2.2 Hz ($J_{4,5e}$), and S_{BX} 7.1 and J_{BX} 8.9 Hz ($J_{4,5a}$). Substantial deviations of first-order spacings from the values of the coupling constants obtained by ABX analysis have also been reported for 100-MHz spectra of 1-thio-β-D-ribopyranose tetraacetate and 1-thio-β-D-galactopyranose pentaacetate[56] (**71**).

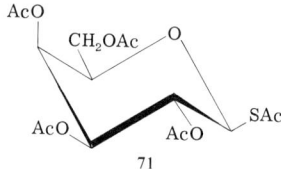

71

a. **Virtual Coupling.**—This confusing term was coined[77] to describe the extra splittings or broadenings that may occur in the signal of a nucleus A if it is coupled to a nucleus B that is strongly coupled (large ratio J_{BC}/δ_{BC}) to a third nucleus C. Under these conditions, the nuclear-energy levels of A are affected by the spin states of C, even when J_{AC} is zero.[59,77] The signal of A is only first-order under conditions of weak coupling of B and C; that is, if[77] $(a) |\delta_{BC}| \gtrsim J_{BC}$ and $|\delta_{BC}| \gtrsim 2J_{AB}$ or if (b) $|\delta_{BC}| \gtrsim \frac{4}{3}J_{BC}$ and $|\delta_{BC}| \sim J_{AB}$. These conditions have been applied in an investigation,[198] at 60 MHz, of the variation of the extra splittings in the H-3 triplets of methyl 4,6-O-benzylidene-α-D-glucopyranoside derivatives with the strength of the coupling (in the sense already described) between H-1 and H-2. For example, in pyridine solution, methyl 2,3-di-O-acetyl-4,6-O-benzylidene-α-D-glucopyranoside showed $\delta_{1,2}$ ~5.4, $J_{1,2}$ 3.6, and $J_{2,3}$ 9.8 Hz, and neither condition (a) nor condition (b) is satisfied. The H-3 signal of this derivative appeared as a complex multiplet containing at least 17 lines. However, in other similar derivatives, these virtual-coupling effects gradually disappeared[198] as $\delta_{1,2}$ was increased by changing the substituent groups at C-2 and C-3.

The foregoing example illustrates an advantage, in the analysis of p.m.r. spectra, that is enjoyed by the synthetic chemist. If he is faced with an over-complex spectrum, he can often simplify it merely by changing the substituents in the molecule; in fact, comparisons of the spectra of a series of similar compounds provide one of the most useful methods for understanding and assigning the spectra.

An unusual example of solvent-dependent, virtual coupling involving a hydroxyl group has been reported for 1,3,5-tri-O-benzoyl-α-D-ribofuranose (72) which, in acetone solution, showed the signals of

the hydroxyl proton, H-2, H-4, H-5, and H-5′ as a complex multiplet, but those of H-1 and H-3 as a quartet and a complex multiplet, respectively.[208] In chloroform, H-1, H-3, and the hydroxyl proton appeared as the doublet, quartet, and separated doublet, respectively, to be expected on a first-order basis.[209] The hydroxyl signal gradually moved downfield toward the H-2 resonance as the chloroform solution was diluted with acetone, and, eventually, the hydroxyl proton and H-2 became strongly coupled, thereby producing additional (non-first-order) complexity in the H-1 and H-3 multiplets that was also reproduced by computation of theoretical spectra[208] by using the FREQINT III program (see Section V, 4a; p. 73).

In contrast, the 100-MHz p.m.r. spectrum of methyl 2-O-acetyl-4,6-O-benzylidene-3-deoxy-3-phenylazo-α-D-glucopyranoside (73) in

(208) J. D. Stevens and H. G. Fletcher, Jr., *J. Org. Chem.*, **33**, 1799 (1968).
(209) L. D. Hall, *Chem. Ind.* (London), 950 (1963).

benzene-d_6 shows additional splitting, but *only* in the outer peaks of the triplet expected for H-6a from first-order considerations.[210] This splitting was described[210] as "partial virtual coupling," and as it was removed by irradiation of H-4, it appears to be due to fairly strong coupling of H-4, H-5, and H-6e. A theoretical analysis of H-5, H-6e, H-4, and H-6a as an *ABMX* sub-system[211] indicated that the strong, central peak of the H-6a signal is not split by virtual coupling, because this peak arises[210] from transitions between H-6a spin-states that are unmixed[212] with those of H-4. This situation occurs only for $J_{4,6a}$ ~0. The unmixed transitions were identified as those that were unaffected by irradiation of H-4. Again, the second-order splitting effects were confirmed by computation of theoretical spectra.[210] These effects have also been observed in the signals of groups of protons;[213] for example, the methyl protons of 6-deoxyaldohexopyranosides.[2]

Virtual coupling is not a special physical phenomenon, but is only a particular manifestation of second-order effects that cannot under any circumstances be subjected to first-order analysis. Thus, the reporting of supposed magnitudes (in Hz units) of virtual coupling for carbohydrates is quite improper, as these spacings have little physical significance and tend to disappear at higher field-strengths[59] if the strongly coupled nuclei are not exactly isochronous. For example, the extra splittings in the H-1 doublet of β-D-galactopyranose pentaacetate that were observed at 60 MHz (because of strong coupling of H-2 and H-3) were not present in a 100-MHz spectrum.[59] True coupling-constants are, of course, independent of field strength, and this fact may be used to distinguish genuine, small, long-range coupling-constants from splittings due to virtual-coupling effects that often occur when the coupling in question is zero. For example, certain combinations of parameters cause the *B* and *X* signals of an *ABX* system to appear as a quartet and sextet (including combination lines), even when J_{BX} is zero. However, at higher fields, a simplification of these multiplets is to be expected as this system tends towards *AMX*.

The conditions under which first-order analysis of spectra is justified have been variously stated as $(a)^{214}$ $\delta/J > 7$, $(b)^{215}$ $\delta/J \geq 10$, and

(210) C. B. Barlow, E. O. Bishop, P. R. Carey, R. D. Guthrie, M. A. Jensen, and J. E. Lewis, *Tetrahedron*, **24**, 4517 (1968).
(211) A. A. Khan, S. Rodman, and R. A. Hoffman, *Acta Chem. Scand.*, **21**, 63 (1967).
(212) E. O. Bishop, *Annu. Rev. NMR Spectrosc.*, **1**, 91 (1968).
(213) F. A. L. Anet, *Can. J. Chem.*, **39**, 2622 (1961).
(214) Ref. 17, p. 91.
(215) R. J. Abraham, "The Analysis of High Resolution NMR Spectra," Elsevier, Amsterdam, 1971, p. 21.

$(c)^{216}$ $4\delta/J^2 > 10$ Hz^{-1}, where the chemical-shift difference (δ) and coupling constant (J) are considered for each and every pair of nuclei in the spin system. For an ABX system, first-order analysis of the X multiplet is (very approximately) a reasonable approximation if the ratio $\delta_{AB}/J_{AB} \geq 3$, but, as this ratio becomes much smaller, the errors become very large.[217]

3. Manual Analysis of Spectra

The manual analysis of alphabetically classified spin-systems has been discussed extensively in many standard texts.[4,5,9,16,17,19,212,215,218] In three articles that are aimed specifically at encouraging the organic chemist to employ second-order methods of analysis more often, theories of the two-spin AB system,[216] the three-spin systems[219] ABC, ABX, ABK, and AB_2, and the four-spin systems[220] ABC_2, ABX_2, ABK_2, $AA'BB'$, and $AA'XX'$ are given, together with an explicit method for the analysis of each type of spin system. A concise procedure for analysis of the ABX system has been given by Becker.[221] The general four- to eight-spin systems, $ABCD$, $ABCDE$, $ABCDEF$, $ABCDEFG$, and $ABCDEFGH$, are probably the ones of most interest in the p.m.r. spectroscopy of carbohydrates, but, unfortunately, explicit manual analyses are not available for these unsymmetric spin-systems,[212] and they must normally be analyzed by iterative methods. Regrettably, many of the spin systems described in the literature possess a high degree of symmetric or magnetic equivalence that is uncommon in carbohydrates. In addition, the manual methods that are available have the disadvantage that a different, and usually complicated, method is required for each type of spin system. For these reasons (and others), manual, second-order analysis of the p.m.r. spectra of carbohydrates is significantly less useful than computerized, iterative analysis.

Deceptively simple ABX patterns[222] are occasionally encountered in the p.m.r. spectra of carbohydrates, particularly at 60 MHz. An extreme example (for which $\delta_{AB} + \frac{1}{2}|J_{AX} - J_{BX}| \ll J_{AB}$) has been reported[62] for 3,4,5,6-tetra-O-acetyl-1,2-dideoxy-1-nitro-D-*arabino*-hex-

---

(216) E. W. Garbisch, Jr., *J. Chem. Educ.*, **45**, 311 (1968).
(217) Ref. 215, p. 71.
(218) For early references, see Ref. 2.
(219) E. W. Garbisch, Jr., *J. Chem. Educ.*, **45**, 402 (1968).
(220) E. W. Garbisch, Jr., *J. Chem. Educ.*, **45**, 480 (1968).
(221) Ref. 17, p. 157.
(222) R. J. Abraham and H. J. Bernstein, *Can. J. Chem.*, **39**, 216 (1961); Ref. 215, pp. 75 and 163.

1-enitol (**74**) in acetone at 60 MHz, for which was observed a doublet

$$\begin{array}{c} \text{HCNO}_2 \\ \parallel \\ \text{CH} \\ | \\ \text{AcOCH} \\ | \\ \text{HCOAc} \\ | \\ \text{HCOAc} \\ | \\ \text{CH}_2\text{OAc} \\ \mathbf{74} \end{array}$$

for the *AB* part that consists of the olefinic-proton resonances (H-1 and H-2). In chloroform-*d* solution, this nitro derivative showed a triplet for the *X* nucleus (H-3) that was interpreted as a doublet that experienced further splitting by coupling with H-4. From analysis of the transition energies and nomograms of the variation of the *X*-line positions with δ_{AB}, it was concluded that coalescence of lines 9 and 11, and 10 and 12, of the *X* pattern can occur (for a certain relationship of δ_{AB} and J_{AB}) only if J_{AX} and J_{BX} have opposite signs. If, in addition, lines 14 and 15 have negligible intensity, the *X* signal then becomes a doublet. For the nitro derivative **74**, J_{AX} and J_{BX} have opposite signs, because they correspond respectively to an allylic coupling over four bonds ($J_{1,3}$ −1.3 Hz) and a vicinal coupling over three bonds ($J_{2,3}$ 4.9 Hz). Thus, the type of deceptively simple pattern observed may provide evidence for the relative signs of coupling constants, as was also found for some dihydroxy derivatives of 1,3-cyclohexadiene.[223] Deceptively simple patterns have been observed, even at 220 MHz, for the H-5 and H-5′ signals[224] (*AB* of an *ABX* sub-spectrum) of 3-*O*-benzyl- and -(*p*-nitrobenzyl) derivatives of 5-deoxy-5-iodo-1,2-*O*-isopropylidene-α-D-xylofuranose[82] (**26** and **27**; see p. 29).

A limited number of more complex spin-systems of carbohydrates have been analyzed by manual methods. For example, H-4, H-5, H-3, H-6, and H-6′ of 3,4,5,6-tetra-*O*-acetyl-1,2-dideoxy-1-nitro-D-*xylo*-hex-1-enitol (**75**) have been analyzed, respectively, as an *ABMRX*

$$\begin{array}{c} \text{HCNO}_2 \\ \parallel \\ \text{CH} \\ | \\ \text{HCOAc} \\ | \\ \text{AcOCH} \\ | \\ \text{HCOAc} \\ | \\ \text{CH}_2\text{OAc} \\ \mathbf{75} \end{array}$$

(223) T. J. Batterham, *Tetrahedron Lett.*, 949 (1969).
(224) The conditions for simplification of the H-5 and H-5′ signals to occur were stated[82] to be

$$\delta_{5,5'} \longrightarrow 0 \text{ and } \tfrac{1}{2}(J_{4,5} - J_{4,5'})/J_{5,5'} \longrightarrow 0$$

sub-system.⁶³ The approximate parameters obtained were refined by computerized, iterative analysis of this five-spin system and then of the complete seven-spin system in which H-1 and H-2 were included.⁶³ The variability of the deviations of the line frequencies from simple first-order rules was demonstrated by an intercomparison of the elements of the Jacobian matrix that were used in the iterative computations and then printed by the computer program LAME⁶³ (see Section V, 4a; p. 75).

The p.m.r. spectrum of 1,6-dideoxy-2,5-O-methylene-D-mannitol (**66**; see p. 65) at 60 MHz has been classified as an $AA'BB'M_3M_3'$ system, but the full complexity of this system was avoided by analysis of the spectrum of the related 1,3,4,6-tetra-O-acetyl-2,5-O-methylene-D-mannitol-$1,1,6,6$-d_4 (**76**) as an $AA'BB'$ system (compare *myo*-

$$\text{AcO} \quad H_A \quad H_{A'} \quad \text{OAc}$$
$$\text{AcOCD}_2 \qquad \qquad \text{CD}_2\text{OAc}$$
$$H_B \qquad \qquad H_{B'}$$
$$\text{O} \qquad \text{O}$$
$$H_X \quad H_X$$
76

inosose-2 phenylosotriazoles,¹⁸⁷ **60**, see p. 64), albeit with some broadening of the BB' signals (H-2 and H-5) by coupling with the deuterium nuclei.¹⁹²

4. Computerized, Iterative Analysis of Spectra

Iterative analysis of spectra by means of a computer is undoubtedly the most useful and general method available. However, the method depends heavily on the ability of the user to compute, initially, a trial, theoretical spectrum that sufficiently well resembles the experimental spectrum for many line-assignments to be made in the latter spectrum.

a. **Computer Programs.**—The earliest programs²²⁵ for n.m.r. analysis, such as FREQINT III and IV, were restricted to computation of a list of line frequencies and intensities from a chosen set of coupling constants and chemical shifts. Depending on the way in which these non-iterative programs are implemented for particular computers, theoretical spectra for spin systems of up to five, or possibly seven, nuclei, each having spin $\frac{1}{2}$, may be calculated.²²⁵ However, some refinement of the values of the chemical shifts and coupling con-

(225) A. A. Bothner-By and C. Naar-Colin, *J. Amer. Chem. Soc.*, **83**, 231 (1961).

stants is possible by making manual, first-order corrections to them, so as to improve the agreement between theoretical and observed spectra.[225] The FREQINT programs were soon modified to give the iterative programs LAOCOON,[226,227] LAOCOON II,[228] and[229] LAOCN3, which, when directed by manual assignments of theoretical lines to peaks in the experimental spectrum, automatically and iteratively compute corrections to the initial parameters so that the theoretical lines are brought into coincidence with the observed lines, according to a criterion of least squares.[228]

Another series of programs (NMRIT–NMREN) has been written[230] in which[231] (a) the NMRIT program is used for computing nuclear energy-levels and theoretical spectra from a variety of sets of trial chemical-shifts and coupling-constants, until a set is found that gives a theoretical spectrum sufficiently close to the observed spectrum that an "assignment" can be made [this assignment consists of identifying the pair(s) of energy levels associated with each observed transition by comparison of the line frequencies and intensities in the theoretical and observed spectra]; (b) the NMREN program computes (by a linear, least-squares procedure) refined, experimental energy-levels that best match the observed line-frequencies for the specified assignment; and (c) the iterative feature of NMRIT adjusts the trial parameters in order to obtain an optimal match between the experimental energy-levels from NMREN and the theoretical energy-levels from NMRIT.

The original capacity of the NMRIT–NMREN programs for analysis of up to eight nuclei of spin $\frac{1}{2}$ has been extended to analysis of up to six *groups* of magnetically equivalent nuclei[231] by using a composite-particle model with spin $>\frac{1}{2}$ to factorize the n.m.r. Hamiltonian matrix into smaller submatrices that can be computed more rapidly than the full matrix (as in the TWOSUM program,[232] for example). This

---

(226) A. A. Bothner-By, C. Naar-Colin, and H. Günther, *J. Amer. Chem. Soc.*, **84**, 2748 (1962).

(227) A. A. Bothner-By and H. Günther, *Discussions Faraday Soc.*, **34**, 127 (1962).

(228) S. Castellano and A. A. Bothner-By, *J. Chem. Phys.*, **41**, 3863 (1964). These authors have compared[229] their struggles in keeping line assignments untangled with the struggle of the Trojan priest Laokoon and his sons with serpents. (LAOCOON is an acronym for **L**east-squares **A**djustment **O**f **C**alculated **O**n **O**bserved **N**.m.r. spectra).

(229) A. A. Bothner-By and S. Castellano, "LAOCN3," a publication of the Mellon Institute, Pittsburgh, 1966.

(230) J. D. Swalen and C. A. Reilly, *J. Chem. Phys.*, **37**, 21 (1962).

(231) R. C. Ferguson and D. W. Marquardt, *J. Chem. Phys.*, **41**, 2087 (1964).

(232) J. C. Martin, University of Alberta, personal communication.

modification has also been applied to the LAOCOON programs to give a non-iterative program[233] (UEANMR II), and later, an improved version[234] (UEAITR) for iterative analysis of up to seven groups of magnetically equivalent nuclei of spin $\frac{1}{2}$. These programs may also be used for nuclei having spin $>\frac{1}{2}$; for example, deuterium. The programs LAOCOON LC[235] and LAOCOONOR[236] have been developed for the anisotropic analysis of oriented nuclei (for example, those in "liquid crystals"), and the programs LACX[237] and LAME[63;237] for matrix factorization according to chemical equivalence, and magnetic equivalence, respectively.

Other programs include EXAN II, a program[238] that obtains non-unique sets of chemical shifts and coupling constants by explicit analysis of the experimental line-positions of ABC spectra; CHEM 3, a program[239] that computes a theoretical spectrum for a three-spin system; and NMDRS, a program[240] for computation of double-resonance spectra.

For an appreciation of the mathematics in these programs, the reader is referred to the literature cited and to a general discussion.[241]

The LAOCOON II and LAOCN3 programs, and their modifications, have proved to be the most popular and useful programs for analysis of the n.m.r. spectra of carbohydrates. It should be emphasized that any one of the iterative programs may be used non-iteratively in order to compute only a theoretical spectrum. This, in itself, is an extremely useful method for confirming spectral analyses, even if these are merely first-order. Later versions of these programs include, for output of the theoretical frequencies and intensities, various options that remove the necessity for tedious, manual plotting of the theoretical spectra. A "stick plot" of the spectrum may be obtained from a line printer, or a plotting routine and line-width function may be used for converting the output data into a spectrum envelope that is digitally stored on a magnetic tape and then drawn automatically by a digital, incremental, $x-y$ plotter (see Fig. 5b).

(233) R. K. Harris and C. M. Woodman, *Mol. Phys.*, **10**, 437 (1966).
(234) R. B. Johannesen, J. A. Ferretti, and R. K. Harris, *J. Magn. Resonance*, **3**, 84 (1970).
(235) R. K. Harris and V. J. Gazzard, *Org. Magn. Resonance*, **3**, 495 (1971).
(236) P. Diehl, C. L. Khetrapal, and H. P. Kellerhals, *Mol. Phys.*, **15**, 333 (1968).
(237) C. W. Haigh, *Annu. Rep. NMR Spectrosc.*, **4**, 311 (1971).
(238) A. A. Bothner-By, S. Castellano, and H. Günther, *J. Amer. Chem. Soc.*, **87**, 2439 (1965).
(239) C. L. Wilkins and C. E. Klopfenstein, *J. Chem. Educ.*, **43**, 10 (1966).
(240) G. Govil and D. H. Whiffen, *Mol. Phys.*, **12**, 449 (1967).
(241) J. D. Swalen, *Progr. Nuc. Magn. Resonance Spectrosc.*, **1**, 205 (1966).

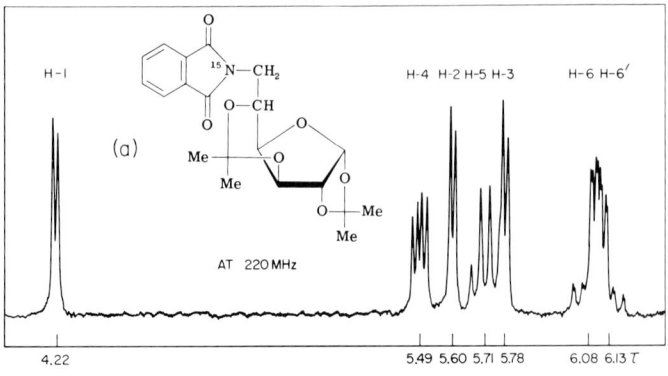

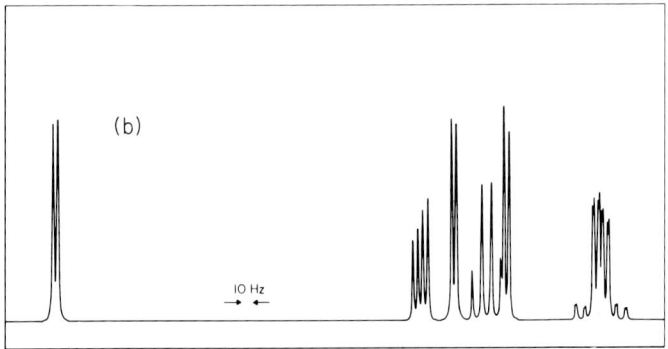

FIG. 5.—Partial, Proton Magnetic Resonance Spectra of 6-Deoxy-1,2:3,5-di-O-isopropylidene-6-phthalimido-α-D-glucofuranose-6-^{15}N (**30**). [(*a*) In solution in benzene-d_6 at 220 MHz; (*b*) computed, theoretical spectrum from iterative analysis.]

If the computer is equipped to provide output of data on a cathode-ray tube, a camera may be used for recording a series of theoretical spectra computed from various sets of parameters, thus simplifying the search for a trial, theoretical spectrum that resembles the experimental spectrum.[242] Output options for n.m.r. programs now include tables of energy levels and error parameters, and lists of progressively and regressively connected transitions that can be correlated with double-resonance experiments (see Part II, Section VI).

The initial chemical-shifts and coupling-constants that are used to calculate a trial, theoretical spectrum before iteration may be obtained by first-order analysis, by an approximate, partial, second-order

(242) R. J. Cushley, Yale University, New Haven, personal communication.

analysis (for example, *ABX* analysis of a sub-system that is really *ABC*), or by estimation from the parameters of similar or model systems.

b. Assignment of Transitions in the Spectra of Carbohydrates.—The most laborious part of the iterative analysis of a spin system by the LAOCOON method is the assignment of some of the numbered, theoretical transitions that are produced in the computation of the trial, theoretical spectrum to the frequencies of the most closely matched peaks in the experimental spectrum, generally by means of a punched card for each transition. It is not necessary to assign all of the theoretical transitions, however, and, in fact, there are often upper and lower limits to the number of these transitions that can be assigned. The LAOCN3 program is designed to handle up to 300 assignments, but, on the other hand, the assignment of too few theoretical transitions (for example, 27 theoretical transitions assigned to 26 peaks) can cause the unassigned transitions to wander in a random way, thereby degrading the theoretical spectrum, instead of improving it.

In iterative analyses of typical six- and seven-spin systems of pentose[201] and hexose[85,196,197,200] derivatives by use of LAOCN3 or UEAITR, it has usually been necessary to assign between 60 and 120 theoretical transitions to between 15 and 30 experimental peaks, although, for the iterative analysis of the seven-spin system of 3,4,5,6-tetra-*O*-acetyl-1,2-dideoxy-1-nitro-D-*xylo*-hex-1-enitol (**75**; p. 72) by means of the LAME program, 408 theoretical transitions were assigned to 43 distinct peaks.[43] As the p.m.r. spectra of carbohydrates are often quite degenerate, because many of the coupling constants possible are near zero, it is usually advantageous to assign all of the peaks observed, so as to provide the maximum amount of experimental input data.

For iterative analysis to be worth while, the accuracy of the experimental line-frequencies should preferably be maximized by scanning the spectrum several times in each direction, in order to give mean values for these frequencies. The intensities of the experimental lines are not used as input data to the programs currently available.

c. Experimental Errors and Variation of Parameters.—The user of a program for iterative analysis of n.m.r. spectra may select those chemical shifts and coupling constants that are allowed to vary during the iterative fitting of the theoretical spectrum to the experimental spectrum. It is obviously appropriate to vary all of the proton chemical-shifts and coupling constants that can be observed as splittings in the spectrum, but it is less certain whether the unresolved coupling-constants should also be varied, or should be allowed to remain zero.

Iterative analyses of the 100-MHz spectra of several methyl 2(or 3)-amino-4,6-O-benzylidene-2(or 3)-deoxy-α-D-glycosides[243] and of methyl 2-O-acetyl-4,6-O-benzylidene-3-deoxy-3-phenylazo-α-D-glucopyranoside[210] by means of the LAOCOON II program have yielded many apparently non-zero values of coupling constants over four or five bonds, for example, $J_{1,3} - 0.16 \pm 0.1$, $J_{4a,6a} - 0.3 \pm 0.1$ to -1.24 ± 0.17, and $J_{2,5} + 0.26 \pm 0.1$ Hz. These values (and others) were believed[210,243] to be genuine, long-range coupling-constants, and the values of $J_{4a,6a}$ and $J_{4a,6e}$ have been interpreted in terms of spatial relationships by means of a simple, molecular-orbital theory.[244]

From iterative analyses of the 100-MHz spectra of 1,2:4,6-di-O-benzylidene-3-O-(methylsulfonyl)-α-D-glucopyranose,[197] 1,2-O-isopropylidene-3,5-O-[(*endo*-methoxy)methylidene]-6-O-*p*-tolylsulfonyl-α-D-glucofuranose[200] (**77**), and 3-O-benzoyl-1,2,4-O-benzylidyne-α-

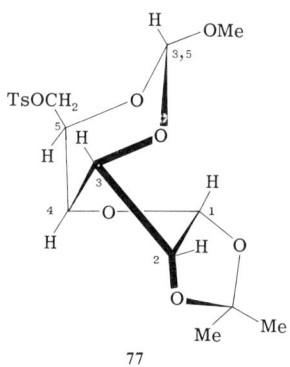

D-ribopyranose[201] by using the LAOCN3 program, the author has obtained many small, apparent values of coupling constants for protons separated by four to six bonds, but is not convinced that these values are necessarily genuine in those cases where no evidence for the couplings can be discerned in the spectrum. In these three analyses, variation of all of the parameters was permitted (six chemical shifts and 15 coupling constants for the six-spin system, and seven chemical shifts and 21 coupling constants for the seven-spin systems).

(243) C. B. Barlow, E. O. Bishop, P. R. Carey, and R. D. Guthrie, *Carbohyd. Res.*, **9**, 99 (1969).
(244) P. R. Carey and R. Ditchfield, *Mol. Phys.*, **15**, 515 (1968).

In other iterative analyses of 6-deoxy-1,2:3,5-di-O-isopropylidene-6-phthalimido-α-D-glucofuranose (**29**; p. 30) and its ^{15}N-labeled derivative[85] (**30**; p. 30) (see Fig. 5; p. 76) by LAOCN3, and of[196] 6-deoxy-1,2:3,5-di-O-isopropylidene-α-D-glucofuranose-d_{12} (**78**) by

CH$_3$
O—CH
D$_3$C—   O
D$_3$C   O
O—CD$_3$
CD$_3$

78

UEAITR, by using magnetic equivalence-factoring for the methyl protons, variation of only geminal and vicinal coupling-constants, and of chemical shifts, was allowed, and the refined, theoretical spectra agreed with the observed spectra about as well as when all of the parameters were varied.

The almost-degenerate nature and finite line-width of the p.m.r. signals of many carbohydrates appear to impose limitations on the accuracy of iterative analysis. For example, computation of theoretical spectra for the seven-spin systems of hexose derivatives for which H-1 is observably coupled only to H-2 shows that each half of the H-1 "doublet" is actually composed of 32 closely spaced, but slightly different, theoretical transitions, of intensity[245] >0.1, that are not usually resolved unless such second-order effects as virtual coupling due to strong coupling of H-2 and H-3 are present. In an iterative analysis, each of these groups of transitions (or a fraction of each of them) can only be reasonably assigned to one experimental frequency, namely, that of the center of one peak, or the other, of the doublet. Therefore, it seems likely that assignments of experimental frequencies to almost-degenerate, theoretical transitions are subject to errors of up to approximately one-half of the width of a peak, and that these errors and those that arise from nonlinear frequency-sweeps can cause the generation of spurious, small "values" of coupling constants for nuclei that are not expected to be coupled. These small errors do not seriously detract from the iterative refinement of approximate first-order values of coupling constants that may change by several Hz during the iterations, that is, by an amount that is much greater than the line width (typically, 0.2–0.6 Hz) divided by two.

(245) Based on a total intensity of 64 for the H-1 signal.

The probable errors of the refined chemical-shifts and coupling-constants are printed in the output of the LAOCOON programs, but these errors are now generally believed[246] to be unreasonably small, and they have sometimes been multiplied by a factor of two[246] or five[247] when reported.

d. Criteria for a Good Analysis.—For an iterative analysis of a p.m.r. spectrum of a carbohydrate to be valid, the following conditions should be satisfied, at least approximately. (a) A good visual fit of the digitally plotted, theoretical spectrum to the experimental spectrum should be obtained; (b) the computed, root-mean-square error of the frequencies of the theoretical lines assigned should be <0.2 Hz (preferably, <0.1 Hz); (c) the computed, probable errors of the parameters should be <0.1 Hz; and (d) the errors in the frequencies of individual, assigned theoretical lines should be <0.4 Hz. If special care is taken in the measurement of line frequencies, it should be feasible to lower these limits.

Additional confirmation of the validity of an analysis (iterative, or first-order) is possible if good agreement between theoretical and experimental spectra is obtained at two different frequencies (for example, at 60 and 100 MHz), and if the assignments of lines are consistent with their connectivity, as determined by double-resonance techniques (see Part II, Section VI).

e. Other Applications.—Non-iterative computations of theoretical spectra have proved to be very useful for confirmation of spectral assignments in manual and first-order analyses of p.m.r. spectra. The FREQINT IV program has been used for the five-spin systems of methyl 3,4-dichloro-4-deoxy-α- and β-D-*glycero*-pent-2-enopyranosides[94] (**79** and **80**), and the CHEM 3 program for confirmation of the

first-order analysis of the H-4, H-5, and H-5' *ABX* sub-system of 1,3,4-tri-*O*-benzoyl-2-deoxy-2-fluoro-D-ribose[248] (**81**). In studies of glyculose derivatives, the chemical shifts and coupling constants of 1,2-dideoxy-5,6-*O*-isopropylidene-D-*glycero*-pent-1-enitol-3-ulose

(246) G. E. Hall and J. D. Roberts, *J. Amer. Chem. Soc.*, **93**, 2203 (1971).
(247) S. Castellano, C. Sun, and R. Kostelnik, *Tetrahedron Lett.*, 5205 (1967).
(248) R. J. Cushley, J. F. Codington, and J. J. Fox, *Carbohyd. Res.*, **5**, 31 (1967).

(**82**) were obtained by second-order analysis, and a comparison of

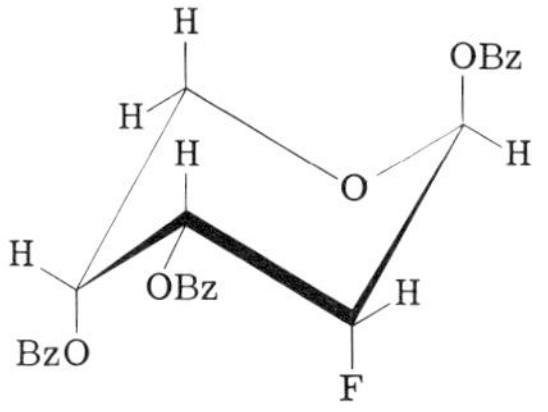

81 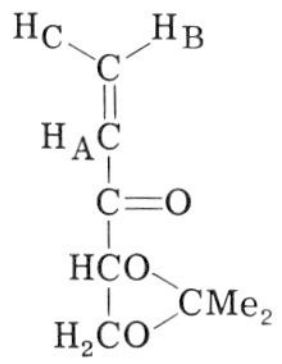 82

computed line-intensities with the spectrum observed showed[249] that $^2J_{1,1'}^{gem}$, and $^3J_{1,2}^{cis}$ have the same sign; also, all assignments in the proton spectra of a series of unsaturated derivatives, for example **83**,

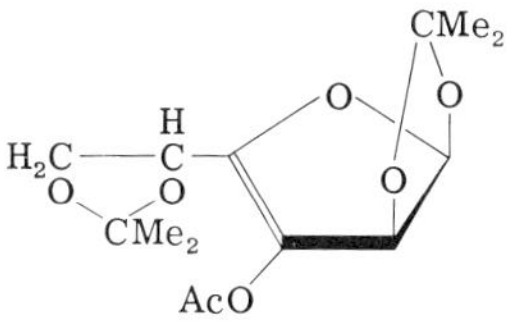

83

of 1,2:5,6-di-O-isopropylidene-D-hexofuranos-3-uloses have been confirmed[250] by LAOCN3.

The LAOCOON II program has been used to calculate theoretical spectra for 1,3,4,6-tetra-O-acetyl-2,5-O-methylene-D-mannitol-*1,1,6,6-d_4* (**76**; p. 73),[192] and in the iterative analysis and simulation of spectra of methyl 2,3- and 3,4-anhydroglycopyranosides[86] (LAOCN3 was also used), the protons of the sugar portion of many nucleosides,[104,105,107,109] for example, 2'-deoxyadenosine[251] (**84**), and adenosine 5'-phosphate[126] (**55**; p. 60).

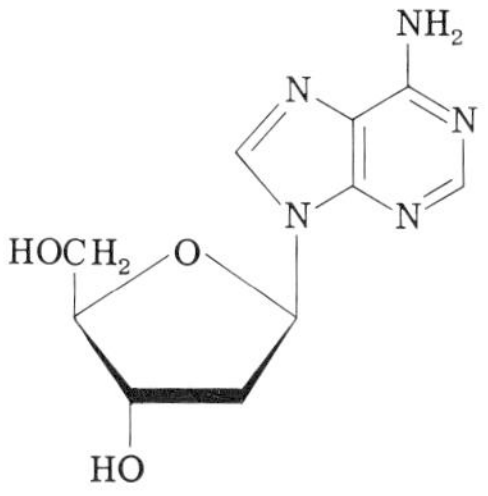

84

(249) D. J. Walton, *Can. J. Chem.*, **47**, 3483 (1969).
(250) K. N. Slessor and A. S. Tracey, *Can. J. Chem.*, **48**, 2900 (1970).
(251) T. J. Batterham, R. K. Ghambeer, R. L. Blakley, and C. Brownson, *Biochemistry*, **6**, 1203 (1967).

The LAOCN3 and TWOSUM programs have been used in the iterative analysis or spectrum simulation of tri-O-acetyl-β-D-arabinopyranosyl fluoride,[252] tri-O-benzoyl-α- and β-D-ribofuranosyl fluoride,[253] and 3,6-anhydro-5-deoxy-5-fluoro-1,2-O-isopropylidene-α-L-idofuranose[254] (85).

The NMRIT and NMREN programs have been employed in the iterative analysis[255] of 9-(4-O-acetyl-2,3-dideoxy-α- and β-D-glycero-pent-2-enopyranosyl)-6-chloropurines (86 and 87) and 4-O-acetyl-1,5-anhydro-3-(6-chloro-9-purinyl)-1,2,3-trideoxy-D-threo-pent-1-enitol (88). These compounds, which were prepared from 6-chloropurine

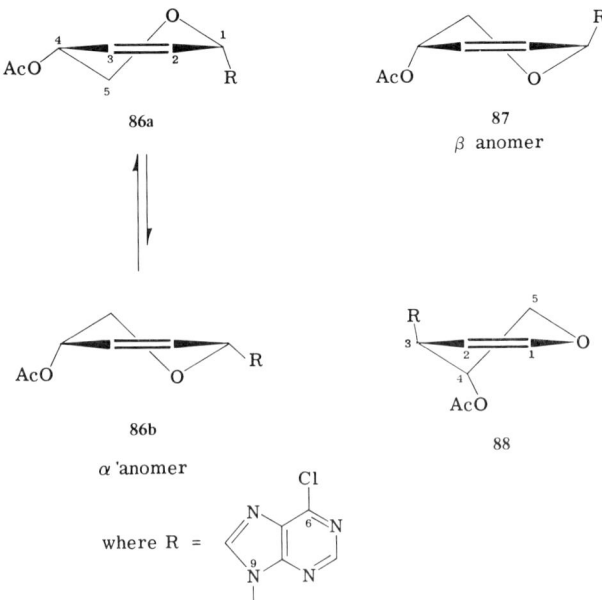

(252) L. D. Hall and J. F. Manville, *Can. J. Chem.*, **47**, 19 (1969).
(253) L. D. Hall, P. R. Steiner, and C. Pedersen, *Can. J. Chem.*, **48**, 1155 (1970).
(254) L. D. Hall and P. R. Steiner, *Can. J. Chem.*, **48**, 451 (1970).

and 3,4-di-O-acetyl-D-xylal, showed, in their p.m.r. spectra, second-order effects that were analyzed by assignment of 50–70 lines.[255] An iterative analysis and spectrum simulation has also been reported for 1-(2,3,4,6-tetra-O-acetyl-β-D-glucopyranosyl)indoline[207] (**89**), and a

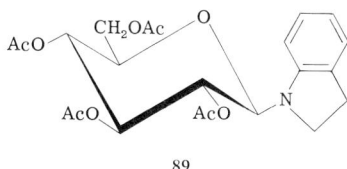

89

simulation of the 220-MHz spectrum of methyl 2-deoxy-β-D-*erythro*-pentofuranoside[80] (**23**; p. 29) has been done.

The chemical shifts and coupling constants obtained from the p.m.r. spectra of carbohydrates by a valid, computerized, iterative analysis may generally be regarded as particularly accurate examples of these parameters.

The indirect method of computing theoretical line-frequencies that is used in such programs as LAOCOON and NMRIT-NMREN is based[229,230,241] on diagonalization of the Hamiltonian matrix of a spin system, followed by calculation of its line frequencies as the differences between the energy levels given by the eigenvalues of the matrix (roots of its determinant), chosen according to a selection rule. This is not the only theoretical approach to calculation of spectra, and a direct method[256] based on super-operators has been implemented,[257] but this type of program is not yet in general use.

A simple guide to the use of iterative computer programs in the analysis of n.m.r. spectra has appeared.[237]

(255) M. Fuertes, G. Garcia-Muñoz, R. Madroñero, and M. Stud, *Tetrahedron*, **26**, 4823 (1970).
(256) C. N. Banwell and H. Primas, *Mol. Phys.*, **6**, 225 (1963).
(257) G. Binsch, *Mol. Phys.*, **15**, 469 (1968).

NON-AQUEOUS SOLVENTS FOR CARBOHYDRATES

By C. J. Moye

Cottee's General Foods Ltd., Sydney, Australia

I. Introduction	85
II. Solubility and Solvation	86
III. Inorganic Solvents	90
1. Liquids	91
2. Salts	95
IV. Organic Solvents	96
1. Protic	96
2. Aprotic	105
V. Special Applications	123
1. Chromatographic Systems	123
2. Nuclear Magnetic Resonance Spectroscopy	124

I. Introduction

The purpose of this article is to bring together the scattered literature dealing with the solubility and reactions of unsubstituted carbohydrates in non-aqueous media. The scope has been limited to unsubstituted carbohydrates to keep it within manageable proportions, and naturally occurring, substituted compounds have therefore been excluded. Only one major review[1] covering the limited area of solubility of sucrose has previously been published, but several articles have appeared on solvents for carbohydrate reactions.[2-4]

Many systematic studies employing non-aqueous solvents have been sponsored by the International Sugar Research Foundation (S.R.F.), which received unpublished reports from the participants. Access to much of this work has now been made possible by the publication of a book[5] sponsored by the Foundation entitled "Sucrose Chemicals." Direct reference to S.R.F. reports as indexed in the book

---

(1) O. K. Kononenko and K. M. Herstein, *Chem. Eng. Data Series*, **1**, 87 (1956).
(2) V. Prey and W. Unger, *Osterr. Chem. Z.*, **61**, 10 (1960).
(3) H. Schiweck, *Zucker*, **14**, 114 (1961).
(4) G. R. Ames, *Chem. Rev.*, **60**, 541 (1960).
(5) V. Kollonitsch, "Sucrose Chemicals," The International Sugar Research Foundation, through V. Kollonitsch, W. S. Cowell Ltd., Ipswich, England (1970).

has been made in the present Chapter, as these reports are available on application. The Foundation has, in fact, played a major role in stimulating interest in the use of non-aqueous solvents for sugars, the most important outcome being the development of the sucrose ester surfactants. Although the Foundation's interest has centered on sucrose, it is clearly possible to extrapolate many of the results directly to other sugars.

An effort has been made to collect information that would not come to light in a literature search; this has met with partial success. The main aim, namely, to determine what non-aqueous solvents have been specifically investigated for their solvating power for carbohydrates, has resulted in such an interesting array of compounds, both inorganic and organic, that the author ventures to hope that further interest in the chemistry of carbohydrates in non-aqueous solvents will be generated.

On succeeding pages, specific solubilities for solute–solvent systems are given where these have been determined, but, as Kononenko and Herstein have pointed out,[1] many literature values are in poor agreement; this is due to a combination of such factors as vastly different equilibration times, impure solvents, and inadequate analytical procedures. The reader should bear this in mind, particularly with regard to older references. Unless otherwise stated, solubility given as % (w/w) or % (w/v) refers to the weight of *solute* dissolved in 100 grams or 100 ml of *solution*.

It should not be forgotten that, even with the most rigorously purified solvent, water can be produced very readily from sugars, and this water can markedly affect the rate and course of a reaction. Undoubtedly, in many of the reactions reported in this article, neither the sugar nor the solvent was prepared in a rigorously anhydrous state, and earlier determinations of solubility may well have suffered from the same deficiency.

II. SOLUBILITY AND SOLVATION

Water is, with few exceptions, an excellent solvent for sugars ranging up to the oligosaccharides and quite a few polysaccharides as well, and the reverse can be said of most of the common organic solvents. It is therefore, not surprising to find that the latter group has been neglected.

On a simple basis of polarity, it would be expected that simple sugars would be more soluble in such polar solvents as alcohols than they actually are. It would appear, from work conducted by Moye and Smythe,[6] that crystal-lattice energy resulting from multiple hydrogen-bond associations plays a vital role in limiting solubility. In terms of such solvents as alcohols, solvation is, therefore, probably not the major factor, but rather the energy input needed to disrupt the crystal lattice. Thus with ethanol, the solubility of sucrose at 80° is only 0.5% (w/w) (solvent), but this is raised to 5.6% at 140°, a significant degree of solvation occurring after lattice disruption. Of greater interest, perhaps, is the unexpected discovery that a wide range of solvents containing the structural grouping R—O—C—C—OH are better solvents for sugars than would be anticipated from structural considerations.[6-12] 2-Methoxyethanol, for example, was found to be a far better solvent for sucrose than diethyl ether or ethanol. Solvents containing this grouping are shown in Fig. 1, with the proposed solvation structure depicted in Fig. 2. It was proposed that the solvents have favorable boiling points (ranging from 120 to 200°) that facilitate disruption of the lattice, and that the particular solvent structure engages in solvation. The high boiling-points are undoubtedly important in relation to input of energy to disrupt the lattice, and this feature might be inferred to be of greater importance than structure, as such solvents as 3-methoxypropanol and 4-methoxybutanol solvate sucrose appreciably at elevated temperatures, even if slightly less efficiently than 2-methoxyethanol. In Fig. 3 are plotted the comparative solubilities of lower alcohols and their ω-methoxy derivatives, as well as the R—O—C—C—OH series investigated. Even more spectacular than for ethanol and the alkoxy alcohols is the change in solubility of sucrose in the derivatives of tetrahydrothiophene 1,1-dioxide ("sulfolanes"; see p. 111, Table I, under Sulfur Heterocycles); a temperature change from 80 to 150° results in a 300-fold change in solubility for sulfolane and 3-methylsulfolane. Here is further evidence of the need to overcome internal lattice-energy in order to break down the crystal structure prior to solvation.

The foregoing discussion is not to suggest that the dipole moment, dielectric constant, and electron availability of the solvent are not

(6) C. J. Moye and B. M. Smythe, *Carbohyd. Res.*, **1**, 284 (1965).
(7) C. J. Moye and B. M. Smythe, Aust. Pat. 261,561; *Chem. Abstr.*, **68**, 14274h (1968).
(8) C. J. Moye and B. M. Smythe, Brit. Pat. 1,019,511.
(9) C. J. Moye and B. M. Smythe, Brit. Pat. 1,019,512.
(10) C. J. Moye and B. M. Smythe, U. S. Pat. 3,219,484; *Chem. Abstr.*, **64**, 8468g (1966).
(11) C. J. Moye and B. M. Smythe, U. S. Pat. 3,290,263; *Chem. Abstr.*, **66**, 55810w (1967).
(12) C. J. Moye, *Proc. 19th Cereal Chem. Conf. Royal Aust. Chem. Inst., Toowoomba, Australia*, 1969.

88   C. J. MOYE

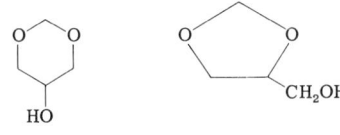

Furfuryl alcohol

R—O—CH$_2$—CH$_2$OH
where R = alkyl or R'—O—CH$_2$—CH$_2$— (R' = alkyl)

Alkoxy alcohols

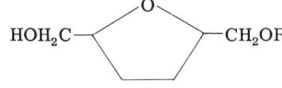

Tetrahydrofurfuryl alcohol

1,3- and 2,3-O-Methylideneglycerol

2,5-Bis(hydroxymethyl)tetrahydrofuran half-ethers
(where R = tetrahydrofurfuryl or alkoxyalkyl)

1,4-Anhydroerythritol ("Erythran")

FIG. 1.—Structures of Solvents Having the General Structure R—O—C—C—OH.

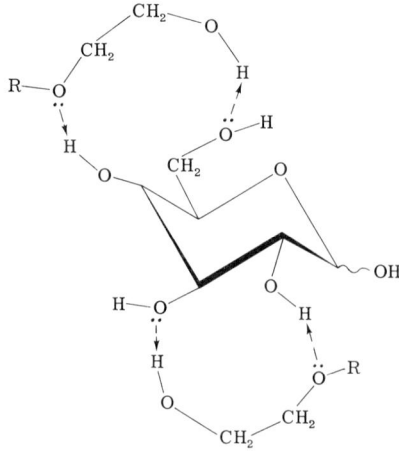

FIG. 2.—Solvation by Alkoxy Alcohols.

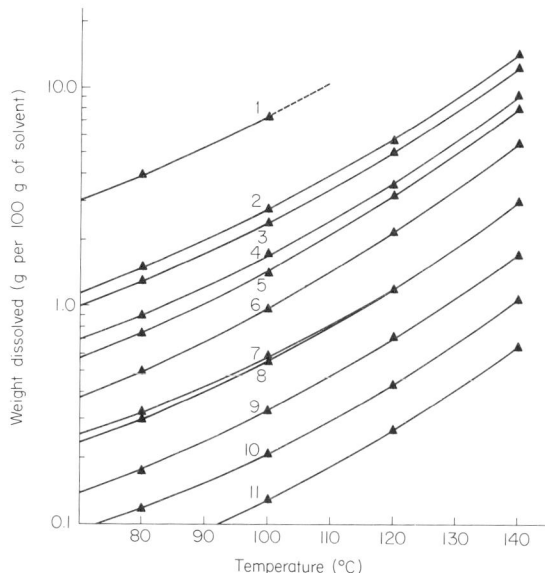

Fig. 3.—Solubility of Sucrose in Lower Alcohols and their Derivatives. (1, Mixed methylideneglycerols; 2, 2-methoxyethanol; 3, tetrahydrofurfuryl alcohol; 4, 3-methoxypropanol; 5, 2-(2-methoxyethoxy)ethanol; 6, ethanol; 7, 2-acetoxyethanol; 8, 4-methoxybutanol; 9, propyl alcohol; 10, butyl alcohol, and 11, 2-butoxyethanol----------Solvent decomposition.)

of paramount importance. These properties *also* play a very important role, as may be judged from the particular effectiveness of such dipolar, aprotic solvents as N,N-dimethylformamide and methyl sulfoxide as solvents for sugars. Parker[13] has discussed the role played by these and related solvents in solvating carbohydrates, which, freed of crystal-lattice associations (these solvents also have high boiling-points), specifically and strongly interact with the exposed, basic atom of the solvent through their dipolar hydroxyl groups. In their review,[1] Kononenko and Herstein examined the work of Orth[14,15] and modified Orth's equation relating solubility to temperature for water as the solvent. They developed the linear equation

$$e = a(b - T)$$

where e is the weight of solvent needed to dissolve 1 g of sucrose

---

(13) A. J. Parker, *Advan. Org. Chem.*, **5**, 5 (1965).
(14) P. Orth, *Bull. Ass. Chim.*, **31**, 94 (1913).
(15) P. Orth, *Bull. Ass. Chim.*, **54**, 605 (1937).

(experimentally determined), and a and b are parameters calculated from their results by the method of least squares.

For those solvents whose dielectric constants were known, a plot of log a *versus* the dielectric constant gave a smooth curve. The significance of this result was not appreciated, but it is interesting that some form of relationship appeared to exist. This result led the authors to suggest that the solubility of sucrose in a non-aqueous solvent could be calculated for any temperature, provided that the solubility at one temperature and the dielectric constant of the solvent were known. Of further interest was the discovery that, for some eight solvents examined, straight lines were obtained for plots of e against T, and these plots, on extrapolation to e_0, gave a range of values averaging out at the value commonly accepted for the melting point of sucrose. As e_0 expresses the situation where solvent is absent, Te_0 might be expected to be a measure of the melting point of the solute and the result could then be regarded as providing experimental support for the correctness of the equation.

The solubility of a sugar in a solvent may be simply and accurately determined by a two-stage procedure. A rough determination of the solubility is first made, by repeatedly observing a series of closed tubes containing increasing weights of sugar for a standard weight (or volume) of solvent, when they are heated under controlled conditions. A bath or oven controlled to ±0.1° is recommended, and the use of finely powdered sugar and rotating tubes helps to decrease the time needed to reach equilibrium. Equilibration can only be determined by repeated observation, and the time interval between observations must be reasonably long (of the order of 3–4 hours as the minimum). A transition range is thus indicated, where tubes in which all of the sugar has dissolved give way to those containing increasing amounts of undissolved sugar. The same process is now repeated, with amounts spanning this narrow range, and an accurate result, expressed as w (sugar)/w or v (solvent), may be obtained without the need for direct assay. Decomposition of the solvent or solute is usually indicated by color changes and the development of u.v. absorbance, particularly at ~280 nm, should furan products have resulted.

III. Inorganic Solvents

Few inorganic solvents have utility as solvents for carbohydrates; however, a number have been investigated, and several are excellent solvents. Observations relating to other solvents suggest areas for interesting future studies.

1. Liquids

Liquid ammonia has considerable potential as a solvent for carbohydrates, as it readily dissolves simple sugars and many polysaccharides. Franklin and Kraus[16] conducted an extensive study of the solubility of a wide range of compounds in liquid ammonia. Amongst these were a number of carbohydrates. D-Glucose, D-fructose, sucrose, lactose, maltose, and raffinose were found to be very soluble, and L-arabinose and D-galactose were described as being moderately soluble. These statements were relative, as specific solubilities were not given. Other early studies confirmed these findings, and, although Sherry,[17] in studies on the mutarotation of sugars in anhydrous solvents, worked at 12% (w/v) sucrose in ammonia, Fitzgerald[18] was able to obtain a 72.4% w/v (solvent) solution that was very viscous but not stated to be saturated. In another early study on the properties and reactions of carbohydrates in liquid ammonia, Miller and Siehrs[19] obtained a solution of 40 g of D-glucose per ml of ammonia. Maltose was the least soluble of the sugars examined, which included L-arabinose, D-fructose, D-galactose, sucrose, and lactose; and glycogen gave an opalescent solution. In the absence of water, the solutions were stable. Recovery of the monosaccharides gave nitrogen-containing, glassy materials, whereas the disaccharides crystallized on evaporation, although all of the ammonia could not be readily removed. Potassium salts were prepared with potassium amide, and the respective sugars could be recovered from them. The monosaccharides gave monopotassium salts and the disaccharides, dipotassium salts. Glycogen also gave a potassium salt.

Liquid ammonia, although not necessarily solubilizing, also has an effect on polysaccharides that can be used to advantage, as evidenced by the increased reactivity of starch that has been pretreated with ammonia.[20-22] It is also possible to methylate cellulose in the presence of liquid ammonia at −70° by use of metallic sodium and methyl iodide,[23,24] so that localized solvation of the carbohydrate chain undoubtedly occurs. Glycogen is, however, soluble, as evidenced by

---

(16) E. C. Franklin and C. A. Kraus, *Amer. Chem. J.*, **20**, 832 (1898).
(17) R. H. Sherry, *J. Phys. Chem.*, **11**, 559 (1907).
(18) F. F. Fitzgerald, *J. Phys. Chem.*, **16**, 651, 656, 660 (1912).
(19) C. O. Miller and A. E. Siehrs, *Proc. Soc. Exp. Biol. Med.*, **29**, 535 (1932).
(20) J. E. Hodge, *Methods Carbohyd. Chem.*, **4**, 279 (1964).
(21) J. E. Hodge, S. J. Karjala, and G. E. Hilbert, *J. Amer. Chem. Soc.*, **73**, 3312 (1951).
(22) G. E. Hilbert, D. W. Olds, and I. A. Wolff, *J. Amer. Chem. Soc.*, **73**, 346 (1951).
(23) E. J. Bourne, K. S. Barclay, M. Stacey, and M. Webb, *J. Chem. Soc.*, 1501 (1954).
(24) K. Freudenberg and H. Boppel, *Ber.*, **71**, 2505 (1938).

the studies of Schmid and coworkers,[25] who conducted studies on the molecular weight of glycogen in liquid ammonia. The metallation of sugars in liquid ammonia is one of the most interesting reactions of this solvent, but the full potential of the resulting "sucrates" cannot be realized because of competing reactions. Muskat[26] performed the first comprehensive study of the properties of "sucrates" prepared in liquid ammonia. He found that alkylation occurs with methyl iodide, and that arylation can be effected with some aryl halides. He also recognized that the halide could react with ammonia, a significant competitive reaction if the "sucrate" is insoluble in ammonia, to give an acid that could decompose the "sucrate" before it could react with the halide. Iodine was found to oxidize the sugar by forming INH_2, the equivalent, in the ammonia solvent system, of hypoiodous acid in aqueous solution. Provided that ammonia was completely removed, acylation occurred in the presence of acid chlorides.

There are numerous publications dealing with metallation reactions in liquid ammonia,[27-39] in particular with the problem of "bound" ammonia, which is difficult to remove from sodium "sucrates," particularly those of the lower degrees of substitution.[33] Prey and coworkers claimed to have solved the problem of removal of this ammonia,[35-37] but Schneider was not able to substantiate their claims.[40] Alternative approaches to the production of "sucrates" by using, for

(25) L. Schmid, E. Ludwig, and K. Pietsch, *Sitzungsber.*, **137** (**2B**), 118 (1928).
(26) I. E. Muskat, *J. Amer. Chem. Soc.*, **56**, 693, 2449 (1934); cf., R. S. Tipson, *Methods Carbohyd. Chem.*, **2**, 150 (1963).
(27) L. Schmid, *Ber.*, **58**, 1966 (1925).
(28) L. Schmid, A. Waschkau, and E. Ludwig, *Monatsh.*, **49**, 102 (1928).
(29) M. Amagasa and N. Onikura, *Kogyo Kagaku Zasshi*, **52**, 2 (1949); *Chem. Abstr.*, **45**, 2414d (1951).
(30) P. C. Arni, W. A. P. Black, E. T. Dewar, J. C. Paterson, and D. Rutherford, *J. Appl. Chem.* (London), **9**, 186 (1959).
(31) W. A. P. Black, E. T. Dewar, J. C. Paterson, and D. Rutherford, *J. Appl. Chem.* (London), **9**, 256 (1959).
(32) W. A. P. Black, E. T. Dewar, and D. Rutherford, *J. Chem. Soc.*, 3073 (1959).
(33) W. A. P. Black and E. T. Dewar, *J. Appl. Chem.* (London), **10**, 134 (1960).
(34) E. Brimacombe, Birmingham University Report to S.R.F., on Project 165.
(35) V. Prey and F. Grundschober, *Monatsh.*, **91**, 1185 (1960).
(36) V. Prey and F. Grundschober, *Monatsh.*, **92**, 1290 (1961).
(37) V. Prey and F. Grundschober, *Z. Zuckerind.*, **12**, 502 (1962).
(38) J. A. Reeder, British Columbia Research Council Reports to S.R.F. on Project 212.
(39) B. D. Jones, Ph. D. Thesis, University of Birmingham, England (1964).
(40) P. Schneider, New York University, Progress Reports 11, 14, and 15 on Project 196 to S.R.F. (Sept. and Dec. 1963; Jan.–Feb. 1964).

example, sodium methoxide,[41] sodium amide,[30,42] or sodium hydride[38,39] in N,N-dimethylformamide or methyl sulfoxide are now superseding the use of liquid ammonia, although, where Purdie's methylation procedure[43] stops short of peralkylation, potassium and liquid ammonia can be exploited to give full substitution, as shown by Irvine and Routledge's work on the permethylation of isosucrose.[44,45] The reactions leading to metallation of sucrose have been well covered.[5] Another interesting reaction of sucrose in liquid ammonia is reductive ammonolysis.[46] Under high temperatures and pressures in the presence of Raney nickel, ethylenediamine and mixed piperazines are produced.

Anhydrous hydrogen fluoride is also an efficient solvent for carbohydrates. Hyman and Katz[47] referred to hydrogen fluoride as capable of solubilizing carbohydrates readily, and stated that cellulose is freely soluble therein. In early studies on the constitution of the solution of cellulose in hydrogen fluoride, it was concluded that the conductivity of the solution is due to the presence of D-glucosyl fluoride.[48,49] This conclusion was later revised[50] when it was found that a product could be obtained that yielded D-glucose on mild hydrolysis. The product was termed a "glucosan." Work by Helferich and coworkers[51–53] also provided support for this conclusion. They found that brief treatment of cellulose and starch gave, in each instance, a product containing very little fluorine. Dalton[54] has attempted to introduce the fluoro group into either the D-glucose or D-fructose molecule by cleavage of sucrose in anhydrous hydrogen fluoride; a nonreducing, readily hydrolyzed, fluorine-containing product was obtained, and it was concluded that some kind of ad-

(41) V. R. Gaertner, *J. Amer. Oil Chem. Soc.*, **38**, 410 (1961).
(42) P. Schneider, New York University Progress Report to S.R.F., on Project 196 (May, 1963).
(43) T. Purdie and J. C. Irvine, *J. Chem. Soc.*, **83**, 1036 (1903).
(44) J. C. Irvine and D. Routledge, *Nature*, **134**, 143 (1934).
(45) J. C. Irvine and D. Routledge, *J. Amer. Chem. Soc.*, **57**, 1411 (1935).
(46) H. B. Haas and P. S. Skell, U. S. Pat. 2,978,451 (1961); *Chem. Abstr.*, **55**, 19968b (1961).
(47) H. H. Hyman and J. J. Katz, "Non-aqueous Solvent Systems," Academic Press, New York, N. Y., 1965, p. 76.
(48) K. Fredenhagen and G. Cadenbach, *Z. Phys. Chem.*, **A146**, 245 (1930).
(49) K. Fredenhagen, *Z. Elektrochem.*, **37**, 684 (1931).
(50) K. Fredenhagen and G. Cadenbach, *Angew. Chem.*, **46**, 113 (1933).
(51) B. Helferich, A Stärker, and O. Peters, *Ann.*, **482**, 183 (1930).
(52) B. Helferich and S. Böttger, *Ann.*, **476**, 150 (1929).
(53) B. Helferich and O. Peters, *Ann.*, **494**, 101 (1932).
(54) L. K. Dalton, personal communication.

dition compound had formed. A 13.6% w/v (solvent) solution of sucrose was used, but this obviously did not approach saturation, as Stattler and coworkers[55] dissolved 475 g of D-fructose and 566 g of D-glucose in 100 g of hydrogen fluoride (in examining the formation of anhydro compounds from these sugars). Two anhydrides, di-D-fructopyranose 2′,1:2,1′-dianhydride and D-fructopyranose D-fructofuranose 2′,1:2,1′-dianhydride, were obtained from D-fructose, and D-glucose yielded a mixture of anhydrides. Micheel and Buller[56] studied ester exchanges with D-glucose, fumaric acid, and 1,3-butanediol polyester in hydrogen fluoride, and Linn has found that sugars can be condensed with aromatic compounds in the presence of hydrogen fluoride, with few side-reactions.[57–59] It was hoped this novel reaction would lead to commercial end-products. Several reviews have appeared in which reactions of carbohydrates in hydrogen fluoride have been covered.[60–62]

Other inorganic compounds examined as possible solvents for sugars include (a) liquid sulfur dioxide, in which sucrose was found to be insoluble,[17] (b) hydrogen sulfide, in which D-glucose was found to give 1-thio-D-glucose and di-D-glucosyl disulfide,[63] (c) dinitrogen tetraoxide, which can act both as a solvent and a reactant in the presence of base to convert D-glucose into D-glucaric acid[64,65] or as a complexing agent to aid dissolution of polysaccharides in N,N-dimethylformamide, and (d) phosphoric acid of various concentrations up to 100%, which was examined as a "less degrading acid than other mineral acids" for the acetylation of cellulose.[66,67]

(55) L. Sattler, F. W. Zerban, G. L. Clark, C. Chia-Chen, N. Albon, D. Gross, and H. C. S. de Whalley, *Ind. Eng. Chem.*, **44**, 1127 (1952).
(56) F. Micheel and M. Buller, *Chem. Ber.*, **101**, 3729 (1968).
(57) C. J. Linn, *Chem. Eng. News*, 84 (1957).
(58) C. J. Linn, U. S. Pat. 2,798,100; *Chem. Abstr.*, **52**, 426b (1958).
(59) C. J. Linn, *Chem. Eng. News*, 29 (1963).
(60) J. Lennard, *Chem. Rev.*, **69**, 625 (1969).
(61) I. J. Goldstein and T. L. Hullar, *Advan. Carbohyd. Chem.*, **21**, 445 (1966).
(62) A. Wagner, in "Friedel–Crafts and Related Reactions," G. A. Olah, ed., Vol. IV, Interscience Publishers, New York, N. Y., 1963, p. 235.
(63) W. G. Overend, University of London Reports to S.R.F. on Project 145 (Dec. 31, 1958; Aug. 31, 1959).
(64) R. A. Bernsten, Ph. D. Thesis, Purdue University (1949).
(65) E. F. Degering, Purdue University Member Report No. 23 to S.R.F. (Oct. 1950).
(66) A. af Ekenstam, *Ber.*, **69**, 549, 553 (1936).
(67) E. Heuser, W. Shockley, A. Adams, and E. A. Grunwald, *Ind. Eng. Chem.*, **40**, 1500 (1948).

2. Salts

Mineral salts have also received attention. Domovs and Freund[68] found the solubility of a wide range of carbohydrates in methanol to be markedly enhanced by the presence of calcium chloride. Mono-, di-, and tri-saccharides in general were found soluble. In the presence of 1.5 molecules of calcium chloride and 12 molecules of methanol per hexose unit, maltose, L-sorbose, melezitose, D-galactose, and sucrose were soluble, whereas D-fructose, raffinose, D-glucose, and lactose were soluble but crystallized out later. With lactose, a crystalline 1:1:4 adduct of β-lactose with calcium chloride and methanol was obtained under anhydrous conditions, whereas an α-lactose complex was obtained in the presence of water, in the molar ratios of 1:1 to 7 of water. Dextran and a "dextrin" were found to be insoluble in this mixed system. Angyal[69] has found that calcium chloride does not interfere with glycosidation reactions. A similar co-solubilization effect occurs with zinc chloride in acetone, where it is evident that the solubility of certain sugars is considerably increased by the presence of zinc chloride, the ensuing reaction with acetone being sufficiently slow to allow this effect to be observed.[69] Further indications that zinc chloride can affect the solubilization of sugars has been obtained by Bourne.[70] In studies of the condensation of acrolein with sucrose, he decided that zinc chloride must function partly as a solvent (as well as a catalyst), as no reaction occurs at a 5:4 ratio of sucrose to zinc chloride, whereas, at a 1:4 ratio, the reaction proceeds. It does not occur in the presence of solvent. Thiocyanates also solubilize carbohydrates,[71] a discovery foreshadowed by the earlier work of Dubosc[72] on the solubilization of cellulose by aqueous calcium thiocyanate. It was found that a fused, 3:1 mixture of potassium and sodium thiocyanates (m.p. 130°) dissolves a moderate amount of sucrose and D-glucose. The sugars were found to be un-ionized, but the solutions discolored after ten minutes at 150°.

In the area intermediate between inorganic and organic solvents, Feuge and coworkers[73] have described the sodium, potassium, and

---

(68) K. B. Domovs and E. H. Freund, *J. Dairy Sci.*, **43**, 1216 (1960).
(69) S. J. Angyal, personal communication.
(70) E. J. Bourne, University of London Report to S.R.F. on Project 127 (Dec. 1960).
(71) T. I. Crowell and P. Hillery, *J. Org. Chem.*, **30**, 1339 (1965).
(72) A. Dubosc, *Rev. Prod. Chim.*, **26**, 507 (1923).
(73) R. O. Feuge, T. J. Weiss, M. L. Brown, and H. J. Zeringue, *J. Amer. Oil Chem. Soc.*, **47**, 56 (1970).

lithium salts of palmitic and oleic acid as catalysts and solubilizers in reactions between molten sucrose and fatty acid esters at elevated temperatures. The nature and yield of the resulting sucrose ester were found to be markedly dependent on the soap or soap mixture used.

IV. ORGANIC SOLVENTS

1. Protic

Alcohols and acids were among the first nonaqueous solvents to be investigated as solvents for carbohydrates. Few systematic studies on their solubility in these solvent systems have been published, however; and, although it is generally appreciated that many monosaccharides can be recrystallized from the lower alcohols, few references to their use have been found. Behr[74] used ethanol as a recrystallizing solvent for D-glucose, Wolfrom and Wood[75] recrystallized DL-glucose from absolute methanol, and Nitsch[76] applied the lower alcohols in preparing D-glucose and D-fructose from sucrose. In preparing D-mannose from ivory-nut shavings, Isbell[77] used methanol and isopropyl alcohol to aid crystallization, but water was present in the system. The most generally useful recrystallizing solvents among the alcohols were, however, found by Moye and Smythe to be those containing the R—O—C—C—OH grouping.[6-12] Hudson[78] referred to a member of this series, namely, 2-methoxyethanol, as a medium for dissolving and recrystallizing sugars, and Payne[79] used several of them in recovery of cane sugar from molasses, but the general nature of the solvent system was not recognized. It would appear that these solvents may provide the best recrystallizing media for sugars, particularly when Kononenko and Herstein's remarks concerning the many solvents they investigated for sucrose are considered; they stated[1] that "pyrazine was the only non-aqueous solvent from which sugar recrystallized readily." The author has found the alkoxy alcohols, tetrahydrofurfuryl alcohol, and related solvents to be particularly useful for recrystallizing a wide range of simple sugars ranging up to trisaccharides.

(74) A. Behr, Ber., **15**, 1104 (1882).
(75) M. L. Wolfrom and H. B. Wood, Jr., *J. Amer. Chem. Soc.*, **71**, 3175 (1949).
(76) E. Nitsch, Austrian Pat. 271,344; *Chem. Abstr.*, **71**, 40555g (1969).
(77) H. S. Isbell, *J. Res. Nat. Bur. Stand.*, **26**, 47 (1941).
(78) C. S. Hudson, *Advan. Carbohyd. Chem.*, **1**, 24 (1945).
(79) J. H. Payne, U. S. Pat. 2,501,914 (1945); *Chem. Abstr.*, **44**, 7079b (1950).

a. **Simple Alcohols.**—The solubility of various sugars in methanol has been determined by a number of authors. Trey[80] found that D-glucose has a solubility of 1.25% w/v (solvent) at 17.5° and 3.19% w/v (solvent) at the boiling point of the solution. During rotation studies, Hudson and Yanovsky[81] found that at 20° α-D-glucose has a solubility of 0.85% w/v, and that, after mutarotation, this increases to 1.6% w/v. Gillis and Nachtergaele[82] spanned the temperature range of 0 to 128.5° and found solubilities ranging from 1.52 to 1,150 g per 100 g of solvent. Other sugars examined[81] included β-D-mannose at 20° (0.78% w/v before and 4.4% w/v after mutarotation), β-D-fructose at 20° (5.2% w/v before and 11.1% w/v after mutarotation),[81] sucrose at 19° (1.18% w/v),[83] and lactose at 19.5° (0.084% w/v).[83] Values at variance with those given for sucrose were published by Gunning,[84] (0.30% w/v at 15°) and Lindet[85] [0.40% w/v (solvent) or 0.50% w/w (presumably at ambient temperature)]. Lindet also found that raffinose has a solubility of 9.5% w/v (solvent) in methanol. In studies on the methyl glycosidation of sugars, Bishop and Cooper[86] found that D-arabinose has low solubility in methanol. Levene and Muskat[87] worked with a 10% w/v (solvent) solution of L-rhamnose in preparing the methyl glycosides.

Rudenko[88] published values for the solubilities of D-glucose [1.52% w/v (solvent) at 19° and 2.26% w/v (solvent) at 50°], D-galactose [0.32% w/v (solvent) at 19° and 0.63% w/v (solvent) at 50°], and D-fructose [18.86% w/v (solvent) at 19° and 22.94% w/v (solvent) at 50°] that agree with those recorded by Hudson and Yanovsky[81] for D-glucose, but disagree with those for D-fructose. In the presence of urea, the solubilities were increased three- to eight-fold, but interaction between the sugar and urea occurred, as shown by u.v. spectral changes. Precipitation with benzene gave a product having a variable composition that, for D-glucose, approached a steady composition at a molar ratio of urea to D-glucose of 5:1.

Fewer solubilities are recorded for ethanol–sugar systems. Trey[80] found the solubility of D-glucose in ethanol to be 0.25% w/v at 20°

(80) H. Trey, *Z. Phys. Chem.*, **18**, 195 (1895).
(81) C. S. Hudson and E. Yanovsky, *J. Amer. Chem. Soc.*, **39**, 1013 (1917).
(82) J. Gillis and H. Nachtergaele, *Rec. Trav. Chim.*, **53**, 31 (1934).
(83) C. A. Lobry de Bruyn, *Z. Physik. Chem.*, **10**, 784 (1892).
(84) J. W. Gunning, *Chem. Ztg.*, **15**, 83 (1891).
(85) M. L. Lindet, *Compt. Rend.*, **110**, 795 (1890).
(86) C. T. Bishop and F. P. Cooper, *Can. J. Chem.*, **41**, 2743 (1963).
(87) P. A. Levene and I. E. Muskat, *J. Biol. Chem.*, **105**, 433 (1934).
(88) N. Z. Rudenko, *Zh. Obshch. Khim.*, **33**, 276 (1963).

and 1.42% w/v at the boiling point of the solution. Levine and coworkers[89] quoted Hudson and Yanovsky[81] as having given the solubility of D-glucose in 100% ethanol at 20° as 1.6–1.7% w/v, but the present author could find no mention of this. The latter authors[81] gave the solubility of α-L-rhamnose hydrate as 8.6% w/v initially, at 20°, rising after mutarotation to 9.5% w/v.

Authors are in better agreement as to the solubility of sucrose in ethanol, with the exception of Scheibler,[90] whose values would be expected to be less accurate, having been obtained by extrapolation from aqueous ethanolic mixtures. Both Schrefeld[91] and Pellet and Pellet[92] gave 0.00% w/v at 14° and 15–16°, respectively, whereas Reber[93] found the solubility to be 0.051% w/v at 25°. Because of the low solubility reported, Moye and Smythe conducted studies over the range of 80 to 140° and found solubilities ranging from 0.5 to 5.6% w/v.[6,8,9] The same authors also made studies with propyl alcohol and butyl alcohol, and a wide range of related alcohols.[6] Their results are recorded graphically in Fig. 3 (see p. 89). The solubility of D-glucose in isobutyl alcohol was found by Trey[80] to be 0.23% w/v (solvent) at the boiling point of the solution.

b. Polyhydric Alcohols.—The solubilities of sugars in ethylene glycol have apparently not been published. Fey and coworkers[94] recorded solubilities for sucrose in propylene glycol and in glycerol at 25° as 1.9% w/v and 5.7% w/v, respectively. Karcz,[95] on the other hand, stated that sucrose is insoluble in glycerol, but this result was not borne out by Strohmer and Stift's extensive studies[96] in re-examining this matter; they made a strong point of the problems encountered with under- and super-saturation, and the difficulties encountered in obtaining pure solvent and pure solute. Their values are in reasonable agreement with Fey's,[94] as were Browne and Randle's.[97] They[97] reported that the solubility of sucrose in glycerol at 20° is 7.0% w/w (solvent). Wahlgren[98] examined "polyethylene glycol 3000 and 400" and stated that the solubility of carbohydrates therein is low. In

(89) M. Levine, J. F. Foster, and R. M. Hixon, *J. Amer. Chem. Soc.*, **64**, 2331 (1942).
(90) C. Scheibler, *Ber.*, **5**, 343 (1872).
(91) O. Schrefeld, *Z. Ver. Deut. Zuckerind.*, **44**, 970 (1894).
(92) H. Pellet and L. Pellet, *Osterr. Ungar. Z. Zuckerind.*, **24**, 41 (1895).
(93) L. A. Reber, *J. Amer. Pharm. Ass. (Sci. Ed.)*, **42**, 192 (1953).
(94) M. W. Fey, C. M. Weil, and J. B. Segur, *Ind. Eng. Chem.*, **43**, 1435 (1951).
(95) M. Karcz, *Österr. Ungar. Z. Zuckerind.*, **23**, 21 (1894).
(96) F. Strohmer and A. Stift, *Osterr. Ungar. Z. Zuckerind.*, **24**, 41 (1895).
(97) F. Browne and D. G. Randle, *Pharm. J.*, **116**, 57 (1926).
(98) S. Wahlgren, *Svensk. Farm. Tidskr.*, **66**, 585 (1962).

order to obtain water-free solutions of carbohydrates in polyhydric alcohols, Merten and Bunge[99] dissolved sugars (ranging from mono- to oligo-saccharides and "dextrins") in a wide range of polyhydric alcohols in the presence of an acid catalyst and an ion exchanger, and removed the water *in vacuo*. To avoid dehydration reactions such as are described later for the alkoxy alcohols, very mild conditions must have been used.

c. **Alkoxy Alcohols.**—An investigation into the solubility of sugars in nonaqueous solvents miscible with water and boiling above 100° led to the discovery, already mentioned, of the ability of compounds containing the R—O—C—C—OH grouping to dissolve sugars. The solubility of sucrose in a wide range of these solvents and related simple alcohols is recorded in Fig. 3 (see p. 89). The solubility of D-glucose and D-fructose in these solvents at temperatures approaching the boiling point was so high that they were regarded as being miscible with the solvents. From these extensive studies emerged a series of papers and patents dealing with the application of the solvents, on the one hand, as purification media, and, on the other, as reaction media.[6-12,100,101]

d. **Other Alcoholic Solvents.**—Rather suprisingly, despite the high solubilities of sugars in amines, little work has been performed on the ω-aminoalkyl alcohols ("alkanolamines"). Hengst[102] patented a process for producing sucrose from sugar beet by extraction with 2-aminoethanol, Lang[103] reported that a 40.6% (by weight) solution of sucrose in 2-aminoethanol is not saturated, and Skell[104] examined the reductive aminolysis of sucrose in 2,2′-iminodiethanol ("diethanolamine"), 2,2′,2″,2‴-(ethylenedinitrilo)tetraethanol and 2,2′,2″,2‴-(propylenedinitrilo)tetraethanol being the major products. Cyclohexanol has been used as a solvent for the catalytic reduction of D-glucose.[105]

e. **Alcohols as Reaction Media.**—The application of alcohols as reaction media is limited to those situations in which the alcoholic group of the solvent either takes part in the reaction, or does not

(99) R. Merten and W. Bunge, Ger. Pat. 1,173,446; *Chem. Abstr.*, **61**, 12184a (1964).
(100) C. J. Moye, Aust. Pat. 272,247 (1964); *Chem. Abstr.*, **69**, 44184s (1968)
(101) C. J. Moye and R. J. Goldsack, *J. Appl. Chem.* (London), **16**, 206 (1966).
(102) G. Hengst, Ger. Pat. 962,599; *Chem. Abstr.*, **53**, 10816f (1959).
(103) Lang, cited in Ref. 1 (see Ref. 33 therein).
(104) P. S. Skell, Pennsylvania State University Summary Report to S.R.F. on Project 86 (Dec. 1958).
(105) A. Ashida, *Nippon Nogei Kagaku Kaishi*, **20**, 621 (1944).

interfere with the course of the reaction by competing with the hydroxyl groups of the sugar.

Where the alcohol functions as a solvent only, it is possible to conduct such reactions as reduction, formation of dithioacetals, oxidation, and amination. Thus, the catalytic reduction of D-glucose in the presence of Raney nickel was achieved by Moye and Smythe[7,9,11] in 2-methoxyethanol, by Ashida and Suzuki[106] in tetrahydrofurfuryl alcohol, and by Ashida[106] in cyclohexanol. The dibutyl dithioacetal of D-glucose was prepared in 2-methoxyethanol,[7,9,11] and an oxidation by bromine[7,9,11] was conducted in 2-(2-methoxyethoxy)ethanol ("diethylene glycol monomethyl ether"). The condensation of reducing sugars with amines to give glycosylamines may be readily accomplished in an alcoholic medium.[107,108] Condensation of D-glucose with butylamine in methanol (followed by decomposition by the action of acetic acid to give 3-deoxy-D-*erythro*-hexulose) has been described,[109] as has its condensation with dodecylamine,[7,9,11] the respective N-substituted D-glucosylamines being obtained. Methanol (and other simple alcohols) has also been used as a solvent for the preparation of various hydrazones,[110,111] and in early studies by Lobry de Bruyn[112] on the nature of the "fructosamin" obtained by decomposition of 2-amino-2-deoxy-D-glucose by the action of methoxide. It has also been used as a cosolvent in the Purdie methylation procedure[43] and as the medium (along with other simple alcohols and nonaqueous media) for the preparation of carbohydrate complexes, a subject beyond the scope of this article, but reviewed by Rendleman.[113]

Reactions of the sugar with the alcohol group of the solvent are confined to glycoside formation, a reaction catalyzed by acids ranging from mineral acids to insoluble ion-exchangers.[114–116] A general review by Shafizadeh[117] describes the earlier work. Bishop and cowork-

(106) A. Ashida and B. Suzuki, Jap. Pat. 172,087; *Chem. Abstr.*, **43**, 7505c (1949).
(107) E. Mitts and R. M. Hixon, *J. Amer. Chem. Soc.*, **66**, 483 (1944).
(108) W. Pigman, E. A. Cleveland, D. H. Couch, and J. H. Cleveland, *J. Amer. Chem. Soc.*, **73**, 1976 (1951).
(109) H. Kato, *Agr. Biol. Chem.* (Tokyo), **26**, 187 (1962).
(110) N. K. Richtmyer, personal communication.
(111) H. H. Stroh and H. Lamprecht, *Chem. Ber.*, **96**, 651 (1963).
(112) C. A. Lobry de Bruyn, *Ber.*, **31**, 2476 (1898).
(113) J. A. Rendleman, *Advan. Carbohyd. Chem.*, **21**, 255 (1966).
(114) B. Helferich and W. Schafer, *Org. Syn., Coll. Vol.*, **1**, 364 (1932).
(115) H. H. Young, U. S. Pat. 2,422,328; *Chem. Abstr.*, **41**, 5753 (1947).
(116) D. F. Mowery, *J. Amer. Chem. Soc.*, **77**, 1667 (1955).
(117) F. Shafizadeh, *Advan. Carbohyd. Chem.*, **13**, 9 (1958).

ers[86,118,119] studied the kinetics of glycosidation in detail, and defined the sequence of steps occurring, and Overend and coworkers[120] examined the reaction of ethanethiol with D-xylose in the presence of N,N-dimethylformamide as cosolvent and found that the diethyl dithioacetal is the main product and that it is subject to further reaction on long standing. The ultimate product is 3-(ethylthio)-2-[(ethylthio)-methyl]furan, which results from an unexpected migration of the ethylthio group. The glycosides of several aldoses with alkoxy alcohols have also been prepared.[100]

The solvent may also react with conversion products of the sugar. Such reactions have been used for preparing ethers of 5-(hydroxymethyl)-2-furaldehyde from ketohexoses, and these products were reduced stepwise[101] to give half ethers of 2,5-furandimethanol and its tetrahydro derivative. The fundamental R—O—C—C—OH structure (see p. 88) is present in these compounds, which, despite their high molecular weights, are all excellent solvents for a range of simple sugars. During studies on the foregoing reaction sequences, articles appeared on (a) the mechanism of formation of 5-(hydroxymethyl)-2-furaldehyde in nonaqueous media, wherein it was shown that acid catalysis (not iodine) is responsible for the conversion of D-fructose by iodine in N,N-dimethylformamide into 5-(hydroxymethyl)-2-furaldehyde[121,122] and (b) radiochemical, thin-layer chromatographic approaches to monitoring the kinetics of such sensitive, complex, chemical reactions.[123,124] The particular system studied in the latter research was the reaction of D-fructose with p-toluenesulfonic acid in tetrahydrofurfuryl alcohol.[124]

Methyl α-D-glucopyranoside is the only product of commercial promise to have thus far emerged from work with protic solvents; it has utility in the preparation of polyurethane foams. Mehltretter and coworkers[125,126] have described the application of mixtures of D-glucosides obtained by the acid-catalyzed reaction of ethylene glycol, 1,2-propanediol ("propylene glycol"), or glycerol with starch,

(118) C. T. Bishop and F. P. Cooper, *Can. J. Chem.*, **40**, 224 (1962).
(119) C. T. Bishop, F. P. Cooper, and V. Smirnyagin, *Can. J. Chem.*, **43**, 3109 (1965).
(120) R. J. Ferrier, L. R. Hatton, and W. G. Overend, *Carbohyd. Res.*, **6**, 87 (1968).
(121) T. G. Bonner, E. J. Bourne, and M. Ruskiewicz, *J. Chem. Soc.*, 787 (1960).
(122) C. J. Moye and Z. S. Krzeminsky, *Aust. J. Chem.*, **16**, 258 (1963).
(123) C. J. Moye, *J. Chromatogr.*, **13**, 56 (1964).
(124) C. J. Moye and R. J. Goldsack, *J. Appl. Chem.* (London), **16**, 209 (1966).
(125) C. L. Mehltretter, F. H. Otey, and B. L. Zagoren, *Ind. Eng. Chem. Prod. Res. Develop.*, **2**, 256 (1963).
(126) C. L. Mehltretter, F. H. Otey, and C. E. Rist, *J. Amer. Oil Chem. Soc.*, **40**, 76 (1963).

in the preparation of rigid polyurethane foams. Glycoside formation can also be used in modifying sucrose-extended, melamine resins by employing methanol, instead of an aqueous medium, as the solvent during formation of the resin.[127]

f. **Acids.**—The lower organic acids (formic to propionic acid) appear to be the only ones thus far studied as carbohydrate solvents, and no specific solubilities have been published. Formic acid has been found[128] to be an excellent solvent for D-glucose, D-fructose, sucrose, lactose (in contrast to pyridine), and raffinose (in contrast to water), and a good solvent for D-xylose, L-rhamnose, and maltose. D-Galactose was stated to be more difficultly soluble. Carr[129] was no more specific in his structure–solubility studies on sugars, "dextrin," starch, inulin, glycogen, and agar; he concluded that, in general, carbohydrates that are readily hydrolyzed and that yield aldehyde and ketone groups are soluble in formic acid. No more-specific information is available for acetic acid. Several authors have referred to the advantages of using acetic acid in the crystallization of sugars,[78, 130-134] and Schiff,[135] in studies on the condensation of sugars with ketones and aldehydes, found that D-glucose and sucrose are readily soluble in glacial acetic acid, with partial crystallization on cooling. The presence of water (2%) gave thick, concentrated syrups in the cold. D-Mannose and D-erythrose tended to give supersaturated solutions, whereas lactose was not so readily soluble and crystallized completely on cooling. D-Xylose was crystallized from very concentrated (95%) aqueous syrups by the addition of acetic acid after its preparation from cottonseed hulls.[136]

g. **Acids as Reaction Media.**—In early studies, Berthelot[137,138]

(127) G. L. Redfern, Yarsley Research Laboratories Ltd., Final Report to S.R.F. on Project 209 (1966).
(128) H. Grossman and F. L. Bloch, Z. Ver. Deut. Zuckerind., **62**, 57 (1912).
(129) R. H. Carr, Science, **69**, 407 (1929).
(130) J. H. Pazur, in "The Carbohydrates; Chemistry and Biochemistry," W. Pigman and D. Horton, eds., Vol IIA, 2nd Ed., Academic Press, Inc., New York, N. Y., 1970, p. 87.
(131) C. S. Hudson and J. K. Dale, J. Amer. Chem. Soc., **39**, 322, footnote 2 (1917).
(132) C. S. Hudson and H. L. Sawyer, J. Amer. Chem. Soc., **39**, 471 (1917).
(133) A. Wernicke and W. Pfitzinger, U. S. Pat. 260,340 (1882).
(134) Soc. Anon des Usines de Melle, Fr. Pat. 621,075 (1927); Chem. Abstr., **25**, 5589 (1931); Ger. Pat. 528,564 (1926).
(135) N. Schiff, Ann., **244**, 20 (1888).
(136) C. S. Hudson and T. S. Harding, J. Amer. Chem. Soc., **39**, 1038 (1917).
(137) M. Berthelot, Compt. Rend., **41**, 452 (1855).
(138) M. Berthelot, Ann. Chim. Phys., **60**, 93 (1860).

heated D-glucose and sucrose with fatty acids in a sealed tube at 120° for two to three days, and obtained complex mixtures. For trehalose, the temperature was raised to 180°. Under these conditions, decomposition of carbohydrate must have occurred, and solubilization in any case would have been minimal. The application of acids (usually acetic) as solvents for reactions of carbohydrates has largely been confined to acylation reactions, in which they have been used in admixture with an acid anhydride. In the absence of a mineral acid, degradation of the carbohydrate is avoided; sulfuric acid is normally used to effect various degrees of acetolysis.[139–142] Thus, cellulose was acetylated by means of a mixture of acetic acid, acetic anhydride, chlorine, and sulfur dioxide,[143,144] and also by use of acetic acid in admixture with trifluoroacetic anhydride; in each instance, the triacetate was formed.[145] The accelerating action of trifluoroacetic anhydride has been reported by a number of authors.[145–148] During one of these studies,[145] a benzoate was prepared by the use of benzoic acid, and maltose was converted into the octapropionate by the use of propionic acid. Cox and coworkers[149] described the preparation of peracylated sucrose by use of acetic acid as the solvent; sucrose acetate isobutyrate may be similarly synthesized by use of mixed anhydrides.[150,151] In nitration studies on dextrans of three ranges of molecular weight [shown to consist almost exclusively of α-D-(1→6)-linked D-glucopyranose residues], Mustafa and coworkers[152] demonstrated that more-effective nitration could be obtained by using nitric acid in acetic acid–acetic anhydride than with nitric acid alone or in admixture with sulfuric acid. Starch, consist-

(139) R. D. Guthrie and J. F. McCarthy, *Advan. Carbohyd. Chem.*, **22**, 11 (1967).
(140) J. H. Pazur, in "The Carbohydrates; Chemistry and Biochemistry," W. Pigman and D. Horton, eds., Vol. IIA, 2nd Ed., Academic Press, Inc., New York, N. Y., 1970, p. 78.
(141) K. Hess and K. Dziengel, *Ber.*, **68**, 1594 (1935).
(142) M. L. Wolfrom and J. C. Dacons, *J. Amer. Chem. Soc.*, **74**, 5331 (1952).
(143) J. C. Irvine and E. L. Hirst, *J. Chem. Soc.*, **121**, 1585 (1922).
(144) E. J. Bourne, M. Stacey, J. C. Tatlow, and J. M. Tedder, *J. Chem. Soc.*, 2976 (1949).
(145) W. L. Barnett, *J. Soc. Chem. Ind.*, **40**, 8T (1921).
(146) C. Hamalainen, R. H. Wade, and E. M. Buras, *Text. Res. J.*, **27**, 168 (1957).
(147) J. M. Tedder, *Chem. Rev.*, **55**, 787 (1955).
(148) H. T. Clarke and C. J. Malm, U. S. Pat. 1,880,808 (1932).
(149) G. J. Cox, J. H. Ferguson, and M. L. Dodds, *Ind. Eng. Chem.*, **25**, 968 (1933).
(150) G. P. Tovey and H. E. Davis, U. S. Pat. 2,931,802 (1960); *Sugar Ind. Abstr.*, **22**, 140 (1960).
(151) Anon, *Chem. Ind. Eng.* (Australia), **11** (4), 34 (1958).
(152) A. Mustafa, A. F. Dawoud, and A. Marawou, *Staerke*, **22**, 17 (1970).

ing of α-D-(1→4)-linked D-glucopyranose residues, is nitrated more effectively than dextran[153] in the nitric acid system, and Mustafa and coworkers considered that their results, and those of earlier workers, could be explained by the more ready gelation of dextran. To the best of the author's knowledge, no connection between the ease of gelation and the nature of the polymeric bond has as yet been demonstrated. Sucrose has been pernitrated (without inversion) by 100% nitric acid in a mixture of acetic acid and acetic anhydride,[154] and the pentanitrates of α- and β-D-glucopyranose were prepared in the same medium.[155]

Incidental reactions that have been reported include the preparation of derivatives of 5-(hydroxymethyl)-2-furaldehyde by reaction of D-fructose in acetic or propionic acid in the presence of the respective anhydride.[156] The condensation of D-glucose with phenol has been effected in acetic acid in the presence of dry hydrogen chloride, prior to resinification,[157] and the reaction of sucrose with thionyl chloride in acetic acid–acetic anhydride produced partially acetylated chlorodeoxysucroses.[158] Sucrose has been condensed with maleic anhydride in acetic anhydride mixed with acetic acid or formic acid, to give solid products having an undetermined structure.[159]

A rather unusual solvent system for sucrose, namely, molten chloroacetic acid, was used by Lorand,[160] who prepared sucrose esters from palmitic anhydride dissolved in this medium. It is unlikely that the glycosidic link in the sucrose is unaffected under these conditions.

h. Phenols.—Lowry and Faulkner[161] conducted studies on the mutarotation of α-D-glucose and tetra-O-methyl-α-D-glucopyranose in both cresol and pyridine. The solubility in cresol was not reported. It was incorrectly stated that D-glucose does not mutarotate in either solvent, but of more significance to the subject under discussion was the observation that, with the tetramethyl ether, a mixture of pyridine and cresol is very much more effective in assisting mutarotation than

---

(153) M. L. Wolfrom, A. Chaney, and K. S. Ennor, *J. Amer. Chem. Soc.*, **81**, 3469 (1959).
(154) J. Honeyman, Report on Project 149 to S.R.F. (1957–1960).
(155) G. Fleury and L. Brissaud, *Compt. Rend.*, **222**, 1051 (1946).
(156) Merck and Co., Inc., Brit. Pat. 925,812; *Chem. Abstr.*, **59**, 12761b (1963).
(157) Herstein Laboratories Reports on Project 82 to S.R.F. (Dec. 1952–May 1953); U. S. Pat. 2,252,725 (1941); *Chem. Abstr.*, **35**, 7420 (1941).
(158) H. I. Szmant, University of Puerto Rico Quarterly Report to S.R.F. on Project 157 (May 18, 1959).
(159) Bjorksten Research Laboratories Reports to SRF on Project 105 (July 10, 1957).
(160) E. J. Lorand, U. S. Pat. 1,959,590; *Chem. Abstr.*, **28**, 4432 (1934).
(161) T. M. Lowry and I. J. Faulkner, *J. Chem. Soc.*, 2883 (1925).

either of the constituents. A similar acceleration was observed by Swain and Brown[162] and Romy[163] for 2-pyridinone, and these results led to drawing of analogies between enzymic catalysis and this type of bifunctional catalysis. The only other study encountered in which a phenol was used as a sugar solvent is that of Theobald,[164(a)] who reinvestigated Hochstetter's work[164(b)] on the condensation of sucrose with diphenyl carbonate in resorcinol. In this work, an ~20% solution of sucrose in resorcinol was used.

2. Aprotic

There are fewer complications in the use of aprotic media as solvents for sugars, and a wide range of such media has been investigated for their potential utility. Several, in particular, methyl sulfoxide, N,N-dimethylformamide, and a number of nitrogenous, heterocyclic compounds, have proved to be exceptional solvents for the lower saccharides.

a. **Hydrocarbons.**—(i) **Aliphatic.**—The only solvents of the aliphatic hydrocarbon type utilized in carbohydrate chemistry appear to be the simple polychlorinated alkanes. Chloroform has been used as a medium in which sugars and starch can be nitrated by dinitrogen pentaoxide,[165] The presence of sodium fluoride lessens problems associated with the formation of nitric acid. The use of this medium has been reviewed.[166] Chloroform and dichloromethane have also been used, in conjunction with pyridine, as media in which sulfates or sulfites of sugars could be formed.[167] For D-xylose, the anomeric hydroxyl group was replaced by chlorine at the same time as the tris(chlorosulfite) was formed.[168] Pyridine, however, tends to suppress undesired chlorination.

(ii) **Aromatic.**—Aromatic hydrocarbons examined include mono- and bi-cyclic compounds and their partial reduction products. The fact that sugars are completely insoluble in benzene and naphthalene was used to advantage by Plato and coworkers[169] in studies of the den-

(162) C. G. Swain and J. F. Brown, *J. Amer. Chem. Soc.*, **74**, 2534 (1952).
(163) P. R. Romy, *J. Amer. Chem. Soc.*, **90**, 2824 (1968).
(164) (a) R. S. Theobald, *J. Chem. Soc.*, 5370 (1961), and references cited therein;
    (b) A. Hochstetter, Ger. Pat. 268,452 (1912); *Chem. Abstr.*, **20**, 1854 (1914).
(165) M. Goldfrank, *Methods Carbohyd. Chem.*, **4**, 291 (1964).
(166) J. Honeyman and J. W. W. Morgan, *Advan. Carbohyd. Chem.*, **12**, 120 (1957).
(167) P. D. Bragg, J. K. N. Jones, and J. C. Turner, *Can. J. Chem.*, **37**, 1412 (1959).
(168) H. J. Jennings, *Can. J. Chem.*, **46**, 2799 (1968).
(169) F. Plato, J. Domke, and H. Harting, *Z. Ver. Deut. Zuckerind.*, **50**, 1009 (1900).

sity of sugar solutions. Other authors have used aromatic hydrocarbons because the products of the reaction were soluble; for example, xylene was used as a dispersion medium for the preparation of sucrose mono-esters[170] and in condensation of alkylene oxides with sucrose.[171] In reduction studies, Ashida[106] found that tetralin is a useful hydrogen-donor, but he had to use N-heterocyclic compounds as solubilizing agents because of the insolubility of D-glucose in tetralin. Toluene has been used as the dispersing medium, with pyridine as the acid acceptor, in the reaction of sucrose with phosgene.[164]

b. **Solvents Based on an Oxygen Function.**—(i) Aliphatic.—Brief mention has been made of the use of esters as solvents for sugars. Hudson[78] stated, without a reference, that ethyl lactate is a better solvent than ethyl acetate for crystallizing some sugars, and Hamalainen and coworkers[172] used amyl acetate as a nonswelling diluent to enable peracetylated fibrous cotton, suitable for textiles, to be produced. Ethyl acetoacetate reacts with D-glucose in ethanol in the presence of fused zinc chloride to give ethyl 5-methyl-2-(D-*arabino*-tetrahydroxybutyl)-4-furoate,[173,174] a reaction that has been reviewed by García González[173] and re-investigated by Dalton.[175] Sucrose is reported to have limited solubility in a mixture of methyl and ethyl carbamates.[176] Lactones, in particular, tetrahydrofuran-2-one, have more utility than esters as sugar solvents. Werner and Hessel[177] described the use of tetrahydrofuran-2-one for a wide range of reactions of carbohydrates, but its utility is probably limited by the fact that solubilities are low at temperatures below 100° [0.26% sucrose w/v (solvent) at 100°] and increasing decomposition of sucrose occurs above this temperature, where the solubility increases[7-11] to 3.0% w/v (solvent) at 130° and 7.8% w/v (solvent) at 140°; its use would, therefore, be limited to the monosaccharides. Tetrahydropyran-2-one has also been used to assist the acylation of cellulose.[178]

Carbohydrates have limited solubility in carbonyl-containing sol-

---

(170) R. S. Aries, U. S. Pat. 2,999,858; *Chem. Abstr.*, **56**, 3557d (1962).
(171) M. de Groote, U. S. Pat. 2,602,051 (1952); *Chem. Abstr.*, **46**, 11662b (1952).
(172) C. Hamalainen, E. M. Buras, S. R. Hobart, and A. S. Cooper, *Text. Res. J.*, **27**, 214 (1957).
(173) F. García González, *Advan. Carbohyd. Chem.*, **11**, 97 (1956).
(174) E. S. West, *J. Biol. Chem.*, **74**, 561 (1927).
(175) L. K. Dalton, *Aust. J. Chem.*, **17**, 117 (1964).
(176) Herstein Laboratories Report No. 112 on Project 82 to the S.R.F. (May 1960).
(177) J. Werner and F. A. Hessel, U. S. Pat. 2,938,898 (1960); *Chem. Abstr.*, **55**, 1468i (1961).
(178) R. C. Blume and F. H. Swezey, *Tappi*, **37**, 481 (1954).

vents, but where solution occurs during or attendant on reaction, the ratio of carbohydrate to "solvent" can be quite high, and it is then difficult, from the information given, to determine whether true dissolution of the sugar occurred. Verhaar[179] determined the solubility of sucrose in acetone to be 0.007% w/v at 30° a value probably far more accurate than Herz and Knock's[180] figure of ~0.23% w/v at 25°. However, the presence of zinc chloride in the medium considerably increases the solubility of the sugar and leads at the same time to the isopropylidene acetals, the formation of which is catalyzed by zinc chloride.[69,181] Phosphoric acid has also been used as the catalyst in this reaction system.[182] In reactions of sugars with paraformaldehyde and paraldehyde, sulfuric acid and zinc chloride are the catalysts usually used, concomitant dissolution and reaction occurring. With D-glucose, paraformaldehyde gives a di-O-methylene derivative,[54] D-galactose gives 4,6-O-ethylidene-D-galactose with paraldehyde,[183] and sucrose is reported to afford a mixture of the tri-O-ethylidene (29%) and penta-O-ethylidene (71%) derivatives, of which the tri-O-ethylidene derivative is said to be the only crystalline, ethylidene cyclic acetal.[181] Tipson[184] has used acid-free chloral as a sugar solvent, but solubilities were not determined.

(ii) **Heterocyclic.**—As mentioned earlier (see p. 89), tetrahydrofurfuryl alcohol is an excellent solvent for mono- and di-saccharides at high temperatures,[6-11] and, although insufficient material was available for determinations of solubility, the same applies to the alkoxyalkyl half ethers of tetrahydro-2,5-furandimethanol examined[6,101] qualitatively as solvents. The solubility of sucrose in tetrahydrofuran[1] is 0.01% (w/v) at 60°. Several authors have examined p-dioxane as a solvent and found it, in contrast to morpholine, to be a very poor solvent. Kononenko and Herstein[1] reported solubilities of sucrose ranging from 0.07% (w/v) at 60° to 0.11% (w/v) at 100°, whereas Helferich and Masamune[185] gave a value of 0.08% (w/w) at 100°. The latter authors also gave solubilities of approximately ten times these for D-galactose and for α- and β-D-glucose, and a value of 5.8% (w/w) for D-fructose at 100°. The oxygen heterocycles have

(179) G. Verhaar, *Arch. Suikerind. Nederland Ned.-Indie*, **1**, 464 (1940).
(180) W. Herz and M. Knock, *Z. Anorg. Chem.*, **41**, 322 (1904).
(181) E. J. Bourne, University of London Report to S.R.F. (final) on Project 159 and part Project 161 (Nov. 1959).
(182) W. L. Glen, G. S. Myers, and G. A. Grant, *J. Chem. Soc.*, 2568 (1951).
(183) D. H. Ball, *J. Org. Chem.*, **31**, 220 (1966).
(184) R. S. Tipson, personal communication.
(185) B. Helferich and H. Masamune, *Ber.*, **64**, 1257 (1931).

little utility as reaction media for sugars. Dalton[54] used a multi-component system containing chloroform and p-dioxane in acetalation reactions, but this system was designed to dissolve the product rather than the starting material. The only encountered report[186] of the use of p-dioxane as a reaction medium is for reductions, with lithium aluminum hydride, of D-glucose, D-fructose, L-arabinose, and D-ribose, in which the concentration of the sugar was low.

c. **Solvents Based on a Sulfur Function.**—(i) **Aliphatic**—With the notable exception of methyl sulfoxide and the related dipropyl compound, no other aliphatic sulfur-containing compounds appear to have been examined as sugar solvents. Methyl sulfoxide[187-189] is an excellent solvent for sugars and polysaccharides of high molecular weight, and it has become popular as a reaction medium for sugars, along with N,N-dimethylformamide, the other well known member of the dipolar, aprotic series of solvents. Kononenko and Herstein[1] found the solubility of sucrose in methyl sulfoxide to be: 41.6% w/w (30°), 49.1% w/w (60°), 51.1% w/w (85°), 58.7% w/w (100°), and 61.6% w/w (110°), and another publication gave the solubility of starch at 20–30° as 2% w/v (solvent) and of sucrose at 20–30° as 30% w/v (solvent) and at 90–100° as 100% w/v (solvent).[188] These appear to be the only values thus far published. Dipropyl sulfoxide is also a good solvent for sucrose, dissolving 22.6% w/w (60°), 38.0% w/w (80°), and 42.0% w/w (100°), with some decomposition occurring at the high temperatures.[190]

One of the more important applications of methyl sulfoxide as a sugar solvent is for n.m.r. spectral studies, where, in conjunction with acetone-d_6, it has been used because exchange reactions are suppressed[191] (see the n.m.r. section, p. 124). Methyl sulfoxide dissolves both amylose and amylopectin,[192] and these can then be separated by the judicious use of co-solvents after dissolution.[193] Ultracentri-

---

(186) H. Endres and M. Oppelt, *Chem. Ber.*, **91**, 478 (1958).
(187) "DMSO, Reaction Medium and Reactant," Chemical Products Division, Crown Zellerbach Corporation, Camas, Washington (1962).
(188) "DMSO Technical Bulletin," Chemical Products Division, Crown Zellerbach Corporation, Camas, Washington (1966).
(189) E. M. Arnett and F. Douty, *J. Amer. Chem. Soc.*, **86**, 409 (1964).
(190) P. Schneider, New York University Progress Report No. 6 to SRF on Project 196 (April 1963).
(191) J. C. Jochims, G. Taigel, A. Seeliger, P. Lutz, and H. E. Driesen, *Tetrahedron Lett.*, 4363 (1967).
(192) S. Tomita and K. Terajima, *J. Sci. Food Agr.*, **21**, Abstr. i-20 (1970).
(193) G. K. Adkins and C. T. Greenwood, *Carbohyd. Res.*, **2**, 217 (1969).

fugation studies have been performed[194] on the resultant solutions in methyl sulfoxide.

Everett and J. F. Foster[195] examined a solution of amylose (from potato starch) in methyl sulfoxide, and were able to separate it into seven fractions, by molecular weight, by precipitation with ethanol. From examination of the fractions by various physical means, particularly limiting-viscosity numbers, it was suggested that the polymer was present in a coiled conformation. From the molecular weights of the fractions, it was concluded that amylopectin was absent. Preliminary solubilization of starch by methyl sulfoxide is claimed[196] to provide the basis for an improved assay for starch by use of glucoamylase. Glycogen is also readily solubilized under mild conditions,[197] and cellulosic fractions can be extracted from wood pulp.[198-200] It is, therefore, apparent that methyl sulfoxide is a general solvent for all classes of carbohydrates.

Methyl sulfoxide has not been so frequently used as N,N-dimethylformamide for carbohydrate reactions, but this is probably due to the longer availability of the latter. Both solvents have drawbacks in that they are not completely inert. Oxidation reactions[201,202] have been performed in methyl sulfoxide, but it is more usual to use derivatives instead of the free sugars. Thus it has been reported that, in the presence of N,N'-dicyclohexylcarbodiimide, methyl sulfoxide is a very mild oxidizing agent that will oxidize most free hydroxyl groups,[201] and Ag^{2+} (argentic picolinate) in methyl sulfoxide or a mixture thereof with bis(2-methoxyethyl) ether oxidizes free sugars (such as D-glucose) and their derivatives.[202] Methyl sulfoxide can also be used as a solvent for oxidations with lead tetraacetate,[203] where acetic acid must be added to prevent oxidation of the solvent, and for periodate oxidations.[204] In the latter oxidation of dextran-500, oxidation was limited, because of internal hemiacetal formation after the primary oxidation step. Reversion reactions have also been conducted in methyl sulfoxide in the presence of dry hydrogen chloride, to give

(194) G. K. Adkins, C. T. Greenwood, and D. J. Hourston, *Cereal Chem.*, **47**, 13 (1970).
(195) W. W. Everett and J. F. Foster, *J. Amer. Chem. Soc.*, **81**, 3459, 3464 (1959).
(196) R. A. Libby, *Cereal Chem.*, **47**, 273 (1970).
(197) R. L. Whistler and J. N. BeMiller, *Advan. Carbohyd. Chem.*, **13**, 289 (1958).
(198) F. A. Abadie-Maumert, *Papeterie*, **79**, 519 (1957).
(199) S. Hossain, *Pulp Paper Mag.* (Canada), **59**, 127 (Aug. 1958).
(200) J. A. McPherson, *Acta Chem. Scand.*, **12**, 779 (1958).
(201) W. W. Epstein and F. W. Sweat, *Chem. Rev.*, **67**, 247 (1967).
(202) J. B. Lee and J. G. Clarke, *Tetrahedron Lett.*, 415 (1967).
(203) C. T. Bishop and V. Zitko, *Can. J. Chem.*, **44**, 1749 (1966).
(204) C. T. Bishop and R. J. Yu, *Can. J. Chem.*, **45**, 2195 (1967).

oligo- and poly-saccharides, which were then degraded back to the starting D-glucose.[205] Esterification reactions in methyl sulfoxide include esterification of monosaccharides[206,207] and formation of sucrose esters by ester exchange.[208-210] Boiling methyl sulfoxide has been investigated as a medium for the preparation of carbonic esters by exchange with diethyl or diphenyl carbonate.[164] Polymerization occurs when a tris-O-(phenoxycarbonyl)sucrose, prepared[211] in pyridine at 0°, is heated at 135° in methyl sulfoxide, and the polymer appears identical with that obtained directly from sucrose and diphenyl carbonate in methyl sulfoxide at 135° in the presence of 1% of sodium hydrogen carbonate. Methyl sulfoxide is also used, in the preparation of ethers, as a medium for the formation of metal derivatives of sugars, but, as it is decomposed by metallic sodium,[45] use of sodium hydride or sodium amide as the base is usual. Sodium amide is reported[212] to be the base best suited for the preparation of the lower sucrates in methyl sulfoxide, and benzylation can be performed efficiently in this medium. Cellulose has been etherified to a low degree of subsituation by 2-halo acids in the presence of methyl sulfoxide,[213] and the properties of cellulose fibers have been modified by N-carboxyanthranilic anhydride in this solvent.[214,215] As with formation of urethans,[216,217] the cellulosic material is rendered more reactive by the swelling action of the solvent, and localized solvation probably occurs. Methyl sulfoxide is more effective than N,N-dimethylformamide in swelling cellulose, and the latter is, in turn, more effective than pyridine.[216-218] Cellulose can also be sulfated by a complex of

(205) F. Micheel and W. Gresser, *Chem. Ber.*, **91**, 1214 (1958).
(206) R. J. Wicker and M. Gates, Brit. Pat. 872,293; *Chem. Abstr.*, **58**, 573b (1963).
(207) R. J. Wicker and M. Gates, Brit. Pat. 872,507; *Chem. Abstr.*, **56**, 4853b (1962).
(208) C. C. Price and W. H. Snyder, *J. Amer. Chem. Soc.*, **83**, 1773 (1961).
(209) D. C. Nelson, U. S. Pat. 3,023,183 (1962); *Chem. Abstr.*, **57**, 11026d (1962).
(210) W. F. Huber and N. B. Tucker, U. S. Pat. 2,812,324 (1957); *Sugar Ind. Abstr.*, **21**, 91 (1959).
(211) L. Hough, J. E. Priddle, and R. S. Theobald, *J. Chem. Soc.*, 1934 (1962).
(212) H. Saeki and T. Iwashige, *Chem. Pharm. Bull.*, **15**, 1803 (1967); see *Carbohyd. Chem.* (Chem. Soc., London), **2**, 31 (1969).
(213) L. Murray and K. Ward, U. S. Pat. 2,987,434 (1961); *Chem. Abstr.*, **55**, 21587g (1961).
(214) W. A. Reeves and R. H. Wade, U. S. Pat. 2,926,063 (1960); *Chem. Abstr.*, **54**, 11502a (1960).
(215) W. R. Adams, U. S. Pat. 3,007,763 (1961); *Chem. Abstr.*, **56**, 491b (1962).
(216) P. Schneebeli, *Compt. Rend.*, **234**, 738 (1952).
(217) P. Schneebeli, *Compt. Rend.*, **236**, 1034 (1953).
(218) S. E. Ellzey and C. H. Mack, *Text. Res. J.*, **32**, 1023 (1962).

TABLE I

Solubility of Sucrose in Tetrahydrothiophene S-Oxides ("Sulfolanes")

Solvent	Solubility (g/g of solvent)				
	80°	100°	120°	150°	180°
"Sulfolane"	0.05	0.17	0.20	16.0	85.0
3-methyl-	0.02	0.07	2.00	7.0	63.0
3,4-dimethyl-	0.30 (85°), 0.37 (100°), 0.50 (110°), 0.52 (120°)				

sulfur trioxide and methyl sulfoxide to a degree of substitution of 1.3–2.0 at ambient temperature in a surprisingly rapid reaction.[219]

Anionic graft-polymerization of paraformaldehyde onto starch and "dextrin" has been effected in methyl sulfoxide solution, polymerization being initiated by the carbohydrate potassium alcoholate formed from the reaction of the carbohydrate with naphthalene potassium, a metallation procedure not previously described for carbohydrates.[220]

(ii) **Heterocyclic.**—Structural analogy with methyl sulfoxide suggests that tetrahydrothiophene S-oxide derivatives ("sulfolanes") should be solvents for sugars, and although, below 100°, the solubility of sucrose in "sulfolane" and its 3-methyl and 3,4-dimethyl derivatives is low, "sulfolane" and 3-methyl "sulfolane" are good solvents for sucrose at temperatures above 150° (see Table I; figures taken from Ref. 5, p. 169).

Some decomposition was noted[1] at higher temperatures in 3,4-dimethyl"sulfolane," implying that the solvent was not pure. The remarkable changes in solubility occurring with the "sulfolanes" above 120°, and the fact that the solubility curves are not parallel, as would have been expected for such similar solvents (see Fig. 2; p. 88), shed some doubts on the accuracy of the figures given, and it could be that inadequate equilibration times were allowed for the determinations, particularly at the lower temperatures. The present author has found that sucrose crystallizes incompletely on cooling from solution in "sulfolane."

"Sulfolane" has found utility as a solvent for the preparation of vinyl ethers of sucrose by transetherification with butyl vinyl ether;

(219) R. L. Whistler, A. H. King, G. Ruffini, and F. A. Lucas, *Arch. Biochem. Biophys.*, **121**, 358 (1967).
(220) S. Sasson and A. Zilkha, *Europ. Polymer J.*, **5**, 315 (1969).

other, related solvents are unstable to the high temperatures and the alkaline conditions used.[221]

d. Solvents Based on a Nitrogen Function.—(i) Aliphatic.—Resembling ammonia, the aliphatic amines are excellent solvents for carbohydrates, but lead to reactions that are well documented elsewhere for the hexoses.[222] In the case of sucrose, lithium sucrate has been prepared by using methylamine as solvent; lithium, unlike sodium, is quite soluble in this solvent.[33] In the period when the work of Sherry,[17] Wilcox,[223] and Fitzgerald[18] on the physical properties of sugars in alkylamines was being conducted, it was not realized that the monosaccharides react with the solvent, and much of the work was not valid. However Fitzgerald[18] reported that sucrose is freely soluble in methylamine and ethylamine, and Wilcox[223] claimed that it was more soluble in isopropylamine, allylamine, and pentylamine than in pyridine. Wilcox worked with solutions of 6.25, 25, and 12.5% [assumed to be as w/v (solvent)], respectively, and these were not stated to be saturated. Prey and Unger[224(a)] dissolved both D-glucose and sucrose in ethylenediamine to perform specific-conductivity measurements, and stated that D-glucose reacts with the solvent. With ethylenediamine, cellulose is known to form complexes[224(b)] which may be decomposed completely by N,N-dimethylformamide. Anderson[225] used triethyl- and tripropyl-amine as solvents for the reaction of 1,2-epoxypropane with sucrose. Wolfrom and coworkers[226] used methylamine gas as a reactant in ethanol as solvent for the reaction of L-arabinose with hydrogen cyanide in a synthesis of 2-deoxy-2-(methylamino)-L-gluconic acid, a compound related to a hydrolytic product from streptomycin.

Other classes of aliphatic, nitrogen compounds examined as sugar solvents include nitriles, urethans, ureas, and amides. Of these, the amides are the most significant commercially, but a less-expensive synthesis of some of the tetrasubstituted ureas could furnish solvents

(221) S. A. Barker, J. S. Brimacombe, J. A. Jarvis, and M. R. Harnden, *J. Chem. Soc.*, 3403 (1963).
(222) J. Staněk, M. Černý, J. Kocourek, and J. Pacák, "The Monosaccharides," Academic Press, Inc., New York, N. Y., 1963, p. 449.
(223) G. N. Wilcox, *J. Phys. Chem.*, **6**, 339 (1902).
(224) (a) V. Prey and W. Unger, *Ann.*, **610**, 154 (1962), (b) L. Segal, *J. Polymer Sci.*, **55**, 395 (1961).
(225) A. W. Anderson, U. S. Pat. 2,927,918; *Sugar Ind. Abstr.*, **22**, 166 (1960).
(226) M. L. Wolfrom, A. Thompson, and I. R. Hooper, *J. Amer. Chem. Soc.*, **68**, 2343 (1946).

of equal utility. Heublein[227] utilized trichloroacetonitrile as a solvent for infrared studies of sugars, solubility being just adequate to allow useful spectra to be obtained. He reported the solubility of D-mannose at 20° to be 0.117% w/w (solvent) and D-galactose to be 0.08% w/w (solvent), and the accuracy of his determination to be ±10–15%. It has been mentioned briefly that sucrose is slightly soluble in a mixture of methyl and ethyl carbamate.[176] Bölcs and Gasselseder[228] synthesized, by a condensation reaction between formaldehyde, acetaldehyde, and urea, compounds of unknown structure that were claimed to have utility as solvents for cellulose. In view of the difficulties in swelling and solubilizing cellulose for chemical modification, this claim might merit re-investigation.

Unsubstituted urea reacts with sugars, and although it dissolves sugars above its melting point, it has little utility as a solvent. For example, urea has been reported[229,230] to dissolve sucrose at 140–160° and react with it. Potentially, tetramethylurea is the most useful solvent of this class of compounds.[231] Brief mention was made of the ability of tetramethylurea to dissolve D-glucose [1.1% w/w (solvent) at 22°] in a review[232] of the preparation and properties of tetramethylurea, but no work on its use as a solvent has been reported. Tetramethylurea structurally resembles N,N-dimethylformamide and dissolves more than 20% of sucrose at elevated temperatures (qualitative observation by the author[233]) and has few of the drawbacks of N,N-dimethylformamide. It is, for example, relatively non-toxic, can be dehydrated readily by simple distillation, is not as reactive as N,N-dimethylformamide, and has a pleasant, mild odor. Tetramethylurea could, therefore, be more useful than N,N-dimethylformamide as a reaction medium for carbohydrates if its synthesis could be simplified. The catalytic transamination of urea with dimethylamine, an approach reported by Lüttringhaus and Dirksen[232] to be unsatisfactory, is worth further investigation. Other tetraalkylureas having short alkyl chains

(227) G. Heublein, *J. Prakt. Chem.*, **30** (4), 82 (1965).
(228) J. Bölcs and H. Gasselseder, Fr. Pat. 691,624 (1930); *Chem. Abstr.*, **25**, P1408 (1931).
(229) Herstein Laboratories Progress Reports to S.R.F., Nos. 11–20 on Project 82 (Mar.–Sept. 1953).
(230) Foster D. Snell Inc., Progress Reports No. 6 (April 1954) and No. 7 (July 1954) to S.R.F. on Project 83.
(231) Available (in 1964) from John Deere Chemical Company, Tulsa 1, Oklahoma, for $20 per pound.
(232) A. Lüttringhaus and H. W. Dirksen, *Angew. Chem. Intern. Ed. Engl.*, **3**, 260 (1964).
(233) C. J. Moye, unpublished observation.

are also liquids that are completely miscible with water and a wide range of organic liquids and, therefore, have a potential similar to that of tetramethylurea as solvents for sugars.[232]

N,N-Dimethylformamide is the best known of the dipolar, aprotic solvents for carbohydrates. Compared with N,N-dimethylformamide, the structurally analogous acetamides, 1-methyl-2-pyrrolidinone, and formamide itself, have not received a great deal of attention. Henglein and coworkers[234] have used formamide as a solvent for trimethylsilylation. Moshy and coworkers[235] studied the mutarotation of D-glucose in formamide, and found it to proceed at a measurable rate. Tucker and Martin[236] have used N,N-dimethylacetamide to prepare sucrose-ester surfactants, and Wicker and Gates[206,207] have conducted sucrochemical reactions, esterifications in particular, in mono- and di-substituted formamides and acetamides, and in 1-methyl-2-pyrrolidinone. No solubilities have been recorded for the foregoing examples, but the solubility of sucrose in N,N-dimethylformamide has been recorded. It was found[1] to be 14.1 (30°), 16.9 (60°), 23.6 (85°), 29.6 (100°), and 42.8% w/w (120°). The cycloamyloses (Schardinger dextrins) are also soluble in N,N-dimethylformamide.[237] Although cellulose does not dissolve in pure N,N-dimethylformamide, it swells, and its reactions are markedly accelerated.[216–218] It has been reported that a mixture of dinitrogen tetraoxide and N,N-dimethylformamide solubilizes cellulose and certain marine polysaccharides.[238,239]

Specific reactions performed on mono- and di-saccharides in N,N-dimethylformamide solution include etherification,[240,241] ethanethiolysis,[120] deoxynucleoside condensations,[242–246] acetal formation[54,247–249]

---

(234) F. A. Henglein, G. Abelsnes, H. Heneka, K. Lienhard, P. Nakhre, and K. Scheinost, *Makromol. Chem.*, **24**, 1 (1957).
(235) H. Jacin, J. M. Slansky, and R. J. Moshy, *J. Chromatogr.*, **37**, 103 (1968).
(236) N. B. Tucker and J. B. Martin, U. S. Pat. 2,831,854 and 2,831,856; *Chem. Abstr.*, **52**, 14669 (1958).
(237) N. Wiedenhof and J. N. J. J. Lammers, *Carbohyd. Res.*, **4**, 319 (1967).
(238) G. G. Allan, P. G. Johnson, Y.-Z. Lai, and K. V. Sarkanen, *Chem. Ind.* (London), 127 (1971).
(239) R. G. Schweiger, *Chem. Ind.* (London), 296 (1969).
(240) R. Kuhn and H. Grassner, *Ann.*, **610**, 122 (1957).
(241) R. Kuhn, H. Trischmann, and I. Löw, *Angew. Chem.*, **67**, 32 (1955).
(242) J. A. Carbon, *Chem. Ind.* (London), 529 (1963).
(243) J. A. Carbon, *J. Amer. Chem. Soc.*, **86**, 720 (1964).
(244) G. Schramm, H. Grötsch, and W. Pollman, *Angew. Chem.*, **73**, 619 (1961).
(245) G. Schramm, H. Grötsch, and W. Pollman, *Angew. Chem.*, **74**, 53 (1962).
(246) G. Schramm, H. Grötsch, and W. Pollman, *Angew. Chem. Int. Ed. Engl.*, **1**, 1 (1962).

in the presence of acid catalysts (where an aldehyde, its polymer, or a vinyl ether can be used to condense with the hydroxyl group), chlorination,[250] urethan formation,[251,252] and the preparation of N-glycosylamino acids and esters.[253] Where condensation reactions result in the liberation of acid, as in trimethylsilylation with chlorotrimethylsilane,[254] esterification with acid chlorides,[255] or sulfonylation by means of sulfonyl chlorides,[256] it is possible to use pyridine or N,N-dimethylaniline as the acid acceptor, or silver oxide can be used to react with the acid anion.[257] In one instance, further reaction of the acid chloride (propionyl chloride) was reported to give 25% of a product having a dimeric substituent. This occurred only in the presence of a base, when O-(2-propionyl)propionyl derivatives resulted instead of the O-propionyl analogs (see Ref. 326). Selective, partial acylation of D-glucose has also been conducted in N,N-dimethylformamide and pyridine by using a range of acylating agents such as acid chlorides, N-acylimidazoles, and mixed anhydrides of fatty acids and O-ethylcarbonic acid.[258] The order of reactivities was found to be HO-6 > HO-1 and HO-2, and the disubstituted product was a mixture of 1,6- and 2,6-diesters. A series of 6-O-(fatty acyl) derivatives of D-glucopyranose was prepared during the study.

The best known application of N,N-dimethylformamide in sucrochemistry is as a solvent for a transesterification reaction leading to sucrose mono- and poly-esters of fatty acids. This reaction has been extensively studied and reviewed,[73,259-269] and will not be further

(247) J. A. Jarvis, Ph. D. Thesis, University of Birmingham, England (1962).
(248) J. A. Jarvis, M. Sc. Thesis, University of Birmingham, England (1960).
(249) S. A. Barker, J. S. Brimacombe, J. A. Jarvis, and J. M. Williams, *J. Chem. Soc.*, 3158 (1962).
(250) M. E. Evans, L. Long, Jr., and F. W. Parrish, *J. Org. Chem.*, **33**, 1074 (1968).
(251) H. Bertsch, E. Ulsperger, and F. Mainas, *J. Prakt. Chem.*, **12**, 102 (1961).
(252) Farbenfabriken Bayer A. G., Brit. Pat. 827,358 (1959); *Sugar Ind. Abstr.*, **22**, 76 (1960).
(253) G. P. Ellis, *Advan. Carbohyd. Chem.*, **14**, 71 (1959).
(254) R. Bentley and N. Botlock, *Anal. Biochem.*, **20**, 312 (1967).
(255) Ref. 5, p. 52 ff.
(256) K. Takuira and S. Honda, *Yakugaku Zasshi*, **87**, 1052, 1248, 1256 (1967).
(257) E. J. Bourne, Final Report on Project 126 to S.R.F. (1958).
(258) E. Reinefeld and H. F. Korn, *Staerke*, **20**, 181 (1968).
(259) "Sugar Esters," Noyes Development Corp., Park Ridge, New Jersey, 1968.
(260) L. Osipow and F. D. Snell, *Intern. Sugar J.*, **59**, 68 (1957).
(261) Sugar Research Foundation, Research Report on "Sucrose Ester Surfactants," New York, N. Y., 1961.

discussed. The esters have wide potential utility, particularly as biodegradable detergents,[269(b)] but their application in food, and related areas where their relatively high cost is not such a disadvantage, has been delayed by toxicity problems caused by the presence of residual N,N-dimethylformamide. As already mentioned, the use of tetramethylurea as a solvent could assist in overcoming this problem.

The reactions of polysaccharides in N,N-dimethylformamide have not been extensively studied. Although the solubility of polysaccharides in N,N-dimethylformamide alone is usually limited, swelling (and, presumably, localized solvation) resulting from contact with N,N-dimethylformamide considerably facilitates the acylation of cellulose.[178,270] As already mentioned,[238,239] the presence of dinitrogen tetraoxide in N,N-dimethylformamide effects dissolution of cellulose and various marine polysaccharides, and the carbohydrates can be esterified readily in this medium. Since dinitrogen tetraoxide is an oxidizing agent for monosaccharides in the presence of a base,[64,65] and in the absence of base its solutions in N,N-dimethylformamide are too acidic for acetalation reactions to be conducted, the scope for this solvent system appears to be limited. N,N-Dimethylformamide has also been used as a reaction medium for the sulfation of starch by the use of a triethylamine–sulfur trioxide complex,[271] and as a medium in which D-fructose was converted by iodine into 5-(hydroxymethyl)-2-furaldehyde.[121] By suppressing the reaction with sodium acetate, it was shown that catalysis was by acid and not by iodine.[122]

(262) L. Osipow, F. D. Snell, C. York, and A. Finchler, *Ind. Eng. Chem.*, **48**, 1459 (1956).
(263) H. B. Haas, F. D. Snell, C. York, and L. I. Osipow, U. S. Pat. 2,893,990 (1959); *Chem. Abstr.*, **53**, 19422c (1959).
(264) H. B. Haas, U. S. Pat. 2,970,142; *Chem. Abstr.*, **55**, 12885g (1961).
(265) R. U. Lemieux and A. G. McInnes, *Can. J. Chem.*, **40**, 2376, 2394 (1962).
(266) E. G. Bobalek, A. P. De Mendoza, A. G. Causa, W. J. Collings, and G. Kapo, *Ind. Eng. Chem., Prod. Res. Develop.*, **2**, 9 (1963).
(267) T. C. M. Davis, U. S. Pat. 2,948,716 (1960); *Chem. Abstr.*, **55**, 1041b (1961).
(268) W. Hagge, G. Matthaeus, and M. Quadulieg, Ger. Pat. 1,040,525; *Chem. Abstr.*, **55**, 1467f (1961).
(269) (a) V. R. Gaertner and E. L. Doerr, U. S. Pat. 2,868,781 (1959); *Sugar Ind. Abstr.*, **21**, 114 (1959); (b) H. J. Zimmer, *Europ. Chem. News*, **3**, 25 (1963).
(270) D. M. Jones, *Advan. Carbohyd. Chem.* **19**, 239 (1964).
(271) R. L. Whistler and W. W. Spencer, *Methods Carbohyd. Chem.*, **4**, 297 (1964).

(ii) **Heterocyclic.**—The widest range of non-aqueous solvents for carbohydrates is to be found amongst the nitrogenous heterocycles, where a range of derivatives of pyridine, piperidine, pyrazine, piperazine, pyrrolidine, quinoline, and morpholine has been investigated as solvents for mono- and di-saccharides; in particular, for sucrose.[1,272-274]

The solubilities of various carbohydrates in nitrogenous, heterocyclic solvents are recorded in Table II, where it may be noted that most of the solubilities recorded are substantial; only that of sucrose in 2,4,6-tris(methoxy)-s-triazine being below 1% w/w at elevated temperatures. Noteworthy is the hundred-fold decrease in the solubility of sucrose in N-methylmorpholine as compared with that in morpholine, which serves to emphasize the role played by hydrogen bonding in this and related systems in facilitating solvation of carbohydrates. It was pointed out by this author[12] that, since 2-alkoxy ethanols and 2-aminoethanol derivatives are good solvents for the lower carbohydrates, general solvation of sugars can be expected for solvents having the structure x—CH_2—CH_2—y, where x and y are OR or NR_2 groups. It further appeared that one of these groups must be unsubstituted. The resultant solvents can then form hydrogen bonds with hydroxyl groups of the sugar, as depicted in Fig. 2 for 2-alkoxy alcohols (see p. 88). The difference in the solubility of sucrose in morpholine and in N-methylmorpholine provides extension, to cyclic systems, of this suggestion. These related structures do not differ markedly from such systems as tetramethylurea and N,N-dimethylformamide, pyrrolidinone and its N-methyl derivative, and the various N-acylated heterocycles listed in Table II (see p. 118). Bearing in mind the structural types just mentioned, it is now possible to predict the ability of many organic molecules to solvate sugars.

Qualitative solubility studies with nitrogenous heterocycles have been conducted by Grossman and Bloch,[128] who examined most of the sugars listed in Table II (see p. 118) and also found that xylose and rhamnose are soluble in pyridine. Vogel[275] reported general solvation of sugars in pyridine and piperidine, and drew attention to the fact that lactose and sucrose are not very soluble in piperidine; by virtue

---

272) S. Komori, M. Okahara, and E. Shinsugi, *Tech. Repts. Osaka Univ.*, **8**, 497 (1958); *Chem. Abstr.*, **53**, 18874c (1959).
273) W. M. Dehn, *J. Amer. Chem. Soc.*, **39**, 1402 (1917).
274) J. G. Holty, *J. Phys. Chem.*, **9**, 776 (1905).
275) H. Vogel, *Ber.*, **70**, 1193 (1937).

Table II

Heterocyclic Solvents

Solute	Solubility (g/100 g of solution) (Temperature)	References
N-Formylpyrrolidine		
Sucrose	30.36 (90°), 37.29 (100°)	272
N-Acetylpyrrolidine		
Sucrose	39.89 (90°), 45.17 (100°)	272
1-Methyl-2-pyrrolidinone		
Sucrose	17.3 (30°), 22.6 (60°), 28.0 (85°), 33.5 (100°), 40.4 (120°)	1
	47.16 (90°), 56.81 (100°)	272
Pyridine		
Sucrose	3.12 (30°), 3.75 (60°), 5.00 (85°), 5.99 (100°), 7.46 (110°)	1
	6.25 (26°)	273
	6.45 (26°)	274
D-Glucose	7.62 (26°)	274
D-Fructose	18.49 (26°)	274
Lactose	2.18+ (26°)	273
Galactose	5.45 (26°)	274
Maltose	98.10 (20–25°)	274
Mannose	29.90 (20–25°)	273
Raffinose	75.00 (20–25°)	273
"Dextrin"	65.44 (20–25°)	273
Pyrazine		
Sucrose	1.95 (85°), 2.23 (97°), 3.04 (107°), 3.95 (120°)	1
2-Methylpyrazine		
Sucrose	0.87 (85°), 1.25 (100°), 1.84 (110°), 2.34 (120°)	1
Mixed pyrazines		
Sucrose	2.02 (85°), 2.46 (100°), 2.73 (110°)	1
N-Formylpiperidine		
Sucrose	6.28 (90°), 8.18 (100°)	272
N-Acetylpiperidine		
Sucrose	7.46 (90°), 9.15 (100°)	272
2-Methylpiperazine		
Sucrose	26.1 (85°), 29.5 (100°), 29.6 (110°), 30.1 (120°)	1
Morpholine		
Sucrose	30.7 (30°), 34.7 (60°), 39.8 (85°), 45.1 (100°), 50.8 (120°)	1
N-Methylmorpholine		
Sucrose	0.37 (85°), 0.38 (100°), 0.56 (110°), 0.72 (120°)	1

TABLE II (Continued)

Heterocyclic Solvents

Solute	Solubility (g/100 g of solution) (Temperature)	References
N-Formylmorpholine		
Sucrose	11.57 (80°), 15.25 (90°), 20.02 (100°)	272
N-Acetylmorpholine		
Sucrose	10.21 (80°), 18.73 (100°)	272
N-Propionylmorpholine		
Sucrose	3.38 (80°), 5.20 (100°)	272
N-Butyrylmorpholine		
Sucrose	1.93 (80°), 2.96 (100°)	272
2,4,6-Tris(methoxy)-s-triazine		
Sucrose	0.32 (140°)	1

of this fact, they can be separated from other sugars. Vogel also pointed out that certain sugars react with piperidine after dissolution. The reactions of primary and secondary amines (and their salts) with aldoses and ketoses have been studied extensively in the intervening time[276-282] and capably reviewed by Anet.[279] In general terms, glycosylamines are formed as the initial step in the reaction of aldoses or ketoses with primary or secondary amines; these products react further, depending on whether the reaction medium is acidic or basic, to give either Amadori rearrangement-products or reductones. 2-Piperidinone and 1-methyl-2-piperidinone have also been used as solvents for sugars,[184] but solubility data were not determined.

Largely because of its availability, pyridine is the most frequently used of the nitrogenous, heterocyclic solvents. In the earliest studies,[283-285] physical properties of solutions of sucrose in

(276) F. Weygand, Ber., 73, 1259 (1940).
(277) K. Heyns, K. W. Pflughaupt, and D. Müller, Chem. Ber., 101, 2807 (1968).
(278) K. Heyns, K. W. Pflughaupt, and D. Müller, Chem. Ber., 101, 2800 (1968).
(279) E. F. L. J. Anet, Advan. Carbohyd. Chem., 19, 213 (1964).
(280) J. E. Hodge and C. E. Rist, J. Amer. Chem. Soc., 75, 316 (1953).
(281) J. E. Hodge and C. E. Rist, J. Amer. Chem. Soc., 74, 1494 (1952).
(282) H. Simon and G. Heubach, Chem. Ber., 98, 3703 (1965).
(283) L. Kohlenberg, J. Phys. Chem., 10, 187 (1906).
(284) E. Cohen and J. W. Commelin, Z. Phys. Chem., 64, 45 (1908).
(285) G. N. Wilcox, J. Phys. Chem., 5, 596 (1901).

pyridine were examined, and interest has continued to the present day[161-163,286-289] on the mutarotation of various sugars in pyridine, and their anomeric composition at equilibrium.[286,287] Mutarotation has been studied[288] in such other non-aqueous solvents as N,N-dimethylformamide, methyl sulfoxide, formamide, and quinoline, and it has been proposed that mutarotation is an autocatalytic process involving a six-membered, bimolecular transition-state.[162,286] (See also, discussion of phenols, p. 104.) Prey and Unger[2] examined the electrical conductivity of D-glucose and sucrose in pyridine, and concluded that the hydrogen atom of the glycosidic hydroxyl group is the most labile. Epimerization of monosaccharides can be conducted in boiling pyridine,[289-291] and this procedure has been successfully used to prepare the isomeric D-*threo*- and D-erythropentuloses from D-xylose and D-arabinose.[291].

Pyridine is widely used as a reaction medium in acylation and related reactions involving, for example, acyl chlorides, anhydrides, sulfonyl chlorides, or chlorosilanes, and has the particular advantage that it can function both as a catalyst and acid acceptor. It is usually the solvent of choice for the trimethylsilylation of alditols and sugars,[292-298] and for various derivatization reactions such as tritylation and sulfonylation.[299-307] The tritylation reaction has been reviewed.[306]

(286) A. S. Hill and R. S. Shallenberger, *Carbohyd. Res.*, **11**, 541 (1969).
(287) T. E. Acree, R. S. Shallenberger, and L. R. Mattich, *Carbohyd. Res.*, **6**, 498 (1968).
(288) A. de Grandchamp-Chaudun, *Ann. Pharm. Fr.*, **26**, 115 (1968).
(289) M. Fedoroňko and K. Linek, *Collect. Czech. Chem. Commun.*, **32**, 2177 (1967).
(290) W. Rehpenning and H. Zinner, *Carbohyd. Res.*, **5**, 176 (1967).
(291) R. S. Tipson and R. F. Brady, Jr., *Carbohyd. Res.*, **10**, 549 (1969).
(292) E. J. Hedgley and W. G. Overend, *Chem. Ind.* (London), 378 (1960).
(293) R. J. Ferrier and M. F. Singleton, *Tetrahedron*, **18**, 1143 (1962).
(294) R. J. Ferrier, *Chem. Ind.* (London), 831 (1961).
(295) R. Bentley, C. C. Sweeley, M. Makita, and W. W. Wells, *Biochem. Biophys. Res. Commun.*, **11**, 14 (1963).
(296) R. Bentley, C. C. Sweeley, M. Makita, and W. W. Wells, *J. Amer. Chem. Soc.*, **85**, 2497 (1963).
(297) B. Smith and O. Carlsson, *Acta Chem. Scand.*, **17**, 455 (1963).
(298) H. H. Sephton, *J. Org. Chem.*, **29**, 3415 (1964).
(299) J. P. Barrette, Ph. D. Thesis, University of Ottawa (1960).
(300) J. P. Barrette and R. U. Lemieux, *Can. J. Chem.*, **38**, 656 (1960).
(301) P. D. Bragg and J. K. N. Jones, *Can. J. Chem.*, **37**, 575 (1959).
(302) Birmingham University Report to S.R.F. on Project 165 (1965).
(303) L. Long, Jr., Final Report to the S.R.F. on Project 141 (July, 1959).
(304) H. J. Roberts, *Methods Carbohyd. Chem.*, **4**, 299, 311 (1964).
(305) Ref. 5, p. 70 and following pages.
(306) B. Helferich, *Advan. Carbohyd. Chem.*, **3**, 79 (1948).
(307) D. H. Ball and F. W. Parrish, *Advan. Carbohyd. Chem.*, **23**, 236 (1968); **24**, 139 (1969).

Pyridine has also been used in the synthesis of urethans[308] and in the formation of carbonic and thiocarbonic esters of simple sugars from phosgene, thiophosgene, or the appropriate chloroformates.[309-311] With polysaccharides (for example, starch and cellulose), such reactions as acylation and phosphorylation are accelerated because of activation by the swelling action of the solvent, even though true solvation does not occur.[312,313] High degrees of substitution may readily be achieved in pyridine solution,[314-323] even in the case of non-conjugated, polyunsaturated, drying-oil acid chlorides,[324,325] but it is also possible to achieve selective, partial esterification.[258] Self-condensation of the acylating agent in the presence of pyridine has been observed in the reaction of propionyl chloride with cellulose.[326] The Kiliani synthesis has been performed in pyridine: Kuhn and Klesse[327] condensed L-arabinose and hydrogen cyanide to give L-glucono- and L-mannono-nitriles, which were converted into the phenylhydrazones of L-glucose and L-mannose.

A wide range of other nitrogenous, heterocyclic solvents has been used as solvents for the preparation of sucrose surfactants by ester exchange. These include N-methyl- and various N-acyl-morpholines,[272,328-330] N-formyl- and N-acetyl-piperdine,[272] N-formyl- and N-acetyl-

---

(308) I. A. Wolff, *Methods Carbohyd. Chem.*, **4**, 301 (1964).
(309) W. N. Haworth and C. R. Porter, *J. Chem. Soc.*, 151 (1930).
(310) L. Hough, Quarterly Reports to S.R.F. on Project 192 (Dec. 1962–June 1963).
(311) L. Hough and J. E. Priddle, *Chem. Ind.* (London), 1600 (1959).
(312) R. L. Lohmar, U.S. Pat. 2,627,516; *Chem. Abstr.*, **47**, 5146i (1953).
(313) R. L. Lohmar, J. W. Sloan, and C. E. Rist, *J. Amer. Chem. Soc.*, **72**, 5717 (1950).
(314) H. A. Goldsmith, *Chem. Rev.*, **33**, 257 (1943).
(315) G. Zemplén and E. D. László, *Ber.*, **48**, 915 (1915).
(316) K. Hess and E. Messmer, *Ber.*, **54**, 499 (1921).
(317) G. Nebbia, *Ann. Chim.* (Rome), **47**, 1280 (1957).
(318) J. H. Schwartz and E. A. Talley, *J. Amer. Chem. Soc.*, **73**, 4490 (1951).
(319) H. Willstaedt and M. Borggård, *Bull. Soc. Chim. Biol.*, **28**, 733 (1946).
(320) M. Zief, *J. Amer. Chem. Soc.*, **72**, 1137 (1950).
(321) R. H. Treadway and E. Yanovsky, *J. Amer. Chem. Soc.*, **67**, 1039 (1945).
(322) Z. A. Rogovin, *J. Polym. Sci.*, **48**, 443 (1960).
(323) A. Fernez and P. J. Stoffyn, *Tetrahedron*, **6**, 139 (1939).
(324) L. Rosenthal, Ger. Pat. 478,127 (1924); *Chem. Zentr.*, Part II, 1071 (1929).
(325) A. E. Rheineck, B. Rabin, and J. S. Long, U. S. Pat. 2,077,371 (1937); *Chem. Abstr.*, **31**, 4144 (1937).
(326) A. K. Sircar, D. J. Stanonis, and C. M. Conrad, *J. Appl. Polym. Sci.*, **11**, 1683 (1967).
(327) R. Kuhn and P. Klesse, *Chem. Ber.*, **91**, 1989 (1958).
(328) T. Hedley and Co., Brit. Pat. 804,197 (1958); *Sugar Ind. Abstr.*, **21**, 45 (1959).
(329) N. B. Tucker, U. S. Pat. 2,831,856 (1958): *Chem. Abstr.* **52**, 14669 (1958).
(330) S. Komori, M. Okahara, and K. Okamoto, *J. Amer. Oil Chem. Soc.*, **37**, 468 (1960).

pyrrolidine,[272] pyrazine,[331] pyridine,[332] and 1-methyl-2-pyrrolidinone.[272,331] The last-mentioned solvent has also been investigated[30,333] as a medium in which to prepare metal "sucrates;" as it is decomposed by metallic sodium, the use of sodium hydride as the metal source has been advocated.[38,39] When pyridine and morpholine were used with metallic sodium, the products were said to resemble more the addition compounds with metal hydroxides.[30]

Lead tetraacetate in pyridine has been used[334] to oxidize sucrose, and a "tetraaldehyde" is produced. The catalytic reduction of D-glucose over Raney nickel has been conducted in pyridine, piperidine, picoline, quinoline, and tetrahydroquinoline solutions.

The pyridine–sulfur trioxide complex serves as a sulfating agent for free sugars,[256,335] but specific syntheses require the use of protected derivatives and subsequent removal of the protecting group.[335] Guisley and Ruoff[336] and Jones and coworkers[167,337] have investigated the preparation of sulfate and sulfite esters of sugars in pyridine and other non-aqueous solvents. Sulfur trioxide, chlorosulfonic acid, sulfuryl chloride, and thionyl chloride were examined, and chloroform and dichloromethane were used as solvent diluents. It was found that the use of pyridine decreases undesired chlorination, which, in the D-glucose derivatives, proceeded at C-4 with configurational inversion.[167] However, with thionyl chloride and sucrose in pyridine at 5–10°, a dichlorodideoxysucrose disulfite was produced.[338] 5-Ethyl-2-methylpyridine was also used as a solvent in these studies.

Other reactions conducted in pyridine solution include the esterification of simple sugars by N-benzyloxycarbonyl derivatives of amino acids in the presence of N,N'- dicyclohexylcarbodiimide[339] to produce essentially the 6-O-aminoacyl derivatives, and formation of carbo-

---

(331) Sugar Research Foundation, Indian Pat. 53,792 (1956); *Sugar Ind. Abstr.*, **20**, 60 (1958).
(332) J. B. Martin, U. S. Pat. 2,831,855 (1958); *Sugar Ind. Abstr.*, **21**, 92 (1959).
(333) Herstein Laboratories Progress Reports Nos. 98–103 to S.R.F. on Project 82 (March–Aug. 1959).
(334) R. C. Hockett and M. Zief, *J. Amer. Chem. Soc.*, **72**, 2130 (1950).
(335) M. J. Harris and J. R. Turvey, *Carbohyd. Res.*, **9**, 397 (1969).
(336) K. Guisley and P. M. Ruoff, Syracuse University Reports to S.R.F. on Project 130 (June–July 1957).
(337) J. K. N. Jones, Queens University Report to the S.R.F. on Project 143 (Sept. 1961).
(338) H. H. Szmant, University of Puerto Rico Reports to S.R.F. on Project 157 (July and Sept. 1959, and Jan. 1960).
(339) N. K. Kochetkov, V. A. Derevitskaya, and L. M. Likhosherstov, *Chem. Ind.* (London), 1532 (1960).

hydrate urethans with alkyl isocyanates or phenyl isocyanate.[216–218,340] Methyl sulfoxide and N,N-dimethylformamide were compared with pyridine, and the ability to assist the reaction was found to depend on the swelling power of the solvents, which followed the order methyl sulfoxide > N,N-dimethylformamide > pyridine.[216–218]

e. **Solvents Based on a Phosphorus Function.**—The usefulness of phosphorus-containing compounds as solvents for sugars has, as yet, been little exploited. The use of hexamethylphosphoric triamide, one of the more powerful members of the dipolar, aprotic group of solvents, is undoubtedly limited by its cost. Although it is extensively used as a solvent for nucleophilic displacement-reactions, it has only been reported once as a solvent for unsubstituted carbohydrates. Feuge and coworkers,[341] in transesterification studies on sucrose, observed lower degrees of transesterification in hexamethylphosphoric triamide than in N,N-dimethylformamide. Dalton[54] and Swan[342] have reported that dimethyl phosphite dissolves appreciable proportions of D-glucose and sucrose, and condensation reactions in the presence of polyphosphoric ester have been reported by Schramm,[244–246] whose claim to have synthesized adenosine in a stereospecific synthesis from D-ribose and adenine was discredited by Carbon.[242,243] A similar claim by de Garilhe and de Rudder[343] for the D-arabinose analog was disproved by Cohen.[344]

V. Special Applications

1. Chromatographic Systems

It is usual for the chromatographic flowing-phases used for unsubstituted sugars to contain water. Several systems have been described that contain minimal proportions of water, and there are undoubtedly many more examples than the following two anhydrous systems. Pastuska[345] separated sugars on borate-modified Kieselgel G by using benzene–acetic acid–methanol in the ratios 1:1:3 and

---

(340) S. Komori and M. Agawa, *Technol. Repts. Osaka Univ.*, **8**, 487 (1958); *Chem. Abstr.*, **53**, 18873 (1959).
(341) R. O. Feuge, T. J. Weiss, M. L. Brown, and H. J. Zeringue, *J. Amer. Oil Chem. Soc.*, **47**, 81A (1970).
(342) J. Swan, personal communication.
(343) M. P. de Garilhe and J. de Rudder, *Compt. Rend.*, **259**, 2725 (1964).
(344) S. S. Cohen, *Progr. Nucleic Acid Res. Mol. Biol.*, **6**, 24 (1966).
(345) G. Pastuska, *Z. Anal. Chem.*, **179**, 427 (1961).

butanone–acetic acid–methanol in the ratios 3:1:1. Otake[346] used two solvent systems incorporating methyl sulfoxide and N,N-dimethylformamide, but the latter system was preferred because impurities present in methyl sulfoxide, even when highly purified, gave a red color with the anthrone developing-reagent. For thin-layer chromatography, methyl sulfoxide–ethyl acetate–water (in the ratios of 60:139:1) and N,N-dimethylformamide–ethyl acetate–acetic acid–water (in the ratios 30:68:1:1) solvent-systems were used, and for column chromatography, various mixtures of N,N-dimethylformamide and ethyl acetate were found to be suitable eluting systems.

2. Nuclear Magnetic Resonance Spectroscopy

The nuclear magnetic resonance (n.m.r.) spectroscopy of carbohydrates was significantly advanced by the introduction of methyl sulfoxide as a solvent that suppresses proton-exchange reactions and permits good resolution of hydroxyl hydrogen resonances, particularly in the presence of acetone-d_6.[191,347] In terms of integration, it is important to have the methyl sulfoxide anhydrous, and this is not easy to achieve. Methyl sulfoxide is also a viscous solvent, a factor which tends to broaden the signal, and operation at somewhat elevated temperatures (>70°) improves the resolution. Inch[348] has reviewed the use of methyl sulfoxide as a n.m.r. solvent for carbohydrates.

Studies by Mackie and Perlin[349] on free sugars in methyl sulfoxide were directed towards the determination of furanose–pyranose equilibria and anomeric composition at equilibrium. It was found that there is a definite difference, according to the configuration, between the tendency of a sugar to adopt the furanoid ring-form in water and in methyl sulfoxide. Casu and Reggiani[350] directed their attention to the anomeric hydroxyl groups of D-glucose and related sugars, and reported the chemical shift and splitting to be characteristic of the configuration of the anomeric linkage. The HO-1 proton of the α anomer had a chemical shift of τ 3.70–4.05 and $J_{1,OH}$ 4.0–4.5 Hz, and the HO-1 proton of the β anomer had τ 3.40–3.68 and $J_{1,OH}$ 6.0–7.0 Hz. Non-anomeric hydroxyl groups of D-glucose and related sugars had OH signals lying in the region τ 5–6, which was taken as characteristic of OH groups free to associate by hydrogen bonding with the

(346) T. Otake, *Proc. Res. Soc. Jap. Sugar Refineries Technologists*, **20**, 60 (1968).
(347) O. L. Chapman and R. W. King, *J. Amer. Chem. Soc.*, **86**, 1257 (1964).
(348) T. D. Inch, *Ann. Rev. N.M.R. Spectrosc.*, **2**, 44 (1969).
(349) W. Mackie and A. S. Perlin, *Can. J. Chem.*, **44**, 2039 (1966).
(350) M. Casu and M. Reggiani, *Tetrahedron*, **22**, 3061 (1966).

solvent. In studies on long-range coupling, Jochims and coworkers[191] found stereospecific coupling for the pentoses and hexoses examined; only axial hydroxyl groups vicinal to axial hydrogen atoms in chair conformations showed coupling.

Although few applications of other non-aqueous solvents for n.m.r. studies of free sugars are to be found, N,N-dimethylformamide, dimethyl phosphite, tetramethylurea, and (at high temperatures) a potassium thiocyanate–sodium thiocyanate melt have some potential. Provided that sugar solubilities are high enough and reaction does not occur, chloral is another solvent that might merit investigation. Any particular advantage over methyl sulfoxide that might be gained by using any of these solvents remains to be explored.

SUGARS SPECIFICALLY LABELED WITH ISOTOPES OF HYDROGEN[*]

By J. E. G. Barnett and D. L. Corina

Department of Physiology and Biochemistry, University of Southampton, England

I. Introduction 128
II. Preparation 128
    1. Reduction by Hydride Reagents. 129
    2. Enzymic Reduction by Reduced Nicotinamide Adenine
        Dinucleotide (NADH) 132
    3. Reduction by Hydrogen. 132
    4. Base-catalyzed Solvent-exchange 133
    5. Enzyme-catalyzed Solvent-exchange 134
    6. Incorporation of Solvent on Oxidation of a Hydrazino Group 135
    7. Addition to a Double Bond. 136
    8. Conversion from Another Labeled Sugar. 137
    9. Methods for Non-specific Labeling 137
III. Radiochemical and Chemical Stability. 138
IV. Localization. 140
    1. Chemical Degradation. 140
    2. Use of Enzymes in Determining the Stereochemistry at a Chiral
        Methylene Group. 141
    3. Physical Methods 143
V. Physical Properties and Their Applications. 147
    1. Nuclear Magnetic Resonance Spectroscopy 147
    2. Infrared Spectroscopy. 149
    3. Mass Spectrometry. 149
    4. Gas–liquid Chromatography 151
VI. Applications in Mechanistic Chemistry 151
VII. Applications in Mechanistic Biochemistry. 155
    1. Solvent Exchange 156
    2. Hydrogen Movements Mediated by Cofactor 166
    3. Isotope Effects 171
VIII. General Applications in Biochemistry 176
    1. Overall Pathway of Enzymic Transformations. 176
    2. Hydrogen Movement within the Cell, and Incorporation into
        Cell Components 177
    3. Incorporation of Monomers into Polysaccharides and Other
        Polymers Containing Carbohydrate 179

[*] We thank Drs. M. Akhtar and D. C. Wilton for helpful discussions during the preparation of the manuscript.

    4. Use as Substrates for Enzymes and for Transport Processes 180
    5. Autoradiography 181
IX. Tables of Known Sugars Specifically Labeled with Isotopes of Hydrogen. . 181
    1. Methods of Preparation 181
    2. Position of Label 182
    3. Commercial Availability 182

I. INTRODUCTION

Although the separation of the stable isotopes of hydrogen was achieved[1] in 1932, and tritium (first isolated[2] in 1934) was found to be radioactive[3] in 1939, it was not until the general availability of mass spectrometry and nuclear magnetic resonance spectroscopy on the one hand, and of scintillation–counting equipment on the other, that deuterium and tritium could be fully exploited. The use of the two isotopes differs considerably, as the former is a stable isotope, generally used at 100% substitution, whereas the latter is used in minute proportions as a tracer. The information that can be obtained by their respective use reflects this difference.

This article will be generally confined to a discussion of sugars in which a hydrogen isotope forms a carbon–hydrogen isotope bond.

II. PREPARATION

The methods of preparing sugars specifically labeled with either deuterium or tritium are similar. However, as tritiated sugars with 100% substitution are not used, the possibility of dilution with un-labeled material makes their synthesis, and, particularly, their isolation, easier. In most of the preparative methods, an isotope effect will be observed, leading to a specific activity of the product that is lower than that expected. This effect does not apply to equilibrium-exchange methods, or when an isotopically pure reagent is used to introduce the deuterium atom. The achievable specific activity of the tritium compounds is limited by their radioactivity, which causes chemical decomposition. Some practical details of methods for the introduction of tritium[4] into sugars have been reviewed.

(1) E. W. Washburn and H. C. Urey, *Proc. Nat. Acad. Sci. U. S.*, **18**, 496 (1932).
(2) M. L. E. Oliphant, P. Harteck, and E. Rutherford, *Proc. Roy. Soc., Ser. A*, **144**, 692 (1934).
(3) L. W. Alvarez and R. Cornog, *Phys. Rev.*, **56**, 613 (1939).
(4) J. E. G. Barnett, *Methods Carbohyd. Chem.*, **6**, 499 (1972).

1. Reduction by Hydride Reagents

Of the methods available for the specific introduction of an isotope of hydrogen into a sugar molecule, the most convenient and widely used is reduction of a sugar aldehyde, ketone, lactone, ester group, or halide by a hydride reagent. This method is particularly useful for tritiated sugars, as sodium borohydride-t of high specific activity can be obtained commercially. Early uses of the method involved the reduction by sodium borohydride-t of D-gluconolactone to give[5] D-glucose-1-t, the reduction of aldoses in pyridine or tetrahydrofuran by lithium borohydride-t to give[6] alditols-1-t, or of ketoses to give[6] alditols-2-t, of aldonolactones to give[7] either aldoses-1-t or alditols-1-t, or both, and the reduction of 1,2-O-isopropylidene-D-glucofuranurono-6,3-lactone with[8] sodium borohydride-t or[9] lithium borohydride-t to give D-glucose-6-t after hydrolysis of the isopropylidene group.

Incorporation of deuterium or tritium at the secondary carbon atoms of the sugar ring depends on the availability of the corresponding ketose derivatives; these may be produced either by biological or chemical oxidation. Biological oxidation is convenient, but limited in scope. However, reduction with sodium borohydride-d or -t of D-$xylo$-5-hexulosonic acid (produced by oxidation of D-glucose with *Acetobacter suboxydans*) led[10] to D-glucose-5-d and[11] D-glucose-5-t, and reduction with sodium borohydride-t of β-D-fructofuranosyl α-D-$ribo$-hexopyranosid-3-ulose, formed by oxidation of sucrose with *Agrobacterium tumifaciens*, led to D-glucose-3-t derivatives.[12] A tritium atom can also be introduced onto C-6 of any unsubstituted D-galactose residue in simple D-galactosides[13,14] or in polysaccharides[14,15] by the use of D-galactose oxidase, which specifically

(5) F. Friedberg and L. Kaplan, *J. Amer. Chem. Soc.*, **79**, 2600 (1957).
(6) H. L. Frush, H. S. Isbell, and A. J. Fatiadi, *J. Res. Nat. Bur. Stand.*, A, **64**, 433 (1960).
(7) H. S. Isbell, H. L. Frush, N. B. Holt, and J. D. Moyer, *J. Res. Nat. Bur. Stand.*, A, **64**, 177 (1960).
(8) G. Moss, *Arch. Biochem. Biophys.*, **90**, 111 (1960).
(9) H. S. Isbell, H. L. Frush, and J. D. Moyer, *J. Res. Nat. Bur. Stand.*, A, **64**, 359 (1960).
(10) W. Mackie and A. S. Perlin, *Can. J. Chem.*, **43**, 2645 (1965).
(11) J. E. G. Barnett and D. L. Corina, *Carbohyd. Res.*, **3**, 134 (1966).
(12) O. Gabriel and G. Ashwell, *J. Biol. Chem.*, **240**, 4123 (1965).
(13) G. Avigad, *Carbohyd. Res.*, **3**, 430 (1967).
(14) J. E. G. Barnett, *Carbohyd. Res.*, **4**, 267 (1967).
(15) A. G. Morell, C. J. A. Van den Hamer, I. H. Scheinberg, and G. Ashwell, *J. Biol. Chem.*, **241**, 3745 (1966).

oxidizes the primary alcohol group to an aldehyde group; this may then be reduced with sodium borohydride-t.

The development of methods[16,17] for the mild, specific oxidation of secondary alcohols has led to the availability of a variety of oxidized sugars;[18] these can be reduced in a similar way.

When a ketose is reduced, both of the theoretically obtainable products are usually formed, but the relative proportions may differ widely. Reduction of 1,2:5,6-di-O-isopropylidene-α-D-*ribo*-hexofuranos-3-ulose by sodium borohydride-t leads almost exclusively (98%) to the D-*allo* (not the D-*gluco*) product.[19] Such stereospecific reductions may aid separation of the products, but where, as in this case, the isomer required is the minor product, separation can be difficult. Then, either an alternative reductant can be tried, or an alternative ketose; for example, methyl β-D-*ribo*-hexopyranosid-3-ulose gives ten times as much D-glucose (relative to D-allose) as does the α-D-glycoside.[20] Finally, the *allo* configuration (D-*glycero* at C-3) may be inverted by p-toluenesulfonylation followed by displacement.[21] If no easy alternative is found, the difficulties of separation may be partially overcome by using a selective chemical reaction, such as the favored lactonization of D-gluconic-5-t acid as compared with that of L-idonic-5-t acid,[11] or by using biological specificity. An elegant example of the latter is the synthesis of D-glucosyl-3-t phosphate (**4**) by reduction of "3-keto-sucrose" (**1**); this gave the D-*allo* (**2**) and D-*gluco* (**3**) reduction products in the ratio of 19:1, but only the D-*gluco* isomer (**3**) was a substrate for sucrose phosphorylase.[12]

Similarly, reduction of an aldehyde may lead to random distribution of the label in the resulting hydroxymethyl group. An isotope introduced into this group can afford either the R or S configuration, and the proportions of these need not be equal (see p. 144).

Reduction of a sugar epoxide with lithium aluminum hydride also gives a mixture of deoxy sugars in which one isomer may preponderate. This method has been successfully used[22] to synthesize a derivative (5) convertible into 2-deoxy-D-*erythro*-pentose-2(S)-d.

A stereospecific reduction of a pseudoglycal has been used to prepare the same isomer.[23] Reduction of methyl 4,6-O-benzylidene-2,3-dideoxy-β-D-*erythro*-hex-2-enopyranoside (6) with lithium aluminum deuteride led to the stereospecific synthesis of 4,6-O-benzylidene-3-deoxy-D-glucal-3(S)-d (7), which, after oxidation with osmium tetraoxide–periodate followed by debenzylidenation, gave 2-deoxy-D-*erythro*-pentose-2(S)-d (2-deoxy-2-deuterio-D-arabinose) (8). Reduc-

(16) K. E. Pfitzner and J. G. Moffatt, *J. Amer. Chem. Soc.*, **87**, 5661 (1965).
(17) P. J. Beynon, P. M. Collins, P. T. Doganges, and W. G. Overend, *J. Chem. Soc.* (C), 1131 (1966).
(18) R. F. Butterworth and S. Hanessian, *Synthesis*, 70 (1971).
(19) J. E. G. Barnett and D. L. Corina, *Biochem. J.*, **108**, 125 (1968); compare D. C. Baker, D. Horton, and C. G. Tindall, Jr., *Carbohyd. Res.*, **24**, 192 (1972).
(20) L. Davis and L. Glaser, *Biochem. Biophys. Res. Commun.*, **43**, 1429 (1971).
(21) H. J. Koch and A. S. Perlin, *Carbohyd. Res.*, **15**, 403 (1970).
(22) S. David, J. Eustache, and C. Rouzeau, *Compt. Rend. Ser. C.*, **270**, 1821 (1970).
(23) B. Radatus, M. Yunker, and B. Fraser-Reid, *J. Amer. Chem. Soc.*, **93**, 3086 (1971).

tion of the α-enopyranoside gave stereospecifically the 2(R) isomer. The absolute stereochemistry of these reductions is interesting, and to explain it, a six-membered complex of type 9 has been proposed.[24]

9

2. Enzymic Reduction by Reduced Nicotinamide Adenine Dinucleotide (NADH)

An alternative to chemical reduction of a carbonyl group is enzymic reduction by use of reduced nicotinamide adenine dinucleotide. This gives specifically the product dictated by the enzyme and, in particular, allows the reduction of one carbonyl group in the presence of another. Thus D-*threo*-2,5-hexodiulose was reduced specifically[25] with NADH-*t* to D-fructose-5-*t*.

3. Reduction by Hydrogen

The earliest method for the specific introduction of hydrogen isotopes into sugars was by reduction with sodium amalgam of the aldonolactone in deuterium oxide to give[26] a sugar deuterated at C-1. This is still the method of choice for the introduction of deuterium into this position, because deuterium oxide is relatively cheap, and easy to use. The same method was initially used[7] for the introduction of tritium at C-1, but the difficulties associated with handling water-*t* of high activity, and the consequent low specific activities produced, have led to its replacement by reduction with borohydride-*t*.

Hydrogenation with hydrogen and a metallic catalyst has been used[27] to convert *myo*-inosose-2 into *myo*-inositol-2-*d*, and to obtain methyl 2-deoxy-β(and α)-D-glucopyranoside-2-*d* from methyl 2-deoxy-2-iodo-β-D-glucopyranoside and from methyl 2-deoxy-2-iodo-α-D-mannopyranoside, respectively.[28] The equatorially attached

(24) B. Fraser-Reid and B. Radatus, *J. Amer. Chem. Soc.*, **92**, 6661 (1970).
(25) G. Avigad and S. Englard, *J. Biol. Chem.*, **240**, 2297 (1965).
(26) Y. J. Topper and D. Stetten, *J. Biol. Chem.*, **189**, 191 (1951).
(27) R. U. Margolis and A. Heller, *Biochim. Biophys. Acta*, **98**, 438 (1965).
(28) R. U. Lemieux and S. Levine, *Can. J. Chem.*, **42**, 1473 (1964).

iodine atom of the D-glucose derivative was reduced with retention of configuration, but the axially attached iodine atom of the D-mannose derivative was reduced with 40% inversion.

4. Base-catalyzed Solvent-exchange

Any carbon–hydrogen bond which can, by ionization, be induced to exchange with the hydrogen of the solvent can, in principle, be isotopically labeled. Unfortunately, the mildly basic conditions required for catalyzing the exchange, usually by enolization, are also those leading to rearrangement. If D-glucose is equilibrated in deuterium oxide containing NaOD, a mixture of 2-C-labeled D-glucose and of D-fructose labeled mainly at C-1, but also at C-3, is produced.[26] Isomerization can be lessened by etherification at O-2. Treatment of 2-O-benzyl-D-arabinose with NaOD in deuterium oxide gave a good yield of 2-O-benzyl-D-arabinose-2-d, which was hydrogenolyzed[29] to D-arabinose-2-d.

When the carbonyl group enolized is at a secondary position, the hydrogen can be introduced into four possible adjacent positions, potentially leading to a mixture of four isomers. These possibilities can be decreased if the sugar is held in a rigid conformation so that enolization can take place in one direction only. For example, 1,6-anhydro-2,3-O-isopropylidene-β-D-*lyxo*-hexopyranos-4-ulose (**10**) can-

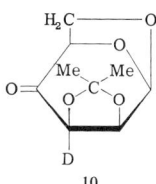

not enolize between C-4 and C-5, as ring strain in the bicyclic system makes achievement of sp² hybridization by C-5 impossible, whereas enolization between C-3 and C-4 is easy. Furthermore, the hydrogen atom on C-3 exchanges in deuterium oxide containing NaOD without change in orientation, because the *cis*-fused 2,3-O-isopropylidene ring is more stable than the *trans*, so that 1,6-anhydro-2,3-O-isopropylidene-β-D-*lyxo*-hexopyranos-4-ulose-3-d was the only product detected.[30,31]

(29) R. U. Lemieux and J. D. Stevens, *Can. J. Chem.*, **44**, 249 (1966).
(30) D. Horton and J. S. Jewell, *Carbohyd. Res.*, **3**, 255 (1966).
(31) D. Horton, J. S. Jewell, E. K. Just, and J. D. Wander, *Carbohyd. Res.*, **18**, 49 (1971).

Base-catalyzed exchanges of this type have generally been used for the preparation of deuterated derivatives, because the reaction can be monitored by n.m.r. spectroscopy, but L-*xylo*- and L-*arabino*-ascorbic-4-*t* acid have been prepared[32] by alkaline exchange of L-*xylo*-ascorbic acid with water-*t*.

5. Enzyme-catalyzed Solvent-exchange

An alternative method for the introduction of the hydrogen of a solvent into a sugar molecule is by the use of enzymes. Most enzymes that ionize a carbon–hydrogen bond to give a proton, or that form a new carbon–hydrogen bond, can be used, particularly aldolases, decarboxylases, and those isomerases that have an enediol intermediate. Other enzymes, such as 3-deoxy-D-*arabino*-heptulosonate synthetase, incorporate hydrogen into the product in a less predictable way (see p. 162). In addition, enzymes that generate a hydride from a dithiol intermediate which can exchange hydrogen with the solvent, such as ribotide reductase (see p. 161), will incorporate hydrogen of the solvent into the product.

The hydrogen atom is always stereospecifically introduced by the enzyme, so that, at a methylene group, only one of the two hydrogen atoms becomes labeled. Thus, D-glucose 6-phosphate ketol isomerase (E.C. 5.3.1.9) incorporates only one deuterium atom from deuterium oxide,[33] or one tritium atom[34] from water-*t*, giving D-fructose-*1*(*R*)-*t* 6-phosphate (**11**). The isomeric D-fructose-*1*(*S*)-*t* 6-phosphate (**12**) can be obtained from D-glucose-*1*-*t* in water,[35] or,

$$\begin{array}{cc}
\text{OH} & \text{OH} \\
| & | \\
\text{HCT} & \text{TCH} \\
| & | \\
\text{C}=\text{O} & \text{C}=\text{O} \\
| & | \\
\text{HOCH} & \text{HOCH} \\
| & | \\
\text{HCOH} & \text{HCOH} \\
| & | \\
\text{HCOH} & \text{HCOH} \\
| & | \\
\text{CH}_2\text{OPO}_3\text{H}_2 & \text{CH}_2\text{OPO}_3\text{H}_2 \\
\mathbf{11} & \mathbf{12}
\end{array}$$

alternatively, the pro-*S* hydrogen atom can be exchanged with water by using[33] D-mannose 6-phosphate ketol isomerase (E.C. 5.3.1.7). These enzymes also give the corresponding 2-labeled aldoses, and this is the most convenient method for preparation[36] of D-glucose-2-*t*.

(32) E. M. Bell, E. M. Baker, and B. M. Tolbert, *J. Label. Compounds*, **2**, 148 (1966).
(33) Y. J. Topper, *J. Biol. Chem.*, **225**, 419 (1957).
(34) I. A. Rose and E. L. O'Connell, *J. Biol. Chem.*, **236**, 3086 (1961).
(35) I. A. Rose and E. L. O'Connell, *Biochim. Biophys. Acta*, **42**, 159 (1960).
(36) J. M. Lowenstein, *Methods Enzymol.*, **6**, 878 (1963).

Aldolases can usually exchange the hydrogen atom lost in the condensation when the appropriate substrate is incubated alone with the enzyme. Thus, 1,3-dihydroxy-2-propanone 1-phosphate stereospecifically exchanges the pro-S hydrogen atom at C-3 with solvent protons in the presence of D-fructose 1,6-diphosphate D-glyceraldehyde lyase (aldolase, E.C. 4.1.2.7).[37-41] Triose phosphate isomerase (E.C. 5.3.1.1) labilizes[38] the pro-R hydrogen atom of 1,3-dihydroxy-2-propanone 1-phosphate, and also introduces a hydrogen atom onto C-2 of D-glyceraldehyde. A combination of aldolase and triose phosphate isomerase has been used to prepare D-glucose-3-t and -4-t (Ref. 42), D-fructose-5-t and D-glucose-5-t 6-phosphate (Ref. 43), and D-fructose-3,4,5-t_3 (Ref. 44).

Decarboxylation *must* lead to incorporation of a proton from the solvent. The oxidative decarboxylation of 6-O-phosphono-D-gluconate gives D-*erythro*-pentulose-*1(S)*-t 5-phosphate (with inversion of configuration).[45] The R-isomer can be obtained[46] from D-ribose 5-phosphate by using D-ribose 5-phosphate ketol isomerase (E.C. 5.3.1.6).

6. Incorporation of Solvent on Oxidation of a Hydrazino Group

When 3-deoxy-3-hydrazino-1,2:5,6-di-O-isopropylidene-α-D-glucofuranose (13) is oxidized[47] in alkaline deuterium oxide, the corresponding deuterated 3-deoxy compound (14) is formed with 94%

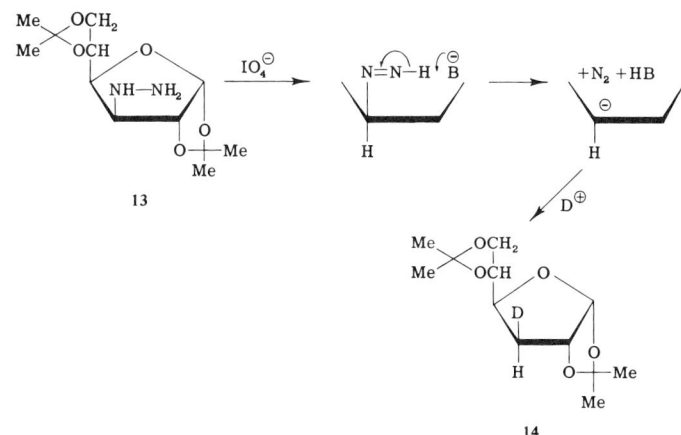

(37) I. A. Rose and S. V. Rieder, *J. Amer. Chem. Soc.*, **77**, 5764 (1955).
(38) B. Bloom and Y. J. Topper, *Science*, **124**, 982 (1956).
(39) W. J. Rutter and K. H. Lung, *Biochim. Biophys. Acta*, **30**, 71 (1958).
(40) I. A. Rose, *J. Amer. Chem. Soc.*, **80**, 5835 (1958).
(41) I. A. Rose and S. V. Rieder, *J. Biol. Chem.*, **231**, 315 (1958).
(42) R. Rognstad, R. G. Kemp, and J. Katz, *Arch. Biochem. Biophys.*, **109**, 372 (1965).

(*References continued on page 136.*)

retention of configuration. At neutrality, the stereospecificity is lost. The proposed[48] mechanism of this type of decomposition is shown.

7. Addition to a Double Bond

An interesting method of introducing tritium into a sugar was used by Lehmann.[49] When an unsaturated sugar is a vinyl ether, it can be hydroborated with diborane-t to give, by anti-Markownikoff addition, a boronated sugar that can then be oxidized to the free sugar by peroxide. Addition to the double bond should theoretically give two isomeric products, but it can actually be quite stereospecific. Treatment of 3-deoxy-1,2:5,6-di-O-isopropylidene-α-D-*ribo*-hex-3-enofuranose (**15**) gave D-galactose-4-t as the sole final product.

Tritium has been introduced[50] onto C-6 of methyl 2,3,4-tri-O-acetyl-6-deoxy-α-D-*arabino*-hex-5-enopyranoside (**16**) by treating it

with silver fluoride in pyridine in the presence of a small proportion of tritiated water. The sugar can add water to form the 6-deoxy-D-hexopyranosid-5-ulose hydrate derivative, and tritium is presumably

(43) V. G. Hauska, H. Kindl, and O. Hoffmann-Ostenhof, Z. *Physiol. Chem.*, **348**, 1273 (1967).
(44) C. Jochmann, P. Rauschenbach, and W. Lamprecht, Z. *Physiol. Chem.*, **349**, 885 (1968).
(45) G. E. Lienhard and I. A. Rose, *Biochemistry*, **3**, 190 (1964).
(46) M. W. McDonough and W. A. Wood. *J. Biol. Chem.*,**236**, 1220 (1961).
(47) D. M. Brown and G. H. Jones, *J. Chem. Soc. (C)*, 252 (1967).
(48) D. J. Cram and J. S. Bradshaw, *J. Amer. Chem. Soc.*, **85**, 1108 (1963).
(49) J. Lehmann, *Carbohyd. Res.*, **2**, 1 (1966).
(50) J. Lehmann, *Carbohyd. Res.*, **4**, 196 (1967).

introduced by equilibration between the unsaturated sugar and this "hydrate." The acetyl groups were also labeled, and contained 99% of the radioactivity.

The classical method of preparing 2-deoxy sugars by hydration of glycals has been adapted[51] for the synthesis of 2-deoxy-D-*arabino*-hexose-2-*t*.

8. Conversion from Another Labeled Sugar

Sometimes, the most convenient method of preparation of the labeled sugar required is by chemical or enzymic modification of a more readily available, or commercially available, sugar. Thus, D-ribose-3-*t* has been prepared from 1,2:5,6-di-*O*-isopropylidene-α-D-allofuranose-3-*t* by way of 1,2-*O*-isopropylidene-α-D-allofuranose-3-*t*, followed by periodate oxidation, and reduction.[52]

9. Methods for Non-specific Labeling

When the specific-labeling methods already described cannot be used, as is often the case with polysaccharides, random-labeling methods must be considered. Those most frequently used are catalytic labeling[53] and the Wilzbach method.[54] These methods, highly successful in other fields, are usually unsuitable for the preparation of labeled carbohydrates, because the conditions used, and, particularly, the free-radical nature of the Wilzbach reaction, cause the generation of highly labeled impurities that are often difficult to separate from the material desired. In addition to the isomerization that may occur when a hydrogen atom is abstracted from the molecule, polysaccharides also suffer radiation damage, particularly at the glycosidic linkage.

a. Gas and Solvent Exchange By Using a Metal Catalyst.—Where the Wilzbach and catalytic methods of labeling have been compared, the latter appears to give the better results,[55,56] and even some specificity.[56] Treatment of *myo*-inositol with a platinum catalyst in deuterium oxide,[57] or of 1L-*chiro*-inositol in water-*t* (Ref. 56), gave label-

(51) J. E. G. Barnett, R. E. Brice, and D. L. Corina, *Biochem. J.*, **119**, 183 (1970).
(52) H. P. C. Hogenkamp, *Carbohyd. Res.*, **3**, 239 (1966).
(53) J. L. Garnett, *Nucleonics*, **20** (12), 86 (1962).
(54) K. E. Wilzbach, *J. Amer. Chem. Soc.*, **79**, 1013 (1957).
(55) N. D. Ferrante, E. A. Popenoe, D. R. Christman, and P. J. Sammon, *Carbohyd. Res.*, **2**, 439 (1966).
(56) S. J. Angyal, C. M. Fernandez, and J. L. Garnett, *Aust. J. Chem.*, **20**, 2647 (1967).
(57) M. R. Stetten and D. Stetten, *J. Biol. Chem.*, **164**, 85 (1946).

ing of the respective inositols. The products of the first reaction were not investigated, but the *chiro*-inositol was shown to give *myo*-inositol, labeled mainly on C-1 and C-6, as the only by-product; this is consistent with favored attack at the axial position, as is found with platinum oxidation of inositols. It has been suggested that the catalyst functions as an oxidation–reduction system.[56]

Heparin has been labeled by this method[55] and also by platinum-catalyzed exchange with tritium gas,[58] in both cases with retention of biological activity, and it is used commercially for labeling C-1 of aldoses.

b. Wilzbach and Related Methods.—In this method for introducing a label, sufficient energy is generated, usually by the natural radiation from decomposition of the tritium itself, to exceed the energy of the carbon–hydrogen bond. Few examples are known where the radiochemical purity of the sugar produced has been rigorously established. 1L-*chiro*-Inositol has been subjected[59] to Wilzbach tritiation, and the products separated by the addition of *myo*-, *muco*-, and *allo*-inositols and fractionation on a cellulose column. All of the separated inositols were labeled. The *chiro*-inositol produced (which was further purified by preparation of a crystalline derivative) was randomly labeled, but the *myo*-inositol formed was labeled only at C-1, as would be expected if the configuration of either C-1 or C-6 of 1L-*chiro*-inositol was inverted during the labeling. Fewer by-products were formed by using crystalline 1L-*chiro*-inositol than with its solution, suggesting that the crystal structure may preserve the necessary configuration during the exchange.[60]

Tritium-recoil labeling, in which the sugar is mixed with a lithium salt and bombarded with neutrons to generate tritium in the mixture, is claimed to give good yields of generally-labeled D-glucose and D-galactose, with few impurities; again, the best results were obtained with crystalline sugars.[61,62]

III. RADIOCHEMICAL AND CHEMICAL STABILITY

Tritium has[63] a half-life of 12.26 years, and so radioactive decay of the label can usually be neglected. However, the radiation emitted

(58) G. H. Barlow and E. V. Cardinal, *Proc. Soc. Exp. Biol. Med.*, **123**, 831 (1966).
(59) S. J. Angyal, C. M. Fernandez, and J. L. Garnett, *Aust. J. Chem.*, **18**, 39 (1965).
(60) S. J. Angyal, J. L. Garnett, and R. M. Hoskinson, *Aust. J. Chem.*, **16**, 252 (1963).
(61) F. S. Rowland, C. N. Turton, and R. Wolfgang, *J. Amer. Chem. Soc.*, **78**, 2354 (1956).
(62) H. Keller and F. S. Rowland, *J. Phys. Chem.*, **62**, 1373 (1958).
(63) W. M. Jones, *Phys. Rev.*, **100**, 124 (1955).

is weak β, and it causes decomposition of tritiated sugars by self-irradiation. The nature of this decomposition with respect to ^{14}C-labeled sugars has been discussed in a previous article in this Series.[64] Because the specific activities of tritium compounds are usually much higher than those of ^{14}C compounds, the problems of storage are much more serious with the former. This problem is obviously of commercial interest, and considerable work on the stability of tritium compounds, undertaken at the Radiochemical Centre, Amersham, England, is described in a pamphlet[65] and a book.[66]

Most tritiated sugars are best stored in aqueous solution, sometimes containing a small proportion of ethanol. Under these conditions, radiolysis usually does not exceed 5 to 10% per year. In the frozen state, between 0° and −196°, the rate of decomposition can be high, presumably due to concentration of the sugar in pockets on freezing.[67] Nucleotides are stored at −20° in 50% ethanol, which acts as a free-radical scavenger,[68] and decompose at the rate of ~12% per year; polymers should be stored as freeze-dried solids.

In addition to radiochemical decomposition, chemical and biological decomposition must be considered, both for tritiated and deuterated compounds. Hydrogen atoms attached to nitrogen, sulfur, and oxygen exchange rapidly with proton-donating solvents, and a hydrogen atom attached to these atoms during synthesis is usually removed during purification, although, occasionally, special precautions, such as repeated addition and evaporation of solvent, must be taken to ensure this removal. Hydrogen atoms attached to carbon also exchange when the hydrogen atom is in an activated position (adjacent to a carbonyl group or a similar electron-withdrawing group). Particular precautions must be taken in the storage and use of such compounds, and, during localization procedures (see p. 140), care must be taken that such labile compounds do not occur as intermediates. Carbon–hydrogen bonds that are chemically stable under the conditions used can be labilized by enzymes, and this possibility should receive consideration when impure enzyme preparations are used.

(64) G. O. Phillips, *Advan. Carbohyd. Chem.*, **16**, 13 (1961).
(65) R. J. Bayley and E. A. Evans, "Storage and Stability of Compounds Labelled with Radioisotopes," Review 7, The Radiochemical Centre, Amersham, England (1968).
(66) E. A. Evans, "Tritium and its Compounds," Butterworth, London, 1966, p. 306 ff.
(67) E. A. Evans, *Nature*, **209**, 169 (1966).
(68) E. A. Evans and F. G. Stanford, *Nature*, **197**, 551 (1963).

IV. LOCALIZATION

Exact localization of the hydrogen isotope is usually required, and it is inadvisable to rely on the method of synthesis, as the label is not always situated entirely at the position predicted.[69] In many mechanistic and biological studies with enzymes, the position of the label alters during the reaction, and, in others, where a proton isotope from the solvent is incorporated, it is important to localize the incorporated isotope, and, if it is at a methylene group, to determine its stereochemistry.

1. Chemical Degradation

Chemical degradation is particularly relevant for the localization of tritium,[70] where a carrier can be added and the molecule specifically degraded, often by periodate oxidation, to give each hydrogen atom uniquely located in a crystalline derivative. Particularly useful derivatives are the dimedone derivative of formaldehyde and the *p*-bromophenacyl ester of formic acid.[71] Tritiated derivatives of 1L-*chiro*-inositol,[56] glycerol,[72] D-xylose,[73] D-ribose,[74] D-mannose,[74] D-glucose,[11,61] D-galactose,[14,62] D-glucose 6-phosphate,[11] and D-glucosyl phosphate[12] are amongst those which have been degraded in this way.

Alternatively, a procedure that specifically removes the tritium by ionization, or oxidation to water-*t*, can be used. Thus, the presence of isotopic hydrogen at C-1 of aldoses was confirmed by oxidation, either by re-isolation of the aldonic acid,[7] or by measurement of the isotope appearing in the water.[26] Other oxidative procedures have involved oxidation to D-glucaric acid,[75] the platinum-catalyzed oxidation[19] of *myo*-inositol-2-*t*, bacterial oxidation[75] of D-glucitol-5-*t*, and osazone formation.[76]

Whenever possible, it is important to test the degradation methods by using authentically labeled molecules, because, sometimes, unforeseen exchanges occur. Thus, it has been reported that only

(69) R. D. Bevill, J. M. Nordin, F. Smith and S. Kirkwood, *Biochem. Biophys. Res. Commun.*, **12**, 152 (1963).
(70) J. E. G. Barnett, *Methods Carbohyd. Chem.*, **6**, 506 (1972).
(71) O. Gabriel, *Anal. Biochem.*, **10**, 143 (1965).
(72) B. Bloom and D. W. Foster, *J. Biol. Chem.*, **239**, 967 (1964).
(73) F. A. Loewus, *Arch. Biochem. Biophys.*, **105**, 590 (1964).
(74) H. Simon, H. D. Dorrer, and K. H. Ebert, *Z. Naturforsch., B*, **18**, 360 (1963).
(75) L. T. Sniegoski and H. S. Isbell, *J. Res. Nat. Bur. Stand., A*, **66**, 29 (1962).
(76) J. S. Schutzbach and D. S. Feingold, *J. Biol. Chem.*, **245**, 2476 (1970).

94% of the label is recovered on formation of p-bromophenacyl formate,[71] that formation of D-*arabino*-hexulose phenylosazone[26] from D-glucose, D-mannose, or D-fructose in water-*d* gives some incorporation of label, and that recrystallization of D-mannose-*1*-*t* phenylhydrazone is accompanied by a small isotope-effect originating in the rate of equilibration between the isomeric forms.[76a]

2. Use of Enzymes in Determining the Stereochemistry at a Chiral Methylene Group

Introduction of a hydrogen isotope by reduction of an aldehyde gives two possible products, resulting from attack on each side of the carbon atom of the carbonyl group. In an asymmetric molecule, such as a sugar, it is improbable that both isomers will be formed in exactly the same proportion.

Because the enzyme surface is stereospecific (usually, even towards the exchange of protons), when a hydrogen isotope is enzymically introduced into a methylene (including a hydroxymethyl) group, the introduction is almost invariably stereospecific, so that only one of the isomers arising from the creation of a new asymmetric center is formed. This property can be used both to determine the proportion of isomers formed in chemical synthesis and to relate the stereochemistry of different enzymes. Such information may be important in mechanistic studies on enzymes (see p. 155). Enzymic methods for assigning configuration at a chiral methylene group may be illustrated by the assignment of configuration to the two 1,3-dihydroxy-2-propanone-3-*t* 1-phosphates formed by exchange with water-*t* in the presence of aldolase and triose phosphate isomerase (see p. 135). The Hanson extension[77] of the Cahn–Ingold–Prelog system of nomenclature[78] is used.

By enzymic dephosphorylation and periodate oxidation, Rose[40] converted the isomeric 1,3-dihydroxy-2-propanone-3-*t* 1-phosphates (**17**), into glycolic-2-*t* acid (**18**), which was oxidized by glycolate dehydrogenase to glyoxylic acid (**19**). By using glycolic-2-*t* acid derived from the aldolase exchange-reaction, the tritium was entirely retained, whereas the tritium in glycolic-2-*t* acid derived from the triose phosphate isomerase exchange was completely lost. It was

(76a) F. Weygand, H. Simon, K. D. Keil, H. S. Isbell, and L. T. Sniegoski, *Anal. Chem.*, **34**, 1753 (1962).
(77) K. R. Hanson, *J. Amer. Chem. Soc.*, **88**, 2731 (1966).
(78) R. S. Cahn, C. K. Ingold, and V. Prelog, *Angew. Chem. Int. Ed. Engl.*, **5**, 385 (1966).

known that L-lactic acid (**20**) is a substrate for glycolate dehydro-

$$\underset{\substack{\text{CH}_2\text{OPO}_3\text{H}_2 \\ }}{\overset{\text{OH}}{\underset{|}{\text{H}_\text{S}-\overset{|}{\text{C}}-\text{H}_\text{R}}}} \overset{\text{aldolase}}{\rightleftharpoons} \underset{\substack{\text{CH}_2\text{OPO}_3\text{H}_2 \\ 17\,S}}{\overset{\text{OH}}{\underset{|}{\text{T}-\overset{|}{\text{C}}-\text{H}}}} \longrightarrow \underset{\substack{\text{CO}_2\text{H} \\ 18\,S}}{\overset{\text{OH}}{\underset{|}{\text{T}-\overset{|}{\text{C}}-\text{H}}}} \overset{\text{glycolate}}{\underset{\text{genase}}{\xrightarrow{\text{dehydro-}}}} \underset{\text{CO}_2\text{H}}{\overset{\text{O}}{\underset{|}{\text{T}-\text{C}\diagup}}}$$
$$19$$

$$\underset{\text{CO}_2\text{H}}{\overset{\text{OH}}{\underset{|}{\text{H}_3\text{C}-\overset{|}{\text{C}}-\text{H}}}} \overset{\text{glycolate}}{\underset{\text{genase}}{\xrightarrow{\text{dehydro-}}}} \underset{\text{CO}_2\text{H}}{\overset{\text{O}}{\underset{|}{\text{H}_3\text{C}-\text{C}\diagup}}}$$
$$20$$

genase, whereas its enantiomer is not; this showed that, if the complex of the enzyme with lactic acid and glycolic acid has the same stereochemistry, it is the R hydrogen atom of glycolic acid that is removed on oxidation. In corroboration, on using L-lactic acid dehydrogenase (E.C. 1.1.1.28) and NADH-t, unlabeled glyoxylic acid was converted into glycolic-2-t acid of the same stereochemistry as that of the acid derived by use of triose phosphate isomerase. It therefore follows that the glycolic-2-t acid and the 1,3-dihydroxy-2-propanone-3-t 1-phosphate derived from the triose phosphate isomerase exchange has the tritium atom attached in the R configuration, and that that from the aldolase exchange is in the S configuration.

The proof of configuration just described relies on the assumption that there is the same relative stereochemistry for the hydrogen atoms in glycolic acid and lactic acid on formation of the enzyme–substrate complexes. The absolute configuration of the glycolate-2-d (**22**) prepared by reduction of glyoxylate-2-d (**21**) with glycolate dehydro-

$$\text{NADH} + \underset{\text{CO}_2\text{H}}{\overset{\text{CDO}}{\underset{|}{|}}} \xrightarrow{\text{glycolate dehydrogenase}} \text{NAD} + \underset{\text{CO}_2\text{H}}{\overset{\text{OH}}{\underset{|}{\text{DCH}}}}$$
$$21 \qquad\qquad 22$$

genase has since been shown,[79] by neutron-diffraction analysis, to be S, confirming the previous assignments.

When one or more molecules have been assigned configuration in this way, molecules having unknown configuration at a labeled hydrogen atom can be enzymically related to the primary standard, if necessary after degradation. Labeled hydroxymethyl groups that

(79) C. K. Johnson, E. J. Gabe, M. R. Taylor, and I. A. Rose, *J. Amer. Chem. Soc.*, **87**, 1802 (1965).

have been degraded to glycolic acid groups and oxidized by glycolate dehydrogenase include D-xylosyl-5-t phosphate,[75] D-fructose-1-t 6-phosphate,[35] and D-erythro-pentulose-1-t 5-phosphate[45,46] (see p. 135). Compounds of known absolute stereochemistry that may be useful in the determination of hydrogen configuration in sugars include ethanol-2-d (Refs. 80 and 81), propionate-2-d (Ref. 82), and malic-3-d acid (Ref. 83). The same hydrogen atom is always labilized by a given enzyme, and deuterium and tritium compounds can be compared; this permits a useful correlation between sugars containing the two isotopes, as the possibility of 100% replacement of hydrogen by deuterium facilitates the use of physical methods in the determination of configuration.

3. Physical Methods

The location of deuterium in sugars is usually most conveniently ascertained by physical methods of analysis. The most important of these is nuclear magnetic resonance (n.m.r.) spectroscopy, a general account of whose application to carbohydrates is given in an earlier Volume of this Series[84] (see also, This Volume, p. 7).

a. **Nuclear Magnetic Resonance Spectroscopy.**—The deuterium nucleus, spin 1, exhibits only very weak coupling with protons, the magnitude of H–D couplings being only about one-seventh of the corresponding H–H couplings, so that proton signals are usually broadened, instead of being split, by deuterium atoms on adjacent carbon atoms. Substitution by deuterium, therefore, simplifies the p.m.r. spectrum, and facilitates assignments, both of the proton substituted and of the adjacent protons. In this way, the identification of the signal caused by the anomeric proton was first made;[81] this was achieved by comparison of the spectra of the penta-O-acetyl-α- or β-D-glucopyranoses and the 1-C-deuterated compounds. Once such assignments have been made with sugars deuterated in known positions, the knowledge can be used to locate deuterium at an unknown position. However, as the position of attachment of the deuterium is usually known, the determination of configuration of

---

(80) R. U. Lemieux and J. Howard, *Can. J. Chem.*, **41**, 308 (1963).
(81) H. Weber, J. Seibl, and D. Arigoni, *Helv. Chim. Acta*, **49**, 741 (1966).
(82) B. Zagalack, P. A. Frey, G. C. Karabatsos, and R. H. Abeles, *J. Biol. Chem.*, **241**, 3028 (1966).
(83) O. Gawron and T. P. Fondy, *J. Amer. Chem. Soc.*, **81**, 6333 (1959).
(84) L. D. Hall, *Advan. Carbohyd. Chem.*, **19**, 51 (1964).

deuterium at a chiral methylene or hydroxymethyl group is generally of much greater importance.

The magnitude of the coupling constants between protons on adjacent carbon atoms is dependent on the angle between them; this was first shown[85] by Lemieux and coworkers, who noted that, for pyranose acetates, the observed splittings were near 8 Hz if the hydrogen atoms were *trans*-diaxial, but only about 3 Hz if the atoms were *gauche* (*a,e* or *e,e*). This result has been generalized[86] to relate the dihedral bond-angle to the magnitude of the coupling (the Karplus equation), but it has been pointed out[87] that the splitting observed depends on several other factors. The development of these concepts with reference to sugars is discussed in Refs. 84 and 87a. In general, assignments for a pyranoid ring can be made with confidence.

If a comparison is made of the coupling constants of the protons of an endocyclic methylene group with a proton on an adjacent carbon atom, for example, at C-5 of a pentopyranose derivative or in certain deoxypyranose derivatives, the two protons can be differentiated. If one of the protons is replaced by deuterium, the orientation of the proton remaining on the methylene group can readily be recognized. This principle is illustrated by the proof of structure of the enantiomers of ethanol-2-*d* given by Lemieux and Howard.[80]

Reduction of 3-O-benzyl-1,2-O-isopropylidene-α-D-*xylo*-pentodialdo-1,4-furanose (**23**) with lithium aluminum deuteride, followed by catalytic debenzylation, gave an unequal R and S mixture (**24**) of 1,2-O-isopropylidene-α-D-xylofuranoses-5-*d* that was converted into a mixture (**25**) of tetra-O-acetyl-β-D-xylopyranose-(R)- and (S)-5-*d*. Inspection of the p.m.r. spectrum showed, for the C-5 proton, two signals that were doublets, with splittings of 4.2 and 8.1 Hz, corresponding to *a,e* and *a,a* coupling, respectively, with the axial proton at C-4.

The signals had relative intensities of 1:1.85, showing that the axial hydrogen atom was the more abundant, and that, therefore, deuterium had been mainly introduced into what was now the equatorial position, having the R configuration. By a series of transformations, the methylene group of the mixture of D-xyloses-5-*d* containing a 30% excess of the R configuration was converted into ethanol-2-*d* (**26**) having a 30% excess of R configuration, and the optical rotation

(85) R. U. Lemieux, R. K. Kullnig, H. J. Bernstein, and W. G. Schneider, *J. Amer. Chem. Soc.*, **80**, 6098 (1958).
(86) M. Karplus, *J. Chem. Phys.*, **30**, 11 (1959).
(87) M. Karplus, *J. Amer. Chem. Soc.*, **85**, 2870 (1963).
(87a) P. L. Durette and D. Horton, *Advan. Carbohyd. Chem. Biochem.*, **26**, 49 (1971).

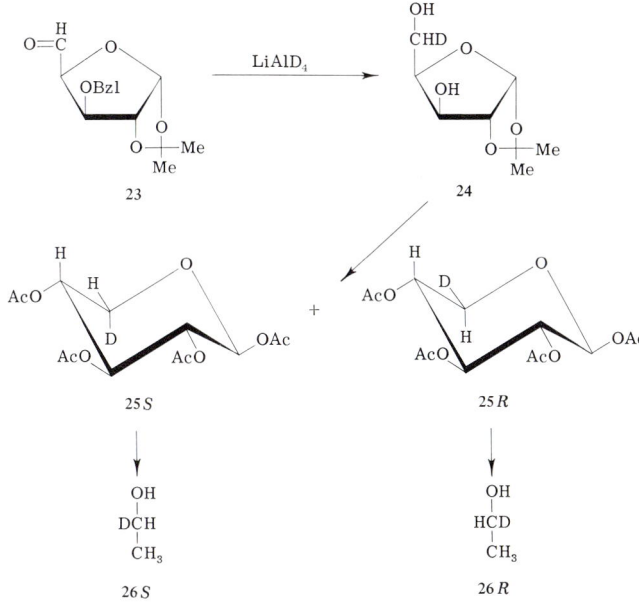

was measured. The corrected dextrorotation for the R form was numerically identical to, and opposite in sign from, the rotation of the levorotatory ethanol-2-d formed[88] by reduction of acetaldehyde-2-d by NADH, showing that the levorotatory isomer is ethanol-2(S)-d.

Although the Karplus equation gives very clear results with pyranoid rings, the assignments for furanoid rings should be regarded as tentative,[89,90] because the furanoid ring is more flexible and the dihedral angles less-clearly delineated.[87a,89,90] The configuration of the deuterium introduced at C-2' of the furanoid ring by the enzymic conversion of adenosine 5'-triphosphate[91] or cytidine 5'-triphosphate[92] in deuterium oxide into the respective 2'-deoxynucleotides has been assigned by p.m.r. spectroscopy. For 2'-deoxycytidine, the difference in coupling of $J_{1',2R'}$ and $J_{1',2S'}$ is very small and completely ambiguous, although the difference in coupling of $J_{2R',3'}$ and $J_{2S',3'}$ is slightly greater and allowed an assignment to be made,[92] but the most convincing evidence was that of an empirical comparison of the

(88) H. R. Levy, F. A. Loewus, and B. Vennesland, *J. Amer. Chem. Soc.*, **79**, 2949 (1957).
(89) R. U. Lemieux and D. R. Lineback, *Ann. Rev. Biochem.*, **32**, 155 (1963).
(90) S. J. Angyal and D. Rutherford, *Ann. Rev. Biochem.*, **34**, 77 (1965).
(91) T. J. Batterham, R. K. Ghambeer, R. L. Blakely, and C. Brownson, *Biochemistry*, **6**, 1203 (1967).
(92) L. J. Durham, A. Larsson, and P. Reichard, *Eur. J. Biochem.*, **1**, 92 (1967).

coupling constants for 2'-deoxycytidine-2'-d, cytidine, 1-(β-D-arabinofuranosyl)cytosine, and 2'-deoxycytidine.[93] Final confirmation of the stereochemistry of this reduction must await comparison of the data for a pyranoid derivative with the clear coupling-data obtained[23] for the pyranose forms of free 2-deoxy-D-*erythro*-pentose-2-d.

b. **Optical Rotation.**—Optical rotation can be extremely useful as a physical constant for comparing samples of deuterated compounds, as in the proof of configuration of ethanol-2-d already described (see p. 144). It has also been used in inferring the stereochemistry of deuterated molecules when the only optical center is that due to the deuterium. Such assignments of configuration should be considered tentative, although, for ethanol-2-d, they proved[88,94] to be correct.

c. **Mass Spectrometry.**—A most important use of mass spectrometry in mechanistic studies is in distinguishing the number of atoms of deuterium present in each molecule. Whereas it is difficult to distinguish a 1:1 mixture of unlabeled molecules with molecules containing two deuterium atoms from molecules containing one atom of deuterium per molecule by some other methods of analysis, mass spectrometry gives an unambiguous answer. It also gives the isotopic abundance of deuterated compounds and, by analysis of the fragmentation pattern, the location of the deuterium atoms in the molecule.

The ability to separate small quantities of isotopically labeled molecules in a mixture may be illustrated by the use of mass spectrometry to distinguish between intermolecular and intramolecular reactions. Some mechanisms proposed for the enzymic NAD-catalyzed isomerization of UDP-D-glucose to UDP-D-galactose involve an intermolecular transfer of hydrogen. This hypothesis has been tested[95] by the simultaneous use of UDP-D-glucose and UDP-D-glucose-d_7 as substrates. Mass-spectrometric analysis of the products, after conversion into hexitol hexaacetates, showed that intermolecular transfer of deuterium had not occurred. A similar study[96] showed that the enzymic conversion of dTDP-D-glucose into dTDP-(6-deoxy-D-*xylo*-hexopyranos-4-ulose) involves an intramolecular transfer (see p. 170).

(93) C. E. Griffin, F. D. Hamilton, S. P. Hopper, and R. Abrams, *Arch. Biochem. Biophys.*, **126**, 905 (1968).
(94) J. H. Brewster, *Tetrahedron Lett.*, 23 (1959).
(95) L. Glaser and L. Ward, *Biochim. Biophys. Acta*, **198**, 613 (1970).
(96) A. Melo, W. H. Elliott, and L. Glaser, *J. Biol. Chem.*, **243**, 1467 (1968).

V. Physical Properties and Their Applications

As tritium is used only in tracer proportions, only the physical properties of deuterated sugars (into which the deuterium is usually introduced at 100% abundance) are of interest. Many of these properties have been discussed in Section IV, as they are very useful for localizing the deuterium.

1. Nuclear Magnetic Resonance Spectroscopy

The deuterium n.m.r. spectrum of 1,2:5,6-di-*O*-isopropylidene-α-D-allofuranose-3-*d* has been measured,[97] but, otherwise, the deuterium nucleus (spin 1) has not received much attention in n.m.r.-spectral studies of carbohydrates. However, replacement of hydrogen by deuterium has important uses in both ^{1}H and ^{13}C n.m.r. spectroscopy. Because D–H couplings are much smaller than the corresponding H–H couplings, replacement of the hydroxyl hydrogen atoms of a sugar by exchange with deuterium oxide is a routine method of simplifying spectra. Similarly, exchange of the acetamido-group hydrogen atom in a peracetylated 2-amino-2-deoxy sugar,[98] or in a benzylidene acetal of a 2-acetamido-2-deoxy sugar,[99] has led, by comparison with the nondeuterated sugars, to the assignment of a signal to the hydrogen atom at C-2. Substitution of hydrogen by deuterium at a carbon atom has also been used to simplify spectra where two signals in a mixture overlap. Analysis of the proportions of the *keto* form of 1,3-dihydroxy-2-propanone 1-phosphate was only possible by using 1,3-dihydroxy-2-propanone-3,3-d_2 1-phosphate.[100]

Deuterium can be substituted for protons attached to carbon, so that the signals of individual protons can be identified. This method has been illustrated (see p. 143) by the identification of the anomeric-proton signal. Although this method of assigning signals has lost some importance since the introduction of double-resonance decoupling for the routine identification of signals, it is still of use in difficult cases[31] and in situations where there is no vicinal coupling. For instance, the easiest way of identifying[98,101] the signal due to

---

(97) J. R. Campbell, L. D. Hall, and P. R. Steiner, *Can. J. Chem.*, **50**, 504 (1972).
(98) D. Horton, J. B. Hughes, J. S. Jewell, K. D. Philips, and W. N. Turner, *J. Org. Chem.*, **32**, 1073 (1967).
(99) B. Coxon, *Tetrahedron*, **21**, 3481 (1965).
(100) G. R. Gray and R. Barker, *Biochemistry*, **9**, 2454 (1970).
(101) D. Horton, W. E. Mast, and K. D. Philips, *J. Org. Chem.*, **32**, 1471 (1967); D. Horton and J. H. Lauterbach, *ibid.*, **34**, 86 (1969).

specific acetoxyl groups in acetylated sugars is by their replacement by acetoxyl-d_3.

In addition to the assignment of configuration to a deuterium atom introduced into a methylene group (see p. 144), two other applications are of interest. By using methyl sulfoxide-d_6 as the solvent, useful spectra of the hydroxyl protons can be obtained. These can quickly be recognized on addition of deuterium oxide, which, by exchange, causes their disappearance. In an investigation of the stability of pyranose and furanose forms of D-glucose and L-arabinose in solution, D-glucose-5,6,6-d_3 and L-arabinose-5,5-d_2 were used.[102] Secondary hydroxyl-group signals appear as doublets, due to coupling with the ring proton. When one of the ring protons is replaced by deuterium, the proton of the hydroxyl group attached to this carbon atom resonates as a singlet; being coupled only to deuterium, the proton signal is merely broadened. Comparison of the spectrum of D-glucose-5,6,6-d_3 with that of D-glucose permitted recognition of the pyranose form and exclusion of the furanose form. The pyranose form (**27**) actually present showed almost identical signals for the

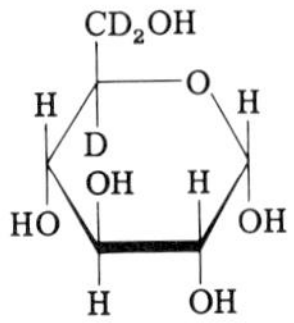

27

secondary hydroxyl groups in the deuterated and non-deuterated sugars, because C-5 does not bear a hydroxyl group, whereas, had the furanose form (**28**) been present, the deuterium atom at C-5 would

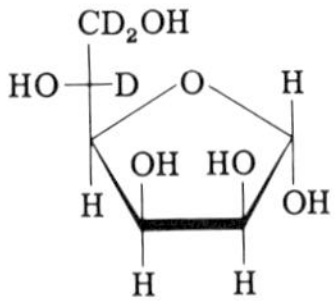

28

have caused one of the hydroxyl-proton signals (that at C-5) to appear as a singlet. This result was confirmed by comparison of the spectra of 1,2-O-isopropylidene-α-D-glucofuranose-5-d and the corresponding nondeuterated compound. L-Arabinose was found to be in the pyranose form when first dissolved, but it was converted into the furanose form.

(102) A. S. Perlin, *Can. J. Chem.*, **44**, 539 (1966).

In a rather similar way, D-glucose-5-d has been used[10] for determining the tautomeric species of D-glucose (pyranose or furanose) involved in glycol cleavage by lead tetraacetate and periodate, respectively. The p.m.r. signal of a proton attached to a carbon atom carrying a formic ester group is observed at low field. In oxidation of a furanose and pyranose, respectively, this signal is due to H-4 or H-5, respectively. When D-glucose-5-d is used, a low-field signal due to the formic ester can be observed only if oxidation of the furanose form occurs. It was found that lead tetraacetate cleaves the furanose form, and periodate cleaves the pyranose form.

Deuterium labeling can also be used[21,103] to identify the natural-abundance ^{13}C signals in sugars. ^{13}C N.m.r. spectra are usually taken with wide-band, proton decoupling, and as, at natural abundance, the ^{13}C nucleus is surrounded by ^{12}C nuclei, discrete singlets are observed for each carbon atom. When deuterium is introduced at a specific carbon atom, the ^{13}C signal is markedly altered, either giving a broad triplet,[104] or effectively causing the signal to disappear.[103]

2. Infrared Spectroscopy

The carbon–deuterium and oxygen–deuterium absorption frequencies lie at about 2160 and 2500 cm^{-1}, respectively,[105] and are readily distinguishable from the corresponding hydrogen vibrations. Specific replacement of hydrogen by deuterium again allows assignments to be made to individual absorption bands, particularly that at C-1 (Ref. 106) and a discussion of the use of deuterium in this way is given in a previous article in this Series.[107] It was found[106,108] that the axial carbon–deuterium stretching frequencies were consistently lower than their equatorial counterparts; for example, the frequencies assigned to these bands in β- and α-D-glucose are 2140 and 2180 cm^{-1}, respectively.

3. Mass Spectrometry

As deuterium has a mass number of 2, compared with a mass number of unity for the proton, it serves as a valuable marker in the

(103) A. S. Perlin, B. Casu, and H. J. Koch, *Can. J. Chem.*, **48**, 2596 (1970).
(104) F. J. Weigert and J. D. Roberts, *J. Amer. Chem. Soc.*, **89**, 2967 (1967).
(105) G. M. Barrow, *J. Chem. Phys.*, **20**, 1739 (1952).
(106) M. Stacey, R. H. Moore, S. A. Barker, H. Weigel, E. J. Bourne, and D. H. Whiffen, *Proc. U. N. Intern. Conf. Peaceful Uses At. Energy, 2nd., Geneva, 1958*, **20**, 251.
(107) H. Spedding, *Advan. Carbohyd. Chem.*, **19**, 23 (1964).
(108) S. A. Barker, R. H. Moore, M. Stacey, and D. H. Whiffen, *Nature*, **186**, 307 (1960).

study of fragmentation patterns. It has been used to aid in the elucidation of the ions produced by mass spectrometry of hexose acetates and pentose acetates,[109] acetates of unsaturated and branched-chain sugars and polyhydric alcohols,[110] permethylated sugars,[111-113] and trimethylsilyl ethers[114,115] of sugars. The deuterium is either attached to the carbon skeleton,[109-111,114,115,115a] or acetoxyl-d_3 (Refs. 109 and 110) and methoxyl-d_3 (Refs. 112 and 113) are specifically substituted for some of the acetoxyl or methoxyl groups.

By analysis[112] of the fragmentation patterns of partially methylated carbohydrates that have subsequently been deuteriomethylated, the disposition of the original methyl groups can be determined. This procedure may have a significant advantage over other methods of identification of partially methylated sugars, and the method has been discussed in a previous article in this Series.[116]

Mass spectrometry has been used for estimating the position and degree of oxidation of oxidized cellulose; this was reduced with sodium borodeuteride, the product hydrolyzed, and the product converted into 1,2:5,6-di-O-isopropylidene-α-D-glucofuranose before study by mass spectrometry and determination of the deuterium isotopic abundance and position.[117] Similar studies have been performed on oxidized derivatives of amylose and starch,[117a] and on 5'-aldehyde analogs of nucleosides.[117b] This method resembles the use[118] of addition of hydrogen cyanide-^{14}C to aldehydes as a method of determining trace quantities of aldehyde, but sodium borohydride-t cannot be used in a similar way because it is not available in 100% isotopic abundance, and reductions with isotopes of hydrogen are subject to unpredictable isotope-effects.

---

(109) K. Heyns and D. Müller, *Tetrahedron Lett.*, 6061 (1966).
(110) A. Rosenthal, *Carbohyd. Res.*, **8**, 61 (1968).
(111) K. Heyns and H. Scharmann, *Tetrahedron*, **21**, 507 (1965).
(112) N. K. Kochetkov and O. S. Chizhov, *Tetrahedron*, **21**, 2029 (1965).
(113) N. K. Kochetkov, N. S. Wulfson, O. S. Chizhov, and B. M. Zolotarev, *Tetrahedron*, **19**, 2209 (1963).
(114) D. C. DeJongh, T. Radford, J. D. Hribar, S. Hanessian, M. Bieber, G. Dawson, and C. C. Sweeley, *J. Amer. Chem. Soc.*, **91**, 1728 (1969).
(115) S. Karay and S. H. Press, *Tetrahedron*, **26**, 4527 (1970).
(115a) D. Horton, E. K. Just, and J. D. Wander, *Org. Mass Spectrom.*, in press.
(116) N. K. Kochetkov and O. S. Chizhov, *Advan. Carbohyd. Chem.*, **21**, 39 (1966).
(117) D. M. Clode and D. Horton, *Carbohyd. Res.*, **12**, 477 (1970); **19**, 329 (1971).
(117a) D. M. Clode and D. Horton, *Carbohyd. Res.*, **17**, 365 (1971).
(117b) D. C. Baker and D. Horton, *Carbohyd. Res.*, **21**, 269 (1972).
(118) J. D. Moyer and H. S. Isbell, *Anal. Chem.*, **29**, 1862 (1957).

4. Gas–liquid Chromatography

Mass-spectrometric analysis of material from biological sources is often performed after direct injection into, and passage through, a gas–liquid chromatographic column. It is usually assumed that the isotopically labeled sugar has the same retention time as the unlabeled material, but this is not necessarily true. It was possible to separate completely the per(trimethylsilyl) derivatives of β-D-glucose from those of β-D-glucose-d_7; the latter have the shorter retention times.[119] By using a technique in which the column effluent was rapidly and repeatedly scanned in the mass spectrometer for two preselected values of m/e, the isotopic abundance in 0.1-μg samples of a mixture of these two derivatives could be determined with an average deviation of 5% from the true value.[120]

VI. Applications in Mechanistic Chemistry

Useful information on the mechanism of chemical reactions can be obtained from the existence of hydrogen isotope-effects, movements of hydrogen atoms (including exchange with solvent), and stereoselectivity in the elimination of hydrogen. Although hydrogen isotopes have not yet found wide application in the elucidation of chemical reaction-mechanisms peculiar to sugars, some examples are of interest and illustrate the type of information that may be obtained.

In an investigation[121] of the mechanism of osazone formation from D-fructose, D-fructose-*1*(S)-*t* and D-fructose-*1*(R)-*t* were converted into the phenylosazone under a variety of conditions that gave different yields. In each experiment, the ratio of the radioactivities in the osazones produced from the two isomers was 3:2. The reaction product was assayed for tritium at intervals during the course of the reaction, up to the point of 100% yield of the osazone, and it was found that the tritium levels for both isomers rose towards the end of the reaction. If elimination of hydrogen is subject to a primary isotope-effect leading to accumulation of tritium in the unchanged starting-material, this behavior would be predicted. The discrimination between the two isomeric forms of D-fructose-*1*-*t* was interpreted as indicating that either a furanose or a pyranose (**29**) form of

---

(119) R. Bentley, N. C. Saha, and C. C. Sweeley, *Anal. Chem.*, **37**, 1118 (1965).
(120) C. C. Sweeley, W. H. Elliott, I. Fries, and R. Ryhage, *Anal. Chem.*, **38**, 1549 (1966).
(121) H. Simon, H. D. Dorrer, and A. Trebst, *Chem. Ber.*, **96**, 1285 (1963).

the hydrazone is first formed, instead of the acyclic form, and that this undergoes elimination. The constant ratio of the radioactivity in the osazone derived from the specifically labeled D-fructoses was attributed to *trans*-elimination of the hydrogen atom and ring-oxygen atom of the α (**29**) and β (**30**) forms of the hydrazone, which may be held stereospecifically in a six-membered ring by hydrogen-bonding.

The further reaction of intermediates of type **31** has been discussed

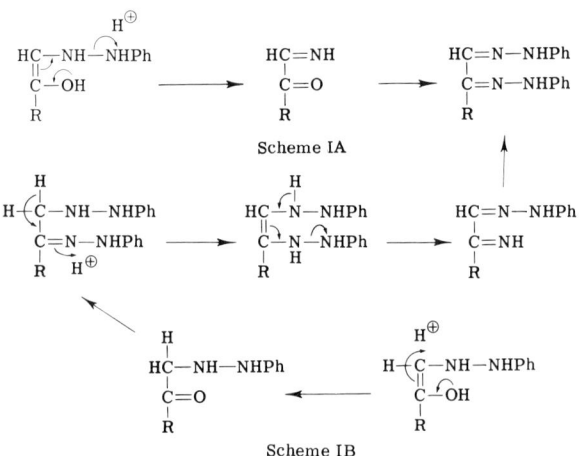

by Weygand and coworkers;[122–124] they suggested two alternative schemes for the conversion of D-glucose phenylhydrazone into the phenylosazone (see Schemes IA and IB).

(122) F. Weygand, *Ber.*, **73**, 1284 (1940).
(123) H. Simon, K. D. Keil, and F. Weygand, *Chem. Ber.*, **95**, 17 (1962).
(124) F. Weygand, H. Simon, and J. F. Klebe, *Chem. Ber.*, **91**, 1567 (1958).

Scheme B requires exchange of the H-1 with the medium, whereas Scheme A does not. Despite a report[5] that stated that there was no loss of tritium from D-glucose-1-t on osazone formation, it seems that about 15% of the hydrogen on C-1 does undergo exchange. Such an exchange is consistent with the operation of mechanism B, with a large isotope-effect, but does not preclude the simultaneous operation of mechanism A.

Solvent exchange may be due to a free-radical mechanism rather than an ionic one. When the reaction of sodium sulfite with methyl 6-deoxy-6-iodo-α-D-glucopyranoside (32) to give methyl 6-deoxy-6-sulfo-α-D-glucopyranoside (34) was conducted[125] in water-t, tritium was incorporated at C-5. This, and the drop in pH at the start of the reaction, suggested that the hex-5-enose 33 was formed by elimination of hydrogen iodide, followed by a free-radical attack of sulfite at C-6 of 33, leading to a free radical at C-5, which then abstracted a proton to afford 34. Addition of sulfite to 33 showed a similar incor-

poration. The large isotope-discrimination against tritium was attributed to a combination of a kinetic and an equilibrium isotope-effect for bisulfite–tritium exchange.

$$HSO_3^- + T^+ \rightleftharpoons TSO_3^- + H^+$$

As an aid to understanding the mechanism of oxidation of aldoses by alkaline iodine,[126,127] bromine,[128] and acidified sodium chlorite,[129]

(125) J. Lehmann and A. A. Benson, J. Amer. Chem. Soc., **86**, 4469 (1964).
(126) H. S. Isbell, L. T. Sniegoski, and H. L. Frush, Anal. Chem., **34**, 982 (1962).
(127) H. S. Isbell and L. T. Sniegoski, J. Res. Nat. Bur. Stand., A, **67**, 569 (1963).
(128) H. S. Isbell and L. T. Sniegoski, J. Res. Nat. Bur. Stand., A, **68**, 145 (1964).
(129) H. S. Isbell and L. T. Sniegoski, J. Res. Nat. Bur. Stand., A, **68**, 301 (1964).

the isotope effects occurring during the oxidation of 1-C-tritiated aldoses have been investigated.

The magnitude of an isotope effect is usually expressed by the ratio k^*/k, where k^* and k are, respectively, the rates for the labeled and unlabeled forms of the reacting molecule. By using isotopes at less than 100% abundance, when k^* is less than k, the labeled reactant is consumed more slowly than the unlabeled one, and the concentration of the labeled form in the residual reactant gradually increases. At the beginning of the reaction, the molar concentration of the isotope in the product is lower than that in the initial reactant, but it increases as the reaction proceeds, reaching that of the original starting-material when the reaction reaches completion or equilibrium. For a first-order (or pseudo-first-order) reaction, the value of k^*/k can be determined,[130,131] if the specific activity of either the reactant or the product is known after a given fraction of the starting material has reacted, by using the equations

(a) $k^*/k = 1 + \log r'/\log(1-f)$
(b) $k^*/k = \log (1 - rf)/\log (1 - f)$,

where f is the fraction of starting material reacted, r is the ratio of the molar specific activity of accumulated product to initial reactant, and r' is the ratio of the residual reactant to that of the initial reactant. By using these equations and measuring the specific activity of (a) the residual reactant by means of a ^{14}C and T double-labeling technique, and (b) the product by measuring the activity of sublimed water-t, Isbell and coworkers showed[126,127] that the ratio k^*/k of 0.14 for oxidation by alkaline iodine of D-glucose-1-t represents the primary isotope-effect of the rupture of the carbon–hydrogen bond at C-1. Oxidation of various aldoses-1-t by slightly acidic bromine-water (in which mutarotation was slow) gave values lying between 0.20 and 0.59. If it is assumed that oxidation takes place only when there is an axial hydrogen of a ring form (35), these isotope effects can be rationalized by invoking rapid oxidation of that anomer that has H-1 axially attached in the more stable chair conformation [for example, β-D-glucose (35)], whereas the anomer having H-1 equatorially attached in the more stable chair conformation [for example, α-D-glucose (36)] is oxidized only slowly, as it is the unstable conformer that reacts the faster. The oxidation of α-D-glucose is, therefore, determined by the rate of mutarotation to β-D-glucose,

(130) W. H. Stevens and R. W. Attree, Can. J. Res., B, **27**, 807 (1949).
(131) A. Ropp, J. Amer. Chem. Soc., **82**, 842 (1960).

which has a secondary isotope-effect of about 0.8. In such aldoses as
D-lyxose, for which both anomeric forms contain considerable proportions of both chair conformations, the isotope effect tends[128]
towards 0.14.

Oxidation of aldoses-*1-t* by acidified sodium chlorite gave[129] k^*/k
values ranging from 0.56 to 0.75, indicating that the rate-determining
step is *not* rupture of the carbon–hydrogen bond. In contrast to oxidation with iodine or bromine, this oxidation is thought to proceed by
way of an *aldehydo* form, and formation of this form was proposed
as the rate-limiting step,[129] because the isotope effect observed is
similar to the values of k^*/k obtained[128] for (*a*) the reduction of D-
glucose-*1-t* with sodium borohydride (0.73), and (*b*) formation of
D-mannose-*1-t* phenylhydrazone (0.83).

VII. Applications in Mechanistic Biochemistry

Most of the sugars specifically labeled with isotopes of hydrogen
that have been synthesized were prepared for use in mechanistic
studies of enzymes.

In the course of enzyme reactions, carbon-bound hydrogen may
exchange between the substrate and a cofactor, an enzyme, the solvent, or a second substrate molecule. When the hydrogen is transferred to a cofactor, it may either be retained, retransferred to the
same or another position on the original substrate molecule, or transferred to another molecule. Usually, transfer to a cofactor, often a
"pyridine nucleotide," occurs as "hydride," and transfer to an enzyme

or solvent is as a proton, but there may be exceptions. In either case, the hydrogen movement is usually absolutely stereospecific, and because the enzyme surface is asymmetric, the hydrogen atoms of a methylene group are differentiated (see p. 141).

Comprehensive reviews dealing with several enzymes acting on sugars, notably aldolases,[132,133] isomerases,[132] epimerases,[132] and ribonucleotide reductase,[134] include information derived from the use of hydrogen isotopes. Reviews on molecular asymmetry in biology,[135] the enzymology of proton abstraction and transfer reactions,[136] and kinetic isotope-effects in enzymic reactions[137] supplement the material discussed in this Section, which is concerned with the ways in which sugars labeled with isotopes of hydrogen have been exploited to give information on mechanisms of enzyme action.

Enzymes catalyzing the same reaction but obtained from different sources may have very different properties, but usually operate by the same mechanism, an exception being D-fructose 1,6-diphosphate lyase (aldolase). Caution must, therefore, be observed in correlating mechanistic evidence derived for enzymes from different sources.

1. Solvent Exchange

a. Hydride and Proton Exchange Related to Enediol or Oxido-reduction Mechanisms.—For most isomerases and epimerases, either oxidation followed by reduction (A), or a keto–enol isomerization (B),

(132) I. A. Rose, *Ann. Rev. Biochem.*, **35**, 23 (1966).
(133) D. E. Morse and B. L. Horecker, *Advan. Enzymol.*, **31**, 125 (1968).
(134) H. P. C. Hogenkamp, *Ann. Rev. Biochem.*, **37**, 225 (1968).
(135) R. Bentley, "Molecular Asymmetry in Biology," Academic Press, New York N. Y., Vol. II, 1970.
(136) I. A. Rose, in "The Enzymes," P. D. Boyer, ed., Academic Press, New York, N. Y., 3rd Edition, 1970, Vol. II, p. 281.
(137) J. H. Richards, in "The Enzymes," P. D. Boyer, ed., Academic Press, New York, N. Y., 3rd Edition, 1970, Vol. II, p. 321.

is[138] the most favored, proposed mechanism, and both mechanisms have been found to operate. For some enzymes, a combination of both of these mechanisms occurs. Other mechanisms that have been proposed[138] include dehydration and rehydration, and carbon–carbon bond-cleavage followed by rearrangement. Generally, hydride transfer might be expected to occur without solvent exchange, whereas proton movement might be expected to be accompanied by solvent exchange.

Enzymes thus far found to catalyze isomerization or epimerization by oxidation and reduction require the cofactor NAD, which is often very tightly bound, not being removed by dialysis, but only by treatment with charcoal. Enzymes for which a keto–enol mechanism has been suggested do not usually involve NAD. The two mechanisms should also be distinguishable by the nonoccurrence or occurrence, respectively, of hydrogen exchange with the solvent.

When D-glucose 6-phosphate and D-fructose 6-phosphate are interconverted by D-glucose 6-phosphate ketol isomerase in either deuterium oxide[33] or water-t (Ref. 34), isotope is incorporated at C-1 of D-fructose 6-phosphate and C-2 of D-glucose 6-phosphate, indicating that the interconversion involves an enediol intermediate, which may arise from an open-chain (**37**) or cyclic (**38**) form of the sugar,

(138) L. F. Leloir, *Advan. Enzymol.*, **14**, 193 (1953).

although, for the yeast enzyme, it seems probable that it is the former.[139] In water-t, as the enzyme distinguishes between the two hydrogen atoms of the hydroxymethyl group, this leads eventually to an equilibrium mixture[34,35] of D-glucose-2-t 6-phosphate and D-fructose-$1(R)$-t 6-phosphate (40). However, when the D-fructose-$1(R)$-t [or $1(R)$-d] 6-phosphate was reconverted[34] into D-glucose 6-phosphate, intramolecular transfer of isotope between C-1 and C-2 was observed, and this transfer was much faster than the exchange reaction. The relative rates were affected by temperature, and, at 60°, 75% exchange was observed, whereas, at 0°, there was 80% retention of isotope. Thus, although the isomerization occurred through an enediol intermediate (39), the proton must have been held by a group (X), on the enzyme, that could exchange with solvent to a limited extent relative to the rate of retransfer, or the proton is supplied partially by this group and partially by the solvent.

Other 1,2-isomerases combine intramolecular transfer and solvent exchange to various extents,[140,141] but, with D-xylose ketol isomerase, solvent exchange was not found,[140] although there is no evidence that the mechanism operates by other than an enediol intermediate. Lack of solvent exchange alone, therefore, cannot be used to exclude the possibility of an enediol intermediate in an enzyme reaction.

When, in a sugar, a hydroxyl group adjacent to a carbonyl group is inverted, as by D-*erythro*-pentulose 5-phosphate 3-epimerase,[46] complete exchange with solvent of the hydrogen on the epimerized carbon atom is usually observed, indicating an enediol mechanism.

$$\begin{array}{c} CH_2OH \\ | \\ -X^{\ominus} \quad H-\overset{|}{\underset{|}{C}}=O \quad HY- \\ H-\overset{|}{C}OH \\ HCOH \\ | \\ CH_2OPO_3H_2 \end{array} \xrightarrow{\begin{array}{c} -XH \\ T_2O \\ -XT \end{array}} \begin{array}{c} CH_2OH \\ | \\ C-O-H^t \quad Y- \\ \overset{|}{C}OH \\ HCOH \\ | \\ CH_2OPO_3H_2 \end{array} \longrightarrow \begin{array}{c} CH_2OH \\ | \\ C=O \quad HY- \\ -X^{\ominus} \quad HO\overset{|}{C}T \\ HCOH \\ | \\ CH_2OPO_3H_2 \end{array}$$

Other epimerases in this class are cellobiose 2-epimerase[141a] and, probably, UDP-2-acetamido-2-deoxy-D-glucose 2-epimerase (which gives[141b] 2-acetamido-2-deoxy-D-mannose and UDP), because both enzymes incorporate tritium from water-t onto C-2 of the products.

In contrast, the UDP-D-glucose 4-epimerases from various sources

(139) M. Salas, E. Vinuela, and A. Sols, *J. Biol. Chem.*, **240**, 561 (1965).
(140) I. A. Rose, E. L. O'Connell, and R. P. Mortlock, *Biochim. Biophys. Acta*, **178**, 376 (1969).
(141) H. Simon, R. Medina, and G. Mullhofer, *Z. Naturforsch.*, B, **23**, 59 (1968).
(141a) T. R. Tyler and J. M. Leatherwood, *Arch. Biochem. Biophys.*, **119**, 363 (1967).
(141b) W. L. Salo and H. G. Fletcher, Jr., *Biochemistry*, **9**, 882 (1970).

require NAD as an obligatory cofactor[142,143] and do not exchange the hydrogen atom at C-4, or any other carbon atom, with the medium; this implies movement of hydrogen as hydride. The mechanism suggested involved[144] the 4-ulose **41**, but, because the enzyme forms a

tightly bound, ternary complex with NAD and the oxidized substrate, this hypothesis is difficult to prove. The presence of enzyme-bound NADH during the epimerization involving *Escherichia coli* enzyme was demonstrated by Wilson and Hogness,[145] but some doubt existed about the identity of the oxidized intermediate.

The stereochemistry of the reaction in which C-4 is oxidized to give intermediate **41** is difficult to envisage, because the hydride abstraction and addition would have to take place from opposite sides of the carbonyl group. Various solutions to this problem have included (*a*) a double binding-site for the substrate, which can transfer from one site to the other as the intermediate 4-ulose,[146] (*b*) use of alternate faces of the NADH molecule (see p. 169),[95] and (*c*) intermolecular transfer between two substrate molecules. The last two suggestions have been disproved by the observation[95] that

(142) E. S. Maxwell, *J. Biol. Chem.*, **229**, 139 (1957).
(143) E. S. Maxwell and H. de Robichon-Szulmajster, *J. Biol. Chem.*, **235**, 308 (1960).
(144) E. S. Maxwell, *J. Amer. Chem. Soc.*, **78**, 1074 (1956).
(145) D. B. Wilson and D. S. Hogness, *J. Biol. Chem.*, **239**, 2469 (1964).
(146) D. C. Wilton, personal communication.

no deuterium is transferred from perdeuterated to non-deuterated substrate in a mixture of reactants (see p. 171). Should alternate sides of the NADH molecule be used, some mixing would occur, as it would were intermolecular transfer between two substrate molecules to occur. It therefore appears that the hydride is abstracted and added, not only at the same position, but at the same position of the same molecule. Davis and Glaser[20] suggested that C-3 is oxidized. If this is so, this 4-epimerase would resemble the D-*erythro*-pentulose 5-phosphate 3-epimerase in using a keto–enol intermediate (**42**), but with proton transfer at C-4 that is completely intramolecular. This theory does not solve the stereochemical problem completely, as it would require movement of the proton to the opposite side of the enediol intermediate, even though movement of hydride is minimized, and subsequent evidence (see p. 168) strongly favors the theory that oxidation occurs at C-4.

UDP-D-glucuronate 5-epimerase also requires NAD,[147] and, for this enzyme, a similar oxidation at the adjacent carbon atom has been proposed,[148] followed by enolization to give the intermediate, which is then reprotonated to give UDP-L-iduronic acid (**43**). However,

occurrence of tritium incorporation could not be demonstrated.[148]

Residues of alginate [β-D-(1→4)-mannuronan or poly-(D-mannuronic acid] are converted into L-guluronic acid residues by D-mannuronan 5-epimerase, with incorporation of tritium from water-*t*. For this epimerization, in which C-4 cannot be oxidized, an enolization of the carbonyl group of the carboxylic acid has been proposed.[148a]

(147) B. Jacobson and E. A. Davidson, *Biochim. Biophys. Acta*, **73**, 145 (1963).
(148) B. Jacobson and E. A. Davidson, *J. Biol. Chem.*, **237**, 638 (1962).

A further example of an epimerase for which a mechanism cannot yet be given is L-*erythro*-pentulose 5-phosphate 4-epimerase, which does not require or contain[149] NAD, and which does not exchange hydrogen or ^{18}O with solvent.[46] The group epimerized is not adjacent to the carbonyl group. The most probable candidates for the mechanism are (*a*) an enediol isomerization involving movement of the double bond from C-2 to C-3, (*b*) hydride transfer to a group on the enzyme, and (*c*) an isomerization involving an aldol cleavage between C-3 and C-4, as shown, giving intermediate **44**.

$$\begin{array}{c}CH_2OH\\|\\C=O\\|\\HOCH\\|\\HOCH\\|\\CH_2OPO_3H_2\end{array} \rightleftarrows \begin{array}{c}CH_2OH\\|\\C=O\\|\\HOCH\\\ominus\\HC=O\\|\\CH_2OPO_3H_2\end{array} \rightleftarrows \begin{array}{c}CH_2OH\\|\\C=O\\|\\HOCH\\|\\HCOH\\|\\CH_2OPO_3H_2\end{array}$$

44

Incorporation of solvent hydrogen into the substrate does not rule out its transfer as hydride if the source of hydride is dithiol, as in the ribonucleotide reductases. In the most understood enzyme of this type, from *Lactobacillus leichmannii*, the function of the dithiol (**45**) is to reduce the cobamide coenzyme (**46**) and provide a source

$$HS \quad SH \xrightarrow{D_2O} D-S \quad S-D \longrightarrow D^\oplus \quad S-S \quad D^\ominus$$

45

of hydride. Because the dithiol freely exchanges with solvent, solvent hydrogen is incorporated into the product.[150–152] The incorporated hydrogen is found only[151] at C-2, and ^{18}O is eliminated from C-2, but[153] not from C-3. When the reaction was conducted in deuterium oxide, either with this enzyme or with the related, iron-containing *E. coli* enzyme, the deuterium atom appeared to be incorporated with overall retention of configuration.[91,92] The mechanism of action of the cobamide-containing enzyme, therefore, appears to involve displacement of the 2'-hydroxyl group by the reduced cofactor (**47**), leading to a cobalt–sugar bond at C-2' (Ref. 154). This bond may be reduced

(148a) B. Larsen and A. Haug, *Carbohyd. Res.*, **20**, 225 (1971).
(149) J. D. Deupree and W. A. Wood, *J. Biol. Chem.*, **245**, 3988 (1970).
(150) A. Larson, *Biochemistry*, 4, 1984 (1965).
(151) M. M. Gottesman and W. S. Beck, *Biochem. Biophys. Res. Commun.*, **24**, 353 (1966).
(152) R. L. Blakely, R. K. Ghambeer, T. J. Batterham, and C. Brownson, *Biochem. Biophys. Res. Commun.*, **24**, 418 (1966).
(153) H. Follmann and H. P. C. Hogenkamp, *Biochemistry*, **8**, 4372 (1969).
(154) M. Akhtar, *Comp. Biochem. Physiol.*, **28**, 1 (1969).

by the 5′-methyl group of the cofactor, which becomes labeled during the reaction,[155] or, alternatively, by hydride from the dithiol, as shown. The latter mechanism would explain the failure[155] of cofactor labeled at C-5′ with tritium to transfer tritium to the substrate, as it does for other cobamide enzymes, and would better account for the stereochemistry observed. Because the enzyme catalyzes a rapid exchange of 5′-tritium to water, the failure to observe transfer to substrate could also be accounted for by rapid exchange preceding reduction by the methyl group of the reduced cobamide.

b. Localization of Proton Movement.—Recognition of proton movement, and determination of the position and stereochemistry of the incorporated atom, can be significant evidence in selecting a particular reaction-mechanism. The need for caution in interpreting lack of solvent-hydrogen exchange, because intramolecular proton transfer can occur, has been illustrated in the previous Section (see p. 158). Incorporation of solvent hydrogen into substrate molecules or product molecules must, on the other hand, be accounted for in any proposed mechanism.

The mechanism of action of 3-deoxy-D-*arabino*-heptulosonate

(155) R. H. Abeles and W. S. Beck, *J. Biol. Chem.*, **242**, 3589 (1967).

7-phosphate synthetase had been tentatively regarded[156] as involving a nucleophilic attack on enolpyruvate phosphate by hydroxide ion, promoting attack on D-erythrose 4-phosphate (48) (Mechanism C).

Mechanism C

$$\begin{array}{c} CO_2H \\ | \\ C-O-P-OH \\ \| \quad | \quad {}^{18}\overset{\ominus}{O}H \\ CH_2 \quad OH \\ H-C{=}O^+ \quad H^{\oplus} \\ HCOH \\ HCOH \\ CH_2OPO_3H_2 \\ \mathbf{48} \end{array} \xrightarrow{\quad\times\quad} \begin{array}{c} CO_2H \\ | \\ C{=}O \\ | \\ CH_2 \\ HOCH \\ HCOH \\ HCOH \\ CH_2OPO_3H_2 \end{array} + \begin{array}{c} O \\ \| \\ HO-P-OH \\ | \\ {}^{18}OH \end{array}$$

Subsequently, it was found[157] that ^{18}O was not incorporated into inorganic phosphate from water-^{18}O, and that tritium from water-t was incorporated into the product. Neither of these observations could be explained by the mechanism just outlined, and the more elaborate mechanism requiring initial formation of an enzyme–pyruvyl complex (Mechanism D) was proposed.[157]

Subsequently some doubt was cast on Mechanism D, because it has been shown[157a] that the condensation occurs with partial retention of the geometric asymmetry of O-phosphono-enolpyruvic-3(Z)-t and -3(E)-t acids (48a), so that, from the former, 3-deoxy-D-*arabino*-heptulosonate-3(S)-t 7-phosphate was the preponderant product. This

Mechanism D

$$\text{Enzyme-}\overset{O}{\overset{\|}{C}}-O^{\ominus}\overset{CO_2H}{\overset{|}{\underset{\substack{C \\ T^{\oplus} \diagup H_Z \diagdown H_E}}{C}}}-O-\overset{O}{\overset{\|}{P}}-OH \longrightarrow \text{Enzyme-}\overset{O}{\overset{\|}{C}}-O-\overset{CO_2H}{\overset{|}{\underset{CH_2T}{C}}}-O-\overset{O}{\overset{\|}{P}}-OH$$

48a   48b

$$\text{Enzyme-}\overset{O}{\overset{\|}{C}}-O-\overset{CO_2H}{\overset{|}{\underset{CHT}{C}}} + HO-\overset{O}{\overset{\|}{P}}-OH \\ \qquad\qquad\qquad OH$$

$$\text{Enzyme-}\overset{O}{\overset{\|}{C}}-O-\overset{CO_2H}{\overset{|}{C}} \\ {}^{18}\overset{\ominus}{O}H \quad \diagdown \overset{\|}{C}HT \longrightarrow \text{Enzyme-}\overset{O}{\overset{\|}{C}}-{}^{18}O^{\ominus} + \begin{array}{c} CO_2H \\ | \\ C{=}O \\ | \\ CHT \\ HOCH \\ | \\ R \end{array} \\ H^{\oplus} \quad O{=}\overset{}{C}-H \\ \qquad\quad R$$

(156) P. R. Srinivasan and D. B. Sprinson, *J. Biol. Chem.*, **234**, 716 (1959).
(157) A. B. DeLeo and D. B. Sprinson, *Biochem. Biophys. Res. Commun.*, **32**, 873 (1968).
(157a) H. G. Floss, D. K. Onderka, and M. Carrol, *J. Biol. Chem.*, **247**, 736 (1972).

observation, together with the discovery[157a] that the tritium incorporated is only about one third of that predicted by Mechanism D, can only be explained in terms of Mechanism D if it is assumed that free rotation of the methyl group in **48b** is restricted by the enzyme during this part of the reaction. Mechanism E would account for the observed stereochemistry of the reaction and for the lack of ^{18}O in the inorganic phosphate recovered, but it does not account for the partial incorporation of tritium, nor for the observed kinetics.[157b]

Mechanism E

$$\begin{array}{c} CO_2H \\ | \\ C-\ddot{O}-P-OH \\ \| \quad | \\ CH_2 \quad OH \\ | \\ H-C=O \quad H^{\oplus} \\ | \\ R \end{array} \longrightarrow \begin{array}{c} ^{18}OH^{\ominus} \quad CO_2H \quad O \\ \searrow \quad | \quad \| \\ C=O-P-OH \\ | \quad \oplus \\ CH_2 \quad OH \\ | \\ HOCH \\ | \\ R \end{array} \longrightarrow \begin{array}{c} CO_2H \quad O \\ | \quad \| \\ H^{18}O-C-O-P-OH \\ | \\ CH_2 \quad OH \\ | \\ HOCH \\ | \\ R \end{array} \longrightarrow \begin{array}{c} O \\ \| \\ HO-P-OH \\ | \\ OH \\ + \\ CO_2H \\ | \\ C=^{18}O \\ | \\ CH_2 \\ | \\ HOCH \\ | \\ R \end{array}$$

When solvent hydrogen is incorporated into substrates or products at a methylene group, the incorporation is usually stereospecific, leading to R or S products (see p. 141). The stereochemistry of the hydrogen introduced is determined by the stereochemistry of the enzyme-bound intermediates. Interconversion of an aldehyde or a ketone by a 1,2-isomerase could have either a *cis*- or *trans*-enediol intermediate. The discovery[34] that the hydrogen atom can be transferred from C-1 to C-2 either partially or totally without exchange suggests that the proton is abstracted and added to the same side of the enediol (**39**, see p. 157). If this hypothesis is correct, the observed stereochemistry of seven isomerases requires that the enediol must be *cis*.[136,140]

Similar considerations suggest that only one basic group is involved in aldolase catalysis.[136] When 1,3-dihydroxy-2-propanone 1-phosphate is incubated with aldolase, a stereospecific exchange with solvent occurs (see p. 135). This exchange shows that carbanion formation precedes condensation with glyceraldehyde 3-phosphate (see also, Ref. 158), and it is found that the new carbon–carbon bond is formed with the same stereochemistry as the departed hydrogen.[40] This situation could be accounted for by the use of the same basic group to abstract the hydrogen atom and then donate it to the oxygen atom, but this theory cannot be experimentally tested, because the oxygen–hydrogen bond is not stable.

In contrast to the aldolases, 6-O-phosphono-D-gluconate dehydrogenase[45] (E.C. 1.1.1.44) and UDP-D-glucuronate decarboxylase[76] (E.C. 4.1.1.b) catalyze decarboxylation with inversion of configuration, giving as products, in water-*t*, D-*erythro*-pentulose-*1*(*S*)-*t* 5-phosphate and UDP-D-xylose-5(*S*)-*t*, respectively.

The idea that enzymically exchanged solvent-protons will always be incorporated stereospecifically had been so widely accepted that, when incorporation of tritium was found to be non-specific, this behavior was cited as evidence for a non-enzymic step.[159] 6-O-Phosphono-D-gluconate hydrolyase (E.C. 4.2.1.12) converts 6-O-phosphono-D-gluconic acid (**49**) into 2-deoxy-6-O-phosphono-D-*erythro*-3-hexulosonate (**50**). When the reaction was conducted in deuterium

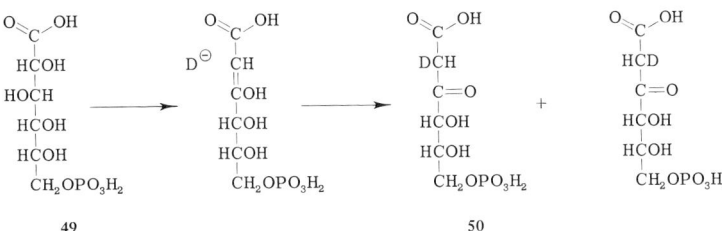

oxide, deuterium was incorporated at C-2, consistent with the dehydration to enol. P.m.r. analysis showed that the atom incorporated was randomly oriented, and it was proposed[159] that the enol was the product of enzyme action and that it spontaneously changed to the ketone, incorporating the solvent proton.

A similar non-specific incorporation might also be expected for the elimination of water from 3-deoxy-L-*glycero*-pentulosonate (**51**) by 3-deoxy-L-*erythro*-pentulosonate hydrolyase (E.C. 4.2.1.h), leading to 2-oxoglutaraldehydic acid (**52**). Reaction in deuterium oxide gave

(157b) M. Staub and G. Dénes, *Biochim. Biophys. Acta*, **148**, 563 (1967).
(158) I. A. Rose, E. L. O'Connell, and A. H. Mehler, *J. Biol. Chem.*, **240**, 1758 (1965).
(159) H. B. Meloche and W. A. Wood, *J. Biol. Chem.*, **239**, 3505 (1964).

incorporation of deuterium at C-3 and C-4 of the product, but both of these were stereospecifically introduced into the pro-R position.[160] In this instance, the reaction must, therefore, be completely enzyme-mediated.

2. Hydrogen Movements Mediated by Cofactor

a. **Intramolecular Transfer.**—Intramolecular movement of hydrogen may occur by transfer as hydride to NAD, as well as by proton movement (see p. 158). In such cases, the transfer is absolute, and there is no exchange with solvent. The most studied enzyme of this type is dTDP-D-glucose oxidoreductase (E.C. 4.2.1.43), which converts dTDP-D-glucose (**53**) into dTDP-6-deoxy-D-*xylo*-hexos-4-ulose (**54**).

The conversion is accomplished with complete retention of the tritium or deuterium originally situated at C-4 of the D-glucose molecule, and analysis of the product showed that tritium was situated only at C-6 of the product.[96,161,162] The hydrogen atom on C-5 exchanges with the solvent,[163] which is consistent with the mechanism proposed.

A similar situation occurs in the formation of UDP-D-apiose (**57**) from UDP-D-glucuronic acid (**55**). This interesting rearrangement requires NAD, and is accompanied by a decarboxylation. The NAD might be expected to oxidize C-4, to facilitate the decarboxylation by formation of a β-keto acid, and give UDP-L-*threo*-pentos-4-ulose (**56**). A 1,3-shift between C-2 and C-4 would give an intermediate that could be reduced by the NADH first formed, to give UDP-D-apiose (**57**). This mechanism is supported by the carbon-atom movement

(160) D. Portsmouth, A. C. Stoolmiller, and R. H. Abeles, *J. Biol. Chem.*, **242**, 2751 (1967).
(161) R. D. Bevill, *Biochem. Biophys. Res. Commun.*, **30**, 595 (1968).
(162) O. Gabriel and L. C. Lindquist, *J. Biol. Chem.*, **243**, 1479 (1968).
(163) K. Hermann and J. Lehmann, *Eur. J. Biochem.*, **3**, 369 (1968).

observed[164] and the predicted transfer of hydrogen from C-4 of UDP-D-glucuronic acid to C-3′ of UDP-D-apiose.[165]

When the hydrogen transferred as hydride to the cofactor is re-transferred to the same carbon atom in the product, the movement is far more difficult to detect. The conversion of D-glucose 6-phosphate (**58**) into 1L-*myo*-inositol 1-phosphate (**61**) occurs by cyclization of the carbon skeleton, with formation of a new bond between C-1 and C-6. When each carbon atom in turn was specifically labeled with tritium, there was complete retention of tritium, even in the presence of added NADH, although there was an apparent, small isotope-effect with D-glucose-5-*t* 6-phosphate.[19] The mechanism proposed for the cyclization[19] was an initial oxidation at C-5 to give NADH and *xylo*-hexos-5-ulose 6-phosphate (**59**), followed by an aldol reaction causing cyclization to 1L-*myo*-inosose-2 1-phosphate (**60**), which is then

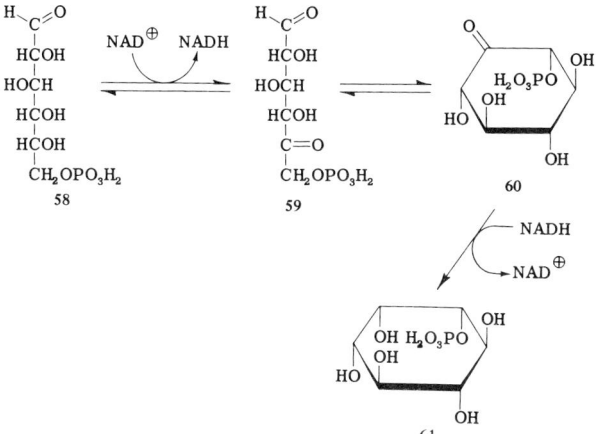

(164) H. Grisebach and U. Doebereiner, *Z. Naturforsch., B*, **21**, 429 (1966).
(165) W. Kelleher and H. Grisebach, *Eur. J. Biochem.*, **23**, 136 (1971).

reduced by the NADH. Substrate, NADH, and enzyme appear to form the familiar, tightly bound, ternary complex that does not exchange intermediates or NADH.

Interruption of such a reaction at the intermediate stage should provide more-positive evidence for the mechanism proposed. Treatment of inositol cyclase with charcoal removed the NAD, and, when the enzyme was reconstituted, its properties were found to have been modified. By using this preparation, hydrogen from NADH-t was transferred to C-5 of both *xylo*-hexos-5-ulose 6-phosphate (59) and L-sorbose 1-phosphate, giving a mixture of D-glucose-5-t 6-phosphate (58) with 1L-*myo*-inositol 1-phosphate (61) or D-glucitol-5-t 6-phosphate, respectively.[166] This shows that the enzyme can catalyze the partial reaction corresponding to reduction of the oxidized intermediate proposed.

Similar experiments have been used to distinguish between the two hexulose derivatives postulated as intermediates in the UDP-D-glucose 4-epimerase reaction, oxidized at C-4 (41) and C-3, respectively (see p. 160). Maitra and Ankel[167] allowed the enzyme from *Escherichia coli* to form the ternary NADH–oxidized sugar–enzyme complex, and then reduced it with sodium borohydride-t. The only products detected were UDP-D-glucose-t and UDP-D-galactose-t (not UDP-D-allose), and the tritium was situated at C-4, not C-3, of the D-glucose residue. It has also been shown[168] that sodium borohydride-t will reduce NAD on the enzyme in the presence of substrate, to give an inactive enzyme containing NADH-t. Activity was restored by incubation with UDP-6-deoxy-D-*xylo*-hexos-4-ulose. Alternatively, the inactive enzyme–NADH-t complex could be formed[168] by incubation of a mixture of UDP-6-deoxy-D-glucose-4-t and UDP-6-deoxy-D-galactose-4-t with the enzyme for 24 hours. These experiments appear to show that the 4-ulose 41, rather than the 3-ulose, is the oxidized sugar intermediate (see p. 159).

A similar method of investigation, particularly applicable when the enzyme can be obtained pure and in sufficient quantity, is to use a substrate analog that can complete only part of the reaction. This method was used[169] for confirming the dTDP-D-glucose oxidoreductase mechanism already discussed. In this system, incubation of the enzyme with dTDP-6-deoxy-D-glucose-4-t (62) (which, after oxidation, cannot re-accept the hydrogen atom on C-6) led to inactive enzyme containing NADH-t. The latter could be isolated

(166) A. Rasheed, D. L. Corina, and J. E. G. Barnett, *Biochem. J.*, **126**, 15p (1972).
(167) U. S. Maitra and H. Ankel, *Proc. Nat. Acad. Sci. U. S.*, **68**, 2660 (1971); but, see also, T. G. Wee, J. Davis, and P. A. Frey, *J. Biol. Chem.*, **247**, 1339 (1972).
(168) G. L. Nelsestuen and S. Kirkwood, *J. Biol. Chem.*, **246**, 7533 (1971).
(169) S. F. Wang and O. Gabriel, *J. Biol. Chem.*, **245**, 8 (1970).

from the impure enzyme, or, alternatively, active enzyme could be regenerated by adding dTDP-6-deoxy-D-*xylo*-hexos-4-ulose (**63**), which became labeled.

A model reaction in which UDP-D-glucose 4-epimerase was shown[20,169a] to form an inactive NADH–enzyme complex on incubation with either D-glucose or D-galactose and UMP has been shown[169b] to be unrelated to the epimerization reaction. The label from D-galactose-*1*-*t* was transferred to NAD, and D-galactonic acid was isolated. Oxidation therefore occurred at C-1, which is normally involved in combination with UDP.

b. Transfer to Cofactor.—Only four cofactors are usually involved in hydrogen transfer, namely, nicotinamide adenine dinucleotide (NAD), nicotinamide adenine dinucleotide phosphate (NADP), flavin adenine dinucleotide (FAD), and flavin mononucleotide (FMN). The active parts of the molecules are, respectively, nicotinamide (**64**) and flavin. The nicotinamide coenzymes catalyze only two-

(169a) H. M. Kalckar, A. U. Bertland, and B. Bugge, *Proc. Nat. Acad. Sci. U. S.*, **65**, 1113 (1970).
(169b) Y. Seyama and H. M. Kalckar, *Biochemistry*, **11**, 36 (1972).

electron transfer (hydride), whereas the flavin-containing coenzymes can catalyze one-electron transfer, and, in particular, oxidations involving molecular oxygen. FAD is involved in the oxidation of D-glucose by D-glucose oxidase (discussed in the Section on isotope effects; see p. 171).

Transfer of hydrogen to NAD or NADP creates a new asymmetric center at C-4 of the pyridine ring. The hydrogen atoms appear on both sides of the plane of the ring; the sides have been designated[170] the A and B sides of the molecule, and the hydrogen atoms as H_A and H_B. The absolute stereochemistry has been elucidated,[171,172] and H_A has been found to correspond to the pro-R hydrogen atom.

The stereochemistry of hydrogen transfer to NAD or NADP for most enzymes catalyzing the oxidation of sugars has been determined. D-Glycerol 1-phosphate dehydrogenase (E.C. 1.1.1.8),[172] triose phosphate dehydrogenase (E.C. 1.2.1.12),[173] D-glucose dehydrogenase (E.C. 1.1.1.47),[174] D-glucose 6-phosphate dehydrogenase (E.C. 1.1.1.49),[175] and 6-O-phosphono-D-gluconate dehydrogenase (decarboxylating) (E.C. 1.1.1.44)[175] have all been shown to transfer hydrogen to the B (pro-S) side of the NAD or NADP. Reduction[176] of D-*threo*-2,5-hexodiulose by an enzyme from *Gluconobacter cerinus* uses the A (pro R) hydrogen atom, as does glycerate dehydrogenase.[177]

The use of dTDP-6-deoxy-D-glucose-4-*t* as a substrate for dTDP-D-glucose oxidoreductase (see p. 169) has permitted the stereochemistry of this normally closed system to be determined.[169] The B side of NAD accepts the hydrogen atom. UDP-D-glucose 4-epimerase has also been shown[168] to use this side of the NAD molecule.

Such an enzyme might accept the hydrogen atom on one side of the NAD, and retransfer hydrogen to the enzyme-bound intermediate from the other face. Use of a substrate having a labeled hydrogen atom in the position to be transferred to NAD would then lead to labeling of the NAD (65), and, if a mixture of labeled and unlabeled molecules were used as substrate, intermolecular transfer of label would occur. The latter transfer may be detected by the use of per-

(170) J. W. Cornforth, G. Ryback, G. Popjack, C. Donninger, and G. L. Schoepfer, Jr., *Biochem. Biophys. Res. Commun.*, **9**, 371 (1962).
(171) J. W. Cornforth, R. M. Cornforth, C. Donninger, G. Popjack, G. Ryback, and G. L. Schoepfer, Jr., *Proc. Roy. Soc., Ser. B*, **163**, 436 (1966).
(172) H. R. Levy and B. Vennesland, *J. Biol. Chem.*, **228**, 85 (1957).
(173) F. A. Loewus, H. R. Levy, and B. Vennesland, *J. Biol. Chem.*, **223**, 589 (1959).
(174) H. R. Levy, F. A. Loewus, and B. Vennesland, *J. Biol. Chem.*, **222**, 685 (1956).
(175) B. K. Stern and B. Vennesland, *J. Biol. Chem.*, **235**, 205 (1960).
(176) S. Englard, G. Kaysen, and G. Avigad, *J. Biol. Chem.*, **245**, 1311 (1970).
(177) F. A. Loewus and H. A. Stafford, *J. Biol. Chem.*, **235**, 3317 (1960).

deuterated substrates: a mixture of perdeuterated and unlabeled substrates is incubated with the enzyme, and the products are analyzed for the presence of species containing one deuterium atom, or one hydrogen atom, by mass spectrometry, as shown for dTDP-D-glucose oxidoreductase.[96] All of the enzymes thus far tested, namely, dTDP-D-glucose oxidoreductase,[96] UDP-D-glucose 4-epimerase,[95] and inositol cyclase,[178] use only one face of NAD, and no mixing has been observed.

3. Isotope Effects

A review of kinetic isotope-effects in enzymic reactions discussed[137] their origin and gave some examples of isotope effects in sugars. The number of stages in an enzyme-catalyzed reaction conspires against the use of isotope effects for the determination of rate-limiting events in enzyme mechanisms. A typical reaction-sequence may be given for the oxidation of D-glucose by D-glucose oxidase.[179] The enzyme (E) contains FAD, which is oxidized by molecular oxygen. G represents D-glucose; Y, D-gluconolactone; and Z, hydrogen peroxide.

(178) W. R. Sherman, M. A. Stewart, and M. Zinbo, *J. Biol. Chem.*, **244**, 5703 (1969).
(179) H. J. Bright and Q. H. Gibson, *J. Biol. Chem.*, **242**, 994 (1967).

E.FAD(ox) + G $\xrightarrow{k_1}$ E.FAD(ox).G $\xrightarrow{k_2}$
E.FAD(red).Y $\xrightarrow{k_3}$ E.FAD(red) + Y
E.FAD(red) + O_2 $\xrightarrow{k_4}$ E.FAD(ox).Z $\xrightarrow{k_5}$ E.FAD(ox) + Z

In addition to the reaction sequence shown, the substrate may have to mutarotate non-enzymically to the form favored for affording the first enzyme–substrate complex. In such multi-stage reactions, the observed isotope-effects for the overall reaction are usually low. The situation is further complicated by possible differences in K_m between a deuterated substrate and the unlabeled substrate, so that, unless a full kinetic analysis of the reaction is performed, the values determined for the apparent isotope-effect may vary.

Direct determination of the kinetic isotope-effect of D-glucose 6-phosphate dehydrogenase (E.C. 1.1.1.49) acting on D-glucose-1-d 6-phosphate, measured by spectrophotometric monitoring of the formation of NADPH, produced an isotope effect of $k_D/k_H = 0.5$, showing that cleavage of the carbon–hydrogen bond is probably a rate-limiting step. However, it was dependent on the concentration of the substrate, and, at low concentrations of substrate, the relative rate of reaction of the deuterated sugar increased.[180]

A further pitfall in the quantitative determination of isotope effects was revealed in a comprehensive study of D-glucose oxidase (E.C. 1.1.3.4). With D-glucose-1-d, a highly variable, overall isotope-effect was observed which depended on the concentration of substrate and oxygen. In the absence of oxygen, the reaction proceeded as far as the formation of E.FAD(red) and D-gluconolactone. By using a stopped-flow system to isolate the stages in the reaction, it was shown that, when D-glucose-1-d was used, k_2 was decreased, giving an isotope effect of $k_D/k_H = 0.06$–0.1, suggesting that the rate-limiting step is the hydrogen transfer. However, the substitution of deuterium for hydrogen was considered to have resulted in a change of the rate-determining step from the dissociation of the enzyme–product complex to the transfer of hydrogen to the cofactor.[180]

Both a two-electron and a one-electron transfer mechanism have been proposed for the reaction,[179] the latter being modelled on the proposed mechanism for the oxidation of D-glucose by bromine (see p. 155).

Although the quantitative aspects of isotope effects are difficult to interpret, qualitative aspects have been of considerable use. Whereas the lack of an isotope effect in an overall enzyme-reaction cannot, in the absence of further kinetic analysis, be used as evidence for any particular mechanism, any isotope effect observed should be explicable by the proposed mechanism.

(180) I. A. Rose, *J. Biol. Chem.*, **236**, 603 (1961).

Small isotope-effects can be detected by double-labeling techniques, in which the carbon skeleton is labeled with ^{14}C, and the ratio of ^{14}C to tritium is measured both in the substrate and the product. Care must be taken in the observation and interpretation of isotope effects determined from the hydrogen-isotope content of the product. Just as in non-enzymic reactions (see p. 154), discrimination against the substrate containing deuterium or tritium leads to an increase in the isotopic content of the substrate, and this decreases the apparent isotope-effect towards the end of the reaction.

This behavior has been observed[181] in the oxidation of a mixture of D-glucose-1-t 6-phosphate and D-glucose-$^{14}C_6$ 6-phosphate by D-glucose 6-phosphate dehydrogenase, where the proportion of the tritiated species remaining in the unreacted product was observed to increase. When 50% of the uniformly ^{14}C-labeled substrate had been oxidized, only 10% of the tritiated substrate had been oxidized, but, when 60% of the former had been oxidized, 20% of the tritiated sugar had been used up.

An interesting consequence of the determination of the apparent isotope-effect by using a mixture of ^{14}C- and tritium-labeled substrates and measuring the labeling in the product is that some isotope-effects which, at first sight, might appear to be detectable, cannot be observed. When randomly labeled D-glucose-6-t 6-phosphate was converted into 1L-*myo*-inositol 1-phosphate (**61**, see p. 167), it was observed[182] that the rate at which tritium was eliminated into water was only half that at which it appeared in the product, showing that there was an isotope effect in the removal of a proton from C-6. Conversely, the T/^{14}C ratio of the product was found[19] to

(181) J. C. Bartley and S. Abraham, *Biochim. Biophys. Acta*, **148**, 563 (1967).
(182) I. W. Chen and F. C. Charalampous, *Biochim. Biophys. Acta*, **136**, 568 (1967).

be 0.5, corresponding to the obligatory loss of one of the hydrogen atoms on C-6, without an isotope effect. This reaction occurs because there are four species in solution, namely, unlabeled D-glucose 6-phosphate, D-glucose-^{14}C 6-phosphate, D-glucose-6(R)-t 6-phosphate and D-glucose-6(S)-t 6-phosphate, and the only one converted at a lower rate is that from which the tritium is eliminated. Therefore, the ratio of the two substrate species giving rise to *labeled* products is unchanged.

Observation of isotope effects is particularly useful with those enzymes that transfer hydrogen to a cofactor and retransfer it back to the same carbon atom. Thus, in the conversion of a mixture of D-glucose-5-t 6-phosphate and D-glucose-^{14}C 6-phosphate into 1L-*myo*-inositol 1-phosphate by inositol cyclase, a tritium/^{14}C ratio of 0.85 was observed[19] in the product; after treatment of the enzyme with charcoal this ratio decreased to 0.65, although there was no exchange with NADH added to the medium. This behavior was interpreted[19] as being caused by an isotope effect that was absent with substrates specifically labeled in the other positions; this hypothesis suggested that hydrogen movement at C-5 was involved in the mechanism (see p. 167).

Retention of tritium, together with an isotope effect (k_T/k_H, 0.32 and 0.42) was also observed[76] when UDP-D-glucuronic-4-t acid (**66**), but

not the corresponding substrates labeled at C-3 and C-5, was used as a substrate for UDP-D-glucuronate decarboxylase from *Cryptococcus laurentii* and wheat germ, respectively. This behavior is consistent with the mechanism proposed, namely, oxidation at C-4 followed by β-elimination of carbon dioxide.

A similar isotope-effect was observed in the conversion of UDP-D-glucose-*4-t* into UDP-D-galactose-*4-t* by UDP-D-glucose 4-epimerase. No exchange of tritium was observed,[69,183] and there was[184] an isotope effect of 0.54. This result was interpreted as indicating a rate-limiting transfer of hydride with UDP-D-*xylo*-hexos-4-ulose as the intermediate. Other sugar enzymes for which isotope effects have been observed are dTDP-D-glucose oxidoreductase,[96] D-glucose 6-phosphate aldo–keto isomerase,[34] D-ribonucleotide reductase,[150] and D-fructose 1,6-diphosphate–glyceraldehyde 3-phosphate lyase.[185]

A rather elegant application of isotope effects was made in distinguishing between the origins of the hydrogen atoms and carbon atoms of D-glyceric acid 3-phosphate formed[186] by D-*erythro*-pentulose 1,5-diphosphate carboxylase (E.C. 4.1.1.f). This enzyme catalyzes the addition of carbon dioxide to D-*erythro*-pentulose 1,5-diphosphate, with simultaneous breakage of a carbon–carbon bond, to give two molecules of glyceric acid 3-phosphate. Conducting the reaction in deuterium oxide results in the formation of glyceric-*2-d* acid 3-phosphate, but this does not permit any conclusions to be made about the site of cleavage of the D-*erythro*-pentulose molecule. If the D-*erythro*-pentulose diphosphate is labeled at C-2 with ^{14}C, the reaction in deuterium oxide results in the formation of four possible species of glyceric-*2-d* acid 3-phosphate (**67, 68, 69, 70**) originating from either end of the D-*erythro*-pentulose 1,5-diphosphate molecule.

$$\begin{array}{cccc}
CO_2H & CO_2H & CO_2H & CO_2H \\
H-{}^{14}C-OH & D-{}^{14}COH & D-{}^{12}COH & H-{}^{12}COH \\
CH_2OPO_3H_2 & CH_2OPO_3H_2 & CH_2OPO_3H_2 & CH_2OPO_3H_2 \\
\mathbf{67} & \mathbf{68} & \mathbf{69} & \mathbf{70}
\end{array}$$

Incubation of the dephosphorylated mixture with glycerate dehydrogenase (E.C. 1.1.1.29), which discriminates against the conversion of glyceric-*2-d* acid into hydroxypyruvic acid, resulted in enrichment with glyceric-*2-d* acid. Enrichment of the deuterated species was paralleled by enrichment of the ^{14}C-labeled material, indicating that they were present in the same molecule (**68**) and that **68** arose from the original D-*erythro*-pentulose-2-^{14}C by cleavage of the C-2–C-3 bond. Substrate tritiated at C-3 also exhibits an isotope effect,[187] measured by the release of tritium into the

(183) B. D. Kohn, R. C. Hoffmann, and P. Kohn, *J. Biol. Chem.*, **238**, 1193 (1963).
(184) G. L. Nelsestuen and S. Kirkwood, *Biochim. Biophys. Acta*, **220**, 633 (1970).
(185) S. V. Rieder and I. A. Rose, *J. Biol. Chem.*, **234**, 1007 (1959).
(186) G. Mullhofer and I. A. Rose, *J. Biol. Chem.*, **240**, 1341 (1965).
(187) F. Fiedler, G. Mullhofer, A. Trebst, and I. A. Rose, *Eur. J. Biochem.*, **1**, 395 (1967).

medium, indicating that there is an initial, rate-limiting, enolization step to **71**, followed by a fast, irreversible carboxylation. As, in the

$$\begin{array}{c} CH_2OPO_3H_2 \\ {}^{14}C=O \\ H-COH \\ HCOH \\ CH_2OPO_3H_2 \end{array} \xrightarrow{H^\oplus, H^\oplus}_{X^\ominus} \begin{array}{c} CH_2OPO_3H_2 \\ {}^{14}C-OH \\ C-OH \\ HCOH \\ CH_2OPO_3H_2 \end{array} \xrightarrow{HO} \begin{array}{c} CH_2OPO_3H_2 \\ C-{}^{14}COH \\ HO^\ominus \\ C=O \\ HCOH \\ CH_2OPO_3H_2 \end{array}$$

71

$$\begin{array}{c} CH_2OPO_3H_2 \\ HO-{}^{14}C^\ominus \\ CO_2H \\ CO_2H \\ HCOH \\ CH_2OPO_3H_2 \end{array} \xrightarrow{D^\oplus} \begin{array}{c} CH_2OPO_3H_2 \\ HO-{}^{14}CD \\ CO_2H \\ CO_2H \\ HCOH \\ CH_2OPO_3H_2 \end{array}$$

absence of carbon dioxide, no exchange with the medium was found at C-3 of D-*erythro*-pentulose diphosphate or at C-2 of glyceric acid phosphate, the presence of carbon dioxide must be necessary for the labilization of H-3, which is immediately followed by carboxylation and hydrolysis.

Isotope effects have also been observed in whole cells. *Acetobacter suboxydans* selectively oxidizes polyhydric alcohols having a specific pattern of hydroxyl groups. It oxidizes D-mannitol at C-2 and C-5. When D-mannitol-*1-t*, D-mannitol-*2-t*, and D-mannitol-*3-t* are oxidized to D-fructose, the unlabeled half of the molecule is oxidized the more rapidly. This result was attributed[188] to a primary isotope-effect at C-3, and a secondary isotope-effect at C-2.

Small, secondary isotope-effects have also been noticed in the hydrolysis of substrates of phosphorylase[189] and lysozyme.[190]

VIII. GENERAL APPLICATIONS IN BIOCHEMISTRY

1. Overall Pathway of Enzymic Transformations

The relative ease of synthesis of tritiated sugars, as compared with that of ^{14}C-labeled sugars, has led to the increasing use of tritiated substrates in the determination of the molecular course of a particular biological transformation. These studies, although limited

(188) L. T. Sniegoski, H. L. Frush, and H. S. Isbell, *J. Res. Nat. Bur. Stand.*, A, **65**, 441 (1961).
(189) J. I. Tu, G. R. Jacobson, and D. J. Graves, *Biochemistry*, **10**, 1229 (1971).
(190) F. W. Dahlquist, T. Rand-Meir, and M. A. Raftery, *Proc. Nat. Acad. Sci. U. S.*, **61**, 1194 (1968).

by the lowered biological stability of a hydrogen isotope, compared with that of carbon, can form a useful guide for more-detailed investigations. Some examples from cyclitol metabolism will be given.

The degradation of *myo*-inositol in higher plants was investigated by the infusion of *myo*-inositol-2-*t* into the plant, leading to the accumulation of D-glucuronic-5-*t* acid and D-xylose-5-*t* as the preponderant products.[191] The position of tritium in the products[192] indicated that the cleavage of the inositol ring had occurred between C-1 and C-6, and, as *myo*-inositol-2-*t* was converted into xylose-5(*R*)-*t* by way of a D-glucuronic-5(*S*)-*t* acid intermediate, the D-glucuronate-D-xylose step was probably catalyzed by UDP-D-glucuronate carboxylase, which is known[73] to decarboxylate with inversion of configuration (see p. 165).

Inositol is formed from D-glucose 6-phosphate by direct cyclization of the chain. D-Glucose 6-phosphate was also shown to be a precursor of deoxyinositols in leaf tissues, as D-glucose-6-*t* 6-phosphate (**72**) was converted[193] into quercitol-2-*t* (L-2,3,5/1,4) (**73**) and[194] viburnitol-4-*t* (L-1,2,4/3,5) (**74**). Labeling in the precursor and the products corresponded to a direct cyclization-mechanism.

Loss of tritium from D-glucose-3-*t* (but not from D-glucose-*1*-*t*) during the biosynthesis of L-streptidine (**77**), which is a component of the antibiotic streptomycin, showed that the 3-guanidination of D-1-deoxy-1-guanidino-*scyllo*-inositol (**75**) may proceed through the 3-ketone intermediate **76**. The enzyme system was specific[195] for the D-isomer (**75**).

2. Hydrogen Movement within the Cell, and Incorporation into Cell Components

In addition to providing information concerning hydrogen movements during a particular enzyme reaction, hydrogen isotopes have

(191) F. A. Loewus and S. Kelly, *Arch. Biochem. Biophys.*, **102**, 96 (1963).
(192) F. A. Loewus, S. Kelly, and E. F. Neufeld, *Proc. Nat. Acad. Sci. U. S.*, **48**, 421 (1962).
(193) H. Kindl, R. Scholda, and O. Hoffmann-Ostenhof, *Phytochemistry*, **6**, 237 (1967).
(194) H. Kindl and O. Hoffmann-Ostenhof, *Phytochemistry*, **6**, 77 (1967).
(195) W. H. Horner and G. A. Russ, *Biochim. Biophys. Acta*, **237**, 123 (1971).

also found application in the study of hydrogen movements in general metabolism. Early studies[196-198] on the metabolic transfer of hydrogen atoms from carbohydrates were concerned with assessing the relative involvement of the pyridine nucleotide cofactors NAD^+ and $NADP^+$ in the biosynthesis of fatty acids.

D-Glucose 6-phosphate:$NADP^+$ oxidoreductase, the first enzyme in the pentose phosphate pathway, transfers H-1 of D-glucose 6-phosphate to $NADP^+$, and 6-O-phosphono-D-gluconate:$NADP^+$ oxidoreductase (decarboxylating) oxidizes 6-O-phosphono-D-gluconate to D-*erythro*-pentulose 5-phosphate with loss of H-3, forming NADPH. Other dehydrogenations in the glycolytic sequences produce NADH.

The H-1 and H-3 atoms of D-glucose, which tend to be transferred to $NADP^+$ instead of to NAD^+, are more readily transferred[199-202] to fatty acids in adipose tissues and micro-organisms than are H-4, H-5, and H-6 of D-glucose[200] and H-2 of glycerol,[197,198] thus supporting the suggestion[203] that NADPH plays a predominantly biosynthetic role in the organism. Changes in the incorporation of tritium from D-

(196) J. M. Lowenstein, *J. Biol. Chem.*, **236**, 1213 (1961).
(197) B. Bloom, *J. Biol. Chem.*, **234**, 2158 (1959).
(198) R. O. Brady, R. M. Bradley, and E. C. Trams, *J. Biol. Chem.*, **235**, 3093 (1960).
(199) R. G. Kemp and I. A. Rose, *J. Biol. Chem.*, **239**, 2998 (1964).
(200) J. Katz and R. Rognstad, *J. Biol. Chem.*, **241**, 3600 (1966).
(201) D. W. Foster and B. Bloom, *J. Biol. Chem.*, **236**, 2548 (1961).
(202) D. W. Foster and B. Bloom, *J. Biol. Chem.*, **238**, 888 (1963).
(203) N. O. Kaplan, M. N. Swartz, M. E. Frech, and M. M. Ciotti, *Proc. Nat. Acad. Sci. U. S.*, **42**, 481 (1962).

glucose-*1-t* into fatty acids in lactating-rat mammary-glands have also been used to indicate changes in the metabolic patterns of D-glucose during development.[196] The general limitations and parameters affecting studies on hydrogen transfers from carbohydrates have been discussed by Katz and Rognstad[204] and others.[179,205–207] In general, any result obtained by the use of D-glucose labeled with hydrogen isotopes is not considered a good guide to the catabolism of the D-glucose skeleton,[208] although approximations of the extent of carbon-chain recycling[209] and the proportions of substrate catabolized by the various metabolic pathways have been derived[210] from measurements of the half-life of tritium loss from tritiated D-glucoses. In mammals, tritium loss from blood D-glucose was shown to be rapid; in plasma, the half-life of tritium loss from tritiated D-glucoses ranged from 22 to 130 minutes, depending on the position of the label.[208,209,211]

3. Incorporation of Monomers into Polysaccharides and Other Polymers Containing Carbohydrate

Although the use of hydrogen isotopes in the study of carbohydrate catabolism is subject to some limitations, it would appear that the study of the incorporation of labeled monosaccharides into polymeric material can generally be evaluated by use of isotopes of hydrogen employed as a general tracer in sugars.[212] Tritium at C-2 and C-4 of D-glucose was readily lost to the solvent during catabolism, probably through isomerase reactions,[204] but the incorporation of labeled D-glucose into such di- and poly-saccharides as sucrose and glycogen resulted in complete retention of hydrogen at C-2 of the D-glucose units.[213] The incorporation of monomer was probably effected by direct transfer of the intact D-glucosyl group. In some circumstances, the amount of hydrogen isotope incorporated is greater than that predicted from the specific activity of the starting material,[185,214] This

(204) J. Katz and R. Rognstad, *J. Biol. Chem.*, **244**, 99 (1970).
(205) W. W. Shreeve, E. Lamdin, N. Oji, and R. Slavinski, *Biochemistry*, **6**, 1160 (1967).
(206) K. J. Matthes, S. Abraham, and I. L. Chaikoff, *Biochim. Biophys. Acta*, **70**, 242 (1963).
(207) J. Katz, R. Rognstad, and R. G. Kemp, *J. Biol. Chem.*, **240**, 1484 (1965).
(208) A. Dunn and S. Strahs, *Nature*, **205**, 705 (1965).
(209) A. Dunn, M. Chenoweth, and L. D. Schaeffer, *Biochemistry*, **6**, 6 (1967).
(210) H. Simon and R. Medina, *Z. Naturforsch.*, *B*, **23**, 326 (1968).
(211) J. Katz and A. Dunn, *Biochemistry*, **6**, 1 (1967).
(212) H. D. Wulf and H. G. Hers, *Eur. J. Biochem.*, **2**, 50 (1967).
(213) H. D. Dorrer, C. Fedtke, and A. Trebst, *Z. Naturforsch.*, *B*, **21**, 557 (1966).
(214) I. A. Rose, R. Kellermeyer, R. Stjernholm, and H. G. Wood, *J. Biol. Chem.*, **237**, 3325 (1962).

result has been explained[175] by occurrence of an isotope effect in the isomerization reactions that occur before catabolism, leading to an increase in the specific activity of the metabolite.

Incorporation of tritium from D-ribose-1-t and D-mannose-1-t into rat-liver glycogen showed that the D-glucose 6-phosphate and D-mannose 6-phosphate aldo–keto isomerases have the same stereochemical characteristics *in vivo* as *in vitro*.[215]

Tritium-labeled sugars have been used in several widely different contexts to measure the biosynthesis, stability, and turnover of polysaccharides. Use of hydrogen isotopes as tracers in the biosynthesis of glycoproteins established (*a*) the sites of immunoglobin glycan biosynthesis,[216,217] (*b*) the kinetics of monosaccharide incorporation,[218,219] and (*c*) the size,[216] nature,[220] and number[15] of the residues present in the glycan components. Measurements of their intracellular half-life have also been made by monitoring the incorporation of tritiated D-galactose and 2-amino-2-deoxy-D-galactose.[221] Attachment of the carbohydrate component to protein was necessary before excretion from the cell occurred.[222] The plasma half-lives of various tritiated ceruloplasmins (native or chemically modified) have also been measured[223,224] Tritiated antigens have been utilized in preparing pure antibodies of high specificity.[216] Constant retention of tritium relative to ^{14}C in the skeleton of hyphal-wall polysaccharides has been taken to signify metabolic stability in the fungus *Aspergillus clavatus*,[225] and *myo*-inositol-2-t has been used for monitoring the kinetics of phosphoinositide formation[226] and the patterns of development[227] in mammals.

4. Use as Substrates for Enzymes and for Transport Processes

Sugars labeled with tritium have been used for the assay of enzyme activity where no other method exists. For instance, UDP-2-acet-

(215) B. Bloom, *J. Biol. Chem.*, **239**, 2102 (1964).
(216) T. Muramatsu and S. G. Nathenson, *Biochem. Biophys. Res. Commun.*, **38**, 1 (1970).
(217) R. M. Swenson and M. Kern, *J. Biol. Chem.*, **242**, 3242 (1967).
(218) F. Melchers, *Biochem. J.*, **119**, 765 (1970).
(219) A. Herscovics, *Biochem. J.*, **117**, 411 (1970).
(220) T. Muramatsu and S. G. Nathenson, *Biochemistry*, **9**, 4875 (1970).
(221) I. Schenkein and J. W. Uhr, *J. Cell Biol.*, **46**, 42 (1970).
(222) D. Schubert, *J. Mol. Biol.*, **51**, 287 (1970).
(223) L. Van Lenten and G. Ashwell, *J. Biol. Chem.*, **246**, 1889 (1971).
(224) C. J. A. Van den Hamer, A. G. Morell, I. M. Scheinberg, J. Hickman, and G. Ashwell, *J. Biol. Chem.*, **245**, 4397 (1970).
(225) D. L. Corina and K. A. Munday, *J. Gen. Microbiol.*, **65**, 253 (1971).
(226) A. Markovitz, J. A. Cifonelli, and A. Dorfman, *J. Biol. Chem.*, **234**, 2343 (1959).
(227) P. A. Weinhold and R. D. Sanders, *Biochemistry*, **10**, 1090 (1971).

amido-2-deoxy-D-glucose-*t* has been used in demonstrating hyaluronic acid synthesis,[227] and D-glucose-*t* in the biosynthesis of UDP-2-acetamido-2-deoxy-D-galactose 4-sulfate.[228] L-Glucitol-*1-t*,[229] D-mannitol-*1-t*, and inulin-*t* (Ref. 230) have been found to be convenient for the measurement of extracellular spaces in adipose and other tissues, and, together with other tritiated hexoses, have been used in studies of intestinal sugar-transport.[231,232]

5. Autoradiography

Because the range of β-emission from tritium is very short, it has proved ideal for the histochemical detection and exact localization of sugars by use of high-resolution autoradiography. After a brief exposure of intestinal tissue to 6-deoxy-D-glucose-*t*, the sugar was shown to be concentrated at the ends of the villi,[233] confirming that only the tips accumulate sugars readily. Subsequently, autoradiography was used for confirming that phlorizin inhibits sugar transport at the villi tips.[234] The method has also been used in studying the concentration and distribution of injected D-ribose-*t* in rat tissues[235] as an approach complementary to chemical methods.

IX. TABLES OF KNOWN SUGARS SPECIFICALLY LABELED WITH ISOTOPES OF HYDROGEN

1. Methods of Preparation

The numbering of the preparative methods corresponds to that of Section II (see p. 127): 1, Reduction by Hydride Reagents; 2, Enzymic Reduction by Reduced Nicotinamide Adenine Dinucleotide; 3, Reduction by Hydrogen; 4, Base-catalyzed Solvent Exchange; 5, Enzyme-catalyzed Solvent Exchange; 6, Solvent Incorporation on Oxidation of a Hydrazino Group; 7, Addition to a Double Bond; 8, Conversion from Another Labeled Sugar; 9, Non-specific Labeling Methods; and an additional group for fully deuterated sugars, namely, 10, Algal Growth in D_2O.

---

(228) M. Tsuji, S. Shimizu, Y. Nakanishi, and S. Suzuki, *J. Biol. Chem.*, **245**, 6039 (1970).
(229) O. B. Crofford and A. E. Renold, *J. Biol. Chem.*, **240**, 14 (1965).
(230) A. M. Goldner, S. G. Schultz, and P. F. Curran, *J. Gen. Physiol.*, **53**, 362 (1969).
(231) J. E. G. Barnett, A. Ralph, and K. A. Munday, *Biochem. J.*, **118**, 843 (1970).
(232) W. F. Caspary and R. K. Crane, *Biochim. Biophys. Acta*, **203**, 308 (1970).
(233) W. B. Kintner and T. H. Wilson, *J. Cell Biol.*, **25**, 19 (1965).
(234) C. E. Stirling, *J. Cell Biol.*, **35**, 605 (1967).
(235) R. P. Goncalves, G. C. Bennett, and C. P. Leblond, *Anat. Record*, **165**, 543 (1970).

Many of the enzyme reactions previously discussed result in the incorporation of isotopes of hydrogen from the solvent or cofactor. Only those labeled compounds synthesized on a preparative scale are included in the Tables, but suitable modification of the other enzyme systems may make them suitable for use in the preparation of other labeled sugars.

2. Position of Label

For tritium-labeled compounds, "general" indicates non-specific labeling at several carbon atoms that are not uniformly labeled and "random" implies a distribution between the two hydrogen atoms of a methylene group, where the proportions of the isotope in the R or S configurations are not known. In Table III (deuterium compounds), "complete" signifies complete replacement of all carbon-bound hydrogen atoms, or of the hydrogen atoms of a methylene group at the position indicated.

3. Commercial Availability

An asterisk (*) against an entry denotes that, at the time of compilation (1971), the labeled sugar was commercially available from one of the suppliers listed. Where known, the method of preparation is given.

(i) The Radiochemical Centre, Amersham, Buckinghamshire, U.K.; (ii) Amersham-Searle Corporation, 2636 S. Clearbrook Drive, Arlington Heights, Illinois 60005, U. S. A.; (iii) New England Nuclear Corporation, 575 Albany Street, Boston, Massachusetts 02118, U. S. A.; (iv) NEN Chemicals GmbH, 6072 Dreieichenhain bei Frankfurt/-Main, Siemensstrasse 1, W. Germany; (v) Merck, Sharp and Dohme Canada Limited, P.O. Box 899, Pointe Claire-Dorval 700, Quebec, Canada; and (vi) Prochem Ltd., Carolyn House, Dingwall Road, Croydon, Surrey, U. K.

SUGARS LABELED WITH ISOTOPES OF HYDROGEN   183

TABLE I

Tritiated Monosaccharides

Sugar	Position of label	Preparative method	References
D-Allitol	1	1	6
L-Allitol	1	1	6
D-Allofuranose			
1,2:5,6-di-O-isopropylidene-	3	1	52
°D-Allose	3	1	12,19
	general	9	236
D-Altritol	1	1	6
D-Arabinitol	1	1	6
α-D-Arabinofuranosyl			
°uridine 5'-pyrophosphate	general	9	
D-Arabinose	1	1	7
	2	1 + 8	188
°	5 random	1 + 8	
°	general	9	
°L-Arabinose	1	1	
D-Arabinonic acid	2	1	188
L-Ascorbic acid	4	4	32
Conduritol	4	8	237
1,3-Dihydroxy-2-propanone	1	1 + 8	6
1-phosphate	3R	5	185,238
	3S	5	37,41
D-Erythritol	1	1	6
L-Erythritol	1	1	6
D-Fructofuranose			
1-deoxy-1-p-toluidino-	1	1 + 8	124
°D-Fructose	1 random	5 + 8	
		1 + 8	124
	5	2	239
		1 + 8	188
°	6 random	8	
	1,6	1 + 8	188
	3,4	1 + 8	188
	general	9	74
6-phosphate	1R	5	35,213
	1S	8	241
	general	5 + 8	242
	3,5	5 + 8	242
	3,4,5	5 + 8	242
	4,5,6	5 + 8	44
	5	5 + 8	43
°D-Fucose	1	1	
D-Fucose	4	1	20

(continued)

TABLE I (continued)

Sugar	Position of label	Preparative method	References
°L-Fucose	1	1	
°L-Fucose	general	9	236
5-phosphate	2	5	140
D-Galactitol	1	1	6
	2	1	6
1,5-anhydro-	6	1	14
α-D-Galactopyranosyl			
°uridine 5′-pyrophosphate	1	1 + 8	
	6 random	1	240
α-D-Galactopyranoside			
methyl	6 random	1	14
°D-Galactose	1	1	7
°	4	1	243
	4	7	49
°	6	1	13,14
°	general	9	61,62,74
°2-acetamido-2-deoxy-	general	9	
°L-Galactose	1	1	
°D-Glucitol	1 random	1	6
	2	1	6
1,5-anhydro-	general	9	236
1,5-anhydro-6-deoxy-	general	9	236
4-O-D-galactopyranosyl-	1	1	6
D-Gluconic acid	2	1	244
6-phosphate	2	8	45,159
α-D-Glucopyranosyl			
thymidine 5′-pyrophosphate	3	1	12
6′-deoxy-	4	1	169
uridine 5′-pyrophosphate	4	1	69
°	6	8	
α-D-Glucopyranose			
2-acetamido-2-deoxy-	general	9	227
2-acetamido-3-O-			
(D-l-carboxyethyl)-2-			
deoxy- (Muramic acid)	general	9	245
α-D-Glucopyranoside			
methyl	3	1	12
	4	1	69
	5	7	49
6-deoxy-6-sulfo-	5	7	125
α-D-Glucopyranosyluronic acid			
uridine 5′-pyrophosphate	3	8	76
	4	8	76
	5	8	76

TABLE I (continued)

Sugar	Position of label	Preparative method	References
°D-Glucose	1	1	5,196
	1	3	123,124
°	2	1	
	2	1 + 8	244
°	3	1	12,19
	3	5 + 8	42,199
°	4	1	243,246,69
	4	5 + 8	42
°	5	1	11
	5	5 + 8	43
°	6 random	1	8,9
	6 random	7	50
	4,6	1	69
	general	9	61,62,240,242,247
1-phosphate	general	9	248
°2-acetamido-2-deoxy-	1	1	
	general	9	245
°2-amino-2-deoxy-	1	1	
°	6 random	8	
	general	9	249
2-(hydroxymethyl)-	general	9	236
°3-O-methyl-	1	1	
	general	9	236
6-deoxy-	4	1	20
	general	9	236
6-phosphate	2	5	34,211
	5	5 + 8	43
	general	9	242
D-Glyceraldehyde			
3-phosphate	1	5 + 8	185
D-Glyceric acid			
3-phosphate	2R	5	250
°Glycerol	1	1	6,72
°	2	1	
L-Gulitol	1	1	6
	2	1	6
β-L-Gulopyranoside			
methyl	5	7	49
D-*gluco*-Heptose			
7-deoxy-	general	9	236
D-*glycero*-D-*galacto*-Heptitol	2	1	6
D-*glycero*-D-*gluco*-Heptitol	2	1	6
D-*glycero*-D-*gulo*-Heptitol	1	1	6
D-*glycero*-D-*ido*-Heptitol	2	1	6
D-*glycero*-D-*talo*-Heptitol	2	1	6

(continued)

TABLE I (continued)

Sugar	Position of label	Preparative method	References
Hexose			
°2-deoxy-D-*arabino*-	1	3	199
°	2	7	51
	general	9	236
Hexulosonic acid			
3-deoxy-D-*erythro*-	3 random	4	159
6-deoxy-L-*lyxo*-			
5-phosphate	1S	5	140
L-Iditol	1	1	6
	2	1	6
β-L-Idopyranoside			
methyl	5	7	49
allo-Inositol	general	9	59
muco-Inositol	general	9	59
myo-Inositol	1	9	59
	1	3	56
°	2	1	
	2	3	27,251
	general	9	252
(−)-Inositol	1,6	3	56
	general	9	59
L-Lyxitol [D-Arabinitol-(5)]	1	1	6
°D-Mannitol	1	1	9
	2	1	6
	3	1 + 8	188
	general	9	74
1,5-anhydro-	general	9	236
6-deoxy-	1	1	6
D-Mannonic acid	2	1	244
α-D-Mannopyranoside			
methyl	5	7	49
°D-Mannose	1	1	7
	1	3	123
°	2	1	253
°	2	1 + 8	244
	general	9	240,242
	2	5	141b,254
°2-acetamido-2-deoxy-	general	9	
α-D-*erythro*-Pentofuranosyl			
2-deoxy-, cytidine 5′-pyro-			
phosphate	2	5	151
2-deoxy-, thymidine 5′-pyro-			
phosphate	5	1	151
Pentose			
°2-deoxy-D-*erythro*-	1	1	
°	general	9	
5-phosphate	general	9	258

TABLE I (continued)

Sugar	Position of label	Preparative method	References
D-erythro-Pentulose			
5-phosphate	1R	5	35,45
	2	5	46
L-erythro-Pentulose	1	1 + 8	6
	5	1 + 8	6
5-phosphate	1R	5	140
D-threo-Pentulose	1	1 + 8	6
	5	1 + 8	6
5-phosphate	3	5	46
L-Psicose	1	1 + 8	6
	6	1 + 8	6
L-Rhamnitol	1	1	6
°L-Rhamnose	1	1	7
°	general	9	
D-Ribitol	1	1	6
β-D-Ribofuranose			
1,2,3,5-tetra-O-acetyl-	general	9	74,256
D-Ribonic acid	2	1	188
D-Ribose	1	1	257
	3	1	52
°	5	1	
	general	9	74,256
L-Ribose			
5-phosphate	2	5	140
L-Sorbose	1	1 + 8	6
	6	1 + 8	6
D-Tagatose	1	1 + 8	6
	6	1 + 8	6
D-Talitol	1	1	6
	2	1	6
L-glycero-Tetrulose	1	1 + 8	6
	4	1 + 8	6
D-Threitol	1	1	6
L-Threitol	1	1	6
D-Xylitol	1	1	6
D-Xylose	1	1	7
	1	8	140
	5	1	9
°	general	9	
L-Xylose	5	1	259
α-D-Xylofuranosyl			
uridine 5'-pyrophosphate	3	8	76
	4	8	76
	5R	8	76

TABLE II

Tritiated Disaccharides and Polysaccharides

Sugar	Position of label	Preparative method	References
α-D-Allosyl β-D-fructoside (Allosucrose)	allose-3	1	12,260
Ceruloplasmin	modified sialic acid	1	223
	terminal galactose	1	15
Chitotriose	general	9	249
°Dextran	general	9	247
Galactan	terminal galactose	1	14
Heparin	general	3	55
	general	9	55,261,262
°Inulin	general	9	
°Lactose	glucose-1	1	7
	general	9	263
Maltose	glucose-1	1	7
Raffinose	galactose-6	1	13
Starch	general	9	248
°Sucrose	fructose-1	8	
	glucose-3	1	12,260
°	6,6' random	8	
	general	8	54,74,247,248

TABLE III

Deuterated Monosaccharides

Sugar	Position of label	Preparative method	References
D-Allose	3	1	20
D-Arabinose	1	1	106,264
	5 complete	1	264
°	complete	10	
2-O-methyl-4-O-formyl-1,3-Dihydroxy-2-propanone	4	8	10
1-phosphate	3S	5	185
	3 complete	5	100
D-Erythrose			
2,3-di-O-formyl-	4 complete	8	10
D-Fructose	1 complete	4 + 8	26
°	complete	10	
6-phosphate	1R	5	26,34
	1S	5 + 8	39
D-Galactose	1	1	106
°	complete	10	

(*continued*)

TABLE III (continued)

Sugar	Position of label	Preparative method	References
°D-Glucitol	complete	10	
°D-Glucose	1	1	106,175
	1	3	106,265
	1	4	26
	2	4	26
	3	1	20,21
	5	1	266
	5,6	1	266
°	6 complete		
°	complete	10	267
3-O-methyl-	5	1	10
	6 complete	1	10
6-phosphate	2	2	33,34
D-Glyceric acid			
2-phosphate	2R	5	250
Glycerol	1	5 + 8	214
°	complete	10	
α-D-*ribo*-Hexofuranose			
3-deoxy-1,2:5,6-di-O-isopropylidene-	3	6	47
Hexopyranoside			
methyl 2-deoxy-α-D-*arabino*-	2	3	28
D-*erythro*-Hexulosonic acid			
3-deoxy-, 6-phosphate	3 complete	5	159
L-Idose	5	1	10
myo-Inositol	2	3	27,268
	4	3	255
	6	3	255
	complete	9	57
D-Mannose	1	1	106
°	complete	10	267
Pentose			
2-deoxy-D-*erythro*-	2	1	22
	2R	1	23
	2S	1	23
°L-Rhamnose	complete	10	267
°D-Ribose	2	1	22
°	3		
	complete	10	267
D-Talopyranose			
1,6-anhydro-	3	4	30
	4	1	30
2,3-O-isopropylidene-	3,4	1,4	31,115a
α-D-Talopyranoside			
methyl	3	4	31
	4	1	31
D-Xylose	1	1	106
	5	1	80
°	complete	10	267

(236) R. K. Crane, R. Drydale, and H. Hawkins, *Atomlight*, **15**, 4 (1960).
(237) G. Legler, *Z. Physiol. Chem.*, **351**, 25 (1970).
(238) T. H. Chiu and D. S. Feingold, *Fed. Proc.*, **26**, 835 (1967).
(239) S. Englard and G. Avigad, *J. Biol. Chem.*, **240**, 2297 (1965).
(240) G. L. Nelsestuen and S. Kirkwood, *J. Biol. Chem.*, **246**, 3828 (1971).
(241) H. Simon, H. D. Dorrer, and A. Trebst, *Chem. Ber.*, **96**, 1285 (1963).
(242) H. Simon, G. Mullhofer, and H. D. Dorrer, *Proc. Symp. Methods of Preparing Stored Marked Molecules, Euratom*, Eur. 1625E, p. 997 (1964).
(243) B. D. Kohn and P. Kohn, *J. Org. Chem.*, **28**, 1037 (1963).
(244) H. S. Isbell, H. L. Frush, C. W. R. Wade, and A. J. Fatiadi, *J. Res. Nat. Bur. Stand.*, A, **73**, 75 (1969).
(245) A. Markovitz, J. A. Cifonelli, and J. I. Gross, *Atomlight*, **16**, 1 (1961).
(246) O. Gabriel, *Carbohyd. Res.*, **6**, 319 (1968).
(247) K. H. Ebert and J. Richter, *Z. Naturforsch.*, B, **22**, 788 (1967).
(248) P. Nordin, H. C. Moser, and J. K. Senne, *Biochem. J.*, **96**, 336 (1965).
(249) T. Guenther, M. Wenzel, and H. Greiling, *Z. Physiol. Chem.*, **326**, 212 (1961).
(250) E. C. Dinovo and P. D. Boyer, *J. Biol. Chem.*, **246**, 4586 (1971).
(251) L. E. Hokin and M. R. Hokin, *J. Biol. Chem.*, **233**, 805 (1958).
(252) T. Meshi and T. Takahashi, *Bull. Chem. Soc. Jap.*, **35**, 1510 (1962).
(253) H. Simon and R. Medina, *Z. Naturforsch.*, B, **21**, 496 (1966).
(254) L. Glaser, *Biochim. Biophys. Acta*, **41**, 534 (1960).
(255) T. Posternak, W. H. Schopfer, D. Reymond, and C. Lark, *Helv. Chim. Acta*, **41**, 235 (1958).
(256) M. Gordon, O. M. Intrieri, and G. B. Brown, *J. Amer. Chem. Soc.*, **80**, 5161 (1958).
(257) R. J. Suhadolnik, T. Uematsu, and R. M. Ramer, *Carbohyd. Res.*, **5**, 479 (1967).
(258) C. Rosenblum, *Nucleonics*, **17** (12), 80 (1959).
(259) S. Schuching and G. H. Frye, *J. Org. Chem.*, **30**, 1288 (1965).
(260) G. A. Marzluf and R. L. Metzenberg, *Anal. Biochem.*, **13**, 168 (1965).
(261) N. D. Ferrante and E. A. Popenoe, *Carbohyd. Res.*, **13**, 306 (1970).
(262) G. H. Barlow and E. V. Cardinal, *Proc. Soc. Exp. Biol. Med.*, **123**, 831 (1966).
(263) T. Meshi and Y. Sato, *Bull. Chem. Soc. Jap.*, **36**, 750 (1963).
(264) K. Heyns and D. Müller, *Tetrahedron*, **21**, 55 (1965).
(265) R. Bentley and D. S. Bhate, *J. Biol. Chem.*, **235**, 1225 (1960).
(266) W. Mackie and A. S. Perlin, *Can. J. Chem.*, **43**, 2921 (1965).
(267) M. I. Blake, H. L. Crespi, V. Mohan, and J. J. Katz, *J. Pharm. Sci.*, **50**, 425 (1961).
(268) T. Posternak, W. H. Schopfer, and D. Reymond, *Helv. Chim. Acta*, **38**, 1283 (1955).

THE USE OF CARBOHYDRATES IN THE SYNTHESIS AND CONFIGURATIONAL ASSIGNMENTS OF OPTICALLY ACTIVE, NON-CARBOHYDRATE COMPOUNDS

By T. D. Inch

Chemical Defence Establishment, Porton Down, Salisbury, Wiltshire, England

I. Introduction 191
II. Use of Carbohydrates in Asymmetric Synthesis 192
  1. Grignard Addition Reactions in Solvents containing
    Carbohydrate Derivatives [Type 1] 193
  2. Reductions with Complexes of Lithium Aluminum Hydride [Type 1] . 195
  3. Addition Reactions of α-Oxo Esters of Carbohydrate Derivatives
    [Type 2] 199
  4. Reduction Reactions of α-Oxo Esters of Carbohydrate Derivatives
    [Type 2] 201
  5. Conjugate-addition Reactions of Grignard Reagents with α,β-
    Unsaturated Carbohydrate Esters [Type 2] 202
  6. Reactions in which Carbonyl Groups in Glycosuloses are Converted
    into New Asymmetric Centers [Type 3] 204
III. Carbohydrates as Sources of Asymmetric Carbon Atoms for the
    Synthesis and Proof of Configuration of Biologically Important,
    Non-carbohydrate Compounds 205
  1. Methods that Utilize Asymmetric Carbon Atoms Already Present in
    Carbohydrates 206
  2. Methods that Require the Creation of New Asymmetric Centers in
    Carbohydrates 213
IV. Stereoselective Synthesis of Asymmetric Sulfoxides 222

I. INTRODUCTION

The scope of chemical investigations of monosaccharides has gradually increased from the time when carbohydrate research was concerned primarily with structural assignments of such commonly occurring sugars as D-glucose and D-mannose until the present, when carbohydrates provide a fertile field for investigations of mechanistic and stereochemical effects (including n.m.r. effects) which may, in many cases, be studied uniquely within a carbohydrate framework. This article attempts to extend the utility of carbohydrates still further by drawing attention to the fact that carbohydrates have considerable practical application in "asymmetric synthesis," provide excellent chiral frameworks for mechanistic studies of asymmetric synthesis, and, in addition (and, perhaps, of more importance) can serve as readily available and easily modified sources of asymmetric centers having known configuration for the synthesis of other biologically important molecules. For example, it is possible to (*a*) create a new asymmetric center within a carbohydrate framework, (*b*) establish the configuration of the new asymmetric center by spectroscopic or other appropriate techniques, (*c*) detach the new asymmetric center from the carbohydrate framework by simple chemical reactions, and (*d*) incorporate it into molecules of biological interest. This article considers first, the use of carbohydrates in asymmetric synthesis, and second, the application of carbohydrates for the synthesis and configurational proof of other biologically important molecules.

II. Use of Carbohydrates in Asymmetric Synthesis

The essence of asymmetric synthesis is the creation of asymmetric centers under the influence of other asymmetric centers in such a way that the resulting enantiomers or diastereoisomers are formed in unequal proportions. Most reactions in asymmetric synthesis that have been described involve the conversion of trigonal carbon atoms into asymmetric, quadrivalent carbon atoms, and this article will be principally concerned with such reactions, although, in many instances, the principles involved may also be applied to asymmetric reactions in which, for example, chiral phosphorus or sulfur atoms are formed. In all reactions in which are formed mixtures of enantiomers having one enantiomer in preponderance, it is possible to describe the stereoselectivity of the reaction in terms of optical yield (optical purity, or enantiomeric yield). The precise significance of these terms has been described in detail elsewhere,[1] but, practically, where at a selected wavelength, $[\alpha]$ is the specific rotation of the reaction product and $[A]$ is the specific rotation of a pure enantiomer, the optical yield $= [\alpha]/[A]$. Thus, the value of the optical yield is a measure of the excess of one enantiomer over the other.

In general terms, asymmetric reactions may be divided into three types. *Type 1. Reactions in which optically active compounds are formed from achiral substrates and chiral reagents, or from achiral substrates and achiral reagents in chiral solvents.* An example of asymmetric synthesis with an asymmetric reagent is the reduction of cyclohexyl phenyl ketone with the Grignard reagent from (+)-chloro-2-methylbutane to give the corresponding alcohol in 25%

(1) M. Raban and K. Mislow, *Topics Stereochem.*, **2**, 199 (1967).

optical yield.[2] An example of an asymmetric reaction in a chiral solvent is the reaction of butanone and phenylmagnesium bromide in (S)-2,3-dimethoxybutane, which afforded 2-phenyl-2-butanol in 17% optical yield.[3] This general class of asymmetric reactions is characterized by the fact that the asymmetric products may be isolated by normal processing without further chemical reactions.

Type 2. Reactions in which new asymmetric centers are created in molecules that already possess one asymmetric center or more and in which the new asymmetric center may be liberated from the parent molecule by simple hydrolytic reactions. The most common reactions of this type are those to which the Prelog rule[4] applies; for example, the addition of methylmagnesium iodide to (−)-phenylpyruvate, followed by hydrolysis to give (−)-atrolactic acid.

Type 3. Reactions in which new asymmetric centers are created immediately adjacent to existing asymmetric centers in molecules that contain one asymmetric center (or more). These are the reactions to which the Cram rules[5] are usually considered to apply. The stereoselectivity of this type of reaction is usually higher than that normally found in reactions of Types 1 and 2. The synthetic utility of reactions of this type has suffered from the disadvantage that the new asymmetric center in non-carbohydrate compounds cannot usually be detached easily from the original asymmetric framework. However, the ready availability of glycosuloses has provided many examples of Cram-type reactions in carbohydrate chemistry, and the fact that the new asymmetric centers may be detached from the original carbohydrate framework has been exploited.

In the following Sections will be given examples illustrating that, in the three types of reaction already listed, where carbohydrates have been used to provide the asymmetric environment, the optical yields obtained were in most instances significantly higher than when other chiral molecules were employed.

1. Grignard Addition Reactions in Solvents containing Carbohydrate Derivatives [Type 1]

It has been demonstrated by independent groups of investigators[6] that asymmetric products can result from reactions between achiral

(2) E. P. Burrows, F. J. Welch, and H. S. Mosher, *J. Amer. Chem. Soc.*, **82**, 880 (1960).
(3) N. Allentof and G. F. Wright, *J. Org. Chem.*, **22**, 1 (1957).
(4) V. Prelog, *Helv. Chim. Acta*, **36**, 308 (1953).
(5) D. J. Cram and F. A. A. Elhafez, *J. Amer. Chem. Soc.*, **74**, 5828 (1952); D. J. Cram and K. R. Kopecky, *ibid.*, **81**, 2748 (1959).
(6) C. Blomberg and J. Coops, *Rec. Trav. Chim.*, **83**, 1083 (1964); H. L. Cohen and G. F. Wright, *J. Org. Chem.*, **18**, 432 (1953).

aldehydes or ketones and achiral Grignard reagents, provided that the solvent is optically active or contains an optically active ether. Generally, the stereoselectivity of such reactions is low and, indeed, the highest optical yield reported[3] was 17% for the formation of 2-phenyl-2-butanol from butanone and phenylmagnesium bromide in benzene containing (S)-2,3-dimethoxybutane. In examples where Grignard reactions were performed in benzene containing permethylated D-mannitol or D-arabinitol, the products obtained exhibited negligible optical activity.[3] However, much higher degrees of stereoselectivity have now been reported[7] for reactions between carbonyl compounds and Grignard reagents in ether or benzene containing suitably substituted carbohydrate derivatives. Thus, (R)-(+)-1-cyclohexyl-1-phenylethanol was obtained in 70% optical yield from the reaction between methylmagnesium bromide and cyclohexyl phenyl ketone in the presence of 1,2:5,6-di-O-isopropylidene-α-D-glucofuranose (1) in ether. The best chemical and optical yields were obtained

1

when 1 molar proportion of the ketone was added at room temperature to a solution of 2 molar proportions of **1** in ether containing the Grignard reagent (3.5 molar proportions). With these proportions, 2 molecules of the Grignard reagent react with the free hydroxyl group of **1**, leaving 1.5 molecules of complexed Grignard reagent available for reaction with the carbonyl compound. When cyclohexylmagnesium bromide was added to acetophenone, or phenylmagnesium bromide was added to cyclohexyl methyl ketone, (S)-(−)-1-cyclohexyl-1-phenylethanol was obtained. The yields and optical yields for these and other reactions performed in the presence of **1** are shown in Table I. The optical yields were generally about 25%, but in some cases were much higher.

Attempts to rationalize the stereochemistry of the product in terms of (a) "bulk" effects or (b) coordinated transition-states for this type of reaction were unsuccessful, although results showing that the Grignard reagent complexes strongly with the glucofuranose deriv-

(7) T. D. Inch, G. J. Lewis, G. L. Sainsbury, and D. J. Sellers, *Tetrahedron Lett.*, 3567 (1969).

OPTICALLY ACTIVE, NON-CARBOHYDRATE COMPOUNDS   195

TABLE I

Stereoselectivity of RMgX–R'COR" Reactions in the Presence of
1,2:5,6-Di-O-isopropylidene-α-D-glucofuranose (1)

RMgX	R'COR"	Product	Yield (%)	Specific rotation (Optical yield, %)	Enantiomeric proportions (R):(S)
MeMgI	PhCOC$_6$H$_{11}$	PhC(OH)(Me)C$_6$H$_{11}$	88	+13.35 (65)	82.5:17.5
MeMgBr	PhCOC$_6$H$_{11}$	PhC(OH)(Me)C$_6$H$_{11}$	95	+14.3 (70)	85:15
PhMgBr	C$_6$H$_{11}$COMe	PhC(OH)(Me)C$_6$H$_{11}$	60	− 5.3 (26)	37:63
C$_6$H$_{11}$MgBr	PhCOMe	PhC(OH)(Me)C$_6$H$_{11}$	50	− 5.8 (28)	36:64
EtMgBr	PhCOMe	PhC(OH)(Me)Et	45	− 4.75 (27)	36.5:63.5
PhMgBr	MeCOEt	PhC(OH)(Me)Et	70	− 1.8 (10)	45:55
MeMgBr	PhCOEt	PhC(OH)(Me)Et	70	+ 4.2 (24)	62:38
EtMgBr	PhCHO	PhCHOHCEt	81	+ 7 (24)	62:38
PhMgBr	EtCHO	PhCHOHEt	60	+ 2.7 (9)	54.5:45.5

ative were provided. Thus, although addition of an equimolar quantity of Grignard reagent to the D-glucofuranose derivative (1) in ether caused only a small change in specific rotation (from −6 to −3°), the addition of a second equimolar quantity of Grignard reagent caused the rotation to change from −3° to −30°. This result suggested that a particularly strong complex between the sugar derivative and the Grignard reagent was formed, as it has also been shown that, whereas the rotation of other chiral ethers is enhanced by complex formation with Grignard reagents, the addition of diethyl ether to such mixtures considerably lowers the rotational enhancement. Perhaps significantly, the rotations of the 3-benzyl and 3-methyl ethers of 1 were not enhanced by addition of an equimolar quantity of Grignard reagent, and optically inactive tertiary alcohols resulted from reactions of RMgX with R'COR" in their presence.

It is not as yet possible to assess the potential synthetic utility of this type of reaction properly, but, clearly, the apparent strength of the Grignard reagent–sugar complex suggests that detailed studies of the nature of this complex may prove rewarding.

2. Reductions with Complexes of Lithium Aluminum Hydride [Type 1]

In 1951, it was suggested[8] that reduction of butanone and 2-hexanone with lithium aluminum hydride in the presence of (+)-camphor afforded optically active alcohols. Although it was later shown[9] that the

(8) A. A. Bothner-By, *J. Amer. Chem. Soc.*, **73**, 846 (1951).
(9) P. S. Portoghese, *J. Org. Chem.*, **27**, 3359 (1962).

optical activity was actually due to residual (+)-camphor, the concept of using complexes of lithium aluminum hydride for asymmetric reduction was further investigated, and it was found that low stereoselectivity could be obtained.[10] Landor and coworkers[11] have demonstrated that, under favorable circumstances, reductions with complexes of lithium aluminum hydride with suitable sugar derivatives may be highly stereoselective.

Landor and coworkers[12] reasoned that, in complexes of lithium aluminum hydride with diols, the equilibrium represented by the equation would lie far to the left, and that reduction of aldehydes and

$$2\left(Li^{\oplus}\left[\begin{array}{c}O\\ \diagdown\\ O\end{array}Al\begin{array}{c}H\\ \diagup\\ H\end{array}\right]^{\ominus}\right) \rightleftharpoons LiAlH_4 + Li^{\oplus}\left[\begin{array}{cc}O & O\\ \diagdown & \diagup\\ & Al\\ \diagup & \diagdown\\ O & O\end{array}\right]^{\ominus}$$

ketones would be effected by the complex rather than by lithium aluminum hydride in the disproportionated mixture, and, consequently, that, were the complex optically active, highly stereoselective, asymmetric reductions would be feasible. When a number of such ketones as pinacolone and acetophenone were reduced with 1:1 complexes of lithium aluminum hydride with methyl 4,6-O-benzylidene-α-D-glucopyranoside (2), 1,2-O-isopropylidene-α-D-glucofuranose (3) and its 3-methyl, 3-ethyl, and 3-benzyl ethers, and 1,2-O-cyclo-

hexylidene-α-D-glucofuranose (4) and its 3-methyl and 3-benzyl ethers, optically active secondary alcohols were formed. The max-

(10) H. Haubenstock and E. L. Eliel, *J. Amer. Chem. Soc.*, **84**, 2363 (1962).
(11) S. R. Landor, B. J. Mitchell, and A. R. Tatchell, *J. Chem. Soc. (C)*, 1822 (1966).
(12) S. R. Landor, B. J. Mitchell, and A. R. Tatchell, *J. Chem. Soc. (C)*, 2280 (1966).

imum optical yield recorded for any reduction was 20%. Perhaps surprisingly, the products from reductions with complexes of **3** had signs of optical rotation opposite to those of the corresponding products from reduction with complexes of **4**, whereas reductions in the presence of the 3-benzyl ethers of both **3** and **4** gave products having the same sign of optical rotation.

All of the reductions with complexes of **2** showed low stereoselectivity. It was considered that the diequatorial, vicinal diol would not form a stable, five-membered, cyclic aluminum complex, and thus, even were a cyclic complex formed, considerable disproportionation would occur, thus lessening the stereoselectivity of the reaction. A number of different reasons were put forward to rationalize the variable results with the D-glucofuranose derivatives. The complex with 3-O-benzyl-1,2-O-cyclohexylidene-α-D-glucofuranose (**5**) gave the highest stereoselectivity, and it was consequently selected for further study; maximum stereoselectivities were obtained with the 1:1 lithium aluminum hydride–sugar complex. This observation was consistent with the formation of a cyclic complex between lithium aluminum hydride and the 5- and 6-hydroxyl groups of the monosaccharide. When acetophenone was reduced with the 1:1 complex of lithium aluminum hydride and **5**, the α-methylbenzyl alcohol (1-phenylethanol) formed in 33% optical yield had the S configuration. From an inspection of molecular models, it was considered that H-1 in the complex was more shielded than H-2 by the 3-O-benzyl group, and, consequently, transfer of H-2 to acetophenone was favored.

Evidence in favor of this theory was claimed[13] when it was found that reduction of acetophenone with the 1:1 complex of lithium aluminum hydride and **5** to which one equivalent of ethanol had been

(13) S. R. Landor, B. J. Mitchell, and A. R. Tatchell, *J. Chem. Soc. (C)*, 197 (1967).

added afforded α-methylbenzyl alcohol having the R configuration. It was suggested that the reversal of configuration of the product occurred because of favored reaction of the most accessible hydrogen atom (H-2) with ethanol, and that it was H-1 that was then transferred preponderantly to acetophenone.

It must be pointed out that, if H-2, not H-1, is the more shielded by the 3-O-benzyl substituent, a product having the S configuration can still be formed in the absence of ethanol, provided that the orientation of the acetophenone to the complex is reversed.

This reasoning does not detract from the basic argument, but merely illustrates that the nature of the complex cannot be deduced on the basis of the experimental evidence available at present, nor from molecular models, as either H-1 or H-2 could be the more shielded, depending on the precise orientation of the 3-O-benzyl substituent.

It was found that the stereoselectivity of reduction was increased by ethanol, and that (R)-1-phenylethanol was obtained in 70% optical yield when acetophenone (12.5 mmoles) was reduced with an ethanol-modified complex of lithium aluminum hydride with 3-O-benzyl-1,2-O-cyclohexylidene-α-D-glucofuranose prepared from the sugar derivative (26 mmoles), lithium aluminum hydride (58 mmoles), and ethanol (110 mmoles).

The reactions described clearly demonstrate the potential synthetic utility of asymmetric reductions with complexes of metal hydrides with sugars, but also indicate that much work is needed before the precise nature of the complexes is fully understood.

Landor's group and other workers[14] have investigated reductions with lithium aluminum hydride complexes of other sugars, but, to date, the 3-O-benzyl-1,2-O-cyclohexylidene derivatives have proved the most satisfactory. It has, however, been shown that complexes with certain amino alcohols give metal hydride complexes that appear very promising for asymmetric reductions[15]; thus, suitable amino sugars may be ideal asymmetric constituents for optically active complexes of lithium aluminum hydride.

3. Addition Reactions of α-Oxo Esters of Carbohydrate Derivatives [Type 2]

Hydrolytic cleavage of the products formed by addition of Grignard reagents to α-oxo esters of optically active alcohols affords optically active acids.

$$PhC(=O)-CO_2R \longrightarrow Ph-\underset{OH}{\underset{|}{C}}(Me)-CO_2R \xrightarrow[\text{(ii) } H^\oplus]{\text{(i) NaOH}} Ph-\underset{OH}{\underset{|}{C}}(Me)-CO_2H$$

This type of reaction has been most widely investigated by Prelog and coworkers, who have established relationships (Prelog's rules) between the configuration of the parent alcohol and the configuration of the derived acid.[4] For example, the stereochemical course of the reaction between methylmagnesium iodide and (−)-menthylphenyl-glyoxylate is most simply seen by reference to formulas **6** and **7**. The

6   **7**

molecule is, by convention, so oriented that the two carbonyl groups are antiparallel (as in **6**) with the smallest group (S) in the alcohol eclipsed by the carbonyl group of the ketone. The reagent R will then approach the carbonyl group of the ketone from the side of the smaller of the two groups remaining in the alcohol portion of the molecule, that is, the side of the medium-sized group M (in **7**). For menthol, the proton on C-3 is the small group S, the methylene group involving C-2 is the medium group M, and the isopropyl group is the large group L. Thus, reaction of (−)-menthylphenylglyoxylate (**8**)

---

(14) O. Červinka and A. Fabryova, *Tetrahedron Lett.*, 1179 (1967).
(15) O. Červinka, *Collect. Czech. Chem. Commun.*, **30**, 1685 (1965).

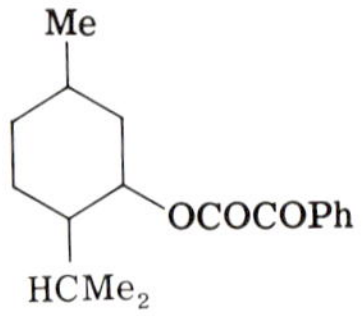

[Structure 8: cyclohexane with Me, OCOCOPh, HCMe₂ substituents]

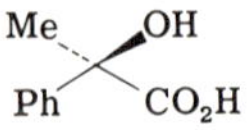

[Structure 9: Ph, Me, OH, CO₂H on central carbon]

with methylmagnesium iodide, followed by hydrolysis, gives preponderantly (R)-(−)-atrolactic acid (9). It should be emphasized that this convention does not necessarily have mechanistic significance.

Prelog's rule has been shown[16] to apply to a number of phenylglyoxylic acid ($PhCOCO_2H$) esters of sugars. Methylmagnesium iodide was added to the phenylglyoxylic acid esters of sugars 10 to 14.

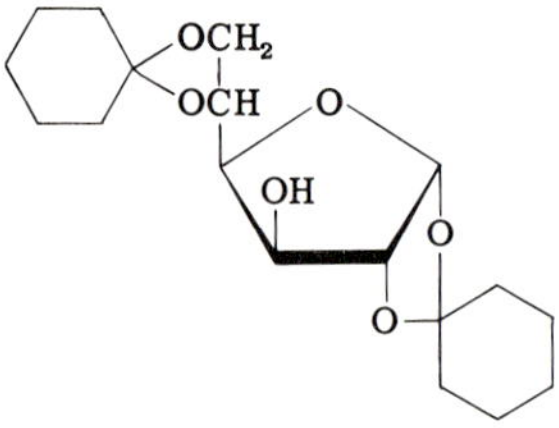

(38%)
10

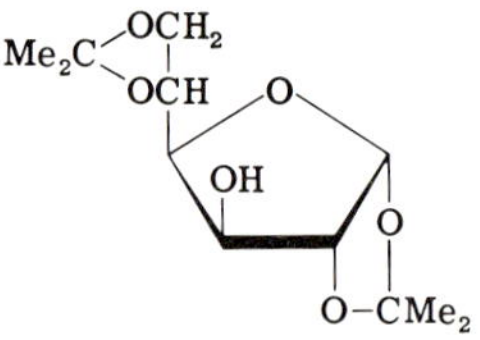

(34%)
11

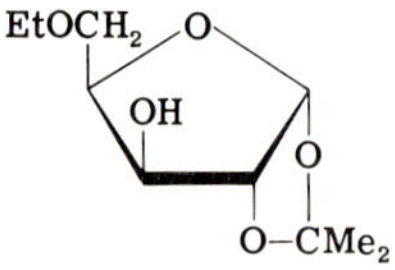

(28%)
12

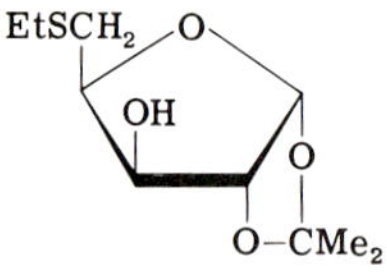

(25%)
13

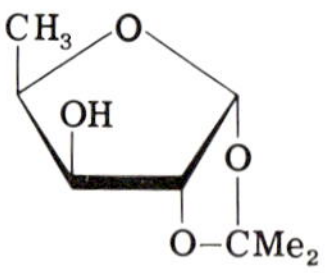

(22%)
14

The percentages given in parentheses are the optical yields of (R)-(−)-atrolactic acid liberated after hydrolysis. In this series, the substituent group on C-4 (for example, in 15) is clearly the large

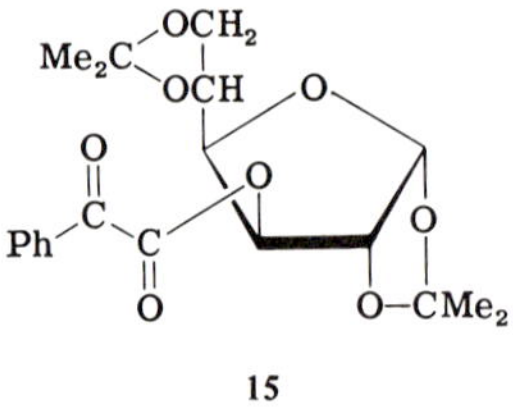

15

(16) M. Kawana and S. Emoto, *Bull. Chem. Soc. Jap.*, **40**, 2168 (1967).

OPTICALLY ACTIVE, NON-CARBOHYDRATE COMPOUNDS 201

substituent L, and the C-2 substituent well removed from the reactive carbonyl group is the medium group M. The fact that, with lessening in the size of the large group, there is a corresponding diminution in the optical yield of the product favors the interpretation that the direction of addition is controlled by factors of steric size rather than by the ability of the Grignard reagent to form coordination complexes with one or more of the oxygen atoms of the sugar. It is also of interest that the addition product of the phenylglyoxylic ester of 1,2:5,6-di-O-isopropylidene-α-D-glucofuranose with cyclohexylmagnesium bromide (an addition reaction in which the steric bulk of the Grignard reagent is appreciably greater than in the other examples) afforded, upon hydrolysis, α-phenylcyclohexaneglycolic acid in 75% optical yield.[17]

The main application of reactions of this type has been for the determination, based on Prelog's rule, of the absolute configuration of alcohols by preparation of optically active atrolactic acid of known configuration through addition of methylmagnesium iodide to the phenylglyoxylic ester of alcohols of unknown absolute configuration, and subsequent reduction. Although any application of Prelog's rule for configurational determinations in carbohydrate chemistry is now unnecessary (with the possible exception of branched-chain sugars) and of doubtful value (because of possible complications from coordination effects), it is possible that reactions of this type, in which the asymmetric alcohol is a sugar, may be a useful synthetic route to hydroxy acids of high optical purity.

4. Reduction Reactions of α-Oxo Esters of Carbohydrate Derivatives [Type 2]

It was demonstrated as early as 1904 that reduction of (−)-methyl-phenylglyoxylate with aluminum amalgam and subsequent hydrolysis afforded optically active (R)-(−)-mandelic acid, although in very low optical yield.[18] Prelog's rule applies to reactions of this type. It has been suggested[19] that Prelog's rules also apply to catalytic reduction of phenylglyoxylic esters of suitable sugar derivatives, although, in these reactions, it was also necessary to consider the effect of the catalyst on the conformation of the phenylglyoxylic ester. Thus, reduction of 1,2:5,6-di-O-cyclohexylidene-3-O-(phenylglyoxylyl)-α-D-glucofuranose in the presence of Raney nickel, and in the presence of

(17) T. D. Inch and R. V. Ley, unpublished results.
(18) A. McKensie, *J. Chem. Soc.*, **85**, 1249 (1904).
(19) M. Kawana and S. Emoto, *Bull. Chem. Soc. Jap.*, **41**, 259 (1967).

5% palladium-on-*basic*-carbon, afforded (R)-(−)-mandelic acid in 44.6 and 22.8% optical yield, respectively, whereas corresponding reduction in the presence of 5% palladium-on-*acidic*-carbon afforded (S)-(+)-mandelic acid in 24.1% optical yield. The stereoselectivity of reactions of this type parallel the reactions described in Section II, 3, in that optical yields decrease as the size of the 4-substituent on the furanose ring is lowered. To account for the difference in the product when basic and acidic catalysts are used, it was suggested that, whereas, with basic catalysts, the carbonyl groups in the glyoxylic ester assume the transoid disposition to which Prelog's rules formally apply, in acidic media, hemiacetal formation of the ester is favored and hydrogen bonding of the ester oxygen atom and hemiacetal hydroxyl group results in a cisoid disposition of the glyoxylic ester. Such rationalizations are not supported by direct evidence for the transition-state conformations.

5. Conjugate-addition Reactions of Grignard Reagents with α,β-Unsaturated Carbohydrate Esters [Type 2]

Optically active alkoxyl groups have been shown to influence the steric course of reactions between Grignard reagents and α,β-unsaturated esters, as well as of those between Grignard reagents and α-oxo esters. Thus, (+)-β-methylhydrocinnamic acid [(+)-3-phenylbutyric acid] was obtained in 6.7% optical yield by addition of phenylmagnesium bromide to (−)-menthyl crotonate, followed by hydrolysis of the adduct.[20] It has been demonstrated that much more highly stereoselective reactions result when the menthyl group is replaced by suitably substituted carbohydrate derivatives.[21] For example, reaction of phenylmagnesium bromide with 3-O-crotonyl-1,2:5,6-di-O-cyclohexylidene-α-D-glucofuranose (16), and subsequent hydrolysis, afforded R-(−)-β-methylhydrocinnamic acid (17) in 33% optical yield

(20) Y. Inouye and H. M. Walborsky, *J. Org. Chem.*, **27**, 2706 (1962).
(21) M. Kawana and S. Emoto, *Bull. Chem. Soc. Jap.*, **39**, 910 (1966).

OPTICALLY ACTIVE, NON-CARBOHYDRATE COMPOUNDS 203

and 12% synthetic yield. In the presence of cuprous chloride, which is known to catalyze 1,4-addition reactions, (−)-β-methylhydrocinnamic acid was formed in 74% optical yield and 58% synthetic yield. On the other hand, reaction of phenylmagnesium bromide with 3-O-crotonyl-5-deoxy-1,2-O-isopropylidene-α-D-xylofuranose in the absence of cuprous chloride, and subsequent hydrolysis, afforded (+)-β-methylhydrocinnamic acid in 16% optical yield, but, in the presence of cuprous chloride, (−)-β-methylhydrocinnamic acid was formed in 58% optical yield.

It was suggested that the coordinated adduct **18** might be the pre-

cursor of 1,2:5,6-di-O-cyclohexylidene-3-O-[(R)-β-methylhydrocinnamoyl]-α-D-glucofuranose, which would result by attack of the phenyl group from the rear of the double bond. In the presence of cuprous chloride, coordination with the double bond also occurs, as illustrated in formula **19**, again with attack of phenyl from the rear side of the double bond with the overall higher optical yield observed.

To explain the formation of (S)-(+)-β-methylhydrocinnamic acid by the reaction of the D-xylofuranose derivative **20** with phenylmagnesium bromide in the absence of cuprous chloride, the coordinated adduct **20** was envisaged as the precursor, with attack from the front side of the double bond now giving the acid having the S-configuration.

From the limited number of examples studied, the validity of the suggestions just outlined is obviously open to question, particularly because there seems no reason why the results for the α-oxo esters should be explained simply by steric effects according to Prelog's rule, but, for the α,β-unsaturated esters, it should have been found necessary to invoke coordination effects in order to rationalize the results. However, the results make it quite clear that carbohydrates provide a unique framework for the study of coordination effects in the steric control of reactions that involve metal complexes.

6. Reactions in which Carbonyl Groups in Glycosuloses are Converted into New Asymmetric Centers [Type 3]

It has been known for many years that addition reactions to carbonyl groups in carbohydrates are highly stereoselective, as in the Kiliani reaction, and examples of addition reactions to ketoses have been reported that are stereospecific.[22,23,23a] The ready availability of derivatives[24] of aldosuloses and diuloses has led to many studies of addition reactions to carbohydrates, and it is apparent that much more work is needed to explain many of the stereochemical "anomalies." For example, in some instances, Grignard reagents and an alkyl(or aryl)lithium react with a ketose to give the same preponderant products, whereas, in others, the preponderant products differ in the configuration of the new asymmetric center.[25-27] In other reactions, it has been shown that different Grignard reagents can react with the same ketose to give preponderant products that differ at the new asymmetric center, although the stereochemical course of a Grignard reaction is usually independent of the nature of the Grignard reagent.[28] As in asymmetric reactions of Types 1 and 2, it is clear that, because of their well defined and easily varied geometry, carbohydrates provide excellent asymmetric frameworks for the study of the factors that control such additions. However, for the purpose of this article, asymmetric reactions of Type 3 will only be discussed in connection with the use of carbohydrates as sources of asymmetric centers for the synthesis of other biologically important molecules.

(22) J. S. Burton, W. G. Overend, and N. R. Williams, *J. Chem. Soc.*, 3433 (1965).
(23) T. D. Inch, R. V. Ley, and P. Rich, *J. Chem. Soc. (C)*, 1683 (1968).
(23a) D. Horton and E. K. Just, *Carbohyd. Res.*, **18**, 81 (1971).
(24) J. S. Brimacombe, *Chem. Brit.*, **2**, 99 (1966); K. Onodera, S. Hirano, and N. Kashimura, *Carbohyd. Res.*, **6**, 276 (1966).
(25) A. A. J. Feast, W. G. Overend, and N. R. Williams, *J. Chem. Soc. (C)*, 303 (1966).
(26) B. Flaherty, W. G. Overend, and N. R. Williams, *J. Chem. Soc. (C)*, 398 (1966).
(27) R. D. Rees, K. James, A. R. Tatchell, and R. H. Williams, *J. Chem. Soc. (C)*, 2716 (1968).
(28) T. D. Inch and G. J. Lewis, *Carbohyd. Res.*, **16**, 455 (1971).

III. CARBOHYDRATES AS SOURCES OF ASYMMETRIC CARBON ATOMS FOR THE SYNTHESIS AND PROOF OF CONFIGURATION OF BIOLOGICALLY IMPORTANT, NON-CARBOHYDRATE COMPOUNDS

The biological properties of many naturally occurring and synthetic molecules that contain an asymmetric center are often very dependent on the absolute configuration of those molecules.[29-31] For example, for some compounds, one enantiomer may be many times more potent than the other, whereas, for other compounds, one enantiomer may have useful therapeutic properties and the other enantiomer may be highly toxic. Similarly, the biological activity of compounds having more than one asymmetric center usually depends both on the absolute and relative configurations of the asymmetric centers in those molecules. Because of these differences, it is necessary to establish the absolute configuration of such isomers and to be able to obtain biologically active, chiral molecules in optically pure forms. A number of examples have been reported in which carbohydrates were used as sources of asymmetric centers for establishing the absolute configuration of biologically active, non-carbohydrate molecules, and a more-limited number of reports have described the use of carbohydrates as synthetic precursors of optically pure, asymmetric compounds required for biochemical evaluation.

The need for *optically pure* asymmetric compounds for biochemical and pharmacological evaluation arises for the following reasons. Where the difference in the potency of enantiomers is large (and many examples are known in which one enantiomer is more than a hundred times as potent as the other), the presence of 1% or less of the active isomer as a contaminant of the less active (or inactive) isomer can account for all or almost all of the activity of the less active (or inactive) isomer,[32] thereby vitiating the value of any results obtained from the more sophisticated type of pharmacological experiment.[33] The difficulties associated with methods for obtaining enantiomers having unequivocal optical purity constitute a major impediment to such studies. As the classical methods of resolution do not necessarily

---

(29) A. F. Casy, in "Medicinal Chemistry," A. Burger, ed., Wiley–Interscience, New York, N. Y., 1970, p. 81.
(30) A. R. Cushny, "Biological Relations of Optically Isomeric Substances," Williams and Wilkins, Baltimore, Md., 1926.
(31) A. H. Beckett and A. F. Casy, *J. Pharm. Pharmacol.*, **7**, 433 (1955).
(32) B. W. J. Ellenbroek, R. J. F. Nivard, J. M. van Rossum, and E. J. Ariens, *J. Pharm. Pharmacol.*, **17**, 393 (1965).
(33) R. W. Brimblecombe, D. M. Green, T. D. Inch, and P. B. Thompson, *J. Pharm. Pharmacol.*, **23**, 745 (1971).

provide optically pure enantiomers[34] and as less than 1% of an optical impurity cannot yet be detected by physical methods of analysis, it is apparent that enantiomers having unequivocal optical purity can at present only be obtained by stereospecific syntheses from optically pure, naturally occurring precursors. Carbohydrates that are cheap, readily available, and optically pure are obvious choices as starting materials for the stereospecific synthesis of many optically pure, non-carbohydrate enantiomers.

The following examples of the use of carbohydrates for the synthesis and configurational assignments of optically pure compounds for pharmacological evaluation, and the examples of the use of carbohydrates only for configurational assignments of other molecules of biological importance, may for convenience be divided into two classes. The first class concerns procedures in which asymmetric centers already present in a carbohydrate are related with, or converted into, other biologically important molecules. The second class consists of procedures in which new asymmetric centers are created within a carbohydrate framework, the configuration of the new asymmetric center is established by reference to other centers having known configuration, and the new asymmetric center is then detached from the carbohydrate framework and related to, or converted into, biologically important molecules. This class of reactions, which usually involves the conversion of carbonyl groups in ketoses into new asymmetric centers, in reactions designated as Type 3 of asymmetric synthesis in Section II,6 (see p. 204), provides a number of examples that illustrate the scope, complexity, and limitations of this type of asymmetric synthesis.

1. Methods that Utilize Asymmetric Carbon Atoms Already Present in Carbohydrates

A number of examples have been described in which carbohydrates have been converted into other biologically active molecules, or in which biologically important molecules, including carbohydrates, have been converted into common products for the purposes of configurational assignment.

a. Relationship of 2-Amino-2-deoxy-D-glucose to L-Alanine.—One of the first examples in which a naturally occurring carbohydrate was

(34) E. L. Eliel, "Stereochemistry of Carbon Compounds," McGraw–Hill, New York, 1962.

OPTICALLY ACTIVE, NON-CARBOHYDRATE COMPOUNDS 207

related to another biologically important molecule was described by Wolfrom and coworkers.[35] 2-Acetamido-2-deoxy-D-glucose (**21**) was converted into the acetylated diethyl dithioacetal **22**, and thence, by desulfurization with Raney nickel, into **23**. O-Deacetylation of **23** afforded **24**, which was oxidized successively with periodate and bromine to give **25**. Compound **25** was also obtained by acetylation of (+)-alanine.

```
      CH₂OH                    H
     ┌──O                    EtSCSEt                CH₃
    /   \                    HCNHAc                HCNHAc
HO  \ OH /  OH               AcOCH                 AcOCH
     \   /                   HCOAc                 HCOAc
      NHAc                   HCOAc                 HCOAc
                             CH₂OAc                CH₂OAc

      21                       22                    23

                       CH₃
                      HCNHAc
                      HOCH
                      HCOH                  CO₂H
                      HCOH                AcHNCH
                      CH₂OH                 CH₃

                        24                    25
```

This series of reactions established that (+)-alanine has the L configuration, and effected a configurational correlation between the standard reference compounds D-glyceraldehyde and L-serine, because C-2 in 2-amino-2-deoxy-D-glucose had been shown to be related to D-glyceraldehyde[36] and (+)-alanine had been related to (−)-serine.[37]

b. Conversion of 2-Amino-2-deoxy-L-glucose into L-(+)-Muscarine.—(+)-Muscarine,[38] the potent cholinergic drug found in *Amanita muscaria* (L), has been synthesized from 2-amino-2-deoxy-L-glucose by the following procedure.[39] 2-Amino-2-deoxy-L-gluconic acid (**26**) was converted into 2,5-anhydro-L-gluconic acid (**27**) by deamination

---

(35) M. L. Wolfrom, R. U. Lemieux, and S. M. Olin, *J. Amer. Chem. Soc.*, **71**, 2870 (1949).
(36) W. N. Haworth, W. H. G. Lake, and S. Peat, *J. Chem. Soc.*, 271 (1939).
(37) E. Fischer and K. Raske, *Ber.*, **40**, 3717 (1907).
(38) P. G. Waser, *Pharmacol. Rev.*, **13**, 465 (1961).
(39) E. Hardegger and F. Lohse, *Helv. Chim. Acta*, **40**, 2383 (1957).

```
      CO₂H
      |
    H₂NCH
      |
     HCOH
      |
     HOCH
      |
     HOCH
      |
     CH₂OH
      26
```

[structure 27: furanose ring with OH, HOCH₂, CO₂H, OH substituents]

26   27

with nitrous acid. The methyl ester of **27**, prepared from **27** by esterification with diazomethane, was converted into the dimethylamide (**28**), and **28** was esterified to the tris(*p*-toluenesulfonate) (**29**). Reduction of **29** with lithium aluminum hydride afforded normuscarine (**30**), which was converted into (+)-muscarine (**31**) by quaternization with methyl iodide. D-(−)-Muscarine has been synthesized from 2-amino-2-deoxy-D-glucose by a similar procedure.[40] The configurational assignment of (+)-muscarine as [tetrahydro-3(*R*)-hydroxy-2(*S*)-methylfur-5(*S*)-ylmethyl]trimethylammonium has had important implications in the formulation of current concepts of structure–activity relationships of cholinergic drugs.[41]

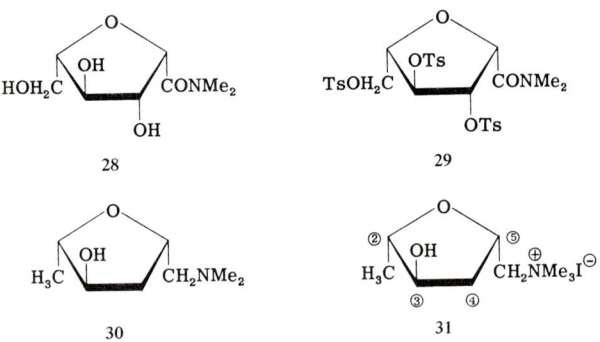

28   29

30   31

c. The Relative Configurations of 4-(Hydroxymethyl)-2-methyl-1,3-dioxolanes.—The isomeric 4-[(dimethylamino)methyl]-2-methyl-1,3-dioxolane methiodides[42] (**32**) have been prepared from 4-(hydroxy-

[structure 32: 1,3-dioxolane with Me₃N⁺H₂C and Me substituents]

32

(40) H. C. Cox, E. Hardegger, F. Kögl, P. Liechti, F. Lohse, and C. A. Salemink, *Helv. Chim. Acta*, **41**, 229 (1958).
(41) A. Bebbington and R. W. Brimblecombe, *Advan. Drug Res.*, **2**, 143 (1965).
(42) B. Belleau and J. Puranen, *J. Med. Chem.*, **6**, 325 (1963); E. Fourneau, D. Bovet, and G. Montezin, *Bull. Soc. Chim. Biol.*, **26**, 134, 516 (1944).

methyl)-2-methyl-1,3-dioxolanes by sequential treatment with
p-toluenesulfonyl chloride, dimethylamine, and methyl iodide, and
by other methods. They are amongst the most potent cholinergic
drugs known. The cholinergic potency of these dioxolanes differs
considerably from isomer to isomer, and depends critically[43] on
the absolute and relative configurations at C-2 and C-4. In order
to establish the relative configuration of dioxolanes formed from the
cis and *trans* isomers separated from the products of reactions between aldehydes and 1-O-*p*-tolylsulfonyl-D- and -L-glycerol, it was
essential to have available, as a reference compound for physicochemical comparisons, a *cis*- or *trans*-2,4-dimethyl-1,3-dioxolane of
unequivocal identity. Thus, D-*cis*-2,4-dimethyl-1,3-dioxolane (**34**)
was prepared from 1,6-anhydro-D-galactose (**33**) by successive application of periodate oxidation, borohydride reduction, *p*-toluene-
sulfonylation, and reduction[44] with lithium aluminum hydride. The
importance of *cis*-2,4-dimethyl-1,3-dioxolane as a reference compound for nuclear magnetic resonance (n.m.r.) studies is now well
established.[45]

d. The Absolute Configuration of S-(–)-3-Piperidinol.—The anticholinergic potency of aminoalkyl glycolates depends[46] on the
absolute configuration of the asymmetric, benzylic carbon atoms in
the acid moiety (see Section III, 2; p. 213), and, in some instances, also
on the configuration of the amino alcohol moiety.[47] In connection
with studies on the relationship between the configuration and the
potency of piperidin-3-yl esters of glycolic acid, the absolute configuration of (–)-3-piperidinol has been established by synthesis[48]
from D-mannitol (**35**), which was converted into the unsaturated

(43) B. Belleau and J. L. Lavoie, *Can. J. Biochem.*, **46**, 1397 (1968).
(44) D. J. Triggle and B. Belleau, *Can. J. Chem.*, **40**, 1201 (1962).
(45) N. Baggett, K. W. Buck, A. B. Foster, M. H. Randall, and J. M. Webber, *J. Chem. Soc.*, 3394 (1965); N. Baggett, K. W. Buck, A. B. Foster, R. Jefferis, B. H. Rees, and J. M. Webber, *ibid.*, 3382 (1965).
(46) E. J. Ariens, *Advan. Drug Res.*, **3**, 235 (1966).
(47) L. H. Steinbach and S. Kaiser, *J. Amer. Chem. Soc.*, **74**, 2219 (1952).
(48) C. C. Deane and T. D. Inch, *Chem. Commun.*, 813 (1969).

derivative **36** by standard procedures, and thence, by successive reduction, acid hydrolysis, selective mono-*p*-toluenesulfonylation, and acetonation, into **37**. Compound **37** was converted into **38** by reaction with sodium azide in *N*,*N*-dimethylformamide, followed by acid hydrolysis. The lactone **39**, formed by sequential oxidation of **38** with sodium periodate and chromium trioxide in pyridine, was converted into **40** by catalytic hydrogenation.[49] Finally, reduction of **40** with lithium aluminum hydride afforded (*S*)-(−)-3-piperidinol (**41**).

e. **The Enantiomeric Acetyl-α- and β-methylcholines.**—Derivatives of 2-(dimethylamino)-1-propanol (**42**) and 1-(dimethylamino)-2-propanol (**43**) have been used for studies of drug–receptor interactions in the central and peripheral nervous systems, and for enzymic studies.[50] For example, benzilic or glycolic acid esters of **42** and **43** antagonize the effects of acetylcholine,[32,51] whereas "acetyl-α-methylcholine" [**44**; (2-acetoxy-1-methylethyl)trimethylammonium iodide,

(49) S. Hanessian and T. H. Haskell, *J. Heterocycl. Chem.*, **1**, 55 (1964); H. Paulsen, *Angew. Chem. Intern. Ed. Engl.*, **5**, 495 (1966).
(50) A. H. Beckett, N. J. Harper, and J. W. Clitherow, *J. Pharm. Pharmacol.*, **15**, 362 (1963).
(51) R. W. Brimblecombe, D. M. Green, and T. D. Inch, *J. Pharm. Pharmacol.*, **22**, 951 (1970).

from **42**] and "acetyl-β-methylcholine" [**45**; (2-acetoxypropyl)trimethylammonium iodide, from **43**] act like acetylcholine in the central and peripheral nervous systems. To avoid use of resolution procedures, the enantiomers of 2-(dimethylamino)-1-propanol and 1-(dimethylamino)-2-propanol have been prepared from carbohydrates, and converted[52] into the enantiomeric "acetyl-α- and β-methylcholines" and also into anticholinergic drugs (see Section III, p. 219). For example, S-(+)-1-(dimethylamino)-2-propanol (**46**) was prepared from 5-O-benzyl-L-rhamnitol (**47**). Compound **47** was converted into the p-toluenesulfonic ester (**48**) by application, in sequence, of periodate oxidation, lithium aluminum hydride reduction, p-toluenesulfonylation, and catalytic hydrogenolysis, and the p-toluenesulfonate **48** was converted into **46** by treatment with dimethylamine. Acetylation of **46**, followed by quaternization, afforded "(S)-(+)-acetyl-β-methylcholine." (R)-(−)-Acetyl-β-methylcholine was similarly prepared from 5-O-benzyl-D-rhamnitol.

```
        CH₂NMe₂          CH₂OH              CH₂OTs
        |                |                  |
        HOCH             HCOH               HOCH
        |                |                  |
        CH₃              HCOH               CH₃
        46               |                  48
                         HOCH
                         |
                         ROCH
                         |
                         CH₃
                         47
```

(R)-(+)-2-(Dimethylamino)-1-propanol (**51**) was prepared from 1,2:3,4-di-O-isopropylidene-5-O-p-tolylsulfonyl-L-rhamnitol (**49**). Treatment of **49** with sodium azide in N,N-dimethylformamide afforded the corresponding 5-azido-5,6-dideoxy-D-gulitol derivative, which was converted into **50** by acid hydrolysis. Sequential application to **50** of periodate oxidation, reduction with lithium aluminum hydride, and methylation gave **51**, which was subsequently converted into "(R)-(+)-acetyl-α-methylcholine." (S)-(−)-Acetyl-α-methylcholine was similarly prepared from 1,2:3,4-di-O-isopropylidene-5-O-p-tolylsulfonyl-D-rhamnitol.

```
     H₂CO                CH₂OH
         \CMe₂           |
     HCO/                HCOH
     |                   |
     HCO                 HCOH              CH₂OH
     |    \CMe₂          |                 |
     OCH  /              HOCH              HCNMe₂
     |                   |                 |
     TsOCH               HCN₃              CH₃
     |                   |                 51
     CH₃                 CH₃
     49                  50
```

52) T. D. Inch and G. J. Lewis, *Carbohyd. Res.*, **16**, 455 (1971).

f. **The Absolute Configuration of (+)-Dethiobiotin.**—(+)-Dethiobiotin (**52**) has been synthesized[53] by using D-glucose as the source of asymmetric centers. As the methyl ester (**53**) of (+)-dethiobiotin has also been derived from (+)-biotin methyl ester (**55**) by desulfurization with Raney nickel, the synthesis of (+)-dethiobiotin also provided a chemical proof of the absolute configuration of biotin (**54**).

```
        H       H                         O
         \     /                          ‖
      Me⟋C —— C⟍CH(CH₂)₄CO₂R          HN⟋C⟍NH
         |     |                          |    |
         HN    NH                        HC —— CH
           ⟍C⟋                            |    |
            ‖                           H₂C    CH(CH₂)₄CO₂R
            O                              ⟍S⟋

        52, R = H                       54, R = H
        53, R = Me                      55, R = Me
```

Methyl 4,6-O-benzylidene-3-deoxy-α-D-*ribo*-hexopyranoside (**56**) was benzoylated, debenzylidenated, and partially *p*-toluenesulfonylated to **57**; this was converted into **58** by reaction with sodium iodide, followed by catalytic reduction. The methanesulfonate of **58** was converted into **59** by reaction with sodium azide in *N*,*N*-dimethylformamide, and **59** was converted into 4-azido-3,4,6-trideoxy-α-D-*xylo*-hexose (**60**) by acetolysis followed by alkaline hydrolysis. Reduction of **60** with borohydride in methanol afforded **61**, which was converted into **62** by successive condensation with acetone, methanesulfonylation, and azide exchange. The 4,5-diazido-3,4,5,6-tetradeoxy-1,2-O-isopropylidene-L-*arabino*-hexitol (**62**) was reduced with hydrogen in the presence of Raney nickel, the resultant diamine was treated with phosgene in the presence of sodium carbonate, and the product was hydrolyzed under acidic conditions to give **63**. The overall yield of **63** from **56** was 4%. The next three reactions (with sodium periodate, the Wittig reaction, and catalytic reduction) were performed without characterization of the intermediate products, and gave (+)-dethiobiotin methyl ester indistinguishable from an authentic sample thereof prepared from (+)-biotin methyl ester.

```
      OCH₂                    CH₂OTs                  CH₃
      /   \                   /    \                  /   \
     /     O                 /      O                /     O
 PhCH      \                /       \               /      \
     \      \              /         \             /        \
      O      OMe         HO           OMe        HO          OMe
      |                   |                       |
      OH                  OBz                     OBz

      56                  57                      58
```

(53) H. Kuzuhara, H. Ohrui, and S. Emoto, *Tetrahedron Lett.* 1185 (1970).

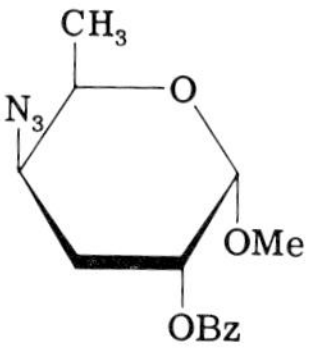

OPTICALLY ACTIVE, NON-CARBOHYDRATE COMPOUNDS 213

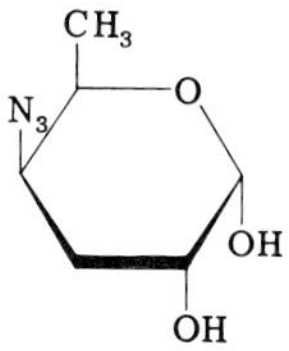

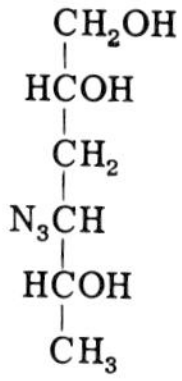

[Structures 59, 60, 61, 62, 63, and reaction scheme 63 → [intermediates] → 53]

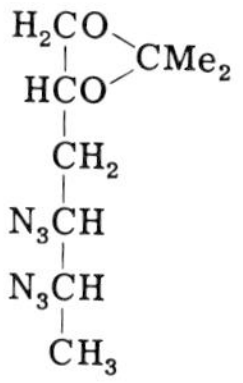

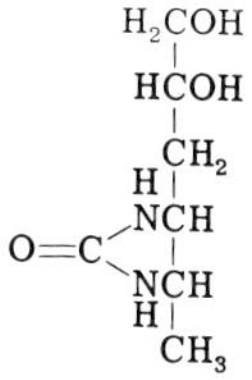

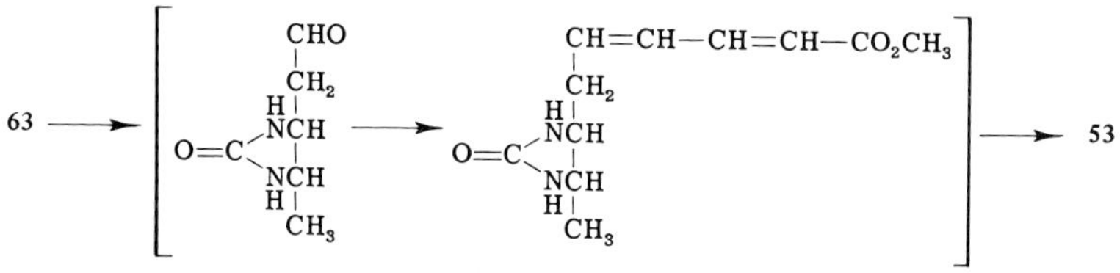

2. Methods that Require the Creation of New Asymmetric Centers in Carbohydrates

a. Asymmetric, Benzylic Carbon Atoms.—There are many examples of highly potent drugs (for example, atropine and adrenaline) that contain an asymmetric, benzylic carbon atom,[54] and it is well known that, in many instances, the biological activity of these drugs depends critically on the absolute configuration of the benzylic carbon atom.[29,46] In order to facilitate pharmacological studies, methods that may have wider applications have been developed for the synthesis of compounds containing asymmetric carbon atoms, with the objective of converting these compounds into atropine-like drugs. Those procedures, used for the synthesis of asymmetric, benzylic carbon atoms within a carbohydrate framework, that involved treatment of derivatives of dicarbonyl sugars with phenylmagnesium bromide illustrate many features of the asymmetric synthesis reactions of Type 3 (see Section II, 6, p. 204) and will be described in detail.

The reaction between phenylmagnesium bromide and 3-O-benzyl-

(54) J. J. Lewis, "Introduction to Pharmacology," Livingstone, Edinburgh, 1965; R. B. Barlow, "Introduction to Chemical Pharmacology," Methuen, London, 1964.

1,2-O-isopropylidene-α-D-xylo-pentodialdo-1,4-furanose (**64**, R = CH$_2$Ph) is highly stereoselective, affording a mixture of the L-*ido* (**65**; R = CH$_2$Ph) and D-*gluco* isomers (**66**; R = CH$_2$Ph) in which

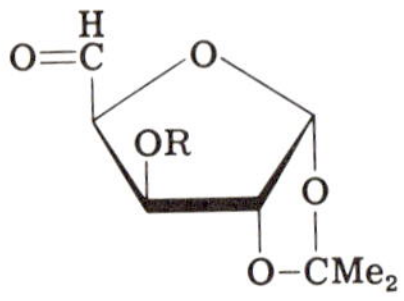

64 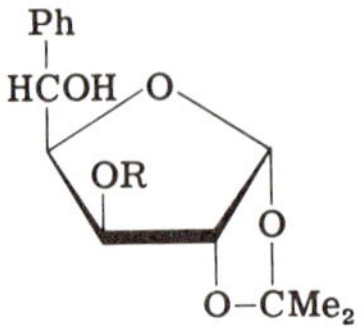   **65** 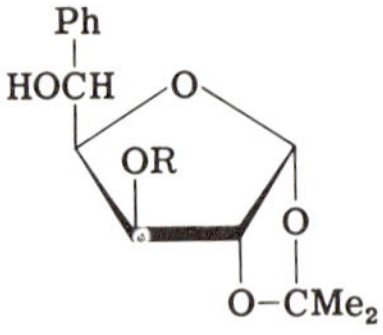   **66**

65 preponderated.[55] The reactions between **67** and phenylmagnesium bromide, and between **68** and methylmagnesium iodide,[22] showed a much higher degree of stereoselectivity than reactions between **64** and methylmagnesium iodide or phenylmagnesium bromide. Thus, **70** was the only product from the reaction of methylmagnesium iodide

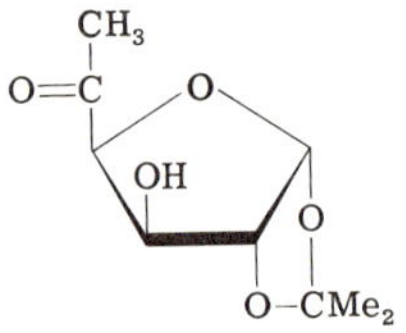

67 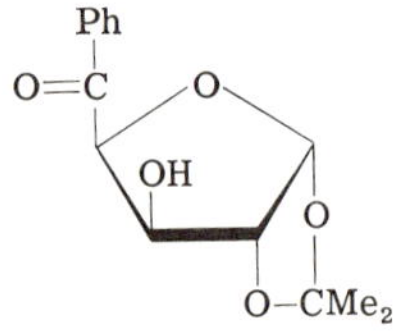   **68**

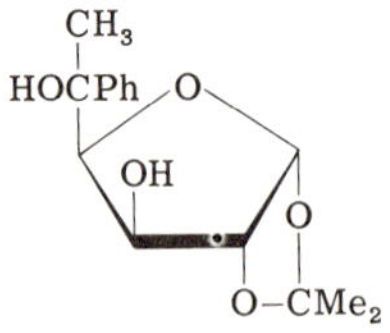

69 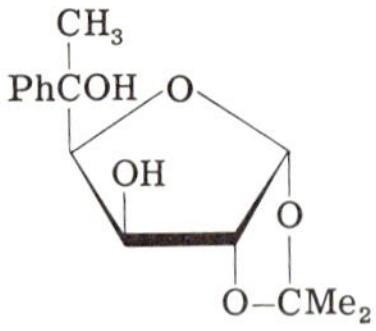   **70**

with **68**; and **69**, the preponderant product from the reaction of phenylmagnesium bromide with **67**, was contaminated with less than 2% of its diastereoisomer (**70**). It was also shown that the reaction between the 3-benzyl ether of **67** and phenylmagnesium bromide is stereospecific, affording only the 3-benzyl ether of **69**. This result is in accordance with the observation that the reaction of **64** (R = CH$_2$Ph) with phenylmagnesium bromide is more stereoselective than the reaction of **64** (R = H) with the same bromide.[55] The Grignard reactions studied were all more highly stereoselective in ether

(55) T. D. Inch, *Carbohyd. Res.*, **5**, 45 (1967).

than in tetrahydrofuran. The fact that the stereoselectivity of reactions of Grignard reagents with dicarbonyl sugar derivatives can be considerably modified by change of solvent or change of a substituent can have important implications in experiments designed to provide one, or both, of the diastereoisomers. That the reactions of the Grignard reagent with **67** and **68** are more stereoselective than those[56] of Grignard reagents and **64** accords with results from reactions of non-carbohydrates, wherein reactions between a Grignard reagent and a ketone have been shown to be more stereoselective than those between a Grignard reagent and an aldehyde.[34]

The formation of **69** and **70** by the reaction sequences illustrated showed that diastereoisomers may conveniently be formed by altering the order of addition of Grignard reagents to the parent aldehyde.

$$64 \ (R = CH_2Ph) \xrightarrow{MeMgI} \xrightarrow{oxidation} \xrightarrow{PhMgBr} 69$$

$$64 \ (R = CH_2Ph) \xrightarrow{PhMgBr} \xrightarrow{oxidation} \xrightarrow{MeMgI} 70$$

In carbohydrate chemistry, as in other branches of organic chemistry, there is considerable difficulty in making unequivocal configurational assignments to asymmetric tertiary alcohols, and the configurational proofs of the structures of **69** and **70** posed the usual problems. Methods for establishing the configuration of specific, branched-chain sugars have been reported,[23,25,28,57] but general procedures have not yet been described. For compounds **69** and **70**, n.m.r. methods permitted only tentative configurational assignments to be made. Thus, the L-*ido* isomer (**70**) was converted into the pyranose tetraacetate (**71**), which was assigned the L-*ido* configuration because it had n.m.r. parameters (chemical shifts and coupling constants) similar to those of 5-C-phenyl-β-D-xylopyranose tetraacetate (**72**), which was obtained from **66**. The n.m.r. spectrum of the pyranose tetraacetate from **69** was similar to that of the pyranose tetraacetate from **65**, and neither spectrum could be analyzed by first-order methods. Thus, the configurational assignment of only one isomer of each of the pairs **65** and **66**, and **69** and **70**, could be made by n.m.r. methods, and no corroboration was forthcoming from n.m.r. analysis of the other isomers.

---

(56) M. L. Wolfrom and S. Hanessian, *J. Org. Chem.*, **27**, 1800 (1962).
(57) R. J. Ferrier, W. G. Overend, G. A. Rafferty, H. M. Wall, and N. R. Williams, *J. Chem. Soc.* (C), 1091 (1968); D. Horton and J. K. Thomson, *Chem. Commun.*, 1389 (1971).

Although probably correct for the examples cited, some of the other arguments used for justifying the structural assignment of **69** as 6-deoxy-1,2-O-isopropylidene-5-C-phenyl-β-D-idofuranose, and of **70** as 6-deoxy-1,2-O-isopropylidene-5-C-phenyl-α-D-glucofuranose, have no real, general value. For example, reports that the steric courses of addition reactions of metal hydrides and of Grignard reagents are similar,[27,58] and the fact that lithium aluminum hydride reduction of **68** affords mainly the D-*gluco* isomer **73**, are consistent with the assignment of the L-*ido* configuration (**70**) to the preponderant product of the reaction of **68** with methylmagnesium iodide. It was also considered that, as the preponderant products from the reaction of methylmagnesium iodide and of phenylmagnesium bromide with **64** could be predicted from a postulated, coordinated transition-state (such as **74**),[22,56] similar transition states for the reaction of phenylmagnesium bromide with **67** and of methylmagnesium iodide with **68** would lead to **69** and **70**, respectively, as the preponderant products.

Examples have been reported that show that reduction with lithium aluminum hydride and Grignard additions do not necessarily

(58) B. R. Baker and D. H. Buss, *J. Org. Chem.*, **31**, 217 (1966).

follow the same steric course, and, moreover, it has also been shown that different Grignard reagents can react with the same ketone to give, as the preponderant product, a compound of different configuration.[28] Thus, whereas reduction of **75** with lithium aluminum hydride afforded only **76**, and, whereas the preponderant product from the reaction of **75** with methylmagnesium iodide was the α-D-glucopyranoside derivative **77** (formed in 72% yield), the preponderant product from the reaction of **76** with phenylmagnesium bromide was the α-D-mannopyranoside derivative (**78**), formed[28] in 62% yield. In contrast to this example, the reactions of **79**, the C-3 epimer of

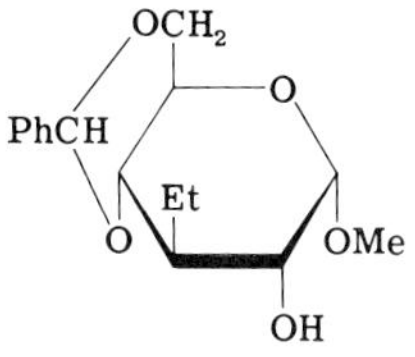

75

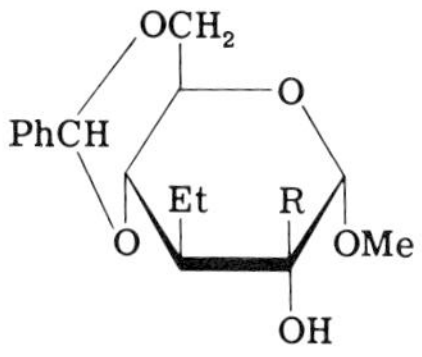

76, R = H
77, R = Me

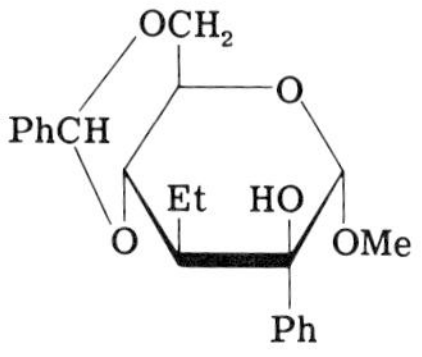

78

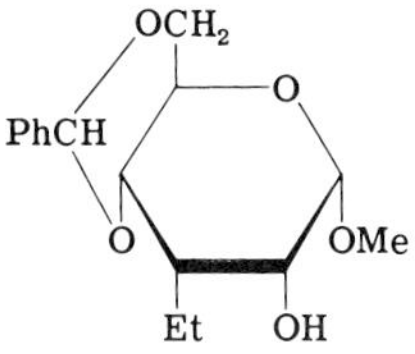

79

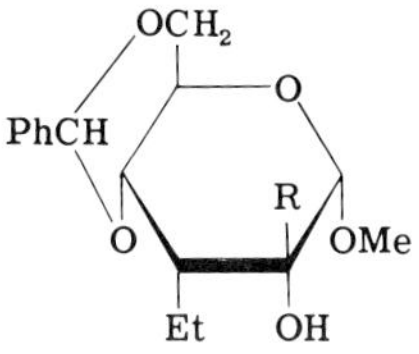

80, R = H
81, R = Me
82, R = Ph

75, with lithium aluminum hydride, methylmagnesium iodide, and phenylmagnesium bromide all followed the same steric course, to give only **80**, **81**, and **82**, respectively. Another example of differences between the steric course of the reactions with Grignard reagents and metal hydrides will be described in Section III, 2(b) (see p. 220). These results illustrate the difficulties that may be encountered if imagined transition-states are used for predicting the favored course of reactions between Grignard reagents and dicarbonyl sugar derivatives, and show that configurational proofs based primarily on comparisons with the configurations of products from related reactions must be considered suspect until actually proved by reliable methods.

The reactions of Grignard reagents with acyclic sugar derivatives have also been investigated as potential routes for the synthesis of compounds containing asymmetric, benzylic carbon atoms.[22] W. A.

Bonner[59] showed that the reaction between phenylmagnesium bromide and 2,3:4,5-di-*O*-isopropylidene-*aldehydo*-D-arabinose (83) is stereoselective and that the preponderant product is the 1-*C*-phenyl-D-*gluco*-pentitol derivative 84. The configuration of 84 was established by its conversion into suitable mandelic acid derivatives of known configuration. Grignard reactions with derivatives of *keto*-ketoses were found much more highly stereoselective than those with derivatives of *aldehydo*-aldoses. Thus, the reaction of 85 with methylmagnesium iodide afforded 86 only, and the reaction between phenylmagnesium bromide and 87 gave mainly 88, with only 8% of the isomer 86. Similarly reactions between cyclohexylmagnesium bromide and 85, and between phenylmagnesium bromide and 89, also proceed with high degrees of stereoselectivity, to give 90 and 91, respectively, as the preponderant, or only, products.[60]

```
                        Ph
                        |
     CHO               HCOH
     |                  |
    OCH                OCH
     |  CMe₂            |  CMe₂
    HCO                HCO
     |                  |
    HCO                HCO
     |  CMe₂            |  CMe₂
    H₂CO               H₂CO

      83                 84


     R                  R'                 R'
     |                  |                  |
     C=O               PhCOH              HOCPh
     |                  |                  |
    OCH                OCH                OCH
     |  CMe₂            |  CMe₂            |  CMe₂
    HCO                HCO                HCO
     |                  |                  |
    HCO                HCO                HCO
     |  CMe₂            |  CMe₂            |  CMe₂
    H₂CO               H₂CO               H₂CO

  85, R = Ph         86, R' = Me        88, R' = Me
  87, R = Me         90, R' = cyclohexyl 91, R' = cyclohexyl
  89, R = cyclohexyl
```

Configurational proofs for 86, 90, 88, and 91 were obtained by n.m.r. analysis of 92 and 93, and 94 and 95. Compounds 92 and 93 were obtained from 86 and 90, respectively, by successive treatment as follows: partial hydrolysis with acid, periodate oxidation, acid hydrolysis, acetonation, and acetylation. Compounds 94 and 95 were similarly prepared from 88 and 91, respectively. The main argument in favor of the structures assigned was that, in 94 and 95, the acetoxymethyl group is shielded by the phenyl substituent and resonates

(59) W. A. Bonner, *J. Amer. Chem. Soc.*, **73**, 3126 (1951).
(60) T. D. Inch, R. V. Ley, and P. Rich, *J. Chem. Soc. (C)*, 1693 (1968).

at very high field (1.72 p.p.m.), whereas, in **92** and **93**, in which no such shielding occurs, the acetoxymethyl group resonates at the more usual frequency (for furanoid derivatives) of 2.1 p.p.m. The consistency of the configurational assignments for **86, 88, 69,** and **70** was demonstrated when the structures of **70** and **86** were correlated by chemical methods.[23]

```
                  R'
                  |
                PhCOH                        CHO                        CMe₂
86                |                          |               Ph    O    O
   H⊕, H₂O      OCH            IO₄⁻        OCH                \  /  \ /
or    ──────→      \  CMe₂   ──────→          \  CMe₂  ──────→   /    \
90                HCO/                       HCO/             R'        O
                  |                          |                  \     /
                 HCO                        HOCPh                 AcO
                  |  \ CMe₂                  |
                H₂CO/                        R'
                                                              92, R = Me
                                                              93, R = cyclohexyl
```

```
              CMe₂
           /\
       R'  O  O
        \ /  /
         \  O
       Ph/
         |
         AcO

     94, R' = Me
     95, R' = cyclohexyl
```

In the foregoing examples, which show that it is relatively easy to form asymmetric centers within a carbohydrate framework, it was possible to obtain optically pure, benzylic carbon atoms, because, even where the reactions were not stereospecific, it was feasible to separate the diastereoisomers by chromatographic methods. The moieties containing the benzylic carbon atoms were readily detached from the carbohydrate framework for subsequent conversion into anticholinergic drugs. Thus, (R)-(−)- and (S)-(+)-2-cyclohexyl-2-phenylglycolaldehyde (**96** and **97**) were obtained from **90** and **91**, respectively, by acid hydrolysis followed by periodate oxidation.[60] Condensation of **96** and **97** with 1-O-p-tolylsulfonyl-D-(or L-)glycerol, chromatographic separation of the *cis* and *trans* isomers on silica gel, and treatment of the isomers with dimethylamine, afforded[61] the eight isomers of the highly potent, anticholinergic drug **98**.

The methyl esters prepared by oxidation of **96** and **97**, respectively, with iodine in methanolic potassium hydroxide were converted into atropine-like drugs by transesterification with 2-(dimethylamino)-

(61) R. W. Brimblecombe, T. D. Inch, J. Wetherell, and N. Williams, *J. Pharm. Pharmacol.*, **23**, 649 (1971).

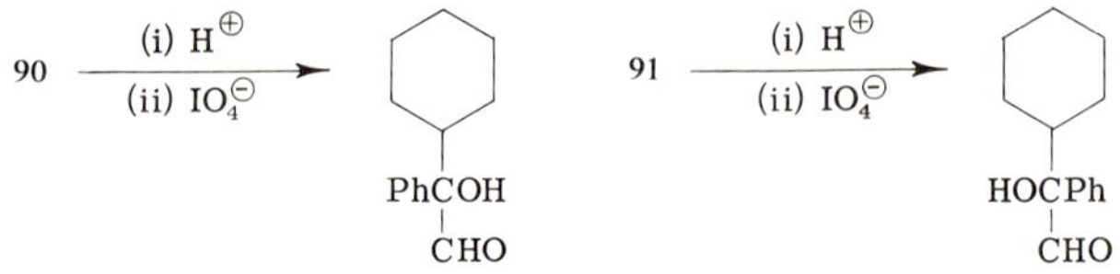

220   T. D. INCH

$$90 \xrightarrow[\text{(ii) IO}_4^\ominus]{\text{(i) H}^\oplus} \begin{array}{c} \text{Cy} \\ \text{PhCOH} \\ | \\ \text{CHO} \\ 96 \end{array} \quad 91 \xrightarrow[\text{(ii) IO}_4^\ominus]{\text{(i) H}^\oplus} \begin{array}{c} \text{Cy} \\ \text{HOCPh} \\ | \\ \text{CHO} \\ 97 \end{array}$$

H₂CNMe₂
   \
    O   O
     \ /
     PhCOH
      |
      Cy

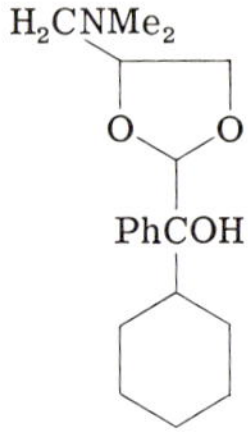

98

ethanol, N-methyl-4-piperidinol, N-methyl-3-piperidinol (see Section III, 1d, p. 209), 2-(dimethylamino)-1-propanol, and 1-(dimethylamino)-2-propanol (see Section III, 1e, p. 210).[33,51]

b. The Absolute Configuration of (+)-Ethanol-1-d.—Optically active ethanol-1-d is produced by alcohol dehydrogenase-catalyzed reduction of acetaldehyde-1-d by reduced nicotinamide adenine dinucleotide (β-NADH) or by similar reduction of acetaldehyde by means of deuterium-reduced β-NAD (see also, This Volume, Chapter 4). The absolute configuration of (+)-ethanol-1-d, knowledge of which was essential to a fuller understanding of the reductions mentioned, was established by the following stereoselective synthesis.[62] The dialdose derivative **64** was reduced with lithium aluminum deuteride in ether, to afford **99**, which was converted into β-D-xylopyranose-5-d tetraacetate (**100**) by successive catalytic

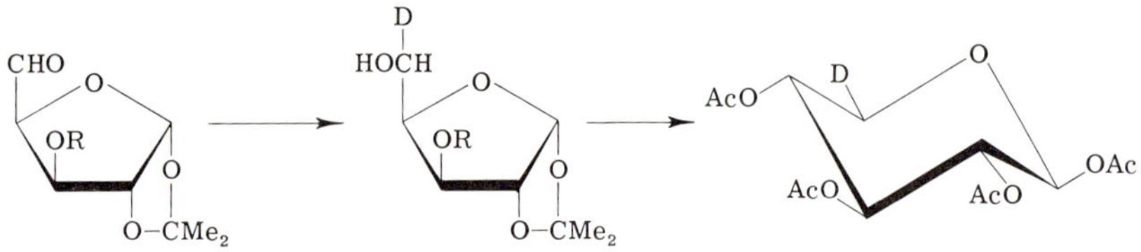

                D
                |
   CHO   HOCH
    \O   \O   AcO   D   O
    OR   →   OR   →
      \O   \O   AcO   OAc
       | | AcO
       O-CMe₂   O-CMe₂
       64   99   100

where R = CH₂Ph.

hydrogenolysis, acid hydrolysis, and acetylation. The intensity of the n.m.r. signals from the protons on C-5 showed that that isomer of

(62) R. U. Lemieux and J. Howard, *Can. J. Chem.*, **41**, 308 (1963).

100 in which the deuterium is equatorially attached was some 30% in excess of the isomer of **100** in which the deuterium is axially attached. In terms of the Cahn–Ingold–Prelog notation, the preponderant isomer of **99** had the *R*-configuration at C-5. The β-D-xylopyranose-5-*d* tetraacetate (**100**) was degraded to (+)-ethanol-*1*-*d* by the procedure illustrated, in which the configuration at C-5 was not affected, thereby establishing the *R*-configuration for (+)-ethanol-*1*-*d*. It is of interest

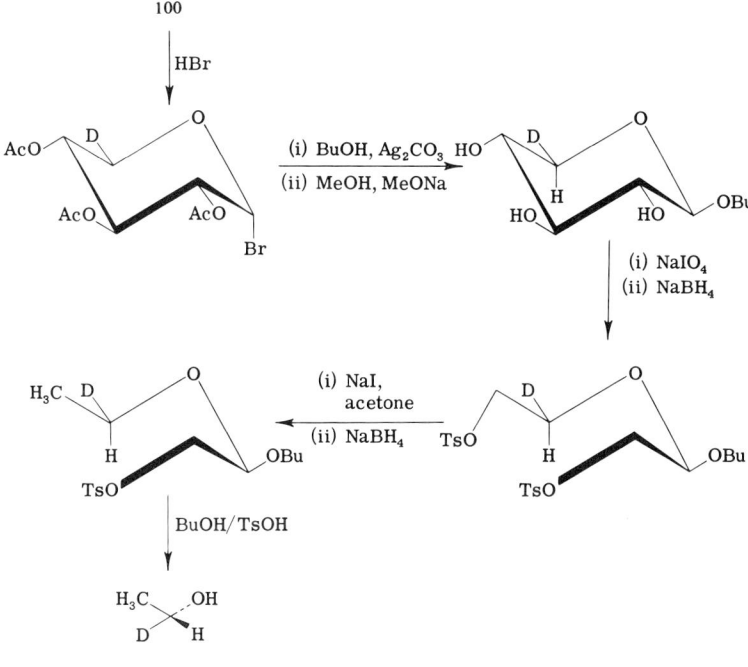

that the steric course of the key step in this configurational proof, namely, the stereoselective reduction of **64**, differed from the steric course of Grignard addition reactions of **64**. For example, it has been demonstrated that the preponderant product from the reaction of methylmagnesium iodide with **64** is the 6-deoxy-L-idose derivative **101**.

IV. STEREOSELECTIVE SYNTHESIS OF ASYMMETRIC SULFOXIDES

Carbohydrates have been envisaged as precursors of naturally occurring, optically active sulfoxides having the general formulas $CH_3SO(CH_2)_nNCS$ and $CH_3SOCH_2CH(NH_2)CO_2H$, and a variety of other types of optically active sulfoxides.[63,64] In principle, the use of carbohydrates for the synthesis of naturally occurring, optically active sulfoxides parallels the use of carbohydrates in the synthesis of biologically active compounds containing only asymmetric carbon atoms. Thus, it has been shown that sulfoxides may be created stereoselectively within a carbohydrate framework,[65-68] and that the sulfoxide configuration may be related to the configuration of the asymmetric carbon atoms within the framework.[65,68-71] However, the conversion of asymmetric sulfoxide centers that have been generated (within a carbohydrate framework) into naturally occurring products remains to be achieved.

For example, the isomeric 1,4-oxathiane S-oxides **102** and **103** have

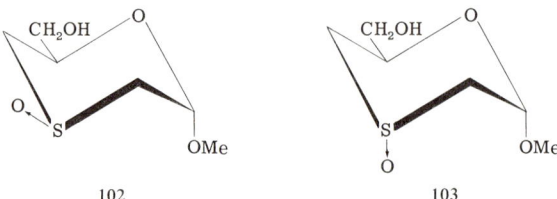

been prepared in the ratio 10:1 from methyl 6-O-trityl-α-D-glucopyranoside by a synthetic procedure involving application, in sequence, of lead tetraacetate oxidation, sodium borohydride reduction, p-

(63) K. W. Buck, F. A. Fahim, A. B. Foster, A. R. Perry, and J. M. Webber, *Abstr. Papers Amer. Chem. Soc. Meeting*, **150**, 18 D (1965).
(64) Q. H. Hasan, Ph. D. Thesis, University of Birmingham, England, 1970.
(65) K. W. Buck, A. B. Foster, A. R. Perry, and J. M. Webber, *Chem. Commun.*, 433 (1965).
(66) A. B. Foster, Q. H. Hasan, D. R. Hawkins, and J. M. Webber, *Chem. Commun.*, 1084 (1968).
(67) K. W. Buck, F. A. Fahim, A. B. Foster, A. R. Perry, M. H. Qadir, and J. M. Webber, *Carbohyd. Res.*, **2**, 14 (1966).
(68) K. W. Buck, A. B. Foster, W. D. Pardoe, M. H. Qadir, and J. M. Webber, *Chem. Commun.*, 759 (1966).
(69) A. B. Foster, T. D. Inch, M. H. Qadir, and J. M. Webber, *Chem. Commun.*, 1086 (1968).
(70) A. B. Foster, J. M. Duxbury, T. D. Inch, and J. M. Webber, *Chem. Commun.*, 881 (1967).
(71) K. W. Buck, T. A. Hamor, and D. J. Watkin, *Chem. Commun.*, 759 (1966).

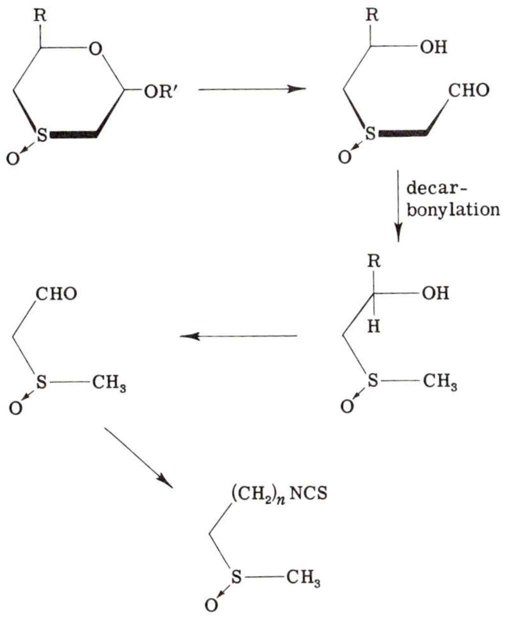

OPTICALLY ACTIVE, NON-CARBOHYDRATE COMPOUNDS 223

Scheme 1

toluenesulfonylation, treatment with sodium sulfide in boiling methanol, hydrogenolysis, and periodate oxidation.[65,67] It was proposed[64] to convert the sulfoxides into isocyanates by a general sequence, such as that illustrated in Scheme 1. The most interesting features of this work are the stereoselectivity of the reactions involved in the conversion of the oxathianes into their S-oxides, and the methods that have been used to establish the configuration of the S-oxides.

It has been shown[66] that, whereas oxidation of **104** with sodium periodate affords a mixture of the axial and equatorial sulfoxides, **102** and **103**, in which the equatorial sulfoxide preponderates (the ratio of **102:103** is 10:1), similar oxidation of **105** affords a mixture of **106** and **107** in which the axial isomer preponderates (ratio of **106:107**,

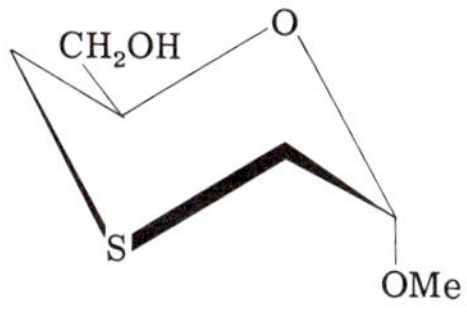

104 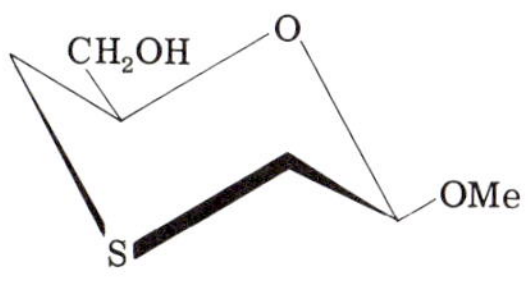   105

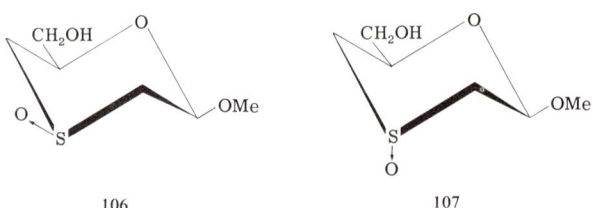

106   107

1:10). On the basis of these and related results, and because variation of the oxidant did not produce the clear-cut differences in the steric course of oxidation observed with thianes, it was suggested[66] that control of the S-oxidation of derivatives of 2-hydroxy-1,4-oxathiane is best achieved by changing the configuration at the anomeric center; an axially attached substituent such as methoxyl or acetoxyl engenders equatorial S-oxygenation, whereas an equatorially attached, anomeric substituent leads to an axial S-oxide.

The ease of reduction of sulfoxides with diborane also depends on the configuration of the sulfoxide.[64] Thus, in such compounds as 106 and 107, reduction of the equatorial sulfoxide is favored, presumably because diborane attacks the less-hindered sulfoxide oxygen atom. However, with the sulfoxide pair 102 and 103, the axial sulfoxide is reduced to a greater extent than the equatorial one. The role of the axially attached substituents on C-2 in promoting reduction of the axial sulfoxides is not yet clear. From such studies, the mechanisms of oxidation of sulfides to sulfoxides, and the reduction of sulfoxides to sulfides, will, perhaps, become sufficiently well elucidated to be of value in establishing the configuration of sulfoxides.

At present, three procedures have been used for ascertaining the configuration at sulfur in a number of cyclic S-oxides. The structure of (2S,6R)-6-(hydroxymethyl)-2-methoxy-1,4-oxathiane has been established by X-ray crystallography[71] and by a chemical method.[65] In the latter method, the solvolysis in water (at pH 7 and 95°) of the methanesulfonates of 102 and of the corresponding axial sulfoxide (103) was studied. It was found that the methanesulfonate of 102 is converted into 103, whereas the sulfoxide 103 is slowly regenerated from its methanesulfonate. Presumably, the equatorial sulfoxide reacts, as illustrated (108), in the $1C$(D) form, by way of an intramolecular displacement followed by attack of water on the sulfur atom of the resultant sulfoxonium salt. Finally, n.m.r. studies on oxathiane S-oxides[68,69] and the more rigid S-oxides of the methyl 3,4-di-O-acetyl-2,6-anhydro-2-thio-α-D-altropyranosides[70] have shown that certain of the n.m.r. parameters are diagnostic of the configuration of S-oxides. Thus, the difference in chemical shift between geminal protons vicinal to the sulfoxide center is larger when the S→O bond is equatorial than when it is axial. Furthermore, the geminal coupling-constants of such protons in compounds in which the S→O bond is equatorial are smaller than when it is axial (and, moreover, smaller than for the parent sulfide). Also, in compounds in which the S→O bond is axial, any protons situated *syn*-axially will resonate at lower field (by ~1 p.p.m.) than when the bond is equatorial.

The aforementioned studies demonstrate quite clearly the potential use of asymmetric reactions at sulfur within a carbohydrate framework, and suggest the possibility that carbohydrates may provide convenient frameworks for studies of asymmetric reactions at nitrogen[67] and phosphorus. Also, just as carbohydrates have been shown to be useful chiral groups for promoting highly stereoselective reactions in oxo esters (see Sections II, 3 and II, 4, pp. 199 and 201). it is probable that carbohydrates could profitably be employed as replacements for the menthyl group, for example, in stereoselective Grignard reactions with sulfinates[72] and phosphonates.[73]

(72) P. Bickart, M. Axelrod, J. Jacobus, and K. Mislow, *J. Amer. Chem. Soc.*, **89**, 697 (1967)
(73) O. Korpiun and K. Mislow, *J. Amer. Chem. Soc.*, **89**, 4784 (1967).

THE WITTIG REACTION IN CARBOHYDRATE CHEMISTRY

BY YU. A. ZHDANOV, YU. E. ALEXEEV, AND V. G. ALEXEEVA

Department of Chemistry, The Rostov State University, Rostov-on-Don, U. S. S. R.

I. Introduction 227
II. General Considerations 228
III. Types of Final Products and Their Utilization in Synthesis 232
    1. C-Glycosylated Unsaturated Hydrocarbons 233
    2. α,β-Unsaturated Carbonyl Sugars 244
    3. α,β-Unsaturated Acids and Their Derivatives................. 253
    4. Alkyl Ald-3-enulosonates 267
    5. Enol, Thioenol, and Enamino Derivatives 270
IV. Anomalous Wittig Reaction With Free and With Partially Protected Sugars . 284
    1. Main Directions of the Reaction 284
    2. Stereochemical Considerations 290
V. Conclusion 292
VI. Further Developments 292

I. INTRODUCTION

Discovered only nineteen years ago,[1] the Wittig reaction has already become one of the fundamental synthetic methods of modern organic chemistry[2(a)–(f)] as a most convenient route to unsaturated compounds. Mild reaction-conditions, together with high, sometimes almost quantitative, yields and the absence of migration of the bond formed are the remarkable features of this method. A number of natural compounds (*e.g.*, carotenoids and methylenic steroids) have been synthesized only by using this reaction, and many others are obtained in better yields by this than by other methods.

As synthetic carbohydrate chemistry is based on the fundamental reactions of general organic chemistry, the Wittig reaction is important in the sugar field. It was first applied[3] to sugars in 1962,

(1) G. Wittig and G. Geissler, *Ann.*, **580**, 44 (1953).
(2) (a) G. Wittig, *Experientia*, **12**, 41 (1956); (b) G. Wittig, *Angew. Chem.*, **68**, 505 (1956); (c) J. Levisalles, *Bull Soc. Chim. Fr.*, 1021 (1958); (d) U. Schöllkopf, *Angew. Chem.*, **71**, 240 (1959); (e) L. A. Yanovskaya, *Usp. Khim.*, **30**, 813 (1961); (f) A. Maercker, *Org. Reactions*, **14**, 270 (1965).
(3) R. Kuhn and R. Brossmer, *Angew. Chem.*, **74**, 252 (1962).

227

when glyceraldehyde was allowed to react with carboethoxymethylenetriphenylphosphorane. However, systematic investigations in this field began in 1963 with work by Zhdanov and coworkers[4] and Kochetkov and Dmitriev.[5] Application of this reaction was then expanded in various directions.[6] It has proved an extremely useful synthesis for numerous, previously inaccessible, unsaturated sugars which were shown to be capable of undergoing a wide range of useful transformations. The versatility of this method has been demonstrated by several syntheses of such biologically important products as sialic acids, 2-deoxy sugars, and sphingosines.

Further development of the Wittig reaction in carbohydrate chemistry is demonstrated by the many publications each year on this topic. Hence, there is now an urgent need to review these reports. In the present article, the authors summarize first the advances in the application of the Wittig reaction in the carbohydrate field through the end of 1970. As the structural variations in the carbonyl compounds employed (*aldehydo* or *keto* sugars, and, to a lesser extent, partially protected or free sugars) are not so great in comparison with the practically unlimited variety of phosphoranes, it seems logical to discuss, in sequence, the structural types of unsaturated sugars resulting, as well as to give a detailed discussion of their properties and synthetic utility. Developments after 1970 are discussed in Section VI (see p. 292).

II. GENERAL CONSIDERATIONS

The Wittig reaction involves the interaction of an oxo compound (an aldehyde or ketone) (**1**) with phosphonium ylides [substituted methylene phosphoranes (**2a,b**)], through an intermediate betaine (**3**), to yield the appropriate alkene (**4**) and triphenylphosphine oxide (**5**), as shown in equation 1.

$$RR_1CO + Ph_3\overset{\oplus}{P}-\overset{\ominus}{C}R_2R_3 \longrightarrow \begin{matrix} RR_1C-\overset{\ominus}{O} \\ | \quad \overset{\oplus}{} \\ R_2R_3C-PPh_3 \end{matrix} \longrightarrow \begin{matrix} RR_1C \\ || \\ R_2R_3C \end{matrix} + Ph_3P=O \qquad (1)$$

1   2a   3   4   5

$$Ph_3P=CR_2R_3$$
2b

(4) Yu. A. Zhdanov, G. N. Dorofeenko, and L. A. Uslova, *Dokl. Akad. Nauk SSSR*, **33**, 3444 (1963).
(5) N. K. Kochetkov and B. A. Dmitriev, *Chem. Ind.* (London), 864 (1963).
(6) Yu. A. Zhdanov and L. A. Uslova, *Khim. Rast. Veshchestv* (Tashkent), **3**, 15 (1968).

Depending on the magnitude of the contribution of the two resonance structures **2a** and **2b**, phosphoranes may be roughly divided into true phosphonium ylides having the preponderant structure **2a**, and resonance-stabilized phosphoranes having the preponderant structure **2b**. The phosphoranes of type **2a** that contain electron-releasing groups R_2 or R_3 are extremely reactive and unstable, whereas the phosphoranes of type **2b** that contain electron-attracting groups R_2 or R_3 may, with some exceptions, react only with aldehydes, being rather unreactive under ordinary conditions.

The Wittig reaction generally results in the preponderant formation of the *trans*-isomer; the yield of the *cis*-isomer is increased with decrease in stability of the phosphorane.

Application of the Wittig reaction in the carbohydrate field is accompanied by certain difficulties. A correct choice of the initial sugar components is the main problem, owing to the basicity of phosphoranes and, especially, to the drastically basic conditions employed with phosphonium ylides (**2a**). It is not surprising, therefore, that protected (acetalated and acetylated) *aldehydo* sugars and resonance-stabilized phosphoranes were used at first,[3–5] although partially protected, and even unprotected, aldoses were shown to be amenable to the reaction with various resonance-stabilized phosphoranes,[7(a)–(i)] thanks to the presence of the carbonyl form in the mobile equilibrium. The latter reactions, however, are extremely complicated (see Section IV, p. 284).

The course of reaction of phosphonium ylides (**2a**) with sugars is unambiguous only when protected *aldehydo* or *keto* sugars are used, although the interaction of (methylthio)methylenetriphenylphosphonium ylide with free sugars has been reported.[8] Such ylides are usually obtained by treatment of the corresponding phosphonium salts with a suitable proton-acceptor, for example, phenyllithium[9(a)–(g)] or sodium

---

(7) (a) V. A. Polenov and Yu. A. Zhdanov, *Zh. Obshch. Khim.*, **37**, 2455 (1967); *Chem. Abstr.*, **69**, 77652 (1968); (b) V. A. Polenov, *Tsesisy Vses. Konf. Khim. Biokhim. Uglevodov, 4th, Moskow-Lvov*, 20 (1967); (c) Yu. A. Zhdanov and V. A. Polenov, *Zh. Obshch. Khim.*, **38**, 1046 (1968); *Chem. Abstr.*, **69**, 97046 (1968); (d) V. A. Polenov, *Mater. Nauch. Konf. Molodykh Uchenych Rost. Obl., Ser. Est. Nauk, 2nd, Rostov-on-Don*, 32 (1968); (e) Yu. A. Zhdanov and V. A. Polenov, *Zh. Obshch. Khim.*, **39**, 119 (1969); (f) *idem, ibid.*, **39**, 1121 (1969); (g) *idem, ibid.*, **39**, 1124 (1969); (h) V. A. Polenov and Yu. A. Zhdanov, *Khim. Biokhim. Uglevodov, Moskow*, 94 (1969); (i) V. A. Polenov, Dissertation, Rostov-on-Don, 1969.

(8) H. J. Bestmann and J. Angerer, *Tetrahedron Lett.*, 3665 (1969).

(9) (a) Yu. A. Zhdanov and V. G. Alexeeva, *Zh. Obshch. Khim.*, **37**, 1408 (1967); (b) V. G. Alexeeva and Yu. E. Alexeev, *Tsesisy Vses. Konf. Khim. Biokhim. Uglevodov, 4th, Moskow-Lvov*, 20 (1967); (c) Yu. A. Zhdanov and V. G. Alexeeva, *Zh. Obshch.*

(*References continued on page 230.*)

amide,[10,11] in an inert atmosphere, the product being used without isolation. It would certainly be desirable to facilitate the reaction by generating the ylides (2a) *in situ* with an alkali metal alkoxide as the proton acceptor. However, this modified method[12] is normally not practicable in the carbohydrate series, although it has been used successfully in certain cases.[13-16] This difficulty arises from the instability of acetalated or alkylated ketoses in basic media.

Thus, permethylated *aldehydo* sugars (6) have been found very susceptible to alkaline conditions, and suffer rapid β-elimination, because of the acidity of the α-proton, with the formation of the enolic derivative (8) through a possible intermediate anion (7), as shown in equation 2. This phenomenon may explain the absence of interaction[17] between the protected dialdose 9 and phenylenedimethylene-bis(triphenylphosphonium) chloride in the presence of lithium ethoxide. Dialdose 9, indeed, is rapidly transformed[18] in an alkaline medium into the unsaturated aldehyde 10 (see also, Ref. 19). Un-

$$\begin{array}{ccc}
\text{CHO} & \text{H}\diagdown\text{C}\!\!\diagup\!\!\text{O}^{\ominus} & \text{CHO} \\
| & \| & | \\
\text{CHOMe} & \text{COMe} & \text{COMe} \\
| \quad \text{OR}^{\ominus} & | \quad -\text{OMe}^{\ominus} & \| \\
\text{CHOMe} \longrightarrow & \text{CHOMe} \longrightarrow & \text{CH} \qquad (2) \\
| & | & | \\
(\text{CHOMe})_n & (\text{CHOMe})_n & (\text{CHOMe})_n \\
| & | & | \\
\text{CH}_2\text{OMe} & \text{CH}_2\text{OMe} & \text{CH}_2\text{OMe} \\
6 & 7 & 8
\end{array}$$

Khim., 38, 1951 (1968); (d) *idem, ibid.*, 39, 405 (1969); (e) V. G. Alexeeva and Yu. E. Alexeev, Khim. Biokhim. Uglevodov, Moskow, 91 (1969); (f) Yu. A. Zhdanov, V. G. Alexeeva, L. A. Uslova, V. A. Polenov, and Yu. E. Alexeev, *Mezhdunar. Simp. Khim. Prir. Soed.*, 7th, Riga, 409 (1970); (g) V. G. Alexeeva, Dissertation, Rostov-on-Don, 1969.

(10) E. H. Williams, W. A. Szarek, and J. K. N. Jones, *Can. J. Chem.*, 47, 4467 (1969).
(11) D. G. Lance, W. A. Szarek, J. K. N. Jones, and G. B. Howarth, *Can. J., Chem.*, 47, 2871 (1969).
(12) T. W. Campbell and R. N. McDonald, *J. Org. Chem.*, 24, 1246 (1959).
(13) A. Rosenthal and L. Nguyen, *Tetrahedron Lett.*, 2393 (1967).
(14) A. Rosenthal and L. (Benzing) Nguyen, *J. Org. Chem.*, 34, 1029 (1969).
(15) (a) M. N. Mirzayanova, L. P. Davydova, and G. I. Samokhvalov, *Dokl. Akad. Nauk SSSR*, 173, 367 (1967); *Chem. Abstr.*, 67, 54368 (1967); (b) *idem, Zh. Obshch. Khim.*, 38, 1954 (1968); (c) *idem, ibid.*, 40, 693 (1970); (d) M. N. Mirzayanova and G. I. Samokhvalov, *Mezhdunar. Simp. Khim. Prir. Soed.*, 7th, Riga, 386 (1970).
(16) G. Baschang and H. Fritz, *Helv. Chim. Acta*, 52, 300 (1969).
(17) Yu. A. Zhdanov, Yu. E. Alexeev, and G. N. Dorofeenko, *Zh. Obshch. Khim.*, 37, 98 (1967).
(18) Yu. E. Alexeev, Dissertation, Rostov-on-Don, 1969.
(19) D. M. Brown and G. H. Jones, *J. Chem. Soc.* (C), 249 (1967).

THE WITTIG REACTION IN CARBOHYDRATE CHEMISTRY 231

saturated compounds of the types **8** or **10** may be degraded further, thus complicating the outcome of the Wittig reaction.

Self-addition of the aldol type is also a general feature of protected glyculoses in basic media.[20(a)–(c)] Acetalated *keto* sugars were shown to be able to undergo other types of transformations under these conditions. Thus, the protected aldose **11** possibly undergoes epimerization at C-4 during the Wittig reaction, to yield its epimer **12**, with concomitant rearrangement into **13** involving participation of the dioxolane ring (see later, p. 273).

It is, therefore, necessary to conclude that use of protected ketoses having alkali-stable protecting groups in the Wittig reaction or its modifications does not guarantee absolute success. Other examples of this kind will be discussed later (see p. 258).

The basicity of the medium is considerably less when resonance-stabilized phosphoranes are employed, resulting generally in a normal course for the Wittig reaction. A direct correlation between the basicity of certain phosphoranes and their reactivity towards 2,4:3,5-di-*O*-benzylidene-*aldehydo*-D-ribose and 2,3,4,5,6-penta-*O*-acetyl-*aldehydo*-D-glucose was not found,[21] thus indicating a steric influence prevalent in the sugar components, in comparison with electronic factors in phosphoranes.

Besides the correct choice of the sugar component, the separation of the final unsaturated sugar from the side-product triphenylphos-

(20) (a) R. Schaffer, *J. Amer. Chem. Soc.*, **81**, 2838 (1959); (b) R. Schaffer and H. S. Isbell, *ibid.*, **81**, 2178 (1959); (c) D. Horton and E. K. Just, *Carbohyd. Res.*, **9**, 129 (1969).
(21) R. E. Harmon, G. Wellman, and S. K. Gupta, *Carbohyd. Res.*, **14**, 123 (1970).

phine oxide (5) is a second problem encountered in the Wittig reaction in the carbohydrate series, since there is no common mode of processing in this respect. The insolubility of compound 5 in light petroleum is often used (see, for instance, Ref. 10), provided that the sugar product has good solubility in this solvent. Precipitation of the sugar product from aqueous alcohol may be also used, as triphenylphosphine oxide possesses good solubility therein.[22,23] These treatments, however, afford only partial separation. Complete purification of the final unsaturated sugars is achieved either by chromatography or by precipitation of remaining triphenylphosphine oxide with anhydrous lithium bromide in dry ether.[23,24] Deacetylation of the sugar product followed by extraction with water is also possible[25] when an acetylated *aldehydo* aldose is the starting material.

The Wittig reaction in the carbohydrate series generally results in the insertion of an unsaturated fragment in various positions of the sugar molecule. This fragment may be simple or derivatized. The latter case is of especial interest, as such derivatization leads to various "modified" sugars capable of further transformations. The derivatives of the first type may be considered to be C-glycosylated alkenes. The second type of final Wittig products includes mainly α,β-unsaturated carbonyl compounds, as well as sugar enol derivatives. These sugar Wittig products may be classified as follows: 1, C-glycosylated, unsaturated hydrocarbons; 2, α,β-unsaturated ketoses; 3, derivatives of α,β-unsaturated aldonic and uronic acids; 4, α-keto-β,γ-unsaturated esters; and 5, sugar enol, thioenol, and enamino derivatives. The abnormal interaction of *p*-methoxybenzoyl and acetylmethylenetriphenylphosphorane with free and with partially protected sugars, examined in detail in our laboratory, will be discussed separately in Section IV (see p. 284).

III. Types of Final Products and Their Utilization in Synthesis

The methylenetriphenylphosphoranes employed in the carbohydrate series are extremely diverse, including, (*a*) the very unstable alkylmethylenephosphonium ylides, (*b*) aryl and alkoxycarbonyl-

(22) See Ref. 17.
(23) Yu. A. Zhdanov, Yu. E. Alexeev, and G. N. Dorofeenko, *Zh. Obshch. Khim.*, **36**, 1742 (1966).
(24) Yu. A. Zhdanov, L. A. Uslova, and G. N. Dorofeenko, *Carbohyd. Res.*, **3**, 69 (1966).
(25) Yu. A. Zhdanov, G. N. Dorofeenko, and L. A. Uslova, *Khim. Obmen Uglevodov, Mater. Vses. Konf. Kh., 3rd, Moscow, 1963*, 67 (1965); *Chem. Abstr.*, **65**, 3946 (1966).

alkoxymethylenephosphoranes of intermediate stability, and (c) a number of resonance-stabilized phosphoranes.

The main interest in unsaturated sugars prepared by the Wittig reaction (and described in this Section) has concerned their synthetic utilization. The pathways of the latter depend on the structure of the unsaturated precursor. In the case of C-glycosylated alkenes, addition to the double bond (mainly hydration and hydrogenation) leads to a branched-chain or long-chain sugar, although correct choice of the reactant to be added may provide a variety of derivatives.

Derivatized unsaturated sugars provide more possibilities for further derivatization, by using both the double bond and the functional group. Thus, hydroxylation of α,β-unsaturated aldonic esters has led, through higher aldonic esters followed by reduction, to the corresponding higher sugars,[26] and indirect hydration of α-keto-β,γ-unsaturated aldonic esters produced[27(a)-(c)] naturally occurring 3-deoxyglyculosonic acids. Acid hydrolysis of enol and enamino derivatives resulted in 2-deoxy sugars[28(a)-(c)] and 3-deoxyglyculosonic acids. [15(a)-(c),16] Thioenol derivatives may be a source of 2-deoxy[8] and branched-chain sugars.[29(a),(b)]

Besides such utilization, certain studies[9(c),(d),17,21,24,29(c)] have been dedicated to a systematic examination of the reactivity of sugar derivatives and phosphoranes at qualitative and even semi-quantitative[21] levels.

1. C-Glycosylated Unsaturated Hydrocarbons

a. Synthesis.—Most of the syntheses next described were performed with unstable phosphonium ylides in an inert atmosphere, the temperature, solvent, and proton acceptor being varied. A series of reactions were conducted with phenyllithium in ether,[9(a)-(g),29(c),30]

(26) N. K. Kochetkov, A. F. Bochkov, B. A. Dmitriev, A. I. Usov, O. S. Chizhov, and V. N. Shibaev,"Khimia Uglevodov," Moscow, 1967, p. 325.
(27) (a) N. K. Kochetkov, B. A. Dmitriev, and L. V. Backinovsky, *Carbohyd. Res.*, **5**, 399 (1967); (b) B. A. Dmitriev, L. V. Backinovsky, and N. K. Kochetkov, *Izv. Akad. Nauk SSSR, Ser. Khim.*, 2341 (1968); (c) N. K. Kochetkov, B. A. Dmitriev, and L. V. Backinovsky, *Carbohyd. Res.*, **11**, 193 (1969).
(28) (a) Yu. A. Zhdanov and V. G. Alexeeva, *Zh. Obshch. Khim.*, **38**, 2594 (1968); *Chem. Abstr.*, **70**, 58175 (1969); (b) Yu. A. Zhdanov and V. G. Alexeeva, *Carbohyd. Res.*, **10**, 184 (1969); (c) Yu. A. Zhdanov, V. I. Kornilov, L. A. Uslova, V. G. Alexeeva, and Lyu Dyc Shoung, *Mezhdunar. Simp. Khim. Prir. Soed., 7th, Riga*, 403 (1970).
(29) (a) J. M. J. Tronchet and J. M. Chalet,*Helv. Chim. Acta*, **53**, 364 (1970); (b) J. M. J. Tronchet and J. M. Bourgeois,*ibid.*, 53, 1463 (1970); (c) Yu. A. Zhdanov and V. G. Alexeeva, *Zh. Obshch. Khim.*, **39**, 112 (1969).
(30) V. G. Alexeeva, *Mater. Nauch. Konf. Molodyh Uchenyh Rost. Obl. Sec. Est. Nauk, 2nd, Rostov-on-Don*, 33 (1968).

although application of sodium amide in ether[10,31,32] and of sodium hydride in methyl sulfoxide[33-35] was also reported. The influence of reaction conditions on the yield of compound **22** was examined.[34]
A series of *C*-glycosylated ethylenes have been prepared by several workers[10,31-35] for different purposes, mainly for the subsequent synthesis of branched-chain nucleosides. Lange and Szarek reported[31] an attempted synthesis of 5,6-dideoxy-1,2-*O*-isopropylidene-α-D-*xylo*-hex-5-enofuranose (**15a**) and its acetate (**15b**) from the *aldehydo*

```
         CH₂                    CH₂                   CH₂
         ‖                      ‖                     ‖
 CHO   O                 CH   O                      CH
     \ /                     \ /                      |
      \                       \                     HCO
      /  OR                   /  OR                     \
     / \                     / \                         CMe₂
    /   \                   /   \                    COCH /
    O    \                  O    \                      |
     \    \                  \    \                    OCH
      O-CMe₂                  O-CMe₂              Me₂C /
                                                      \
                                                       OCH₂
    14a R = H              15a R = H                16
    14b R = Ac             15b R = Ac
```

sugars (**14a**) and (**14b**), respectively. Although 1,2-*O*-isopropylidene-α-D-*xylo*-pentodialdo-1,4-furanose (**14a**) is known to exist as a dimer,[36(a),(b)] it reacts with Grignard reagents.[37(a),(b)] Coupling between **14a** and methylenetriphenylphosphonium ylide was, however, unsuccessful both by application of an alkoxide method and by use of pregenerated ylide. The desired product (**15a**) was isolated, together with its acetate (**15b**), in extremely low yields by the reaction of acetate (**14b**) with pregenerated ylide in ether. Such results are not unexpected, because of the unacceptability of hydroxyl compounds and their acetylated derivatives in reactions with phosphonium ylides.[2(b)] Increased yields of alkenes may be expected with acetalated *aldehydo* sugars as starting materials. Indeed, interaction of 2,3:4,5-di-*O*-isopropylidene-*aldehydo*-L-arabinose with methylenetriphenylphosphonium ylide led[33] to 1,2-dideoxy-3,4:5,6-di-*O*-isopropylidene-L-*arabino*-1-hexene-3,4,5,6-tetrol (**16**) in about 40% yield. Lance and Szarek also reported[31] the preparation of methyl 4,6-*O*-benzylidene-

(31) D. G. Lance and W. A. Szarek, *Carbohyd. Res.*, **10**, 306 (1969).
(32) W. A. Szarek, J. S. Jewell, J. Szczerek, and J. K. N. Jones, *Can. J. Chem.*, **47**, 4473 (1969).
(33) J. M. J. Tronchet, E. Doelker, and B. Baehler, *Helv. Chim. Acta*, **52**, 308 (1969).
(34) A. Rosenthal and M. Sprinzl, *Can. J. Chem.*, **47**, 3941 (1969).
(35) A. Rosenthal and M. Sprinzl, *Carbohyd. Res.*, **16**, 337 (1971).
(36) (a) R. Schaffer and H. S. Isbell, *J. Amer. Chem. Soc.*, **79**, 3864 (1957); (b) T. D. Inch, *Carbohyd. Res.*, **5**, 53 (1967).
(37) (a) D. Horton, J. B. Hughes, and J. M. J. Tronchet, *Chem. Commun.*, 481 (1965); (b) T. D. Inch, *Carbohyd. Res.*, **5**, 45 (1967).

THE WITTIG REACTION IN CARBOHYDRATE CHEMISTRY 235

2,3-dideoxy-3-C-methylene-α-D-*threo*-hexopyranoside (**18**), in 60% yield, from methyl 4,6-O-benzylidene-2-deoxy-α-D-*threo*-hexopyranosid-3-ulose (**17**).

An analogous methyl 2,3-dideoxy-3-C-methylene-α-L-*erythro*-hexopyranoside (**20**) was obtained[10] from methyl 4,6-O-benzylidene-2-deoxy-α-L-*erythro*-hexopyranosid-3-ulose (**19**); it was used in the synthesis of olivomycose (see p. 242). There was also reported[32,34] the preparation of 3-deoxy-1,2:5,6-di-O-isopropylidene-3-C-methylene-α-D-*ribo*-hexofuranose (**22**) from 1,2:5,6-di-O-isopropylidene-α-D-*ribo*-hexofuranos-3-ulose (**21**) in about 60% yield, followed by transformations into branched-chain sugar derivatives, including

nucleosides.[34] Synthesis of 5-O-benzyl-3-deoxy-1,2-O-isopropylidene-3-C-methylene-α-D-*erythro*-pentofuranose (**24**) from 5-O-benzyl-1,2-O-isopropylidene-α-D-*erythro*-pentofuranos-3-ulose (**23**) was similarly realized.[35] The reaction of 1,2:3,4-di-O-isopropylidene-α-D-*galacto*-hexodialdo-1,5-pyranose (**25**) with ethylidenetriphenylphosphonium ylide produced[31] 6,7,8-trideoxy-1,2:3,4-di-O-isopropylidene-α-D-*galacto*-oct-6-enose (**26**) in 72% yield. This compound was later shown[11] to be the *cis* isomer.

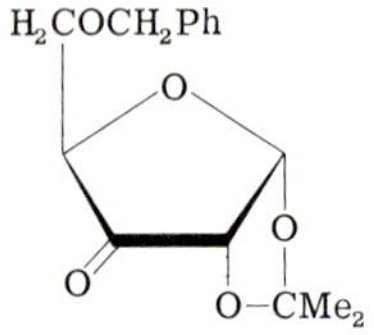

                    23 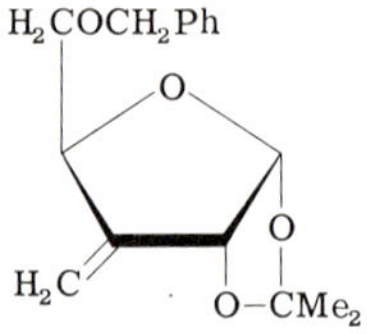   24

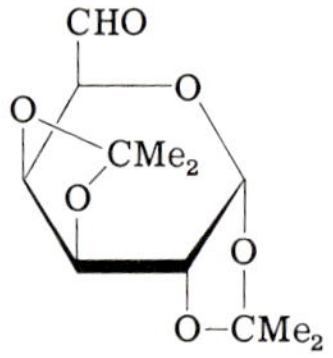

                    25 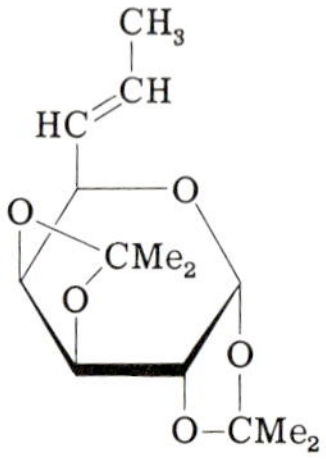   26

More-detailed examination of the reaction between alkylidene(and arylmethylene)triphenylphosphonium ylides and *aldehydo* sugars was conducted[9(a)–(g),30] in our laboratory with 2,3:4,5-di-*O*-cyclohexylidene-*aldehydo*-D-xylose (and -L-arabinose), resulting in the preparation of 1-*C*-alkyl-3,4:5,6-di-*O*-cyclohexylidene-1,2-dideoxy-D-*xylo*(and L-*arabino*)-hex-1-enitol (**27a–27d** and **28a–28d**, respectively[9(d)]), together with 1-*C*-aryl-3,4:5,6-di-*O*-cyclohexylidene-1,2-dideoxy-D-*xylo*(and L-*arabino*)-hex-1-enitol (**29a–29c**[9(g)] and **30a–30d**,[9(a),(c)] respectively) as a

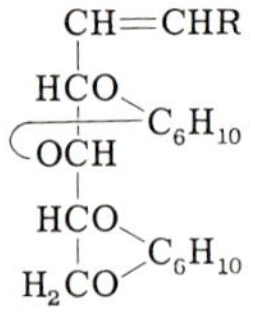

27a R = Me
27b R = *n*-C$_4$H$_9$
27c R = *n*-C$_5$H$_{11}$
27d R = *n*-C$_6$H$_{13}$

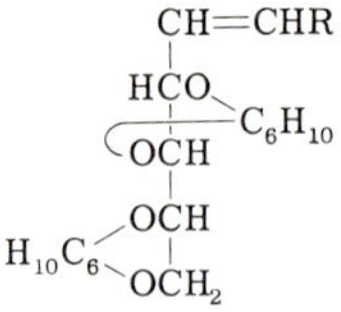

28a R = Me
28b R = *n*-C$_3$H$_7$
28c R = *n*-C$_5$H$_{11}$
28d R = *n*-C$_6$H$_{13}$

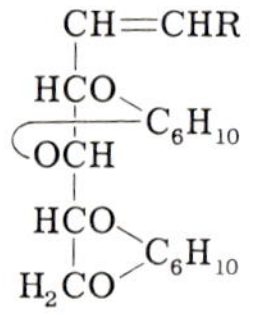

29a R = Ph
29b R = *p*-NO$_2$C$_6$H$_4$
29c R = *p*-MeOC$_6$H$_4$

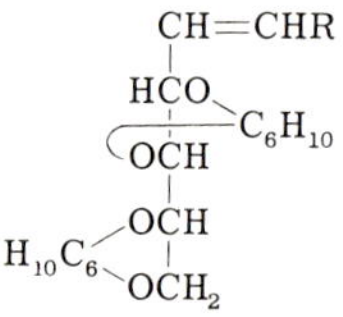

30a R = Ph
30b R = *p*-NO$_2$C$_6$H$_4$
30c R = *p*-MeOC$_6$H$_4$
30d R = 1-naphthyl

mixture of geometrical isomers, with a significant preponderance of *trans* isomer (see later). Acetals **28c** and **28d** were then converted[9(d)] into the free, unsaturated alditols (**31a** and **31b**) by heating with aqueous acetic acid.

```
    CHR                           R
    ‖                             ‖
    CH                            CH
    |                             |
   HCOH                        (CHOAc)₄
    |                             |
   HOCH                          CH₂OAc
    |
   HOCH                    32a  D-galacto,
    |                             R = 10-anthranyl-
   CH₂OH                          idene
31a  R = n-C₅H₁₁           32b  D-galacto,
31b  R = n-C₆H₁₃                 R = 9-fluorenylidene
                           32c  D-gluco,
                                 R = furfurylidene
```

Use of anthranylidene- and furfurylidene-triphenylphosphoranes, as well as of the fluorenylidene derivative, permits introduction of the respective fragment into the sugar molecule. This possibility has been realized[25,38,39] in syntheses of the C-substituted, unsaturated alditols **32a** and **32b** (Ref. 38), and **32c** (Refs. 25 and 39). The possibility of insertion of long-chain hydrocarbon groups into sugar molecules was realized by Gigg and coworkers[40] in connection with phytosphingosine syntheses. The appropriate phosphonium ylides were allowed to react with the *aldehydo*-aldoses **33** and **35** containing a protected amino group; these had been prepared by multistage procedures. The resulting C-alkylated, unsaturated products **34** [Ref. 40(a),(b)] and **36** [Ref. 40(c)] were finally transformed into C_{18} and C_{20} phytosphingo-

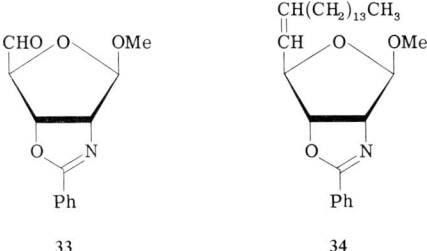

33   34

(38) Yu. A. Zhdanov, L. A. Uslova, and G. N. Dorofeenko, *Zh. Vses. Khim. Obshchest.*, **10**, 600 (1965); *Chem. Abstr.*, **64**, 3671 (1966).
(39) Yu. A. Zhdanov, G. N. Dorofeenko, and L. A. Uslova, *Zh. Obshch. Khim.*, **35**, 181 (1965).
(40) (a) R. Gigg, C. D. Warren, and J. Cunningham, *Tetrahedron Lett.*, 1303 (1965); (b) J. Gigg, R. Gigg, and C. D. Warren, *J. Chem. Soc.* (C), 1872 (1966); (c) J. Gigg and R. Gigg, *ibid.*, 1876 (1966); (d) R. Gigg and C. D. Warren, *ibid.*, 1879 (1966).

sines, providing indirectly the first synthetic proof of the structure assigned to sphingosine. The enantiomer of C_{20} phytosphingosine was synthesized,[40(d)] starting from D-galactose, through an ethylenic inter-

```
     CHO                    CH=CH(CH₂)ₙCH₃           CH=CH(CH₂)₁₃CH₃
      |                           |                        |
   BzlOCH                      BzlOCH                   HCOBzl
      |                           |                        |
   BzlOCH                      BzlOCH                   HCOBzl
      |                           |                        |
   Phth—CH                     Phth—CH                  BzlOCH
      |                           |                        |
   CH₂OBzl                     CH₂OBzl                  CH₂OBzl

     35                     36 where n = 11, 13           37
```

Bzl = CH₂Ph;   Phth = [phthalimido group]

mediate (37). A stereospecific synthesis of dihydrosphingosine has been recorded[41] that involves the reaction of the amino aldehyde 38 with the Wittig reagent prepared from tetradecyltriphenylphosphonium bromide and subsequent transformations of the resulting alkene (39).

```
                                            CH=CH(CH₂)₁₂CH₃
   CHO  O                                    |    O
    \  / \                                    \  / \
     \/   \                                    \/   \
     /\    O                                   /\    O
    /  \  /                                   /  \  /
PhCH₂O₂CNH  O—CMe₂                       PhCH₂O₂CNH  O—CMe₂
        38                                          39
```

The syntheses of sphingosines just described have a substantial advantage in comparison with those previously reported, because carbohydrates offer a variety of starting materials having known absolute configuration and, therefore, the necessity of stereoisomer separation does not arise.

Dienic hydrocarbon residues may also be introduced into a sugar molecule. Certain C-glycosylated dienes (40a–40d) and (41a–41d) were synthesized[29(c),30] by the reaction of 2,3:4,5-di-O-cyclohexylidene-aldehydo-D-xylose (and L-arabinose) with appropriate phosphonium ylides. It may be noted that compounds 40d and 41d are the first examples of C-glycosylated terpenes.

(41) E. J. Reist and P. H. Christie, *J. Org. Chem.*, **35**, 3521 (1970).

THE WITTIG REACTION IN CARBOHYDRATE CHEMISTRY 239

```
       CH=CH—CH=CR'R"                    CH=CH—CH=CR'R"
       |                                  |
       HCO                                HCO
         \                                  \
          C₆H₁₀                              C₆H₁₀
         /                                  /
       OCH                                OCH
       |                                  |
       HCO                                OCH
         \                                  \
          C₆H₁₀                 H₁₀C₆
         /                                  /
       H₂CO                               OCH₂
```

40a R' = R" = H   41a R' = R" = H
40b R' = H, R" = Me   41b R' = H, R" = Me
40c R' = H, R" = Ph   41c R' = H, R" = Ph
40d R' = Me,   41d R' = Me,
    R" = (CH₂)₂CH=CMe₂   R" = (CH₂)₂CH=CMe₂

b. Physical Properties.—Electronic absorption data for the C-arylated, unsaturated hexitols 30a–30c are similar to those for simpler structural analogs, as is shown in Table I. Compounds 29a–29c,

TABLE I
Ultraviolet Absorption[a] of Alkenes 30a–30c
and Their Structural Analogs[9(c)]

Compound	λ max (nm)	$\epsilon_{max} \times 10^5$
Styrene[42]	244	1.2
30a	244	0.9
p-Nitrostyrene[43]	289	1.56
30b	300	1.1
p-Methoxystyrene[44]	275	1.9
30c	257	0.87

[a] The data were obtained in methanol, except those for p-nitrostyrene, which were obtained in hexane.

belonging to the D-*xylo* series, give the same values. Certain discrepancies in wavelength and molar extinction values for each pair of analogs may be explained by the influence of the sugar moiety.

In addition to these results, the ultraviolet (u.v.) spectral data for dienes 41a–41d, given in Table II, are of interest for comparison with the values calculated by use of the Woodward rule,[45] assuming the identity of the contributions of the sugar residue and of the alkyl

(42) O. H. Wheeler and C. B. Covarrubias, *Can. J. Chem.*, **40**, 1244 (1962).
(43) P. P. Shorygin, M. F. Shostakovsky, E. P. Prilezhaeva, T. N. Shkurina, L. G. Stolarova, and A. P. Genich, *Izv. Akad. Nauk SSSR, Ser. Khim.*, 1571 (1961).
(44) C. N. R. Rao, "Ultraviolet and Visible Spectroscopy: Chemical Applications," Butterworth, London, 1961.
(45) R. B. Woodward, *J. Amer. Chem. Soc.*, **63**, 1123 (1941); **64**, 72 (1942).

TABLE II
Ultraviolet Absorption of Dienes 41a–41d[29(c)]

Compound	λ_{max} (nm)		$\epsilon_{max} \times 10^5$
	Obs.	Calc.	
41a	224	222	1.9
41b	285	281	2.0
41c	230	227	2.2
41d	240	238	2.4

groups.[46] Good agreement between the observed and calculated wavelengths was, therefore, observed. Better agreement would be attained were the value of 8–10 nm adopted for the sugar residue, instead of 5–7 nm as for the alkyl group.

The infrared spectra of C-alkylated, unsaturated sugars are characterized by alkenic bond absorption at 5.8–6.05 μm; this shifts[9(c)] to the 6.05–6.23 μm region for C-arylated analogs, with simultaneous increase in the intensity. Both types of compound showed[9,33] *trans*-alkene absorption at 10.3–10.4 μm, indicating a preponderance of *trans*-isomers in the final mixtures. The infrared spectra of dienes **40c** and **41c**, having aromatic substituents, proved to be uninterpretable[29(c)] owing to the masking influence of the aromatic absorption. Dienes **40a, 40b, 41a,** and **41b** showed normal dienic absorption at 6.23 μm (weak) and 6.6 μm (strong).

Methylenic protons in such furanose derivatives as compound **22** resonate at τ 4.4–4.6 p.p.m. in their n.m.r. spectra, the alkenic bond showing a deshielding influence on the proton at C-2, which resonates at ~5 p.p.m. as compared with its normal resonance at[32,34,35,47] ~5.5 p.p.m. The terminal vinyl group in **26** shows[31] a three-proton multiplet at 4.0–4.7 p.p.m., a similar spectrum being observed[33] for compound **16**.

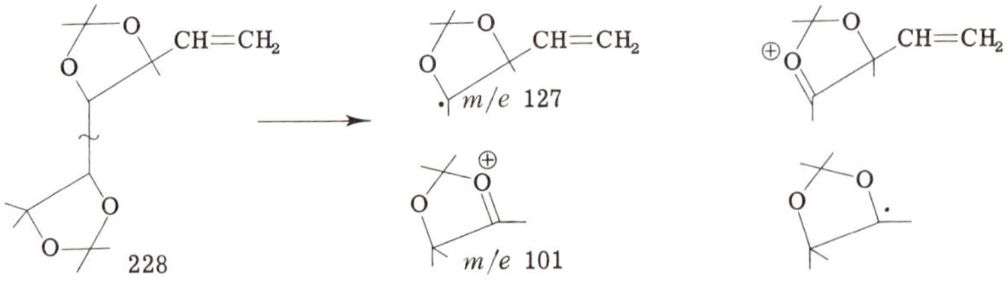

(3)

16

(46) J. C. D. Brand and G. Eglinton, "Applications of Spectroscopy to Organic Chemistry," Oldbourne, London, 1965.
(47) J. S. Jewell and W. A. Szarek, *Tetrahedron Lett.*, 43 (1969).

Compound **16** was also shown[33] to undergo a so-called "h" rupture[48] during chemical ionization, as two mass peaks (at m/e 101 and 127) were observed. This rupture may be represented by equation 3.

The reported[9(d)] surface activity of the C-alkylated hexitols **28c** and **28d**, as well as of the hydrogenated (**44**) and hydroxylated (**46**) derivatives, is of marked interest. This activity was found[9(g)] to increase with lengthening of the hydrocarbon chain.

c. Chemical Properties.—Addition of bromine has been used for a preliminary test for alkenes, but no dibromides have been isolated. Catalytic hydrogenation was employed in a series of studies. Rosenthal and Sprinzl reported,[34] for example, that hydrogenation of olefin **22** proceeded stereoselectively, to yield 3-deoxy-1,2:5,6-di-O-isopropylidene-3-C-methyl-α-D-allofuranose (**42**). The stereoselectivity of the reaction is undoubtedly caused by the shielding effect of the 1,2-O-isopropylidene group, resulting in the preponderant orientation

of alkene molecules on the catalyst surface. Hydrogenation of unsaturated derivatives **34, 36,** and **37** was one of the stages in a synthesis of sphingosine.[40] Similar hydrogenation of the 1-C-alkylated hexenitol **28d** afforded[9(g)] 2,3:4,5-di-O-cyclohexylidene-1-deoxy-1-C-heptyl-L-arabinitol (**43**), which was transformed into the unprotected alditol **44** by acid hydrolysis.

Hydroxylation of the olefinic bond is also a convenient route for utilization of the alkenes being considered. Treatment of alkene **26** with aqueous alkaline potassium permanganate resulted in formation of compound **45**, the structure of which was established by a degradative procedure[11]. It followed from this result that the initial alkene (**26**) was the *cis* isomer. Treatment of alkene **28d** with iodic acid in the presence of a catalytic amount of osmium tetraoxide, with propyl alcohol as the solvent, led[9(g)] to a mixture of two alditols (**46**). Direct hydration of the olefinic bond in C-glycosylated alkenes

---

(48) O. S. Chizhov, L. S. Golovkina, and N. S. Wulfson, *Carbohyd. Res.*, **6**, 138, 143 (1968).

```
      CH₃
       |
    HOCH
       |
    HOCH            CH₂(CH₂)₄CH₃
      ⟋—O⟍           |
     O     ⟍         CHOH
      ⟍  CMe₂        |
       O⟋            CHOH
         ⟍           |
          O          HCOH
          |          |
        O—CMe₂       HOCH
                     |
                     HOCH
                     |
                     CH₂OH

         45             46
```

seems to be impossible, owing to the instability of these compounds in acid media. The employment of the oxymercuration–demercuration procedure, or of hydroboration, however, provided an indirect avenue to such conversion. Thus, methyl 4,6-O-benzylidene-2-deoxy-3-C-methyl-α-L-*arabino*-hexopyranoside (**47**) was obtained[10] by addition of mercuric acetate to alkene **20** with subsequent demercuration by sodium borohydride, and the product was used for the synthesis of olivomycose (2,6-dideoxy-3-C-methyl-L-*arabino*-hexose). The configuration of C-3 of compound **47** clearly indicates the stereoselectivity of the addition of mercuric acetate in this case; it can be explained by a predominant attack of the reagent on the favored conformer (**48**). This procedure was also used[9(g)] for obtaining

```
       CHPh
       ⟋—O⟍
      O      ⟍
       ⟍  OCH₂   OMe
        ⟍  CH₃  ⟋
         ⟍—⟋
          |
          HO

           47
```

```
        O            OMe
      ⟋   ⟍        ⟋
    Ph      ⟍    ⟋
              O⟍  CH₂  O
                    ↑

           48
```

1-deoxy-1-C-methyl-L-glucitol and -L-mannitol (**49a**) from alkene **28a**. The sequence of mercuric acetate addition was established by the fact that the mixture of compounds (**49a**) consumed four moles of periodate per mole, as the alternative structure (**49b**) would correspond to only three moles of periodate consumption. Surprisingly, application of this procedure to alkene **24** failed.[35]

Hydroboration of alkene **22** was realized[49] by treatment with a large excess of diborane in anhydrous tetrahydrofuran, followed by oxidation with alkaline hydrogen peroxide. This procedure resulted in the formation of 3-deoxy-3-C-(hydroxymethyl)-1,2:5,6-di-O-iso-

(49) A. Rosenthal and M. Sprinzl, *Can. J. Chem.*, **47**, 4477 (1969).

THE WITTIG REACTION IN CARBOHYDRATE CHEMISTRY 243

```
   CH₂CH₃              CH₃
   |                   |
   CHOH                CHOH
   |                   |
   HCOH                CH₂
   |                   |
   HOCH                HCOH
   |                   |
   HOCH                HOCH
   |                   |
   CH₂OH               HOCH
                       |
                       CH₂OH

    49a                 49b
```

propylidene-α-D-allofuranose (**50**), and **50** was transformed by a multistage synthesis into a desired branched-chain nucleoside.[49] The stereoselectivity of this reaction, caused by the shielding effect of the 1,2-O-isopropylidene group, was observed to a lesser extent when hydroboration of the structurally related alkene **24** was

[Structures 50, 51a, 51b, 51c shown]

50: Me₂C(OCH₂/OCH) ring with HOCH₂ and O-CMe₂
51a: H₂COCH₂Ph ring with CH₃, HO, O-CMe₂
51b: H₂COCH₂Ph ring with OH, CH₃, O-CMe₂
51c: H₂COCH₂Ph ring with H₂COH, O-CMe₂

performed. It gave[35] a mixture of 5-O-benzyl-1,2-O-isopropylidene-3-C-methyl-α-D-ribofuranose (**51a**), 5-O-benzyl-1,2-O-isopropylidene-3-C-methyl-α-D-xylofuranose (**51b**), and the expected, anti-Markownikov product, 5-O-benzyl-3-deoxy-3-C-(hydroxymethyl)-1,2-O-isopropylidene-α-D-ribofuranose (**51c**) in the ratios of 5:7:88.

The introduction of a nitro group into the alkenes under consideration may be achieved by addition of nitryl iodide (INO_2). In the case of alkene **22**, this addition, followed by reduction with sodium borohydride, led[32] to 3-deoxy-1,2:5,6-di-O-isopropylidene-3-C-(nitromethyl)-α-D-allofuranose (**52**). Light-induced addition of 1,3-dioxolane to alkenes **15a** and **22** has also been reported.[47] The resulting adducts, **53** and **54**, respectively, are of value as they contain a potential *aldehydo* function.

2. α,β-Unsaturated Carbonyl Sugars

a. Synthesis.—Lengthening of the carbon chain of a carbohydrate by the =CH—COR fragment (where R = H, alkyl, or aryl), as well as detailed examination of the compounds resulting, was realized by Zhdanov and coworkers,[22,23,24,50,51-57] starting from acetylated and acetalated *aldehydo* sugars and, in particular, from 3-O-benzyl-1,2-O-cyclohexylidene-α-D-*xylo*-pentodialdo-1,4-furanose[22] (**55**) and its 3-O-benzoyl analog.[23] The formation of two isomers was observed in general, with strong preponderance of the *trans* isomers.

A convenient method of preparation of formylmethylenetriphenylphosphorane[58] made possible a synthesis of α,β-unsaturated *aldehydo*

---

(50) Yu. A. Zhdanov, Yu. E. Alexeev, and G. N. Dorofeenko, *Zh. Obshch. Khim.*, **38**, 231 (1968); *Chem. Abstr.*, **69**, 27681 (1968).

(51) Yu. A. Zhdanov, L. A. Uslova, L. P. Leskina, and O. A. Gavrilenko, *Zh. Obshch. Khim.*, **40**, 666 (1970).

(52) L. A. Uslova, Dissertation, Rostov-on-Don, 1965.

(53) Yu. A. Zhdanov, G. N. Dorofeenko, G. A. Korolchenko, and A. E. Osolin, *Zh. Obshch. Khim.*, **36**, 492 (1966).

(54) Yu. A. Zhdanov, G. N. Dorofeenko, and L. A. Uslova, *Dokl. Akad. Nauk SSSR*, **160**, 339 (1965).

(55) Yu. A. Zhdanov, L. A. Uslova, G. N. Dorofeenko, and G. I. Kravchenko, *Zh. Obshch. Khim.*, **36**, 1025 (1966).

(56) Yu. A. Zhdanov and V. A. Polenov, *Carbohyd. Res.*, **16**, 466 (1971).

(57) Yu. A. Zhdanov, Yu. E. Alexeev, and G. N. Dorofeenko, *Zh. Obshch. Khim.*, **37**, 2635 (1967).

(58) S. Trippett and D. M. Walkers, *J. Chem. Soc.*, 1266 (1961).

sugars. 3-O-Benzyl-1,2-O-cyclohexylidene-5,6-dideoxy-α-D-*xylo*-hept-5-eno-dialdo-1,4-furanose (**56**) was thus prepared[22] starting from *aldehydo* sugar **55**. Preparations of 4,5:6,7-di-O-cyclohexylidene-

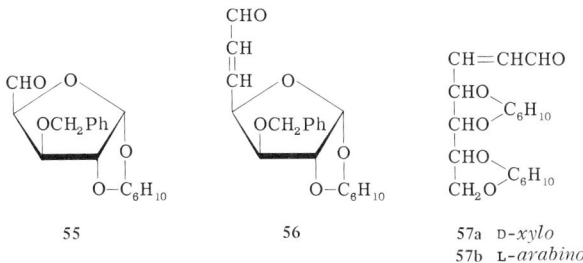

55

56

57a D-*xylo*
57b L-*arabino*

2,3-dideoxy-*aldehydo*-D-*xylo*(and L-*arabino*)-hept-2-enoses (**57a** and **57b**, respectively), starting from 2,3:4,5-di-O-cyclohexylidene-*aldehydo*-D-xylose (and -L-arabinose) were also reported.[29(c)] Similar interaction of acetylmethylenetriphenylphosphorane with certain *aldehydo* sugars led to the preparation of the corresponding α,β-unsaturated ketoses, such as 3-O-benzyl-1,2-O-cyclohexylidene-5,6,8-trideoxy-α-D-*xylo*-oct-5-eno-1,4-furanos-7-ulose[18] (**58a**), together with its benzoyl analog[50] (**58b**), 1,3,4-trideoxy-D-*gluco* (and D-*galacto*)-non-3-enulose 4,5,6,7,8-pentaacetates (**59a** and **59b**), respectively,[9(f),24,25,39] and 5,7:6,8-di-O-benzylidene-1,3,4-trideoxy-D-*ribo*-oct-3-enulose[21] (**60**).

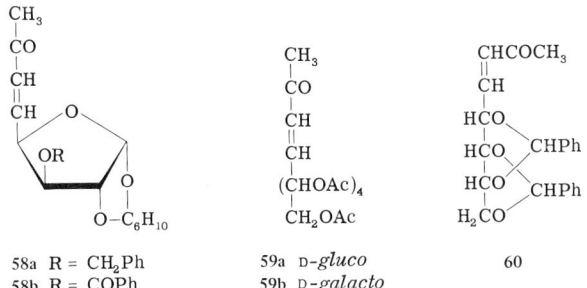

58a R = CH₂Ph
58b R = COPh

59a D-*gluco*
59b D-*galacto*

60

D-Glyceraldehyde was also shown[53] to react with acetylmethylenetriphenylphosphorane to form 1,3,4-trideoxy-D-*glycero*-hex-3-enulose (**61**). Similar condensation with L-sorbose pentaacetate led[51] to the branched-chain, α,β-unsaturated ketose **62**.

Product **59a** was isolated[52] as crystalline (minor) and syrupy (major) isomers, without determination of their geometrical configuration. Surprisingly, the failure of reaction between acetylmethylenetriphenylphosphorane and *aldehydo*-D-glucose pentaacetate has been reported.[21]

Coupling of *aldehydo* sugars with aroylmethylenetriphenylphosphoranes affords a practically unlimited source of various 1-C-aryl-α,β-unsaturated aldoses. *p*-Substituted 2,3-dideoxy-1-C-phenyl-D-*glycero*-pent-2-enoses (**63**) were thus obtained,[53] starting from D-glyceraldehyde, in 10–30% yields.

Syntheses of 4,6:5,7-di-*O*-benzylidene-2,3-dideoxy-1-*C*-phenyl-D-*ribo*-hept-2-enose[21] (**64**) and 4,5:6,7-di-*O*-cyclohexylidene-2,3-dideoxy-1-*C*-phenyl-L-*arabino*-hept-2-enose[9(g)] (**65**) were also reported.

A series of compounds having the general structures **66** and **67** have also been synthesized, as outlined in Tables III and IV, respectively.

Product **66** (D-*galacto*, R = Ph) was then deacetylated[24] to give 2,3-dideoxy-1-C-phenyl-D-*galacto*-oct-2-enose (**68**), which has been

TABLE III
1-C-Aryl α,β-Unsaturated *aldehydo*-Octose Pentaacetates (66)

Configuration	R in 66	Yield (%)	References
D-*gluco*	Ph	74	24,25,54
	p-NO$_2$-C$_6$H$_4$	60	24,25
	p-MeOC$_6$H$_4$	high	24,25,54
D-*manno*	p-NO$_2$C$_6$H$_4$	35	24
D-*galacto*	Ph	79	24,25,54
	p-NO$_2$C$_6$H$_4$	79	24,25,54
	p-MeOC$_6$H$_4$	70	24,25,54
	p-EtOC$_6$H$_4$	54	24,25,54
	p-BrC$_6$H$_4$	65	24,55
	1-naphthyl	50	24,55
	3,4-(MeO)$_2$C$_6$H$_3$	77	24,55
	2,3,4-(MeO)$_3$C$_6$H$_2$	33	24,55
	2-furyl	69	24,55
	2-thienyl	70	24,55

shown by periodate oxidation and by infrared spectral data to exist in the acyclic form (see later).

Ketones are known to be usually unreactive towards resonance-stabilized phosphoranes, excluding ethoxycarbonylmethylenetriphenylphosphorane.[2(f)] L-Sorbose pentaacetate was, however, reported[51] to react, although incompletely, with acetylmethylenetriphenylphosphorane and certain aroylmethylenetriphenylphosphoranes, giving branched-chain, olefinic sugars (69) in about 25% yield.

$$\begin{array}{c} AcOH_2C-C=CHCOR \\ | \\ AcOCH \\ | \\ HCOAc \\ | \\ AcOCH \\ | \\ CH_2OAc \end{array}$$

69   R = Ph or p-MeOC$_6$H$_4$

TABLE IV
7-C-Aryl-1,2-O-cyclohexylidene-5,6-dideoxy-α-
D-*xylo*-hept-5-enodialdo-1,4-furanoses (67)

R' in 67	R in 67	Yield (%)	References
H	Ph	55	23
	p-NO$_2$C$_6$H$_4$	65	23
COPh	p-NO$_2$C$_6$H$_4$	83	23
CH$_2$Ph	Ph	58	17
	p-NO$_2$C$_6$H$_4$	99	17
	p-MeOC$_6$H$_4$	94.5	17
	3,4-(MeO)$_2$C$_6$H$_3$	91	17

Zhdanov and Polenov have reported[56] a synthesis of the first sugar phosphorane (**70**), which was shown to react with *p*-nitro-and *o*-hydroxy-benzaldehyde, forming unsaturated ketoses **71** and **72**, respectively. Attempted reactions with *p*-dimethylamino-, *p*-hydroxy-, and 2,4-dihydroxy-benzaldehyde were unsuccessful. The formation of the dienic ketose **72** is undoubtedly caused by β-elimination of a methoxyl group during treatment of the reaction mixture with aqueous sodium hydroxide during isolation.

70   **71**   **72**

b. Physical Properties—Electronic absorption spectra have been observed for some of these compounds. Aldehydes **57a** and **57b** absorb[29(c)] at ~240 nm (ε 16,000). The different absorptions of ketones **65** (λ_{max} 253 nm, ε 7,500) [Ref. 9(g)] and **67** (R= Ph, R' = CH$_2$Ph) (λ_{max} 244 nm, ε 12,900) may possibly be due to different ratios of geometrical isomers in the two compounds, the absorption of the latter being similar to that of the analogous acrylophenone (λ_{max} 242 nm, ε 27,000). This absorption band varies within the range 260–300 nm for acetylated derivatives (**66**), depending on the substituent on the aromatic nucleus.[54] Ketone **71** was found[56] to absorb at 218 nm (ε 4,300) and 310 nm (ε 8,000). A bathochromic shift of the first absorption band to 256 nm (ε 6,300) was observed[56] for ketone **72**, as expected.

The infrared absorption spectra of α,β-unsaturated ketoses were examined thoroughly. All of these compounds have two absorption bands in common, corresponding to a conjugated-carbonyl (5.93–6.01 μm) and a conjugated alkene bond (6.09–6.18 μm), the positions and intensities of these bands depending on the substituents on the —CH=CHCO fragment. Thus, aldehydes **56, 57a**, and **57b** show conjugated-carbonyl absorption at 5.93 μm, although the second absorption band changes from 6.17 μm in **57a** and **57b** to 6.09 μm in **56**. Ketoses **67** have the first absorption band at 5.95 μm, but, for compound **67** (R = 3,4-(MeO)$_2$C$_6$H$_2$), it is observed at 6.01 μm. The second absorption band is sensitive to the types of aromatic substituents, and varies in the region of 6.09–6.16 μm. Acetylated ketoses **66**, besides having in common bands at 6.0 and 6.15 μm, show[24] an acetyl absorption at 5.7 μm. A free octose (**68**) absorbs[24] at 5.92 μm and 6.10 μm.

c. **Chemical Properties.**—All of the compounds being considered decolorize bromine in carbon tetrachloride or acetic acid, although at different rates. The respective dibromides are usually unstable, and have been identified[24] only for some acetylated derivatives (**66**, R = Ph, *p*-methoxyphenyl, *p*-ethoxyphenyl, and *p*-nitrophenyl) having the D-*galacto* configuration. The rate of addition of bromine was qualitatively observed[52] to depend on the substituents on the aromatic nucleus, and decreased with increasing electronegativity. The stability of these dibromides correlates with the electronegativity of the substituents in the same way. Interestingly, compounds **66** turned out to be unreactive towards Woodward hydroxylation and the action of lead tetraacetate.[52] Formation of an adduct on interaction of ketone **67** (R' = CH$_2$Ph, R = 3,4-dimethoxyphenyl) and sodium metabisulfite in acetic acid was observed.[18] The expected structure of the adduct, represented by formula **73**, follows from a shift in the infrared carbonyl absorption from 6.01 μm for the initial compound to 5.98 μm, and the appearance of sulfonate group absorption at 8.09 and 9.1 μm.

The reactivity of the carbonyl group in α,β-unsaturated dialdoses was studied in more detail. The interaction with phenylhydrazine and its derivatives led to different results. 3-O-Benzyl-1,2-O-cyclohexylidene-5,6-dideoxy-α-D-*xylo*-hept-5-enodialdo-1,4-furanose (**56**) gave[57] a (2,4-dinitrophenyl)hydrazone. Action of (2,4-dinitrophenyl)-hydrazine on derivatives **67** led[18] to complex mixtures. In the case of the acetylated derivatives (**66**), as well as of branched-chain derivatives (**69**), the interaction with (2,4-dinitrophenyl)hydrazine resulted in the formation of the expected hydrazones (Refs. 24 and 51, respectively). The interaction with phenylhydrazine was, however, proved to proceed by two paths in this case. Only the nitro derivatives (**66**, R = *p*-nitrophenyl) and (**69**, R = *p*-nitrophenyl) were shown[24,51] to form the expected hydrazones. The phenyl derivative **66** (D-*galacto*, R = Ph) and *p*-methoxyphenyl derivative **69** (R = *p*–MeOC$_6$H$_4$) afforded substituted pyrazolines (**74**) (Ref. 24) and (**75**) (Ref. 51), respectively, as a result of subsequent intramolecular addition of a hydrazine fragment to the double bond. A strongly electronegative group, either in the sugar or in the hydrazine fragment, obviously prevents this addition, owing to significant lowering of the basicity of the hydrazine fragment.

```
           Ph                           C₆H₄OMe-p
          /                            /
   H₂C—C                         H₂C—C
       ‖                              ‖
       N                              N
       /                              /
   H₂C—N                      AcOH₂C—C—N
       \                              \
        Ph                             Ph
   HCOAc                          AcOCH
   AcOCH                          HCOAc
   AcOCH                          AcOCH
   HCOAc                          CH₂OAc
   CH₂OAc

       74                             75
```

Several reactions have been carried out[50] on the carbonyl group with 3-O-benzoyl-1,2-O-cyclohexylidene-5,6,8-trideoxy-α-D-*xylo*-oct-5-eno-1,4-furanos-7-ulose (**58b**). Treatment of this compound with aluminum isopropoxide (the Meerwein–Ponndorf–Werley reduction) led to 3-O-benzoyl-1,2-O-cyclohexylidene-5,6,8-trideoxy-α-D-*xylo*-oct-5-eno-1,4-furanose (**76**), together with debenzoylated product.

Reaction with phenyllithium afforded[50] the appropriate tertiary alcohol (**77**). The branched-chain ethyl uronate **78** was obtained[50] by the Reformatsky reaction. After continuous storage of **58b** in methanolic hydrogen chloride, and subsequent fractionation on alumina, the dimethyl acetal (**79**) was isolated.[18]

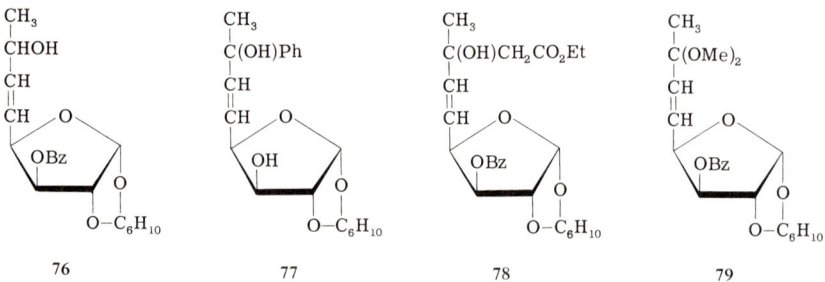

Debenzoylation of **58b** with sodium methoxide in methanol unexpectedly resulted[50] in the formation of the non-carbonyl, unsaturated product (**80**), obviously arising as the result of intramolecular attack of intermediate alkoxide on the olefinic bond.

With the objective of synthetic utilization of α,β-unsaturated dialdoses, the Wittig reaction with certain compounds of this series

80

was examined in our laboratory. When 3-O-benzyl-1,2-O-cyclohexylidene-5,6-dideoxy-α-D-xylo-hept-5-enodialdo-1,4-furanose (**56**) was allowed to react with resonance-stabilized phosphoranes, several dienic carbonyl compounds, **81a–81e**, were obtained[57] in high yields.

81a R = OMe   81d R = 3, 4 -(OMe)$_2$C$_6$H$_3$
81b R = Ph   81e R = H
81c R = p-BrC$_6$H$_4$

Peaks in the electronic absorption spectra of these compounds, together with those of appropriate structural analogs are given in Table V. The aromatic dienes **81b–81d**, as compared with the structurally related monoalkene **67** (R = Ph, R' = CH$_2$Ph), show the expected bathochromic shifts, with values depending on the substituent on the aromatic nucleus. The dienic aldehyde **81e** formed a (2,4-dinitrophenyl)hydrazone[57] having an electronic spectrum (λ_{max} 390 nm; ϵ 27,500) similar to that[61] of the (2,4-dinitrophenyl)hydrazone of

TABLE V
Electronic Absorption Spectra of Dienic Carbonyl Compounds
(81a–81e) and Their Structural Analogs

Compound	λ_{max} (nm)	ϵ_{max} x 10^5	Solvent
Methyl sorbate[59]	258	2.77	ethanol
81a	256	1.57	methanol
67, R = Ph, R' = CH$_2$Ph	244	1.29	methanol
81b	292	1.37	methanol
81c	297	0.68	methanol
81d	287	1.20	methanol
2,4-Hexadienoic aldehyde[60]	271	2.75	ethanol
81e	268	0.90	methanol

(59) D. Peters, *J. Chem. Soc.*, 1832 (1960).
(60) E. R. Blout and M. Fields, *J. Amer. Chem. Soc.*, **70**, 189 (1948).
(61) E. L. Pippen and M. Nonaka, *J. Org. Chem.*, **23**, 1580 (1958).

2,4-hexadienoic aldehyde (λ_{max} 390 nm, ϵ 36,700). Interestingly, the molar rotations of dialkenes **81b, 81d,** and **81e** are considerably greater than that of the corresponding monoalkenes,[57] in accordance with Brewster's assumptions.[62]

Dialkenes **40b, 41b,** and **41c** were resynthesized[29(c)] starting from the α,β-unsaturated aldehydes **57a** and **57b** and the appropriate alkyl- and aryl-methylenetriphenylphosphoranes, as shown in equation 4.

$$\begin{array}{c}\text{CHO}\\|\\\text{CH}\\\|\\\text{CH}\\|\\\text{CHO}\\|\\\text{CHO}\\|\\\text{CHO}\\|\\\text{H}_2\text{CO}\end{array}\!\!\!\!\!\!\text{C}_6\text{H}_{10} \quad \xrightarrow{\text{Ph}_3\text{P}=\text{CHR}} \quad \begin{array}{c}\text{CH}=\text{CHR}\\|\\\text{CH}\\\|\\\text{CH}\\|\\\text{CHO}\\|\\\text{CHO}\\|\\\text{CHO}\\|\\\text{H}_2\text{CO}\end{array}\!\!\!\!\!\!\text{C}_6\text{H}_{10} \qquad (4)$$

57a and 57b   R = H, Ph   40b, 41b, and 41c

Interaction of 4,5:6,7-di-*O*-cyclohexylidene-2,3-dideoxy-1-*C*-phenyl-L-*arabino*-hept-2-enose (**65**) with phenylmethylenetriphenylphosphorane was accompanied[9(g)] by the formation of triphenylphosphine, instead of the expected triphenylphosphine oxide, thus indicating the abnormal character of this reaction. This result may be interpreted as involving possible addition of the phosphonium ylide to the alkenic bond, with subsequent stabilization of the intermediate betaine **82** through elimination of triphenylphosphine, and closure of the three-membered ring[2(f)] with formation of the cyclopropane derivative **83**, as shown in equation 5.

$$\begin{array}{c}\text{G}-\text{CH}\overset{\ominus}{=}\text{CHCOPh}\\|\\\text{HC}-\overset{\oplus}{\text{PPh}_3}\\|\\\text{Ph}\end{array} \quad \xrightarrow{-\text{PPh}_3} \quad \begin{array}{c}\text{H}\quad\text{H}\\|\quad|\\\text{G}-\text{C}-\text{C}-\text{COPh}\\\diagdown\;\diagup\\\text{CH}\\|\\\text{Ph}\end{array} \qquad (5)$$

82   83

The crystalline sugar derivative obtained in this reaction (in 40% yield) had an electronic absorption spectrum having λ_{max} 246 nm. This value lies between the corresponding values for acetophenone (λ_{max} 240 nm) and the initial ketose **65** (λ_{max} 253 nm), and may, therefore, be due to absorption by the cyclopropane fragment. Structure **84** for this compound follows from these considerations.

(62) J. H. Brewster, *J. Amer. Chem. Soc.*, **81**, 5475 (1959).

```
        Ph
        |
        CO
      HC\
       | >CHPh
      HC/
       |
       HCO\
       |   >C₆H₁₀
      (OCH
        |
        OCH
   H₁₀C₆<  |
        OCH₂

        84
```

3. α,β-Unsaturated Acids and Their Derivatives

The simultaneous insertion of an alkene bond and a carboxyl function into a sugar molecule is very attractive for carbohydrate chemistry in view of the further useful derivatization possible. This problem has been successfully solved by employment of the Wittig reaction, as well as its phosphonate modification (see p. 259). The compounds thus far synthesized are not numerous, although their transformations have been studied in detail and have resulted in a new method for the carbon-chain extension of carbohydrates.[26]

a. Synthesis.—The first syntheses[3–5] of this kind were also the first examples of realization of the Wittig reaction in carbohydrate chemistry, as already mentioned. Synthesis of ethyl 2,3-dideoxy-4,5-O-isopropylidene-D-*glycero*-pent-2-enonate (**85**) from 1,2-O-isopropylidene-D-glyceraldehyde and ethoxycarbonylmethylenetri-

```
       CO₂Et
        |
        CH
        ||
        CH
        |
       HCO\
        |   >CMe₂
      H₂CO/

        85
```

phenylphosphorane was reported by Kuhn and Brossmer.[3] Further development of this mode of interaction involved the use of both acetylated and acetalated *aldehydo*-aldoses and of partially protected or even free monosaccharides, as well as of phosphoranes having ethoxycarbonyl, amido, or nitrile functions.

Ethyl 2,3-dideoxy-D-*gluco*-oct-2-enonate pentaacetate (**86a**) (Refs. 4, 5, 21, 25, 39 and 63) and its D-*galacto* analog (**86b**) (Refs. 25, 39, and 63) were reportedly obtained starting from *aldehydo*-D-glucose and *aldehydo*-D-galactose pentaacetates. Syntheses of ethyl 2,3-di-

CO₂Et   CO₂Et
| |
CH   CH
|| ||
CH   CH
| |
(CHOAc)₄ (CHOAc)₃
| |
CH₂OAc   CH₂OAc

86a D-*gluco*   87a D-*arabino*
86b D-*galacto* 87b L-*arabino*

deoxy-D-*arabino*-hept-2-enonate[5] (**87a**) and its enantiomer[63] (**87b**) were also reported.

Acetalated *aldehydo*-aldoses also proved to be suitable starting materials for the synthesis to be considered, as may be illustrated by the successful synthesis of ethyl 2,3-dideoxy-4,5:6,7-di-*O*-isopropylidene-L-*arabino*-hept-2-enonate[64] (**88**) and ethyl 4,6:5,7-di-*O*-benzylidene-2,3-dideoxy-D-*ribo*-hept-2-enonate[21] (**89**).

CO₂Et   CO₂Et
| |
CH   CH
|| ||
CH   CH
| |
HCO\   HCO\
     \CMe₂   \CHPh
 OCH/   HCO/
| |
     OCH   HCO\
Me₂C\   \CHPh
     OCH₂   H₂CO/

     88   89

Free and partially protected aldoses were also shown to interact with alkoxycarbonylmethylenetriphenylphosphoranes when heated in *N,N*-dimethylformamide. Ethyl 2,3-dideoxy-D-*glycero*-pent-2-enonate[53] (**90**), ethyl 2,3-dideoxy-D-*arabino*-hept-2-enonate[65] (**91a**), ethyl 2,3-dideoxy-D-*ribo*-hept-2-enonate[65] (**91b**), ethyl 2,3-dideoxy-D-

	CH=CHCO₂Et
	HCOH
CO₂Et	CO₂Et	CO₂Et		HCOH
CH	CH	CH		HOCH
CH	CH	CH		HCOH
HCOH	(CHOH)₃	(CHOH)₄		HCOH
CH₂OH	CH₂OH	CH₂OH		CH₂OH
90	91a D-*arabino*	92a D-*gluco*		93
	91b D-*ribo*	92b D-*galacto*		

(63) N. K. Kochetkov and B. A. Dmitriev, *Izv. Akad. Nauk SSSR, Ser. Khim.*, 274 (1966); *Chem. Abstr.*, **64**, 19734 (1966).
(64) B. A. Dmitriev, N. E. Bayramova, A. A. Kost, and N. K. Kochetkov, *Izv. Akad. Nauk SSSR, Ser. Khim.*, 2491 (1967); *Chem. Abstr.*, **69**, 77662 (1968).
(65) N. K. Kochetkov and B. A. Dmitriev, *Dokl. Akad. Nauk SSSR*, **151**, 106 (1963).

gluco-oct-2-enonate[65] (**92a**), ethyl 2,3-dideoxy-D-galacto-oct-2-enonate[63,65] (**92b**), and ethyl 2,3-dideoxy-D-glycero-D-gulo-non-2-enonate[65] (**93**) have been prepared, starting from the corresponding aldoses, in about 45% yield. The lowering of yield is caused by the formation of cyclic by-products (see later). A *trans* configuration for these products is most probable, and was definitely established[66] for **92b** by its degradative oxidation into fumaric acid. To prevent cyclization, D-galactose tetraacetate was used.[63] The resulting product (**94**) was then acetylated, to afford **86b** in 80% yield.

```
     CO₂Et              CO₂R              CO₂Me
      |                  |                  |
      CH                 CH                 CH
      ||                 ||                 ||
      CH                 CH                 CH
      |                  |                  |
    HCOAc             HCNHAc             HCNHAc
      |                  |                  |
    AcOCH              HOCH               HOCH
      |                  |                  |
    AcOCH              HCOH               HCO
      |                  |                    \
    HCOH               HCOH               HCOH  CHPh
      |                  |                    /
    CH₂OAc             CH₂OH              H₂CO

      94              95a  R = Me            96
                      95b  R = Et
```

Use of a 2-amino-2-deoxyaldose as the starting material provides a convenient avenue to the appropriate amino derivatives. Thus, methyl 4-acetamido-2,3,4-trideoxy-D-gluco-oct-2-enonate (**95a**) and its ethyl ester (**95b**) were obtained[67,68] in ~30% yield starting from 2-acetamido-2-deoxy-D-glucose. This reaction was found to be accompanied by formation of a cyclic by-product, resulting in lowering of the yield. Partial prevention of this process was achieved by employment of 2-acetamido-4,6-O-benzylidene-2-deoxy-D-glucopyranose, resulting in preparation of the respective derivative (**96**) in 50% yield.

The branched-chain derivatives **97** and **98** were prepared[51] from L-sorbose pentaacetate and 2,3:4,5-di-O-isopropylidene-*aldehydo*-L-arabinose, respectively. In the latter case, a branched-chain phosphorane, namely methoxycarbonyl(methoxycarbonylmethyl) methylenephosphorane, was employed.

---

(66) N. K. Kochetkov and B. A. Dmitriev, *Izv. Akad. Nauk SSSR, Ser. Khim.*, 1405 (1965); *Chem. Abstr.*, **63**, 18273 (1965).

(67) B. A. Dmitriev and N. K. Kochetkov, *Izv. Akad. Nauk SSSR, Ser. Khim.*, 2483 (1967).

(68) (a) B. A. Dmitriev and N. E. Bayramova, *Tesisy Vses. Konf. Khim. Biokhim. Uglevodov, 4th, Moskow-Lvov,* 11 (1967); (b) B. A. Dmitriev and N. E. Bayramova, *Khim. Biokhim. Uglevodov, Moscow,* 58 (1969).

```
        CO₂Me              CO₂Me
        |                  |
        CH                 CH₂
        ‖                  |
        C—CH₂OAc           C—CO₂Me
        |                  ‖
        AcOCH              CH
        |                  
        HCOAc              HCO
        |                     \CMe₂
        AcOCH              OCH/
        |                  
        CH₂OAc         Me₂C   OCH
                           \  |
                            OCH₂

         97                  98
```

A series of interesting derivatives was prepared[21,69] starting from phosphoranes having maleimido, amido, and nitryl functions. These products were 2,4:3,5-di-O-benzylidene-1-deoxy-1-(2,5-dioxo-3-pyrrolylidene)-D-ribitol (**99a**), its N-phenyl derivative (**99b**), 2,4-O-benzylidene-1-deoxy-1-(2,5-dioxo-3-pyrrolylidene)-D-ribitol (**99c**), 4,6:5,7-di-O-benzylidene-2,3-dideoxy-D-*ribo*-hept-2-enonamide (**100**), 2,3-dideoxy-D-*gluco*-oct-2-enonamide 4,5,6,7,8-pentaacetate (**101**), 4,6:5,7-di-O-benzylidene-2,3-dideoxy-D-*ribo*-hept-2-enononitrile (**102**), and 2,3-dideoxy-D-*gluco*-oct-2-enononitrile 4,5,6,7,8-pentaacetate (**103**). Derivatives **99a–99c** were prepared for use in the

```
    RN        O              HN       O
      \    //                  \    //
       \  /                     \  /
        \/                       \/
    O                        O
        |                        |
        CH                       CH
        |                        |
        HCO\                     HCO\
        HCO  CHPh                HCOH  CHPh
        HCO/                     HCO/
        |                        |
        H₂CO                     CH₂OH

      99a R = H                  99c
      99b R = Ph
```

synthesis of the antibiotic showdomycin.[21] Attempted coupling of triphenylphosphoranylidenesuccinimide and its N-phenyl derivative with *aldehydo*-D-ribose and *aldehydo*-D-glucose acetates failed, as it gave intractable mixtures.[21,69] Attempts to interact D-ribose and D-glucose with these phosphoranes were also unsuccessful.[69]

It was also of interest to synthesize derivatives of α,β-unsaturated acids by using modified alkoxycarbonylmethylenetriphenylphosphoranes, $Ph_3P{=}CHXCO_2R$. This approach led to synthesis of ethyl 2,3-dideoxy-2-phenyl-D-*gluco*-oct-2-enonate 4,5,6,7,8-pentaacetate[25]

(69) R. E. Harmon, G. Wellman, and S. K. Gupta, *Carbohyd. Res.*, 11, 574 (1969).

(104) and of a series of 2-bromo derivatives[68,70] (**105–108b**), the latter being used for a convenient preparation of sugar aziridines (see later).

Condensation of alkoxycarbonylmethylenetriphenylphosphoranes with partially protected dialdoses having one free aldehyde group provided a direct avenue to various uronic acids. Two derivatives of this type, namely, butyl 3-O-benzoyl-1,2-O-cyclohexylidene-5,6-dideoxy-α-D-*xylo*-hept-5-eno-1,4-furanuronate[23] (**109a**) and methyl 3-O-benzyl-1,2-O-cyclohexylidene-5,6-dideoxy-α-D-*xylo*-hept-5-eno-1,4-furanuronate[18] (**109b**) were synthesized. An analogous deriva-

(70) B. A. Dmitriev, N. E. Bayramova, and N. K. Kochetkov, *Izv. Akad. Nauk SSSR, Ser. Khim.*, 2691 (1967); *Chem. Abstr.*, **69**, 27687 (1968).

tive (110), synthesized from aldehyde 33, was used in a synthesis[71] of phytosphingosine.

```
        CHCO₂R                    CHCO₂Et
        ‖                         ‖
        CH   O                    CH   O    OMe
         \  / \                    \  / \  /
          \/   \                    \/   \/
          /\    \                   /\
         OR'     \                 O    N
                  O                 \  //
                  |                  C
                  O—C₆H₁₀            |
                                     Ph
     109a  R = Bu,
           R' = Bz                   110
     109b  R = Me,
           R' = CH₂Ph
```

A. S. Jones and coworkers[72] attempted a synthesis of 5'-(carboxymethyl)-5'-deoxyuridine (111) as a nucleotide analog, starting from 2',3'-O-isopropylideneuridine-5'-aldehyde (112) and ethoxycarbonylmethylenetriphenylphosphorane. Surprisingly, no reaction occurred. When the phosphorane was generated *in situ* by means of sodium ethoxide, a mixture of five products was obtained. Four of them were

```
     CH₂CO₂H                                          CHCO₂Et
     |                                                ‖
     CH₂  O   B            CHO  O   B                 CH   O   B
      \  / \ /              \  / \ /                   \  / \ /
       \/   \/                \/   \/                    \/   \/
       /\                     /\                        /\
      HO  OH                 O    O                    O    O
                              \  /                      \  /
                              CMe₂                      CMe₂

       111                     112                      113
```

```
              O   B                              O   B
         ____/ \_/                          ____/ \_/
RO₂CCH₂—CH        \              RO₂CCH=CH        \
         \        /                        \      /
          _____/                          \____/
          /\                                   |
         O   O                                 OH
          \ /
          CMe₂
                       114a  R = Et            115a  R = Et
                       114b  R = H             115b  R = H

              O
              ‖
               ╱‾‾╲
              |    NH
where B =     |    |
               ╲__╱
              N    O
              |
```

proved to be the unsaturated derivatives 114a, 114b, 115a, and 115b; they obviously arise as a result of (a) allylic rearrangement of the expected ester 113 through the intermediate anion 116, or (b) elimi-

(71) J. Gigg, R. Gigg, and C. D. Warren, *J. Chem. Soc.* (C), 1882 (1966).
(72) P. Howgate, A. S. Jones, and J. R. Tittensor, *Carbohyd. Res.*, 12, 403 (1970).

THE WITTIG REACTION IN CARBOHYDRATE CHEMISTRY 259

nation of the isopropylidene group in this anion, as shown in equation 6.

Partial saponification of the ester group in the basic medium also occurred. An alternative route to **115a** and **115b** consists in an initial β-elimination of the isopropylidene group in aldehyde **112**, similar to that for aldehyde **9** (see p. 231), to give the α,β-unsaturated aldehyde **117**, which could react with the Wittig reagent, leading to **115a** as shown in equation 7.

$$113 \xrightarrow{\text{EtO}^{\ominus}} \text{EtO}_2\text{CCH=CH} - \overset{\text{116}}{\underset{}{\text{[structure]}}} \xrightarrow[(a)]{\text{HOR}} 114a \quad (6)$$

$$\downarrow (b)$$

$$\text{EtO}_2\text{CCH=CH} - \overset{}{\underset{}{\text{[structure]}}} \xrightarrow[-\text{Me}_2\text{CO}]{\text{HOR}} 115a$$

$$112 \xrightarrow{\text{EtO}^{\ominus}} \text{OHC} - \overset{117}{\underset{}{\text{[structure]}}} \longrightarrow 115a \quad (7)$$

Although ethoxycarbonylmethylenetriphenylphosphorane is known to react with ketones in the presence of benzoic acid or under drastic conditions,[2(f)] the introduction of the ethoxycarbonylmethylene group into a sugar molecule at a branching point has been achieved by use of the phosphonate modification[73,74] of the Wittig reaction. Rosenthal and coworkers reported[13,14] the condensation of protected glyculose **21** with phosphonoacetic acid trimethyl ester in the presence of potassium *tert*-butoxide, resulting in the formation of the unsaturated, branched-chain derivative **118** as a mixture of the *cis* and *trans* isomers

(73) L. Horner, H. Hoffman, and H. G. Wippel, *Chem. Ber.*, **91**, 61 (1958).
(74) W. S. Wadsworth, O. E. Schupp, E. Y. Seus, and Y. A. Ford, *J. Org. Chem.*, **30**, 680 (1965).

in the ratio of 1:3. The respective pyranose derivative (119) was prepared[75] in a similar way. Synthesis of the nitrile derivative (120) was also reported.[76]

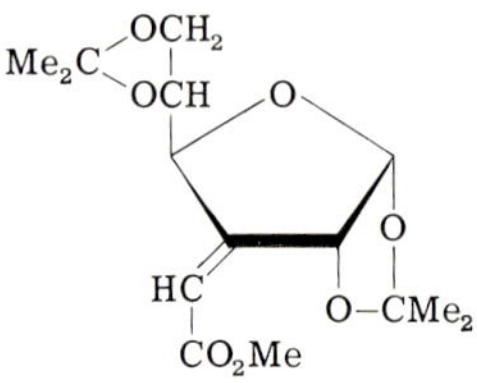

118 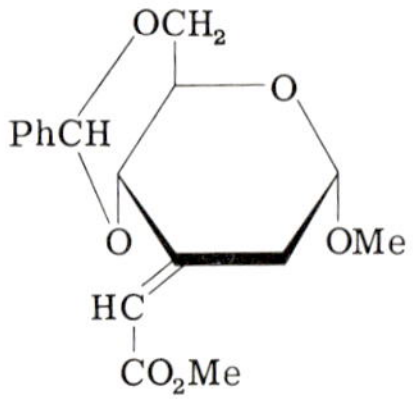   119 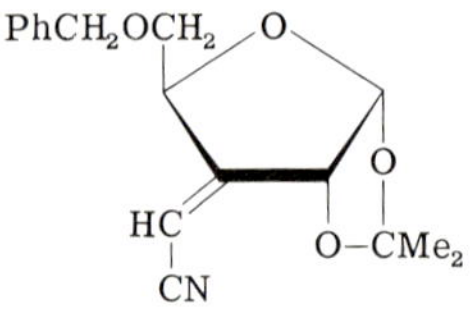   120

The introduction of a dihydroxyphosphinylmethyl grouping into a sugar molecule by using diphenyltriphenylphosphoranylidenemethylphosphonate (120a) is of extreme interest for the eventual synthesis of modified nucleotides having a C–P bond. Jones and Moffatt[77] have shown that the phosphorane 120a reacts smoothyl with 4'-C-formyl-2',3'-O-isopropylidene derivatives of uridine and adenosine, yielding 5'-deoxy-5'-(diphenylphosphinylmethyl)uridine (120b) (58% yield) and an analogous adenosine derivative (120c) (33% yield), respec-

120a

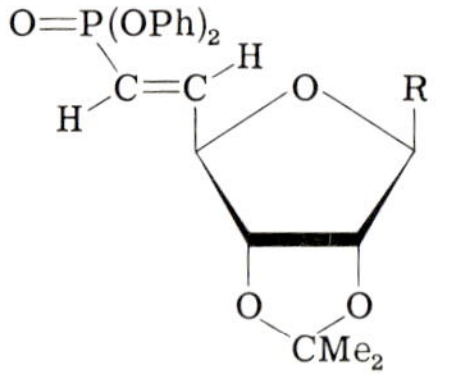

120b R = Uracil-1-yl
120c R = Adenin-9-yl

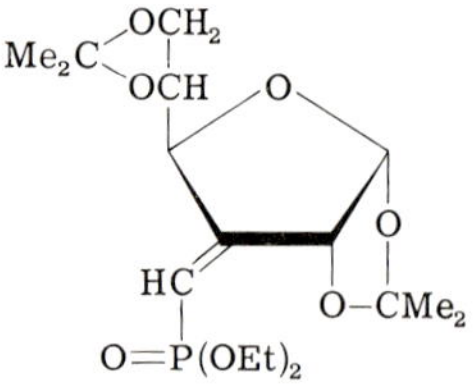

120d

tively. Interaction of 1,2:5,6-di-O-isopropylidene-α-D-ribo-hexofuranos-3-ulose (21) with the carbanion obtained from tetraethylmethylenediphosphate and butyllithium led[78] to a branched-chain, unsaturated sugar phosphonate (120d) (81% yield), used further for preparation of modified nucleotides.

b. Physical Properties.—Electronic absorption spectra have been reported[72] only for the "unexpected" derivatives 114b and 115b. Compound 114b has a u.v. spectrum typical of uridine derivatives (λ_{max} 261 nm; ε 18,700). The bathochromic shift for 115b (λ_{max} 268.5 nm)

(75) A. Rosenthal and P. Catsoulacos, Can. J. Chem., 46, 2869 (1968).
(76) A. Rosenthal and D. A. Baker, Tetrahedron Lett., 397 (1969).
(77) G. H. Jones and J. G. Moffatt, J. Amer. Chem. Soc., 90, 5337 (1968).
(78) H. P. Albrecht, G. H. Jones, and J. G. Moffatt, J. Amer. Chem. Soc., 92, 5511 (1970).

was interpreted as due to the appearance of an additional chromophore.

The compounds under consideration usually show three characteristic infrared absorption bands. The first, near 5.8 μm, corresponds to conjugated ester group absorption, which may be masked by acetyl absorption at ~5.7 μm in the case of acetylated derivatives. The second, near 6.01 μm, and the third, near 10.25 μm, are due to *trans*-alkene bond absorption. The bromo derivatives **105–108b** show ester absorption at 5.68–5.78 μm and olefinic absorption at longer wavelength (6.09 μm and 10.2–10.4 μm, respectively).[70] Ester **110** shows[71] a normal, conjugated ester absorption at 5.79 μm, but absorption at 6.07 μm may be caused either by the conjugated alkene bond or by carbon–nitrogen double-bond absorption. The pyrrolylidene derivatives **99a**, **99b**, and **99c** show[69] two carbonyl peaks, at 5.66 and 5.79 μm, corresponding to free and conjugated carbonyl groups, respectively. The third band, at ~5.88 μm, has also been ascribed[69] to carbonyl absorption, although it may correspond to alkene absorption. The branched-chain derivative **98** shows[51] carbonyl absorption at 5.88 μm.

A few data on the n.m.r. spectroscopy of these compounds have been reported. Thus, the spectrum of bromo derivative **107** exhibits[70] a vinylic, C-3 proton signal at τ 3.0 p.p.m. as a doublet ($J_{3,4}$ 8 Hz). In the "unexpected" derivative **114b**, the vinylic C-5' proton resonates at τ 4.9–5.1 p.p.m. as a triplet. A high-field signal at τ 7.24 p.p.m. (doublet) corresponds to the C-6' methylenic group, because irradiation at τ 7.24 caused collapse of the H-5' triplet to a singlet, and irradiation at τ 5.1 caused the collapse of the H-6 doublet to a singlet. In the spectrum of the dialkene **115b**, vinylic protons resonate at τ 3.88 (doublet, H-3'), 3.28 (doublet, H-5'), and 3.94 p.p.m. (doublet, H-6') with the coupling constant $J_{5',6'}$ 16 Hz, thus indicating the *trans* configuration[72] of the C-5'–C-6' alkenic bond.

c. **Chemical Properties.**—Examination of the chemical properties of the α,β-unsaturated aldonic acid esters was undertaken in connection with their synthetic utilization. All of the subsequent transformations of these compounds began from addition reactions to the alkenic bond, such as hydrogenation,[13,14,67,71,75,76] addition of ammonia,[18,64,70] and hydroxylation.[63,66,67,79] Intramolecular addition of hydroxyl, resulting in anhydridation, was also observed.[63,67]

Hydrogenation of the amino compound **95a** led to methyl 4-acetamido-2,3,4-trideoxy-D-*gluco*-octonate (**121**), its structure being established by degradation[67] to *N*-acetylglutamic acid (**122**). Hydrogenation

(79) N. K. Kochetkov and B. A. Dmitriev, *Izv. Akad. Nauk SSSR, Ser. Khim.*, 670 (1964).

of the unsaturated ester **110** to afford the saturated derivative **123** was accomplished. The ester group in **123** was then reduced to a hydroxymethyl group, and the resulting alcohol was, in turn, transformed into the corresponding iodide. The latter was used for carbon-chain extension in phytosphingosine synthesis.[71]

```
CO₂Me        CO₂H              CO₂Et
 |            |                  |
CH₂          CH₂                CH₂
 |            |                  |
CH₂          CH₂                CH₂   O
 |            |                    \ / \ OMe
HCNHAc       HCNHAc
 |            |
HOCH         CO₂H                  O   N
 |                                   \ /
HCOH                                  
 |                                    
HCOH          122                    Ph
 |
CH₂OH
  121                                 123
```

Hydrogenation of the branched-chain derivatives **118** and **120** to give 3-deoxy-1,2:5,6-di-*O*-isopropylidene-3-*C*-(methoxycarbonyl-methyl)-α-D-allofuranose[13,14] (**124**) and the corresponding nitrile[76] (**125**) was also reported. The D-*allo* configuration of these compounds follows from their n.m.r.-spectral data. Thus, in compound **125**, the proton at C-2 resonates as a triplet, indicating *cis* orientation of H-1, H-2, and H-3, as, in the n.m.r. spectrum of 1,2-*O*-isopropylidene-α-D-xylofuranose, coupling between H-2 and H-3 is absent, leading[76] to a doublet for H-2. Hydrogenation of the branched-chain, pyranoid derivative **119**, followed by acid hydrolysis of the product, resulted in the formation[75] of methyl 2,3-dideoxy-3-*C*-[(methoxycarbonyl)methyl]-D-*ribo*-hexopyranoside (**126**). The stereoselectivity of the hydrogenation may be explained by the shielding effect of the 1,2-*O*-isopropylidene group in **118** and **120**, and of the 4,6-*O*-benzylidene group in **119**.

```
     OCH₂              OCH₂                  CH₂OH
Me₂C<   |         Me₂C<  |                    | 
     OCH  O             OCH  O                 \ O
      \ /                \ /                HO /   \ OMe
       O                  O                     \ /
       |                  |                     CH₂
      CH₂  O-CMe₂        CH₂ O-CMe₂              |
       |                  |                    CO₂Me
      CO₂Me              CN

       124                125                    126
```

Studies on the addition of ammonia to α,β-unsaturated esters were undertaken in order to develop a novel synthesis of 3-amino-2,3-

dideoxy sugars. Heating of ester **88** with ammonia at 100° during 40 hr, followed by acetylation, gave[64,68] 3-acetamido-2,3-dideoxy-4,5:6,7-di-O-isopropylidene-L-*arabino*-heptonamide (**127**), in excel-

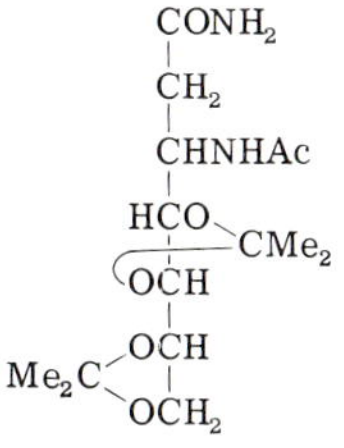

```
       CONH₂              CONH₂              CONH₂
        |                  |                  |
       CH₂                CH₂                CH₂
        |                  |                  |
       CHNHAc             AcHNCH             HCNHAc
        |                  |                  |
       HCO                HCO                HCO
         \CMe₂              \CMe₂              \CMe₂
       OCH                OCH                OCH
        |                  |                  |
       OCH                HOCH               HOCH
   Me₂C  |                  |                  |
       OCH₂              CH₂OH              CH₂OH

       127                128a               128b
```

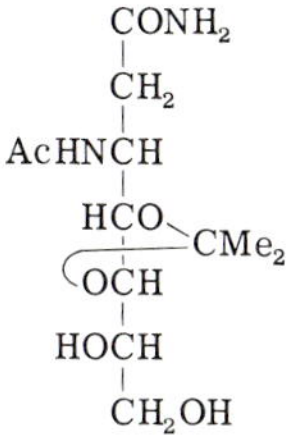

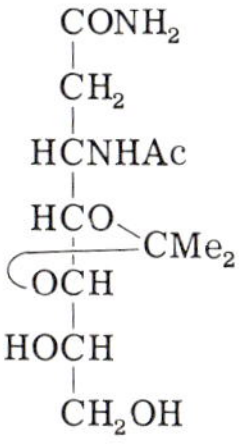

lent yield, as a mixture of the C-3 epimers. This mixture was separated into the two epimers (**128a** and **128b**) after partial hydrolysis with acid. Epimer **128b** would, by the Cram rule, be expected to preponderate as shown. Indeed the preponderant epimer was transformed[64] into

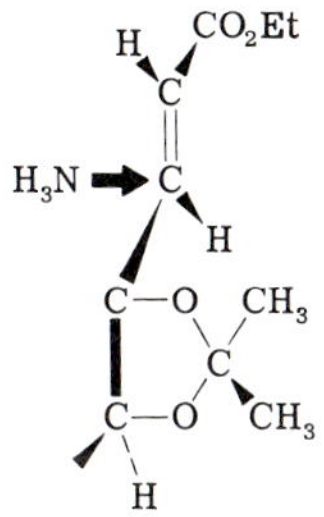

The predominant
attack of ammonia
on the double bond
in compound **88**.

N-benzoyl-L-aspartic acid (**129**), thus confirming its L-*manno* configuration (**128b**). Surprisingly, the addition of ammonia to the uronate **109b** proceeded under the usual conditions, to give[18] the corresponding amino derivative (**130**).

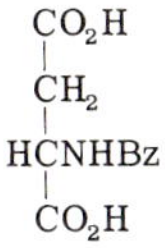

```
   CO₂H              CO₂Me
    |                 |
   CH₂               CH₂
    |                 |
  HCNHBz             CHNH₂
    |                  \O
   CO₂H           OCH₂Ph/
                       \    O
                        \   |
                         O-C₆H₁₀
   129                130
```

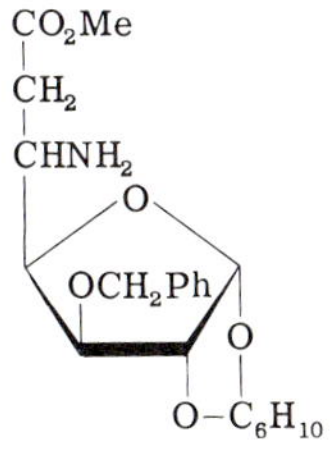

The interaction of such bromo derivatives as **105** with ammonia and amines may provide a convenient method for the preparation of sugar aziridines. The ester **107** was shown[68,70] to react with benzylamine, to produce aziridine **131** as a single isomer (in only 25% yield, because of some simultaneous deacetylation). When the acetalated bromo derivative **105** was allowed to react with ammonia, the aziridine **132** was obtained,[68,70] also as a single isomer, in ~90% yield. Ethyl 2,3-(N-benzylepimino)-2,3-dideoxy-L-*glycero*-L-*galacto*-heptonate 4,5,6,7-tetraacetate (**131**) has the *trans* orientation of the substituents on the epimino ring, because, in its n.m.r. spectrum, signals at τ 7.45 (doublet) and 7.70 p.p.m (quartet) corresponding to H-2 and H-3 resonances were observed,[70] with a coupling constant of 2.8 Hz. In *cis*-substituted aziridines, the neighboring protons are coupled with J ~6.6 Hz. Furthermore, the acetoxyl group on C-4 of **131** exhibits a higher-field signal, at τ 8.45 p.p.m., as compared with those of other acetoxyl groups, due to a shielding effect of the N-benzyl group. Deuterated **132** gives an H-2 signal at τ 7.5 p.p.m. as a doublet, with $J_{2,3}$ 2.8 Hz, and an H-3 quartet at τ 7.6 p.p.m. that confirms its *trans* configuration at the epimino ring. Aziridines **131** and **132** are, therefore, the first examples of sugar *trans*-aziridines.[68(b)]

131   132

Proof of the absolute configuration at C-2 in these compounds seems to be difficult. Taking into account the stereospecificity of the addition of ammonia, as well as the *trans* configuration of the initial bromo derivatives, it may be proposed that C-3 has the D configuration. The L configuration of C-2 follows from the *trans*-epimino configuration of these compounds. The predicted configuration was partially confirmed for **132** by its conversion into L-glucose through successive hydrolysis with acid and oxidation of the resulting α-amino acid with ninhydrin. The formation of L-glucose was,

THE WITTIG REACTION IN CARBOHYDRATE CHEMISTRY 265

indeed, expected in this case, as this treatment usually results in inversion of the configuration of C-3.[68(b),70]

As already noted (see p. 255), the interaction of alkoxycarbonylmethylenetriphenylphosphorane with free sugars is accompanied by the formation of by-products. The structure of the by-product was established[63] in the case of D-galactose; based on its stability to hydrogenation, consumption of one mole of periodate per mole, and the δ-lactonic infrared absorption at 5.7 μm, the by-product corresponded to structure 133. This product may arise as a result of intramolecular attack of the 6-hydroxyl group on the alkenic C-3 atom, with simultaneous lactonization. This conversion is probably catalyzed by phosphorane or triphenylphosphine oxide, as heating of the α,β-unsaturated ester 92b with these reagents in N,N-dimethylformamide led[63] to the formation of 133.

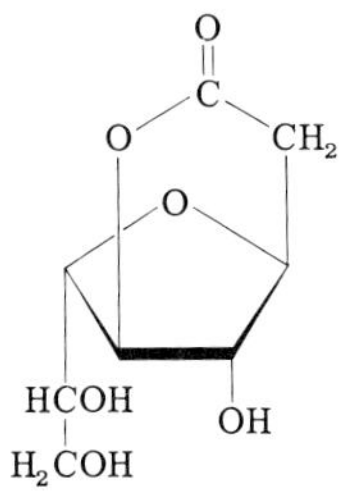

133

Lactonization is not obligatory, as the byproduct formed in the reaction of 2-acetamido-2-deoxy-D-glucose with methoxycarbonylmethylenetriphenylphosphorane was proved[67] to have structure 134. This compound rapidly consumes one mole of periodate per mole, and shows signals at τ 7.95 p.p.m. (doublet) and τ 6.15 p.p.m. in its n.m.r. spectrum, corresponding to the α-methylenic and the ester group, respectively. Use of 2-acetamido-4,6-O-benzylidene-2-deoxy-D-*arabino*-hexose instead of the unprotected sugar does not prevent

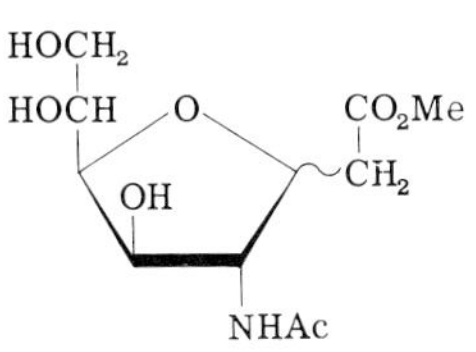

134 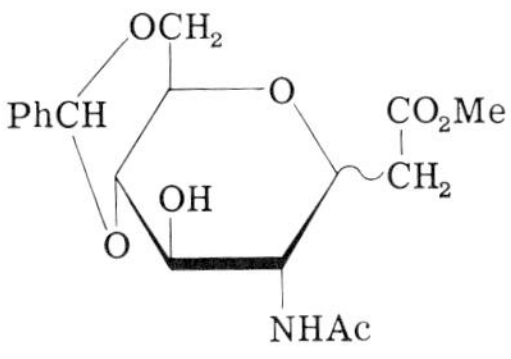   135

anhydridation and results in the formation of the anhydro pyranoid derivative **135**. Its n.m.r. spectrum[67] shows a doublet for the α-methylenic group at τ 6.8 p.p.m., and two close singlets for the ester group, at τ 6.5 and 6.45 p.p.m.

Kochetkov and Dmitriev have reported[66,79] an investigation on the hydroxylation of certain α,β-unsaturated aldonic esters. The hydroxylation was performed by treatment of these compounds with chloric or iodic acid (or their salts) and a catalytic amount of osmium tetraoxide in a suitable solvent. The resulting diastereoisomers were then separated as their cadmium salts.

Thus, hydroxylation of ethyl 2,3-dideoxy-D-*galacto*-oct-2-enonate (**92b**) led[66] to a mixture of D-*threo*-L-*ido*-octonic acid (**136a**), D-*threo*-L-*galacto*-octonic acid (**136b**), and its 1,4-lactone (**136c**), with preponderance of the two latter. After separation, aldonic acid **136b** was converted into lactone **136c**, which was reduced to D-*threo*-L-*galacto*-octose. The stereoselective hydroxylation, with preponderant formation of the D-*threo* fragment, may also be explained by *cis* addition of osmium tetraoxide in a situation similar to that shown on p. 263.

```
      CO₂H              CO₂H              C=O
      HCOH              HOCH              HOCH ⎤
      HOCH              HCOH              HCOH ⎥
      HCOH              HCOH              HCO ─┘
      HOCH              HOCH              HOCH
      HOCH              HOCH              HOCH
      HCOH              HCOH              HCOH
      CH₂OH             CH₂OH             CH₂OH

      136a              136b              136c
```

Hydroxylation of the acetamido derivative **96** proceeded[67] with the formation of a complex mixture, from which methyl 4-acetamido-6,8-O-benzylidene-4-deoxy-D-*threo*-L-*galacto*-octonate (**137**) was

```
      CO₂Me
      HOCH
      HCOH
      HCNHAc
      HOCH
      HCO⎫
      HCOH⎬CHPh
      H₂CO⎭

      137
```

isolated in 50% yield, thus indicating similar stereoregulation of addition of osmium tetraoxide.

4. Alkyl Ald-3-enulosonates

a. **Synthesis.**—Preparation of alkyl ald-3-enulosonates by coupling of (alkoxyoxalyl)methylenetriphenylphosphoranes (138) with suitable *aldehydo*-aldoses is of great interest in the chemistry of sugars, as it leads to biologically important 3-deoxyglyculosonic acids. Reported first by Zhdanov and Uslova,[80] this mode of interaction was further developed by Kochetkov and coworkers[27,81] who have devised a convenient method of transformation of these esters into 3-deoxyglyculosonic acids.

(Methoxyoxalyl)- (138, R = Me) and (ethoxyoxalyl)-methylenetriphenylphosphorane (138, R = Et) were prepared[80] from the appropriate 3-bromopyruvates through their quaternization by triphenylphosphine, and subsequent treatment of the resulting quaternary salts with ammonia.

Condensation of phosphorane (138, R=Me) with *aldehydo*-D-galactose pentaacetate afforded[80] methyl 3,4-dideoxy-D-*galacto*-non-3-enulosonate 5,6,7,8,9-pentaacetate (139e) in 42% yield. Extension of this interaction to include other *aldehydo*-aldoses led[27,81] to the preparation of analogous derivatives (139b–139d). The reactivity of phosphorane (138, R = Me) was found[81] to be lowered, in comparison with that of ethoxycarbonylmethylenetriphenylphosphorane; this was demonstrated by moderate yields (40–60%) and failure of the reaction with free sugars. Phosphorane (138, R = *tert*-Bu) was more reactive, as it gave high yields of 139b, 139c, and 139f.

$$Ph_3P=CHCOCO_2R$$
$$138$$

$$\begin{array}{c} COCO_2R \\ | \\ CH \\ || \\ HC \\ | \\ (CHOAc)_n \\ | \\ CH_2OAc \end{array}$$

139a n = 3, L-*arabino*, R = Me
139b n = 3, L-*arabino*, R = CMe$_3$
139c n = 3, D-*arabino*, R = CMe$_3$
139d n = 4, D-*gluco*,   R = Me
139e n = 4, D-*galacto*, R = Me
139f n = 4, D-*galacto*, R = CMe$_3$

(80) Yu. A. Zhdanov and L. A. Uslova, *Zh. Obshch. Khim.*, **36**, 1211 (1966); *Chem. Abstr.*, **65**, 18670 (1966).

(81) B. A. Dmitriev, N. E. Bayramova, A. B. Backinovsky, and N. K. Kochetkov, *Dokl. Akad. Nauk SSSR*, **173**, 350 (1967); *Chem. Abstr.*, **67**, 54381 (1967).

b. **Physical Properties.**—Ester infrared absorption in the compounds to be discussed is usually masked by acetyl absorption, resulting in the appearance of a complex absorption band at 5.69–5.72 μm, the first value obviously corresponding to an absorption of the alkoxycarbonyl group. The second band, at 5.88–5.91 μm, is undoubtedly caused by the conjugated keto group, and the third band, at 6.07–6.12 μm, as well as absorption at 10.25 μm, confirmed the presence of a conjugated *trans* alkene bond.[80,81]

In the n.m.r. spectra of compounds **139e** and **139f**, the olefinic protons H_α and H_β give a doublet at τ 3.50 p.p.m. ($J_{\alpha,\beta}$ 16 Hz) and a quartet at τ 3.20 p.p.m. ($J_{\beta,\alpha}$ 16 Hz, $J_{\beta,H\text{-}5}$ 4 Hz); the large value of the coupling constant $J_{\alpha,\beta}$ thus confirmed the *trans* configuration of the alkene bond.[81]

c. **Chemical Properties.**—As already noted, the alkyl ald-3-enulosonates **139a–139f** were synthesized for subsequent conversion into 3-deoxyglyculosonic acids. This problem could not be solved by such direct methods as hydration, or lactonization in the presence of mineral acid, because of the instability of the final products in acid media. Kochetkov and coworkers applied[27] an indirect approach, including the conversion of these compounds into the corresponding enamino lactones, followed by acid hydrolysis under mild conditions.

The first stage of this transformation consists in acid hydrolysis of these compounds, with retention of the protecting groups. The *tert*-butoxy esters **139b, 139c,** and **139f** proved to be the most suitable for this purpose, owing both to increased yields (see earlier) and rapid hydrolysis by trifluoroacetic acid. The resulting acids (**140a–140c**) were then allowed to react with methyl *p*-aminobenzoate[27(a)] or with aniline.[27(b),(c)] The interaction of acid **140c** with methyl *p*-amino-

```
        CO₂H                              Meo₂C—H₄C₆—HN    O
        |                                                \\
        CO                                                 N—C₆H₄—CO₂Me
        |                                   ═╱
        CH
        ‖                                   HCOAc
        HC                                  |
        |                                   AcOCH
        (CHOAc)ₙ                            |
        |                                   AcOCH
        CH₂OAc                              |
                                            HCOAc
   140a  n = 3, L-arabino                   |
   140b  n = 3, D-arabino                   CH₂OAc
   140c  n = 4, D-galacto                      141
```

benzoate led[27(a)] mainly to the undesired enamino lactam **141**. This phenomenon was explained by an initial addition of the amine to the conjugated alkene bond. The resulting γ-amino acid (**142**) could then give rise to lactam **143a**, which interacts in its enolic form (**143b**) with the other molecule of amine, as shown in equation 8.

$$
\begin{array}{cccc}
CO_2H & C=O & C=O & \\
CO & CO & COH & \\
| & | & \| & \xrightarrow{ArNH_2} \mathbf{141} \quad (8)\\
CH_2 & CH_2 & CH & \\
CHNHAr & CH-NAr & CH-NAr & \\
\mathbf{142} & \mathbf{143a} & \mathbf{143b} &
\end{array}
$$

To prevent addition of the amine to the alkene bond, methyl *p*-aminobenzoate hydrochloride was used,[27(a)] resulting in the formation of the expected enamino lactone (**144**), which was shown to

$$
\text{AcOH}_2\text{C}-\underset{\text{OAc}}{\overset{\text{H}}{\text{C}}}-\underset{\text{H}}{\overset{\text{OAc}}{\text{C}}}-\underset{\text{H}}{\overset{\text{OAc}}{\text{C}}}-\underset{\text{OAc}}{\overset{\text{H}}{\text{C}}}\diagdown\!\!\!\!\diagup\!\!\!\!\!\! =\text{O} \\
\text{NHC}_6\text{H}_4\text{CO}_2\text{Me}
$$
144

consist of two epimers. In this case, lactonization occurs first, and the resulting lactone (**145a**) reacts with the amine in its enolic form (**145b**) to produce the enamino lactone (**144**), as shown in equation 9. An alternative route through the Schiff base **146** is also possible.

$$
\mathbf{144} \longleftarrow
\begin{array}{c} C=O \\ C=NAr \\ CH_2 \\ OCH \end{array}
\xleftarrow{ArNH_2}
\begin{array}{c} C=O \\ CO \\ CH_2 \\ HCO \end{array}
\rightleftharpoons
\begin{array}{c} C=O \\ COH \\ \| \\ CH \\ HCO \end{array}
\xrightarrow{ArNH_2} \mathbf{144} \quad (9)
$$
$\quad\quad\quad\quad\mathbf{146} \quad\quad\quad\quad \mathbf{145a} \quad\quad\quad \mathbf{145b}$

Reaction of the acid **140a** with aniline hydrochloride led[27(b)] to the formation of two epimeric enamino lactones (**147a** and **148a**) in the ratio of 1:1; these were separated by fractional recrystallization.

Similar reaction of the enantiomeric acid (**140b**) resulted[27(c)] in the formation of the two epimeric enamino lactones **147b** and **148b**. The enamino lactones **144, 147b**, and **148a** were then hydrolyzed[27] to afford the corresponding 3-deoxyglyculosonic acids.

147a L-*arabino*
147b D-*arabino*

148a L-*arabino*
148b D-*arabino*

5. Enol, Thioenol, and Enamino Derivatives

It has been shown in the preceding Sections that the reactivity of various carbonyl-containing phosphoranes provides a direct avenue to introduction of the carbonyl function in a sugar molecule. There is, however, an alternative mode of such transformation that consists in introduction of such carbonyl precursors as enol, thioenol, or enamino groupings, provided that appropriate ylides (**149**) are available and that the final products may be converted into carbonyl compounds under mild conditions.

$$Ph_3 \overset{\oplus}{P} - \overset{\ominus}{\underset{X}{C}} - R$$

149

a. Synthesis.—An attempted interaction of methoxymethylenetriphenylphosphonium ylide (**149**, R = H; X = OMe) with 2,3:4,5-di-*O*-isopropylidene-*aldehydo*-D-arabinose, followed by acid hydrolysis, led[82] to an extremely complex mixture containing only 6% of the expected 2-deoxy-D-*arabino*-hexose, together with D-glucose, D-mannose, and, possibly, D-fructose. This process was explained as proceeding through two concurrent routes of decomposition of the intermediate betaine, with elimination either of triphenylphosphine

(82) B. A. Dmitriev, N. N. Aseeva, and N. K. Kochetkov, *Izv. Akad. Nauk SSSR, Ser. Khim.*, 1342 (1968); *Chem. Abstr.*, **69**, 77644 (1968).

oxide (the classical route) or of triphenylphosphine, these compounds, indeed, being observed in the final mixture.

Zhdanov and Alexeeva[28(a)(b)] reported, however, a successful preparation of 3,4:5,6-di-O-cyclohexylidene-1-O-methyl-1-hexene-D-xylo(and L-arabino)-1,3,4,5,6-pentols (150a and 150b, respectively) in ~25% yields by similar coupling of the appropriate aldehydo-aldose with ylide 149 (R = H; X = OMe), prepared from methoxymethyltriphenylphosphonium chloride in ether at low temperature with use of phenyllithium as the proton acceptor. An analogous compound (151) was obtained[33] by Tronchet and coworkers in ~19% yield, sodium hydride being the proton acceptor used. Use of p-tolyloxymethylenetriphenylphosphonium ylide (149, R = H; X = O-C$_6$H$_4$-CH$_3$-p) in reaction with 2,3:4,5-di-O-isopropylidene-aldehydo-L-arabinose allowed the corresponding vinyl ether[71] (152) to be obtained in 67% yield.

$$\begin{array}{cccc}
\text{CHOMe} & \text{CHOMe} & \text{CHOMe} & \text{CHOC}_6\text{H}_4\text{CH}_3 \\
\| & \| & \| & \| \\
\text{CH} & \text{CH} & \text{CH} & \text{CH} \\
\text{HCO}\!\!\diagdown\!\!\text{C}_6\text{H}_{10} & \text{HCO}\!\!\diagdown\!\!\text{C}_6\text{H}_{10} & \text{HCO}\!\!\diagdown\!\!\text{CMe}_2 & \text{HCO}\!\!\diagdown\!\!\text{CMe}_2 \\
\text{OCH} & \text{OCH} & \text{OCH} & \text{OCH} \\
\text{HCO}\!\!\diagdown\!\!\text{C}_6\text{H}_{10} & \text{H}_{10}\text{C}_6\!\!\diagup\!\!\text{OCH} & \text{Me}_2\text{C}\!\!\diagup\!\!\text{OCH} & \text{Me}_2\text{C}\!\!\diagup\!\!\text{OCH} \\
\text{H}_2\text{CO} & \text{OCH}_2 & \text{OCH}_2 & \text{OCH}_2 \\
\\
\text{150a} & \text{150b} & \text{151} & \text{152}
\end{array}$$

Searching for other 2-deoxy sugar precursors that might be obtained in high yields, attention was paid to (methylthio)methylenetriphenylphosphonium ylide (149, R = H, X = SMe). Tronchet and coworkers reported[29(a),(b),83,84] the interaction of this ylide, prepared from the corresponding phosphonium salt and sodium hydride, with various aldehydo and keto sugars. The methylthio derivative 153 was prepared in this way from 2,3:4,5-di-O-isopropylidene-aldehydo-L-arabinose, in 87% yield, as a mixture of the cis and trans isomers, with a preponderance of the former. Similar treatment of 2,5-anhydro-3,4-O-isopropylidene-D-ribose resulted in the isolation[83(b)] of 3,6-anhydro-4,5-O-isopropylidene-1-S-methyl-1-thio-1-hexene-D-ribo-1,3,4,5,6-pentol as the cis (154a) and trans (154b) isomers. Analogous

(83) (a) J. M. J. Tronchet, S. Jaccard-Thorndal, and B. Baehler, Helv. Chim. Acta, 52, 817 (1969); (b) J. M. J. Tronchet, N. Le-Hong, and F. Perret, ibid., 53, 154 (1970).
(84) J. M. J. Tronchet, J. M. Bourgeois, and B. Baehler, Helv. Chim. Acta, 53, 368 (1970).

derivatives (**155a–155c**) having a potential *aldehydo* function were prepared[83(b)] from 1,2-*O*-isopropylidene-3-*O*-methyl (and 3-*O*-benzyl)-α-D-*xylo*-pentodialdo-1,4-furanose. In the former case, the isomers were separated, and **155a** was shown to preponderate.

```
    CHSMe
    ‖
    CH                                          CHSMe
    |                       O    CH=CHSMe       ‖
    HCO                    ╱  ╲                 CH   O
     ╲ CMe₂                                     |   ╱ ╲
    ╱ OCH                                       OR
    ╲O                                          |     O
     OCH                                        |    ╱
Me₂C╱    |                 O   O                O–CMe₂
    ╲OCH₂                   ╲ ╱
                            CMe₂                155a  R = Me, cis
      153                                       155b  R = Me, trans
                        154a cis isomer         155c  R = CH₂Ph
                        154b trans isomer
```

Coupling of ylide **149** (R = H, X = SMe) with certain protected *keto* sugars proved to be more complex. Methyl 2,3-*O*-isopropylidene-6-*O*-methyl-α-D-*lyxo*-hexopyranos-3-uloside led[29(a)] to compounds **156a** and **156b** (in the ratio of 3:7), and **156c**, in only 20% overall yield. A ketose dimer (of uncertain structure, but of double molecular weight) was also observed in the reaction mixture, thus indicating concurrent aldol self-addition of the initial ketose. This circumstance explains the formation of the unexpected product **156c** as a result of C-5 epimerization of the initial ketose during enolization caused by the basic medium.

```
       CH₂OMe                            CH₂OMe
       ╱─O                               ╱─O
  H   ╱   CMe₂                      MeS ╱    CMe₂
   ╲ C                                 ╲C
  MeS  ╲O  O╱                       H   ╲O   O╱
            ╲                                 ╲
             OMe                               OMe

       156a                              156b

                       ╱─O
                      ╱  CH₂OMe
                MeS  ╱
                  ╲ C
                 H  ╲     CMe₂
                     ╲O   O╱
                          ╲
                           OMe

                      156c
```

Coupling of ylide **149** (R = H, X = SMe) with 1,2:5,6-di-*O*-isopropylidene-α-D-*xylo*-hexofuranos-3-ulose was shown[29(b)] to proceed without epimerization, giving rise to the respective, isomeric thiovinylic ethers (**157a** and **157b**). The same interaction with 1,2:5,6-di-

O-isopropylidene-α-D *ribo*-hexofuranos-3-ulose (**21**) proved[29(b),84] to be quite complex, affording, besides the expected products (**158a** and **158b**), their C-5 epimers (**157a** and **157b**), as well as an "abnormal" product (**159**) of uncertain structure.

157a   157b

158a   158b   159

The interaction of 1,2-O-isopropylidene-α-L-*glycero*-tetrofuranos-3-ulose and 5-deoxy-1,2-O-isopropylidene-β-D-*threo*-pentofuranos-3-ulose with ylide **149** (R = H, X = SMe) leading to branched-chain derivatives (**159a**) and (**159b**), respectively, has also been reported.[85]

159a   159b

Bestmann and Angerer have briefly reported[8] that treatment of D- (and L-)arabinose, D-glucose, and L-rhamnose with ylide **149** (R = H, X = SPh) in methyl sulfoxide produced the sugar thioenol derivatives **160a–160d** in 40–70% yields. As products **160a** and **160b** were converted into 2-deoxy-D-(and L-)*arabino*-hexoses in high yield,

(85) J. M. J. Tronchet, J. M. Bourgeois, R. Graf, and J. Tronchet, *Compt. Rend.*, **269**, 420 (1969).

this mode of preparation may provide a most convenient, two-stage synthesis of 2-deoxy sugars..

CHSPh
‖
CH
|
(CHOH)$_n$
|
CH$_2$R

160a n = 3, D-*arabino*, R = OH
160b n = 3, L-*arabino*, R = OH
160c n = 4, D-*gluco*,   R = OH
160d n = 4, L-*manno*,   R = H

The Wittig reaction with ylide **149** (R = CO$_2$Et; X = OEt) was also applied to certain *aldehydo*-aldoses by Samokhvalov and coworkers[15] in order to find more convenient paths to 3-deoxyglyculosonic and sialic acids. This ylide was generated *in situ* from the corresponding phosphonium salts by means of sodium ethoxide in ethanol at −20°. Such an approach was first attempted[15(a)] by starting from 2,3:4,5-di-*O*-isopropylidene-*aldehydo*-D-arabinose, and the expected ethyl 3-deoxy-2-*O*-ethyl-4,5:6,7-di-*O*-isopropylidene-D-*arabino*-hept-2-enonate (**161**) was formed in about 70% yield.

CO$_2$Et
|
COEt
‖
CH
|
OCH
⎯⎯⎯CMe$_2$
HCO
|
HCO
⎯⎯⎯CMe$_2$
H$_2$CO

161

Similar reactions with 2-acetamido-2-deoxy-3,4:5,6-di-*O*-isopropylidene-*aldehydo*-D-glucose and 3-acetamido-2,4,5,6-tetra-*O*-acetyl-3-deoxy-*aldehydo*-D-*glycero*-D-*galacto*-heptose were also successful, leading to ethyl 4-acetamido-3,4-dideoxy-2-*O*-ethyl-5,6:7,8-di-*O*-isopropylidene-D-*gluco*-oct-2-enonate[15(b)] (**162**) (not isolated) and ethyl 5-acetamido-3,5-dideoxy-D-*glycero*-D-*galacto*-non-2-enonate[15] (**163**), respectively; in the latter case, deacetylation occurred because of the basic medium employed.

An analogous approach to 3-deoxyglyculosonic acids would be realized were an ylide **149** (R = CO$_2$Et; X = NR'$_2$) available. Nevertheless, this problem was solved in an original way. The condensation of 2,4-*O*-ethylidene-D-erythrose with the sodium salt of ethyl (diethyl-

THE WITTIG REACTION IN CARBOHYDRATE CHEMISTRY 275

```
      CO₂Et              CO₂Et
      |                  |
      COEt               COEt
      ‖                  ‖
      CH                 CH
      |                  |
      HCNHAc             HCOH
      |                  |
      OCH                AcHNCH
        \  CMe₂          |
      HCO⁄               HOCH
      |                  |
      HCO                HCOH
      |   \ CMe₂         |
      H₂CO⁄              HCOH
                         |
                         CH₂OH
      162                163
```

phosphono)piperidinoacetate (**164**) has been found to produce[16] the enamino lactone (**165b**) in 70% yield, without isolation of the intermediate enamino ester (**165a**).

```
         O     Na⊕           CO₂Et              OCH₂
         ‖                   |                  /    \O
  (EtO)₂P–C–CO₂Et         C–N⟨   ⟩         MeCH⟨    ⟩
         |⊖                  ‖                  \    /=O
         N                   CH                  O
        ⟨ ⟩                  |                       N–⟨  ⟩
                             HCO
                             |  \
                             HCOH  CHMe
                             |    /
                             H₂CO

        164                165a                  165b
```

Further development of this method will possibly provide a most convenient synthesis of 3-deoxyglyculosonic acids, and, especially, sialic acids, as the configuration of the final product has already been fixed in the enamino lactone of type **165b**, in contradistinction to the analogous enamino lactones (for example, **144**), which are formed as a mixture of epimers.

b. Physical Properties.—Electronic absorption data have been reported[29(b),83(b)] only for certain thioenol derivatives. Two absorption bands, at λ_{max} 233 nm (ϵ 3,400) and 244 nm (ϵ 4,200) were observed[83(b)] for the *cis* isomer **154a** having a terminal double bond. The *trans* isomer **154b** absorbs[83(b)] at λ_{max} 228 nm (ϵ 6,440) and 239 nm (ϵ 5,880). The structurally related thioenols **157a**, **157b**, **158a**, and **158b** show one absorbtion band in the region of 249–256 nm, with the molar extinctions of 7,500 to 9,500 depending on the configuration of the double bond and its steric environment. The "abnormal" product **159** shows[84] absorption for a terminal double bond at λ_{max} 226 nm (ϵ 7,690).

Infrared absorption data for the compounds considered have been reported more often. Absorption of such vinyl ethers as **150a** at 5.9–6.9 μm corresponds to the vinylic fragment.[28(a),(b),33,86] Additional bands in the spectrum of **151** at 10.7 μm (Ref. 33) and of **152** at 10.18 μm (Ref. 86) show the presence of *trans* isomers. Alkene absorption of vinylthio ethers was observed[29(a),(b),83,84] between 6.1 and 6.2 μm, the upper limit corresponding[83(b),84] to absorption by a terminal olefinic bond in such compounds as **154a** or **159**.

The infrared spectum of ester **161** is characterized[15(a)] by ester (5.81 μm) and vinyl (6.07 μm) bands. Similar absorptions at 5.74 μm and 5.98 μm of the amino ester **163** [Ref. 15(c)] may also be caused by the acetamido group, and their identification is, therefore, under question.

Nuclear magnetic resonance spectral data for the vinylthio ethers **153–159**, reported[29(a),(b),83,84] by Tronchet and coworkers, are listed in Table VI. Several general conclusions follow from these data. (*a*) The alkenic hydrogen atom nearest to the sugar residue in such compounds as **153** and **155c**, having a terminal vinylthio group, resonates at higher field as compared with the neighboring one. The resonance of the latter is similar to that of the vinylic hydrogen atom in such compounds as **156a**, having a branching vinylthio group. (*b*) The more distant hydrogen atom of a terminal vinylthio fragment resonates at higher field for a *cis* isomer as compared with the *trans* isomer. A neighboring alkenic hydrogen atom shows the reverse correlation (compare, for example, the pair of isomers **154a** and **154b**). (*c*) Protons of the methylthio group of the terminal vinylthio fragment resonate at somewhat higher field in comparison with that of the branching vinylthio fragment. (*d*) The ring-hydrogen atom nearest to the methylthio group in the pairs of branched-chain isomers **156a–156b**, **157a–157b**, and **158a–158b** resonates at lower field in comparison with the normal resonance of a distant methylthio group, thus indicating its deshielding effect.

It should be noted, in addition, that the resonances of the vinylic protons and the methoxyl group in the vinyl ether **151** are observed[33] at considerably lower field as compared with the vinylthio derivatives, and this indicates a shielding effect of the sulfur atom.

Interestingly, vinylthio derivatives of branched-chain furanoses (**158a** and **158b**) having the D-*ribo* configuration show[29(b)] an additional coupling-constant, $J_{2,4}$, having a value of about 1.5 Hz, that is absent

(86) M. F. Shostakovsky, N. N. Aseeva, and A. I. Polyakov, *Izv. Akad. Nauk SSSR, Ser. Khim.*, 892 (1970).

TABLE VI
Proton Magnetic Resonances of Vinylthio Fragments
and Their Environment in Vinylthio Derivatives (153–159)[a]

Compound	H-1	$J_{1,2}$	H-2	H-3	H-3'	$J_{2,3'}$	$J_{3',4}$	H-4	H-4'	$J_{4',5}$	H-5	SMe	References
153 (cis)	3.84°d	10	4.48°qq									7.76s	83(a)
153 (trans)	3.61°d	15.5	4.68°qq									7.79s	
154a	3.84°dd	9.5	4.55°dd									7.71s	83(b)
154b	3.64°dd	15	4.75°dd									7.78s	
155a												7.70s	83(b)
159	3.59°dd	14.3	4.33°dd								4.32°dd[b]	7.73s	84
156a				4.87d					3.86°d	~1.0	5.61ddd	7.70s	29(a)
156b				5.38d					3.82°d	~1.5	5.37m	7.68s	
157a					3.72°t	1.1	1.4	5.42dd				7.67s	
157b			5.18d		3.58°d	0	1.3	5.36dd				7.66s	29(b)
158a			4.84dt		3.46°t	1.6	1.8	5.38m				7.63s	29(b)
158b			4.97m		3.59°t	1.5	1.8	5.09m				7.73s	

[a] Chemical shifts (τ) refer to chloroform-d and are given in p.p.m.; coupling constants (J) in Hz; alkene protons are indicated by an asterisk. [b] Chemical shift for H-6 is 3.79 p.p.m. (doublet); d, doublet; m, multiplet; q, quartet; s, singlet; t, triplet.

for similar derivatives (**157a** and **157b**) having the D-*xylo* configuration; this circumstance is useful for configurational assignments.

Mean statistical conformations of the last four compounds were roughly estimated,[29(b)] based on the allylic coupling-constants $J_{2,3}$ and $J_{3,4}$ (see Table VI). According to the values of these coupling constants, the angles between the plane of the alkenic bond and C–H-2 and C–H-4 are equal to ~50° in **158a** and **158b**, corresponding to a planar furanoid ring. For compound **157a**, these angles are equal to about 40°, thus suggesting a preponderance of the E^3 conformation. In compound **157b**, the angle for H-2 is equal to 30° and to 45°, corresponding to a conformation intermediate between E^1 and T_0^1. The rotational properties of **157b** also reflect its different conformation. Indeed, the specific rotations of the pair of isomers **157a** and **157b** differ greatly (+216° and −163°, respectively); for the other pair, **158a** and **158b**, the difference in specific rotations is only 35°. On the other hand, the optical rotatory dispersion spectra of these compounds show[29(b)] a Cotton effect at about 250 nm that is negative for **157b** and positive for three other related compounds. A similar, considerable difference in specific rotations was also observed [29(a)] for the pair of pyranose derivatives **156a** and **156b** (+16° and +247°, respectively), also indicating their conformational difference.

c. **Chemical Properties.**—As already noted (see p. 270), the enol, thioenol, and enamino derivatives to be discussed are of great interest primarily in view of their conversion into 2-deoxy carbonyl compounds. This transformation has generally been realized by acid hydrolysis for the enol and enamino derivatives, and by treatment with mercuric chloride for the thioenol derivatives.

The rapid hydrolysis of vinyl ethers with ethereal perchloric acid, reported[87] earlier, was successfully employed[28] to transform the enol derivatives **150a** and **150b** into 3,4:5,6-di-O-cyclohexylidene-2-deoxy-*aldehydo*-D-*xylo*-(**166a**) and -L-*arabino*-hexose (**166b**), respectively,

(87) S. Levine, *J. Amer. Chem. Soc.*, **80**, 6150 (1958).

in about 90% yield. The cyclohexylidene groups were then removed by hydrolysis with 50% aqueous acetic acid, to yield the respective 2-deoxyaldoses. Similar treatment of the enolic ester **161** led[15(a)] to ethyl 3-deoxy-4,5-O-isopropylidene-D-*arabino*-heptulosonate (**167a**). Partial elimination of the isopropylidene groups in this case demonstrates the lower stability of the isopropylidene derivatives in ethereal perchloric acid in comparison with that of the cyclohexylidene derivatives. It is noticeable that, in the infrared spectrum of the keto ester **167a**, a band at 6.07 μm was observed,[15(a)] besides carbonyl absorption at 5.8 μm, indicating partial enolization. On heating **167a** under vacuum, lactonization occurs,[15(a)] resulting in formation of lactone **167b**, which also shows partial enolization.

Lactonization may also occur during acid hydrolysis. For example, ester **162** (not isolated) was directly converted[15(b)] by ethereal perchloric acid into lactone **168** in 76% yield. The structure of **168** was confirmed[15(b)] by the consumption of one mole of periodate per mole without liberation of formic acid, as well as by infrared absorption at 5.78 (C=O) and 6.06 μm, the latter band possibly being caused both by enolic and acetamido groupings. Similarly, the 1,4-lactone of N-acetylneuraminic acid (**169**) was shown to be formed[88] on analogous treatment of ester **163**. Lactonization was also shown to occur at the condensation stage, as illustrated by isolation[16] of the enamino lactone **165** instead of the expected ester (**165a**). Lactone **165b** was then quantitatively hydrolyzed[16] by Dowex-50 (H$^+$) cation-exchange resin to 3-deoxy-D-*erythro*-hexulosonic acid (**170a**), or esterified to ethyl 3-deoxy-D-*erythro*-hexulosonate (**170b**) by means of ethanolic hydrogen chloride.

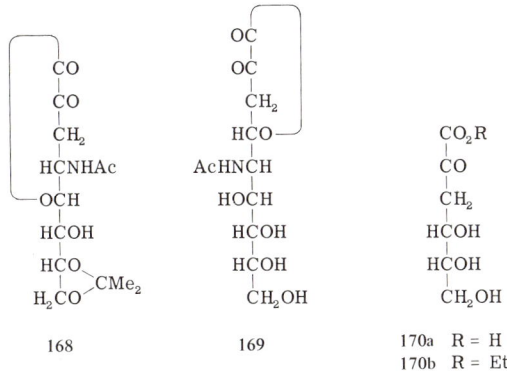

(88) M. N. Mirzayanova, Dissertation, Moscow, 1969.

An attempted hydrolysis of the aromatic vinyl ester **152** by 50% aqueous acetic acid failed,[86] and resulted in the formation of some unsaturated products due to allylic rearrangements of the kind discussed. The indirect transformation of **152** into the respective carbonyl compound[86] consisted in its methoxymercuration, followed by sodium borohydride reduction of the intermediate adduct (**171**), and acid hydrolysis. Methoxymercuration of **152** was found[86] to be regioselective, as 2-deoxy-L-*arabino*-hexose was the main final product of these transformations, together with a small proportion of L-arabinose. The formation of the latter was explained[86] by alternative attack of the methoxymercuration reagent on C-1, followed by elimination of a mercury-containing fragment.

$$\begin{array}{c}\text{OMe}\\|\\\text{CHOC}_6\text{H}_4\text{CH}_3\\|\\\text{CHHgCl}\\|\\\text{HCO}\diagdown\\\phantom{\text{HCO}}\diagup\text{CMe}_2\\\text{OCH}\\|\\\phantom{\text{Me}_2\text{C}}\text{OCH}\\\text{Me}_2\text{C}\diagdown\phantom{\text{O}}|\\\phantom{\text{Me}_2\text{C}}\diagup\text{OCH}_2\end{array}$$

171

An intermediate dimethyl acetal (**172**) was isolated[83(a)] in 63% yield by similar methoxymercuration of ether **151** followed by sodium borohydride reduction; the product was then hydrolyzed to 2-deoxy-L-*arabino*-hexose. The same acetal (**172**) was formed[83(a)] by reaction

$$\begin{array}{c}\text{CH(OMe)}_2\\|\\\text{CH}_2\\|\\\text{HCO}\diagdown\\\phantom{\text{HCO}}\diagup\text{CMe}_2\\\text{OCH}\\|\\\phantom{\text{Me}_2\text{C}}\text{OCH}\\\text{Me}_2\text{C}\diagdown\phantom{\text{O}}|\\\phantom{\text{Me}_2\text{C}}\diagup\text{OCH}_2\end{array}$$

172

of vinylthio ether **153** with mercuric chloride in the presence of mercuric oxide in methanol, followed by borohydride reduction; this indicated initial formation of vinyl ether **151**, which then underwent methoxymercuration. Similar treatment of vinylthio ethers in water leads directly to carbonyl compounds, with no acetal formation. This mode of treatment was used[8] by Bestmann and Angerer to convert vinylthio ethers **160a–160d** into the respective 2-deoxy sugars.

Different behavior of vinylthio ethers on treatment with mercuric salts that depends on the solvent may be explained by an initial formation of cation **173** as a result of electrophilic attack of HgX on C-1. In alcohol, this cation then gives vinyl ether **174**, which undergoes further alkoxymercuration. In water, enol **175** is formed and is immediately transformed into carbonyl compound **176**, as shown in equation *10*.

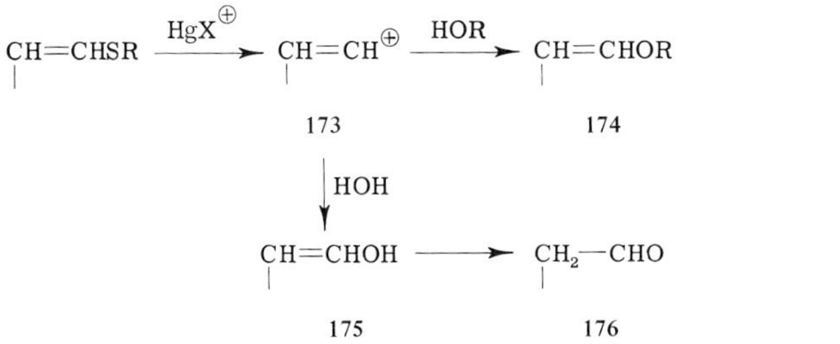

$$\begin{array}{c} \text{CH=CHSR} \xrightarrow{\text{HgX}^\oplus} \text{CH=CH}^\oplus \xrightarrow{\text{HOR}} \text{CH=CHOR} \\ 173 174 \\ \downarrow \text{HOH} \\ \text{CH=CHOH} \longrightarrow \text{CH}_2\text{—CHO} (10) \\ 175 176 \end{array}$$

Thioenol derivatives may also serve as a source of dideoxy sugars when a hydrogenation–desulfurization procedure is used. Thus, the geometrical isomers **154a** and **154b** led[83(b)] to 3,6-anhydro-1,2-dideoxy-4,5-*O*-isopropylidene-D-*ribo*-hexitol (**177**) when treated with Raney nickel. A unique vinylthio ether (**159**) gave in this way[84] the related derivative **178**. Branched-chain thioenol derivatives react in the same way, leading to branched-chain sugars. Thus, the mixture of geometrical isomers (**156a** and **156b**) produced[29(a)] methyl 4-deoxy-2,3-*O*-isopropylidene-4-*C*-methyl-6-*O*-methyl-α-D-talopyranoside (**179a**) and

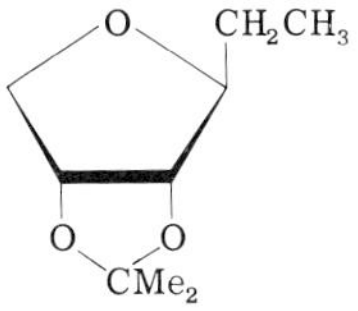

177

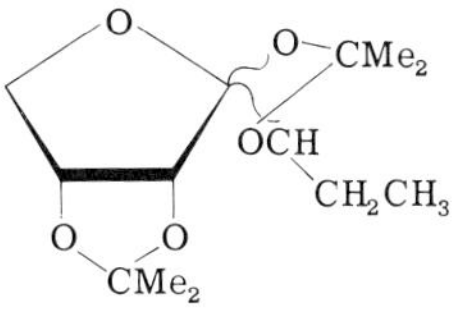

178

its D-*manno* epimer (**179b**) in practically equal amounts, thus indicating the absence of stereoselectivity. The configurations of these epimers follow from their n.m.r. spectral data;[29(a)] the small

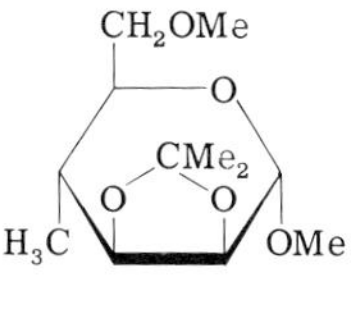

179a

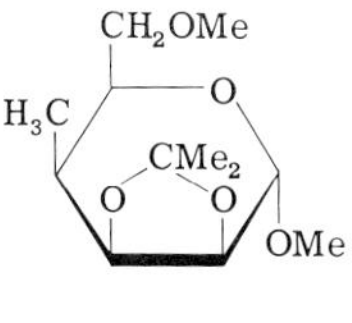

179b

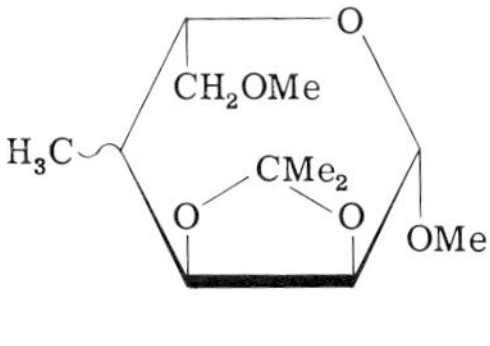

179c

coupling (1 Hz) between H-1 and H-2 for both epimers indicates the α configuration (e,e orientation), and the small coupling (4 Hz) between H-3 and H-4 for **179b** corresponds to the a,e orientation. The mixture of epimers (**179c**) obtained[29(a)] by hydrogenation–desulfurization of **156c** was not separated.

Interesting results were obtained[29(b)] with thiovinylic derivatives of furanoses on treatment with Raney nickel. Isomers **157a** and **157b** gave, in this case, not only the expected 3-deoxy-1,2:5,6-di-O-isopropilidene-3-C-methyl-α-D-gulofuranose (**180a**) and its D-*galacto* epimer (**180b**) in the ratio of 11:4, but the unsaturated product **180c**, the latter in low yield. Analogous isomers **158a** and **158b** having the D-*ribo* configuration gave only the one, expected product (**181**), thus showing a high stereoselectivity due to the 1,2-O-isopropylidene group.

The branched-chain, C-methyl derivative **181a** was obtained in a similar way,[85] together with a small proportion of the C-3 epimer, starting from the glyculose **159a**. It should be noted that these transformations result in branched-chain sugars having at the branching point a configuration epimeric with that which would have been obtained by the Grignard synthesis.

The configuration of the 3-C-methyl derivatives **180a, 180b,** and **181** was determined[29(b)] from empirical rules correlating the values of the coupling constants $J_{1,2}$ and $J_{2,3}$ with the configuration of the furanose moiety.

TABLE VII
Deshielding Effect of the Methylsulfonyl Group[29(a),(b)]

Compound	H-2[a]	H-3[a]	H-5[a]
156b		5.38	5.37
182		5.43	4.53[b]
157a	4.94	5.42	
183	4.29[b]	5.25	
158a	4.84	5.38	
184a	4.13[b]	5.34	
158b	4.97	5.09	
184b	4.86	4.24[b]	

[a] Chemical shifts in τ (chloroform-d).
[b] Change in chemical shift caused by the methylsulfonyl group.

Oxidation of the vinylthio derivatives is also of interest. When compounds **156b, 157a, 158a,** and **158b** were treated with osmium tetraoxide and hydrogen peroxide, the respective sulfonyl derivatives **182** [Ref. 29(a)], **183, 184a,** and **184b** [Ref. 29(b)] were obtained. These derivatives proved to be useful for confirming the geometrical configuration of the unsaturated fragment in the initial compounds, because the methylsulfonyl group has a considerable deshielding effect on the nearest ring-proton, as shown in Table VII.

IV. ANOMALOUS WITTIG REACTION WITH FREE AND WITH
PARTIALLY PROTECTED SUGARS

Successful interaction of alkoxycarbonylmethylenetriphenylphosphoranes with free sugars to give α,β-unsaturated esters (see p. 255) stimulated further progress in this direction, although the formation of such by-products as **133** showed that this interaction is more complex than desired. Indeed, careful examination of the coupling of (p-methoxybenzoyl)- and acetyl-methylenetriphenylphosphorane with free and with partially protected sugars, performed[7(a)–(i)] in our laboratory, demonstrated that this process is extremely complex and leads to the formation of various unexpected products, mainly (a) five- and six-membered anhydro derivatives and (b) C-glycosylated furans.

1. Main Directions of the Reaction

Anhydro-C-(p-methoxyphenyl)aldoses and anhydro-1,3-dideoxyketoses were found to be compounds usually produced on heating free and partially protected aldoses with p-methoxybenzoyl- and acetyl-methylenetriphenylphosphoranes, respectively, in N,N-dimethylformamide. Many of them were isolated by thin-layer chromatography on alumina, although, on several occasions, these compounds were available only as mixtures of the five- and six-membered anhydro derivatives, because of their similar chromatographic mobilities. Five-membered anhydro derivatives are favored, but exclusive formation of six-membered anhydro derivatives was observed when five-membered ring-closure was impossible. Only L-rhamnose on reaction with acetylmethylenephosphorane formed a six-membered derivative, **196b**.

TABLE VIII
Five-membered Wittig Products (188) from D-Glucose and its Derivatives

No. of compound	R	R_1	R_2	R_3	R_4	Yield (%)	References
a	p-OMeC$_6$H$_4$	H	H	H	H	15	7(a)
b	p-OMeC$_6$H$_4$	H	CH$_2$Ph	H	H	11	7(e)
c	CH$_3$	H	H	H	H	35	7(f)
d	CH$_3$	H	Me	H	H	65	7(g)
e	CH$_3$	H	Me	Me	Me	15	7(g)
f	CH$_3$	Me	Me	Me	Me	25	7(g)
g	CH$_3$	H	H	H	Tr	30	7(g)

The five-membered anhydro derivatives discussed next have been isolated (yields are given in parentheses): 4,7-anhydro-1,3-dideoxy-D-*ribo*-octulose[7(i)] (**185**) (21%) from D-ribose, 3,6-anhydro-2-deoxy-1-*C*-(*p*-methoxyphenyl)-D-*xylo*-heptose[7(c)] (**186a**) (18%) and 4,7-anhydro-1,3-dideoxy-D-*xylo*-octulose[7(f)] (**186b**) (40%) from D-xylose, 3,6-anhydro-2-deoxy-1-*C*-(*p*-methoxyphenyl)-D-*galacto*-octose[7(a)] (**187**) (6.5%) from D-galactose, and a series of compounds (**188**) derived from D-glucose and from partially protected D-glucofuranose (see Table VIII), possibly as mixtures of epimers.

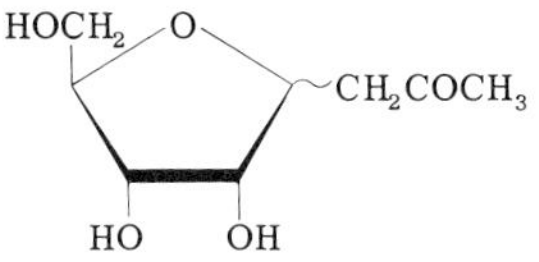

185

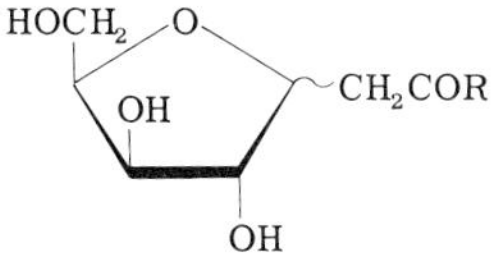

186a R = *p*-MeOC$_6$H$_4$
186b R = Me

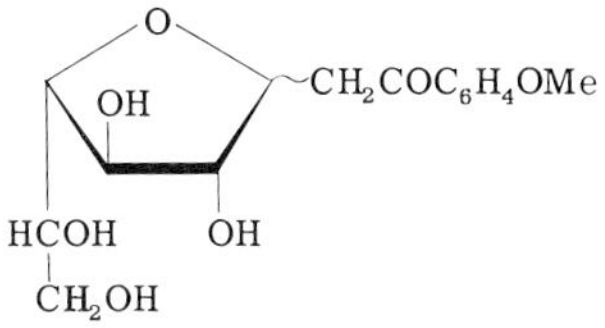

187

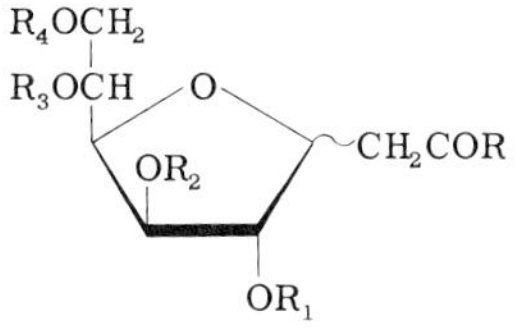

188

On several occasions, separation of epimers was, however, achieved. Thus, 3,6-anhydro-2-deoxy-4,5:7,8-di-*O*-isopropylidene-1-*C*-(*p*-methoxyphenyl)-D-*glycero*-D-*talo*-(and D-*glycero*-D-*galacto*)-octose (**189a** and **190a**, respectively) were isolated[7(e)] after 2,3:5,6-di-*O*-isopropylidene-α-D-manofuranose had been allowed to react with (*p*-methoxybenzoyl)methylenetriphenylphosphorane. Similar interaction with acetylmethylenetriphenylphosphorane resulted[7(g)] in the isolation of 4,7-anhydro-1,3-dideoxy-5,6:8,9-di-*O*-isopropylidene-D-*glycero*-D-*talo*-(and D-*glycero*-D-*galacto*)-nonulose (**189b** and **190b**, respectively).

The configurational assignments for these epimers are based[7(g),(i)] on (*a*) the greater values of specific rotation for the "α" epimers (**189a** and **190a**) as compared with those of the "β" epimers (**189b** and **190b**), provided that the isorotation rules are applicable in this case, and (*b*) the greater chromatographic mobilities of the "β" epimers in comparison with those of the "α" epimers, as the more compact "β" epimers would be expected to be the more mobile. Moreover, the melting point of **189a** is higher than that of **190a**, the same correlation being observed for true anomers. Similar considerations led to the conclusion[7(f)] that the two products of the reaction between D-galactose and acetylmethylenephosphorane are 4,7-anhydro-1,3-dideoxy-D-*glycero*-L-*gluco*(and D-*glycero*-L-*manno*)-nonulose (**191a** and **191b**, respectively).

189a R = *p*-MeOC$_6$H$_4$
189b R = Me

190a R = *p*-MeOC$_6$H$_4$
190b R = Me

191a   191b

The structure of these compounds follows from their spectra and from the results of periodate oxidation. Electronic (273–279 nm) and infrared (5.9–6.03 μm) absorptions of C-(*p*-methoxyphenyl) derivatives[7(c),(e)] are similar to those of *p*-methoxyacetophenone (λ_{max} 272 nm, and 5.98 μm). The infrared spectra of anhydro-1,3-dideoxyglyculoses are characterized[7(f),(g)] by carbonyl absorption at ~5.88 μm. In certain compounds (such as **189a**), the size assigned to the sugar ring was based on the fact that only one mode of ring closure was possible. In other cases, this problem was solved by periodate oxidation. Derivatives similar to **185** (that is, having no terminal glycol grouping) consume one mole of periodate per mole without liberation of formic acid. Such derivatives as **187**, from the malondialdehyde derivative **192**, undergo overoxidation[7(a)] resulting in consumption of six moles of periodate per mole, with liberation of three moles of formic acid, as shown in equation *11*. When R = Me, further overoxidation of the intermediate ketoaldehyde **193** was observed[7(f)] due to increased activity of the methylene group, although the expected overall consumption of nine moles of periodate per mole, with liberation of six moles of carboxylic acid, was not achieved.

Careful examination of the reaction between D-galactose and acetylmethylenetriphenylphosphorane led[7(f)] to isolation of a product having a relatively high chromatographic mobility and lacking i.r.

$$\begin{array}{c}
\text{H}_2\text{C}\!-\!\!\text{CH}\!-\!\!\overset{\displaystyle O}{\underset{\displaystyle\text{HO}\quad\text{OH}}{\diagdown\!\!\!\!\!\!\diagup}}\!\!-\!\text{CH}_2\text{COR} \xrightarrow[-\text{HCHO}]{2\,\text{IO}_4^{\ominus}}
\begin{array}{c}\text{CHO}\\|\\\text{HC}\!-\!\text{O}\!\diagdown\!\!\text{H}\diagup\text{CH}_2\text{COR}\\|\quad\quad\quad\text{C}\\\text{CHO}\quad|\\\quad\quad\quad\text{CHO}\end{array}\\
\text{HO}\quad\text{OH}\qquad\qquad\qquad\qquad\qquad\qquad\quad\mathbf{192}
\end{array}$$

$$\Big\downarrow \text{IO}_4^{\ominus}$$

$$\begin{array}{c}\text{COR}\\|\\\text{CH}_2\\|\\\text{CHO}\end{array} \xleftarrow[-\text{HCO}_2\text{H}]{\text{IO}_4^{\ominus}} \begin{array}{c}\text{COR}\\|\\\text{CH}_2\\|\\\text{CHOH}\\|\\\text{CHO}\end{array} \xleftarrow[\substack{-\text{CO}_2\\-2\text{HCO}_2\text{H}}]{2\,\text{IO}_4^{\ominus}} \begin{array}{c}\text{CHO}\\+\\\text{HO}\!-\!\text{C}\!-\!\text{O}\!\diagdown\!\text{H}\diagup\text{CH}_2\text{COR}\\+\quad\quad\quad\text{C}\\\text{CHO}\quad|\\\quad\quad\quad\text{CHO}\end{array} \quad (11)$$

193

when R = Me

$$\Big\downarrow \text{IO}_4^{\ominus}$$

$$\text{OHC}\!+\!\text{CH(OH)}\!+\!\text{COCH}_3 \xrightarrow{2\,\text{IO}_4^{\ominus}} 2\,\text{HCO}_2\text{H} + \text{CH}_3\text{CO}_2\text{H}$$

where R = p-MeOC$_6$H$_4$ or Me

absorption in the 5.98–6.66 μm region. The periodate oxidation data for this compound proved to be identical to those of **191a** and **191b**. Storage of this compound in aqueous solution led to a mixture of **191a** and **191b**. The structure of the internal acetal **194** is in accordance with these observations, as well as with the elemental analysis data. An analogous acetal was isolated[7(f)] as the acetate (**195**) during the separation of product **186b**.

194   **195**

Six-membered anhydro derivatives are less likely to be formed, as already mentioned (see p. 284). Thus 3,7-anhydro-2,8-dideoxy-1-C-(p-methoxyphenyl)-L-*manno*-octose (**196a**) was isolated[7(c)] in only 5% yield when L-rhamnose reacted with (*p*-methoxybenzoyl)-methylenetriphenylphosphorane. An analogous interaction of L-rhamnose with acetylmethylenetriphenylphosphornae led,[7(f)] however, to 4,8-anhydro-1,3,9-trideoxy-L-*manno*-nonulose (**196b**) in 25% yield. The corresponding five-membered anhydro derivative was not detected. Exclusive formation of such six-membered anhydro derivatives as 3,7-anhydro-6,8-*O*-benzylidene-2-deoxy-1-*C*-(*p*-methoxyphenyl)-D-*gluco*-nonulose[7(i)] (**197a**) and 4,8-anhydro-7,9-*O*-benzylidene-1,3-dideoxy-D-*gluco*-octulose[7(g)] (**197b**), as well as of appropriate disaccharide derivatives[7(e)] (**198a** from maltose and **198b** [Ref. 7(e)] from lactose) is naturally conditioned by the absence of an alternative mode of ring closure.

196a R = *p*-MeOC$_6$H$_4$
196b R = Me

197a R = *p*-MeOC$_6$H$_4$
197b R = Me

198a R = α-D-galactosyl
198b R = β-D-galactosyl

The spectral characteristics[7(c),(e),(f),(g),(i)] of six-membered anhydro compounds are similar to those of their five-membered analogs (see p. 286), as would be expected. The results obtained by periodate oxidation corresponded to the structures proposed.

The second direction of the Wittig reaction with free and with partially protected sugars is the formation of furan derivatives, this direction being realized with (*p*-methoxybenzoyl)methylenetriphenylphosphorane only in certain cases. Thus, model experiments with glycolaldehyde and (*p*-bromobenzoyl)methylenetriphenylphosphorane led[7(a)] to 2-(*p*-bromophenyl)furan (**199a**) having a melting point identical with the reported value. Similar interactions of (*p*-methoxybenzoyl)methylenephosphorane with glycolaldehyde and glyceraldehyde led[7(a)] to isolation of substituted furans **199b** and **200**, respec-

199a R = Br
199b R = OMe

200

tively. Of the other sugars, only D-ribose,[7(c)] D-glucose[7(a)] and its 3-methyl and 3,5-6-trimethyl ethers,[7(e)] and D-galactose[7(a)] were found to form the *C*-glycosylated furans **201**, **202a–202c**, and **203**, respectively.

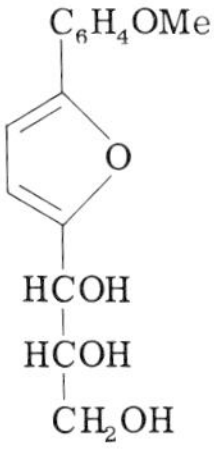

Structures

201
C_6H_4OMe-furan
HCOH
HCOH
CH_2OH

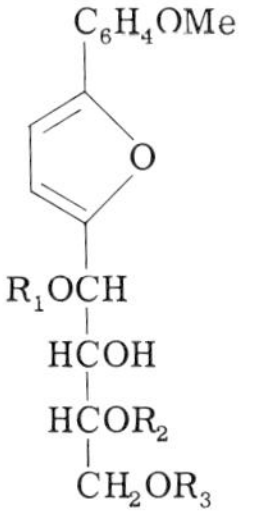

202a $R_1 = R_2 = R_3 = H$
202b $R_1 = Me, R_2 = R_3 = H$
202c $R_1 = R_2 = R_3 = Me$
C_6H_4OMe-furan
R_1OCH
HCOH
$HCOR_2$
CH_2OR_3

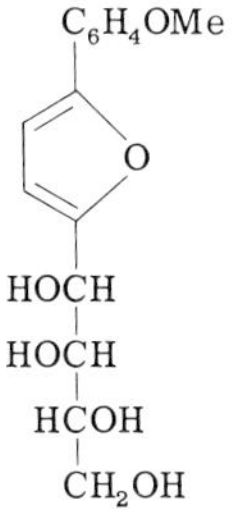

203
C_6H_4OMe-furan
HOCH
HOCH
HCOH
CH_2OH

Only L-rhamnose was found[7(f)] to form a furan derivative (**204**) with acetylmethylenetriphenylphosphorane, and this was obtained in negligible yield.

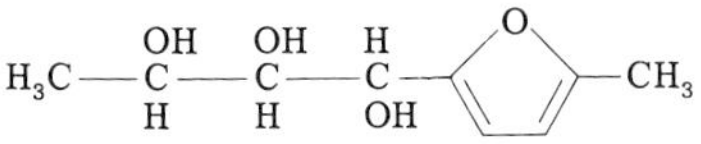

204

$$H_3C-\underset{H}{\overset{OH}{C}}-\underset{H}{\overset{OH}{C}}-\underset{OH}{\overset{H}{C}}-\text{furan}-CH_3$$

The electronic spectra of these furan derivatives are characterized by a broad absorption band at 280–287 nm with an inflection at 295–308 nm. Absence of absorption in the region of 5–6.21 μm was observed in their infrared spectra.

Coupling of glycolaldehyde with (p-methoxybenzoyl)methylenetriphenylphosphorane produced[7(a)] an unexpected product having electronic and infrared spectra similar to those of p-methoxyacetophenone. This compound was readily converted into the furan derivative **199b** and may, therefore, be the cyclopropane derivative **205**.

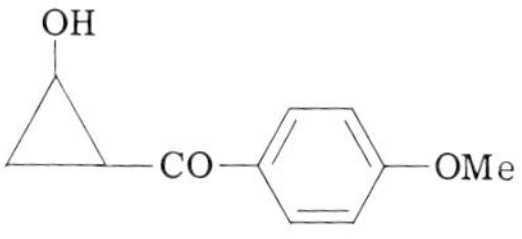

205
cyclopropane(OH)—CO—C₆H₄—OMe

Another unique compound was found[7(g)] among the products of the reaction between 3,5,6-tri-O-methyl-D-glucose and acetylmethylenephosphorane. An inflection at 222 nm in the electronic spectrum, and strong infrared absorption at 5.8 μm, indicated an α,β-unsaturated, cyclic ketone structure.[89] When this compound was allowed to react with methylmagnesium bromide, the product showed[7(g)] an absence

(89) A. N. Elizarova, "Khimia cyclopentenonov," Moscow, 1966.

of the inflection at 222 nm, together with retention of the infrared carbonyl absorption at 5.8 μm, thus indicating 1,4-addition of the Grignard reagent. Steric hindrance to 1,2-addition is indicated by these observations, as well as by the appearance of one hydroxyl group in the final product (data from elemental analyses); the results demonstrate the specific position of one methoxyl group in this compound. Based on these facts, structures **206** and **207** were proposed[7(g)] for the compound isolated and for the product of its interaction with the Grignard reagent, respectively.

H   Me
  H
  Me
O
MeO HCOMe
     |
     HCOMe
     |
     CH₂OMe

206

H   Me
       Me
        H
O
HO   HCOMe
     |
     HCOMe
     |
     CH₂OMe

207

The only "normal" Wittig products [**61** and **63** (R = OMe)] from the reaction of glyceraldehyde with acetyl- and (p-methoxybenzoyl)-methylenephosphorane, respectively, were identified during these investigations [Refs. 7(a) and 7(f), respectively].

2. Stereochemical Considerations

The absence of the "normal" Wittig products from the mixture of compounds arising in the reactions of (p-methoxybenzoyl)- and acetyl-methylenetriphenylphosphorane with free and partially protected aldoses (excluding glyceraldehyde) is the most remarkable feature of these processes. This circumstance therefore raises the question as to whether abnormal Wittig products are formed as a result of subsequent transformations of intermediate, "normal" Wittig products, or arise independently at the stage of betaine formation. Reasonable assumptions in that regard were made.[7(i)] The first point of view seems to be unacceptable, because no "normal" Wittig products were detected in the final mixtures, even on performing the reactions at lowered temperature.[7(i)] Moreover, an analogous compound (**68**) retains its acyclic form, even in the strongly alkaline medium used for deacetylation, as already indicated (see p. 247), and, therefore, would not be expected to be transformable in the presence of phosphoranes.

The second hypothesis[7(i)] is in better agreement with the results observed. Indeed, consideration of the three favored betaines (**208a**–

THE WITTIG REACTION IN CARBOHYDRATE CHEMISTRY 291

208a 208b 208c

where G = sugar residue, and R = p-MeOC$_6$H$_4$ or Me

208c) formed in an initial stage of the reaction leads to a conclusion about different pathways for their subsequent transformations. Thus, betaine **208b** would be expected to undergo facile cyclization into **209**, followed by dehydration, and elimination of triphenylphosphine oxide to form the furan derivative **211**, as shown in equation *12*.

$$(12)$$

Taking into account that the intermediate ion **210** may be stable only when R = OMeC$_6$H$_4$-p, the formation of furan derivatives (**211**) with (p-methoxybenzoyl)methylenetriphenylphosphorane, and the lack of such formation with acetylmethylenetriphenylphosphorane, becomes clear.

Betaines **208a** and **208c** are incapable of formation of furan derivatives, because of an unfavorable orientation of the carbonyl fragment and the nearest hydroxyl group. These betaines may, however, be sources of anhydro derivatives, as well as **208b**, if it is assumed that elimination of triphenylphosphine oxide is accompanied by intramolecular attack of a distant hydroxyl group on the β-carbon atom, as shown.

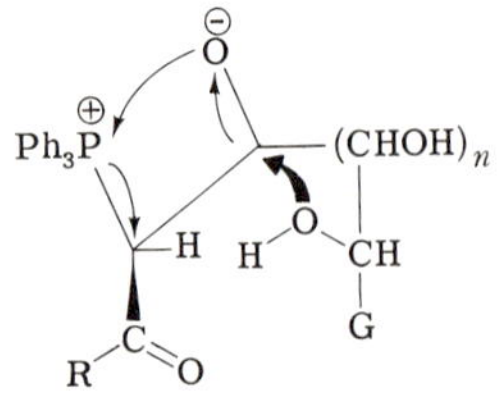

Possible mode of formation
of cyclic derivatives

V. CONCLUSION

It may be concluded from the previous Sections of this Chapter that the Wittig reaction has now become one of the general synthetic methods of modern carbohydrate chemistry, a reaction of exceptional versatility. The further development of this area is still in progress, including both the theoretical and practical aspects. It would be desirable to investigate such problems as, for instance, correlation between the ratio of *cis* and *trans* Wittig isomers and the structure of the initial sugar, as these observations may throw light upon the reaction mechanism. It would also be of great interest to examine the Wittig reaction further with free and partially protected sugars, as this would provide valuable information about the detailed behavior under these conditions, as well as affording syntheses of unusual sugar and furan derivatives. Furthermore, serious account should be taken of the implications of new ylides and carbanions as contributing to further progress in this field.

VI. FURTHER DEVELOPMENTS

Another example in the *C*-glycosylated, unsaturated hydrocarbon series is methyl 2-deoxy-3,4-*O*-isopropylidene-2-*C*-methylene-β-D-*erythro*-pentopyranoside (213), obtained[90] in 55% yield by the reaction of methyl 3,4-*O*-isopropylidene-β-D-*erythro*-pentopyranosid-2-ulose (212) with methylenetriphenylphosphorane; *n*-butyllithium was used as a proton acceptor. Subsequent reduction of 213 with 10% palladium-on-charcoal led to a 7:1 mixture of the 2-epimeric methyl 2-deoxy-2-*C*-methylpentopyranosides (214 and 215, respectively). In the n.m.r. spectrum of 214, the H-1 resonance appeared as a doublet at τ 5.62 p.p.m., exhibiting a $J_{1,2}$ coupling constant of 8 Hz, thus indicating H-2 to be axial and, therefore, the D con-

(90) A. Rosenthal and M. Sprinzl, *Can. J. Chem.*, **48**, 3253 (1970).

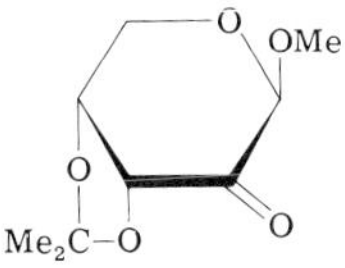

212

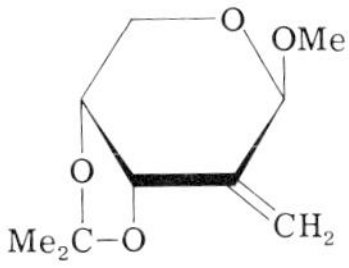

213

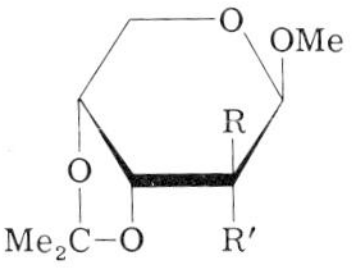

214 R = H, R' = Me
215 R = Me, R' = H

figuration at C-2. The H-1 resonance for **215** was observed at τ 5.51 p.p.m., with $J_{1,2} = 3$ Hz, thus according with the L configuration at C-2. The noticeable preponderance of the D-*ribo* epimer (**214**) shows that the 3,4-O-isopropylidene group causes more steric hindrance during the orientation of the alkene **213** on the catalyst surface than does the methoxyl group at C-1. An attempted synthesis of nucleosides based on the branched-chain sugars **214** and **215** was unsuccessful.

The synthesis of 6,7-dideoxy-1,2:3,4-di-O-isopropylidene-α-D-*galacto*-hept-6-enopyranose (**216**), starting from 1,2:3,4-di-O-isopropylidene-α-D-*galacto*-hexodialdo-1,5-pyranose (**25**) with subsequent addition of diethoxymethylsilane to form 6,7-dideoxy-1,2:3,4-di-O-isopropylidene-7-(diethoxymethylsilyl)-α-D-*galacto*-heptopyranose (**217**), has also been reported.[91]

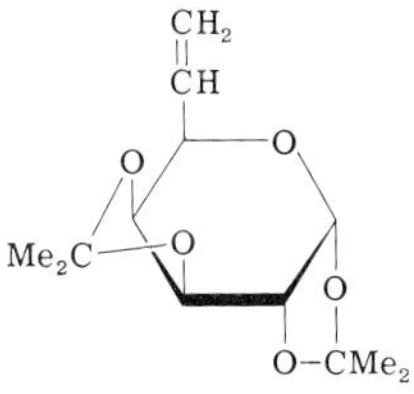

216

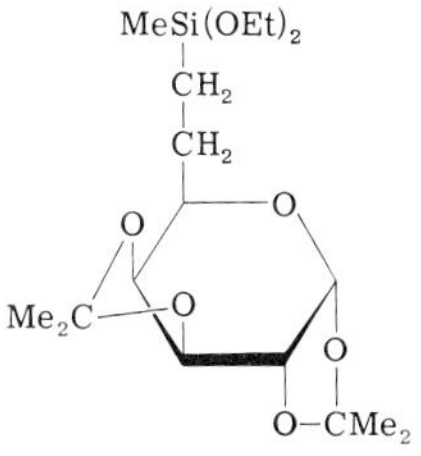

217

The interaction of the *aldehydo* derivatives **14a, 14b**, and **55** with methylenetriphenylphosphorane has been re-investigated[92] with the use of a 1:1 potassium *tert*-butoxide–*tert*-butanol complex in the reaction mixture. This modification affords significantly increased yields of the final alkenes (**15a, 15b**, and their 3-O-benzyl analogs).

In further research on the carbon-chain extension of monosaccharides by reaction of free sugars with (alkoxycarbonylmethylene)-triphenylphosphoranes, Kochetkov and coworkers have reported[93]

(91) J. Lehmann and H. Schäfer, *Chem. Ber.*, **105**, 969 (1972).
(92) N. Baggett, J. M. Webber, and N. K. Whitemore, *Carbohyd. Res.*, **22**, 227 (1972).
(93) B. A. Dmitriev, A. Ya. Cherniak, and N. K. Kochetkov, *Zh. Obshch. Khim.*, **41**, 2754 (1971).

the reaction of D-xylose with (ethoxycarbonylmethylene)triphenylphosphorane, which yields the expected α,β-unsaturated ester **218** (58%), together with the furanoid lactone **219** (24%). An application of the general procedure described in Section III (see p. 266) to compound **218** led to D-*glycero*-L-*galacto*-heptose (**220**) and D-*glycero*-L-*ido*-heptose (**221**) in the ratio of 5:1; the latter was previously unknown.

A convenient approach to carbon-chain branching in monosaccharides is the condensation[51,94] of *aldehydo*-aldoses with the branched-chain phosphoranes[95] **222a** and **222b**. The expected branched-chain derivatives **223a** and **223b**, together with the products (**223d** and **223c**) of condensation of these phosphoranes with acetaldehyde, have been thus obtained.[94] The latter pair gave a racemic mixture (**224**) of 2-deoxy-DL-streptonolactones, after hydroxylation followed by saponification.

By utilizing the tendency of 1,3-dibromoacetone to undergo monosubstitution upon reaction with triphenylphosphine, Zhdanov and Uslova have synthesized[96] (bromoacetylmethylene)triphenylphos-

---

(94) Yu. A. Zhdanov and L. A. Uslova, *Zh. Obshch. Khim.*, **41**, 1844 (1971); *Chem. Abstr.*, **76**, 14841 (1972).
(95) H. J. Bestmann and H. Schultz, *Chem. Ber.*, **95**, 2925 (1962).
(96) Yu. A. Zhdanov and L. A. Uslova, *Zh. Obshch. Khim.*, **42**, 759 (1972).

THE WITTIG REACTION IN CARBOHYDRATE CHEMISTRY 295

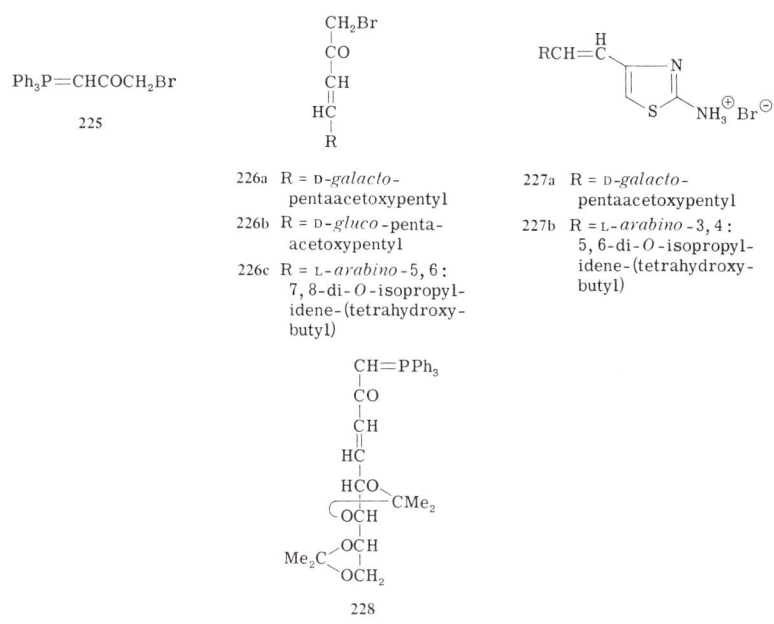

Ph₃PCH=CCH₂CO₂R with CO₂R	CO₂R, CH₂, C—CO₂R, HC-R'	CH₂, C(OH)CO₂H, CH, CH₃
222a R = Me	223a R = Me, R' = D-*gluco*-pentaacetoxypentyl	224
222b R = Et	223b R = Et, R' = D-*galacto*-pentaacetoxypentyl	
	223c R = R' = Me	
	223d R = Et, R' = Me	

phorane (**225**), and have shown that this phosphorane normally reacts with *aldehydo*-aldoses to give 1-bromo-1,3,4-trideoxyglyc-3-enuloses (**226a–226c**). The carbonyl (5.90–5.95 μm) and alkene (6.05–6.09 μm) infrared absorptions of these derivatives are displaced somewhat from those of α,β-unsaturated ketoses (see Section III, 2; p. 248), owing to the presence of bromine. The bromo derivatives **226a** and **226c** were shown[96] to react with thiourea with formation of the C-glycosylated aminothiazoles **227a** and **227b**. Reaction of compound **226c** with triphenylphosphine, following deprotonation, permitted synthesis[96] of the new sugar phosphorane **228**. The phosphorane **228** was shown by chromatography to interact with certain reactive aldehydes.

The introduction of the phosphonate grouping into a sugar molecule (see Section III, 3; p. 260) has been further investigated by Paulsen and coworkers.[97] Thus, the phosphonate phosphorane **120a** was shown to react with 2,3,4,5,6-penta-*O*-acetyl-*aldehydo*-D-glucose, 2,3:4,5-di-*O*-isopropylidene-*aldehydo*-D-arabinose, and 2,4-*O*-ethylidene-*aldehydo*-D-erythrose, leading to the unsaturated phosphonates **229**, **230a**, and **231**, respectively, in moderate yields.

```
O=P(OPh)₂
   |
   CH              O=P(OR)₂
   ||               |
   HC              CH
   |               ||
   HCOAc           HC
   |               |
   AcOCH           OCH
   |               |―CMe₂
   HCOAc           HCO
   |               |
   HCOAc           HCO
   |               |―CMe₂
   CH₂OAc          H₂CO

   229             230a  R = Ph
                   230b  R = Et
```

231

Difficulties were encountered in separation of the products from the side-product (triphenylphosphine oxide). Much better results were obtained when the phosphonate modification was applied, starting from the carbanion **232**, as, in this case, the side-product is the water-soluble diethyl phosphate. Indeed, the phosphonate **230b** was thus obtained[97] in 72% yield.

$$(\text{EtO})_2\overset{O}{\overset{||}{P}}-\overset{\ominus}{C}H-\overset{O}{\overset{||}{P}}(\text{OEt})_2$$

232

233a R = Ph
233b R = Et

The yields of phosphonates **231** and **233b**, arising from 2,4-*O*-ethylidene-*aldehydo*-D-erythrose by coupling with the phosphorane **120a** and carbanion **232**, were only 10% and 30%, respectively. The low yields possibly result because the parent *aldehydo*-aldose exists as a dimer.[98] The formation of the cyclic phosphonate **231** through elimination of phenol, and its failure to be formed from the ethyl analog **233b**, is of interest; the difference in behavior may be explained on the basis of decreased electron-density on phosphorus in the intermediate **233a** as compared with that of **233b**, a property that facilitates an attack of the free hydroxyl group.

(97) H. Paulsen, W. Bartsch, and J. Thiem, *Chem. Ber.*, **104**, 2545 (1971).
(98) R. Schaffer, *J. Amer. Chem. Soc.*, **81**, 2838 (1959).

The structures of the sugar phosphonates obtained were confirmed[97] by n.m.r -spectral data. These compounds are characterized by $J_{1,P}$ values of ~20 Hz. The second phosphorus coupling-constants $J_{2,P}$ could be determined only for **230b**, **231**, and **233b**, and were shown to depend on the geometrical configuration of the alkene bond, being equal to ~20 Hz for the *trans*-alkenes **230b** and **233b**, and 46.5 Hz for the *cis*-alkene **231**. In general, the H-1 signal appeared as a complex octet through coupling with P, H-2, and H-3. Large $J_{1,2}$ values (~17 Hz) for all compounds except **231** confirmed the *trans* configuration of the double bond.

The unsaturated phosphate **230b** was also used[97] for an investigation of some addition reactions. Hydrogenolysis led to a product presumed to be 1,2-dideoxy-1-(diethylphosphono)-3,4:5,6-di-*O*-isopropylidene-D-*arabino*-hexitol (**234**), and coupling with ethanol in the presence of sodium ethoxide resulted in the formation of the adduct **235**; the structures of these products have not been firmly established. Epoxidation of the double bond in **230b** was unsuccessful, presumably because of the electron-attracting properties of the phosphonate grouping and the sugar moiety.

```
    O=P(OEt)₂          O=P(OEt)₂                 CH₂
        |                  |                    O⟨  |
       CH₂                CH₂                      CH
        |                  |                        \O
       CH₂                CHOEt                      \
        |                  |                          OR
       OCH                OCH
        |  ⟩CMe₂           |  ⟩CMe₂
       HCO                HCO                         O-CMe₂
        |                  |
       HCO                HCO
        |  ⟩CMe₂           |  ⟩CMe₂
       H₂CO               H₂CO

        234                235              236a  R = H
                                            236b  R = CH₂Ph
```

Synthetic utilization of sugar epoxides with the use of known Wittig modifications has been shown to be very useful.[92,99] Thus, the alkene **15a** and its 3-*O*-benzyl analog were resynthesized,[92] albeit in moderate yields, by the interaction of epoxides **236a** and **236b**, respectively, with triphenylphosphine in boiling *N,N*-dimethylformamide.

Very significant results have been reported[99(a),(b)] by Meyer zu Reckendorf and Kamprath-Scholtz, who investigated the interaction of methyl 2,3-anhydro-4,6-*O*-benzylidene-α-D-allopyranoside (**237**) with (ethoxycarbonylmethyl)diethylphosphonate in the presence of sodium hydride. The resultant cyclopropane derivative (**238a**) was

---

(99) (a) W. Meyer zu Reckendorf and U. Kamprath-Scholtz, *Angew. Chem.*, **80**, 152 (1968); *Angew. Chem. Intern. Ed. Engl.*, **7**, 142 (1968); (b) *Chem. Ber.*, **105**, 673 (1972); (c) *ibid.*, **105**, 686 (1972).

converted, without isolation, into the corresponding acid (**238b**) which, by subsequent treatment with diazomethane, gave the methyl ester **238c**. The latter was a key compound, both for structural analysis and for subsequent transformations.

The structure of methyl 4,6-O-benzylidene-2,3-dideoxy-2,3-C-(methoxycarbonylmethylene)-α-D-*manno*-hexopyranoside (**238c**) was clearly established by its n.m.r. spectrum. The cyclopropane ring-protons (H-2, H-3, and H-7) resonated at τ 8.17, 7.88, and 8.32 p.p.m., respectively. The small $J_{1,2}$ and $J_{3,4}$ values (0.0 Hz and 1.9 Hz, respectively) indicate the L-*erythro* configuration of the cyclopropane

237	238a R = CO$_2$Et 238b R = CO$_2$H 238c R = CO$_2$Me 238d R = CHO	238e R = CH$_2$OH 238f R = CH$_2$NH$_2$ 238g R = NH$_2$ 238h R = CN
		239

ring. This conclusion was confirmed by comparison of $J_{1,2}$ (2.5 Hz) for the parent epoxide (**237**) with that of its D-*manno* analog ($J_{1,2}$ = 0 Hz). The $J_{2,3}$ value of compound **238c** was 9.1 Hz, indicating a *cis* orientation of H-2 and H-3. The other coupling constants ($J_{2,7}$ and $J_{3,7}$) for the cyclopropane ring-protons are sufficiently large (~5 Hz) as to indicate the *trans* orientation of H-2, H-3, and H-7 and, therefore, the sterically unhindered, *exo* orientation of the methoxycarbonyl group.

The inversion of configuration of the three-membered ring during the interaction of the epoxide **237** with the (ethoxycarbonylmethyl)-diethylphosphonate carbanion is not unexpected, as this phenomenon was also observed[100] when α-styrene oxide was allowed to react with the same carbanion. The mechanism proposed[101] for this inversion may be extended to the present example.

The key compound **238c**, an example of a functionalized cyclopropane sugar, was then subjected to various chemical transformations. The *aldehydo* derivative **238d** was obtained[99(b)] from compound **238c** by way of the corresponding *p*-tolylsulfonylhydrazone, by the Mac-Fadyen–Stevens reaction.[102] Reduction of the ester **238c** with lithium

---

(100) I. Tomoskozi, *Angew. Chem.*, **75**, 294 (1963); *Tetrahedron*, **19**, 1969 (1963).
(101) A. W. Johnson, "Ylid Chemistry," Academic Press, New York, 1967, p. 226.
(102) C. Niemann, R. N. Lewis, and J. T. Hays, *J. Amer. Chem. Soc.*, **64**, 1678 (1942).

aluminum hydride led[99(c)] to the carbinol **238e**. The latter was then converted[99(c)] into the amine **238f** through the respective sulfonic ester, by successive substitution by azide ion and reduction of the azide. The oxime of aldehyde **238d** gave[99(b)] the nitrile **238h** when it was treated with pyridine-acetic anhydride. The *p*-tolylsulfonylhydrazone of aldehyde **238d** was successfully converted[99(b)] into the first example of a cyclobutene sugar derivative (**239**), obtained in 49% yield, by heating with potassium *tert*-butoxide in toluene. An attempted synthesis[99(c)] of the cyclopropylamine **238g** by Curtius degradation of the azide of acid **238b** was, however, unsuccessful, presumably because of its instability.

GLYCOENZYMES: ENZYMES OF GLYCOPROTEIN STRUCTURE

By John H. Pazur and N. N. Aronson, Jr.

Department of Biochemistry, The Pennsylvania State University, University Park, Pennsylvania

I. Introduction 301
II. Types of Glycoenzymes 305
   1. Carbohydrate Hydrolases 305
   2. Nucleic Acid Hydrolases 307
   3. Other Hydrolases 308
   4. Oxidoreductases 309
III. Chemical and Physical Structure 309
   1. Criteria of Purity 309
   2. Protein Component 314
   3. Protein–Carbohydrate Linkages 318
   4. Carbohydrate Component 323
   5. Conformational Structure 327
IV. Biosynthetic Pathways 328
   1. Cellular Locale, Reactions, and Mechanisms 328
   2. Regulation of Synthesis 333
V. Biological and Structural Significance 337

I. Introduction

With the many improvements in methods for the isolation and characterization of enzymes, it has become evident that many of them contain carbohydrate residues covalently linked in their molecular structure. Enzymes of this type are, therefore, glycoproteins and the term *glycoenzyme* has been proposed as appropriately descriptive of this group.[1] Glycoenzymes have been found in a wide variety of biological materials, including plant tissues, animal organs, and microbial extracts. Glycoenzymes often occur in multimolecular forms, and evidence is accumulating that such forms are isoglycoenzymes, differing only in the carbohydrate portion of the molecules. The hydrolase group of enzymes contains the largest number of glycoenzymes,

(1) J. H. Pazur, D. L. Simpson, and H. R. Knull, *Biochem. Biophys. Res. Commun.*, **36**, 394 (1969).

and examples from this group include yeast invertase,[2,3] fungal amylase[4,5] and glucoamylase,[6,7] bovine and porcine pancreatic ribonuclease,[8,9] bovine pancreatic deoxyribonuclease,[10,11] bovine and rat-liver β-D-glucosiduronase,[12,13] pineapple bromelain,[14,15] and broad-bean α-D-galactosidase.[16] Examples from the oxidoreductase group are plant peroxidase,[17,18] fungal D-glucose oxidase,[19,20] fungal chloroperoxidases,[21,22] and bovine-liver D-amino acid oxidase.[23] Glycoenzymes found in other enzyme groups are few in number and have not been investigated in great detail.

The commonest monosaccharide constituents of the carbohydrate moieties of glycoenzymes are D-mannose and 2-acetamido-2-deoxy-D-glucose. Other monosaccharides encountered are D-glucose, D-galactose, D-xylose, L-arabinose, L-fucose, and sialic acid. Various combinations of these monosaccharides occur in the different glycoenzymes. The monosaccharides may occur either as oligomeric or polymeric chains, attached to specific amino acid residues of the polypeptide chains of the enzymes. Whereas some glycoenzymes possess a single carbohydrate chain attached to a specific amino acid residue of the protein, others contain many such chains attached to many amino acid residues. Information on the molecular structure of the carbohydrate component of the glycoenzymes is still incomplete but, in general, the monosaccharide residues are glycosidically joined as in

(2) M. Adams and C. S. Hudson, *J. Amer. Chem. Soc.*, **65**, 1359 (1943).
(3) N. P. Neumann and J. O. Lampen, *Biochemistry*, **6**, 468 (1967).
(4) H. Hanafusa, T. Ikenaka, and S. Akabori, *J. Biochem.* (Tokyo), **42**, 55 (1955).
(5) J. F. McKelvy and Y. C. Lee, *Arch. Biochem. Biophys.*, **132**, 99 (1969).
(6) J. H. Pazur, K. Kleppe, and E. M. Ball, *Arch. Biochem. Biophys.*, **103**, 515 (1963).
(7) J. H. Pazur, H. R. Knull, and A. Cepure, *Carbohyd. Res.*, **20**, 83 (1971).
(8) T. H. Plummer, Jr., and C. H. W. Hirs, *J. Biol. Chem.*, **238**, 1396 (1963).
(9) R. L. Jackson and C. H. W. Hirs, *J. Biol. Chem.*, **245**, 624 (1970).
(10) B. J. Catley, S. Moore, and W. H. Stein, *J. Biol. Chem.*, **244**, 933 (1969).
(11) J. Salnikow, S. Moore, and W. H. Stein, *J. Biol. Chem.*, **245**, 5685 (1970).
(12) B. V. Plapp and R. D. Cole, *Arch. Biochem. Biophys.*, **116**, 193 (1966).
(13) P. D. Stahl and O. Touster, *J. Biol. Chem.*, **246**, 5398 (1971).
(14) T. Murachi, M. Yasui, and Y. Yasuda, *Biochemistry*, **3**, 48 (1964).
(15). J. Scocca and Y. C. Lee, *J. Biol. Chem.*, **224**, 4852 (1969).
(16) P. M. Dey and J. B. Pridham, *Biochem. J.*, **113**, 49 (1969).
(17) H. Theorell and A. Akeson, *Arkiv Kem. Mineral. Geol.*, **16A**, 8.
(18) Y. Morita and K. Kameda, *Bull. Agr. Chem. Soc. Jap.*, **23**, 28 (1959).
(19) J. H. Pazur, K. Kleppe, and A. Cepure, *Arch. Biochem. Biophys.*, **111**, 351 (1965).
(20) B. E. P. Swoboda and V. Massey, *J. Biol. Chem.*, **240**, 2209 (1965).
(21) D. R. Morris and L. P. Hager, *J. Biol. Chem.*, **241**, 1763 (1966).
(22) L. P. Hager, D. R. Morris, F. S. Brown, and H. Eberwein, *J. Biol. Chem.*, **241**, 1769 (1966).
(23) H. Yamada, P. Gee, M. Ebata, and K. Yasunobu, *Biochim. Biophys. Acta*, **81**, 165 (1964).

typical glycans. The interglycosidic bonds of the carbohydrate chains in many glycoenzymes have the α-D (or L) configuration. Structures that are highly branched are often encountered in the carbohydrate chains of glycoenzymes.

Considerable attention has been given to determining the types of linkages between the carbohydrate and protein moieties of glycoenzymes. It is now well established that two types of linkage, the N-glycosyl and the O-glycosyl, are the major ones linking these moieties in the glycoenzymes. The N-glycosyl linkage is situated between the reducing end of the carbohydrate chain and an L-asparagine residue of the protein, whereas the O-glycosyl linkage is found between the carbohydrate moiety and the hydroxyl group of a serine or threonine residue in the protein. Preliminary reports have claimed the occurrence of an ester linkage between the reducing end of the carbohydrate chain and an aspartic or glutamic acid residue of the protein, but these reports have not yet been fully documented.

Glycoproteins are important constituents of virtually all cells, and a knowledge of biosynthetic pathways and factors regulating the biosynthesis of these compounds is required before the role of glycoproteins in cellular processes can be understood. Although experimental data establishing biosynthetic pathways for glycoenzymes are incomplete, it appears probable that they are synthesized *via* pathways similar to those delineated for glycoproteins, for which much information exists. For purposes of discussion, the biosynthetic pathway may be divided into three distinct phases: *phase 1*, when assembly of the polypeptide chain takes place on the ribosomes of the rough endoplasmic-reticulum, *via* the normal reactions of protein synthesis; *phase 2*, during which the bridge-carbohydrate residues are attached to certain amino acid residues of the polypeptide chain *via* transferases acting in conjunction with appropriate nucleoside 5'-(glycosyl pyrophosphates), perhaps while the polypeptide chain is still attached to the ribosomes in the rough endoplasmic-reticulum; and *phase 3*, when, in the region of the smooth endoplasmic-reticulum and in the Golgi particles, the remaining carbohydrate residues are attached to the incomplete glycoenzyme. The role of nucleoside 5'-(glycosyl pyrophosphates) as donors of the monosaccharide residues in the various steps of these reactions has been amply documented. Furthermore, the transferases catalyzing these reactions have been detected in a variety of tissues in which glycoprotein synthesis occurs. These enzymes are bound, for the most part, in the endoplasmic reticulum and in the Golgi particles.

The mechanism whereby the addition of carbohydrates to the polypeptide chain is controlled is not yet fully understood. It appears probable that the control is of the secondary type, mediated by the concentrations and structural features of the donor and acceptor molecules, and also by the concentrations and specificities of the transferases. However, it is possible that there exists a messenger ribonucleic acid that carries a triplet code for the insertion of glycosylated amino acid residues at appropriate points of the growing chain. If, indeed, such messenger molecules do exist, this aspect of glycoprotein synthesis must be under direct genetic control.

Questions relating to function at the molecular level of the carbohydrate residues in glycoproteins and glycoenzymes have been of special interest to investigators. A hypothesis has been advanced that the carbohydrate portion of the molecule is involved mechanistically, or as a "marker," for the transport of these molecules from the site of synthesis in the cell to the site of action or of need. As the transport of glycoproteins and glycoenzymes often involves passage through cellular membranes, the process could well be facilitated by an interaction of the carbohydrate chain of the glycoprotein or glycoenzyme with a carrier substance in the membrane. This interaction could lead to a conformational change in the complex, allowing the passage of the glycoprotein or glycoenzyme through the membrane. The idea of a "marker" function for the carbohydrate arises from observations that the entry or secretion of the molecules into cellular sub-particles is markedly influenced by alterations in the carbohydrate side-chains of glycoproteins. Although such functions in the transport process seem quite plausible for the carbohydrate residues, experimental verification of these suggestions is difficult to achieve.

It has long been known that glycoproteins are less susceptible to proteolysis and to denaturation than are other types of proteins. Accordingly, a suggestion has been advanced that the carbohydrate residues function as protective agents for the protein moiety of glycoproteins and glycoenzymes. This suggested role of the carbohydrate in maintaining the tridimensional structure of the enzyme is indeed attractive. In a number of glycoenzymes, the carbohydrate residues appear to be attached to amino acid residues situated at the surface of the molecule. Such residues might interfere with the interaction of a proteolytic enzyme with the glycoprotein or glycoenzyme and thereby inhibit its hydrolysis. Resistance to denaturation by heat may also be attributable to such steric effects. For example, a polypeptide fold in the native molecule may be held in position through the strategic location of carbohydrate residues along the chain, and the molecular transformations necessary for denaturation might thus be hindered. If, during biosynthesis, the carbohydrate residues become attached to the protein prior to the folding of the latter into its characteristic tridimensional structure, the carbohydrate residues could thereby influence the manner in which the protein molecule folds into this structure and could also impart stability to the folded molecule.

It is the purpose of this article to assemble information on the major types of glycoenzymes, their chemical structures, their physical properties, and the pathways of their biosynthesis, and to discuss and evaluate theories pertaining to the functions, at the molecular and cellular levels, of the carbohydrate residues in glycoenzymes.

II. Types of Glycoenzymes

1. Carbohydrate Hydrolases

In early work on the purification of β-D-glucosidase[24,25] and invertase,[2,26] it was found by reducing-sugar determinations that highly purified samples of these enzymes contain carbohydrate. However, it was not then appreciated that such carbohydrate might constitute an integral structural element of the enzyme. On the contrary, it was generally assumed that the carbohydrate was a contaminant in the enzyme, and had no structural or biological significance. An early suggestion[24] that the carbohydrate might function as a "holding group" for the substrate was not taken seriously. Later reports[27,28] claimed that the carbohydrate in invertase preparations could be removed from the enzyme by adsorption procedures. However, it is now well established that invertase is indeed a glycoprotein, and the carbohydrate component of the enzyme has been extensively investigated.[29,30] The carbohydrate-free preparations may have resulted from degradation of the invertase during isolation. The enzyme β-D-glucosidase has been found to occur in multimolecular forms,[31] but whether all of these forms contain carbohydrate is not yet known.

Other carbohydrate hydrolases that contain carbohydrate moieties

(24) B. Helferich, W. Richter, and S. Grunler, *Ber. Verhandl. Sächs. Akad. Wiss. Leipzig, Math.-phys. Kl.*, 89, 385 (1937).
(25) B. Helferich and W. W. Pigman, *Z. Physiol. Chem.*, 259, 253 (1939).
(26) M. Adams and C. S. Hudson, *J. Amer. Chem. Soc.*, 60, 982 (1938).
(27) E. H. Fischer and L. Kohtes, *Helv. Chim. Acta*, 34, 1123 (1951).
(28) E. H. Fischer, L. Kohtes, and J. Fellig, *Helv. Chim. Acta*, 34, 1132 (1951).
(29) N. P. Neumann and J. O. Lampen, *Biochemistry*, 8, 3552 (1969).
(30) V. H. Greiling, P. Vogele, R. Kisters, and H. Ohlenbusch, *Z. Physiol. Chem.*, 350, 517 (1969).
(31) J. Schwartz, J. Sloan, and Y. C. Lee, *Arch. Biochem. Biophys.*, 137, 122 (1970).

TABLE I
Carbohydrate Content of Some Glycoenzymes

Enzyme	Biological source	Carbohydrate (%)	Monosaccharides	References
alpha-Amylase	*Aspergillus oryzae*	3	Man, GlcNAc, Gal, Xyl, Ara	4,5,33
Glucoamylase I	*Aspergillus niger*	14	Man, Glc, Gal	6,7
Glucoamylase II	*Aspergillus niger*	23	Man, Glc, Gal	6,7
Invertase (external)	yeast	50	Man, GlcNAc	29
β-D-Glucosiduronase	bovine liver	6		12,42
β-D-Glucosaminidase	*Aspergillus oryzae*	5	Man, GlcNAc	39
α-D-Galactosidase I	broad bean	25		16
α-D-Galactosidase II	broad bean	3		16
Ribonuclease B	bovine pancreas	9	Man, GlcNAc	8
	porcine pancreas	35	Man, GlcNAc, Gal, Fuc, Sialic acid	9
Deoxyribonuclease A	bovine pancreas	3	Man, GlcNAc	10,11
Deoxyribonuclease B	bovine pancreas	4	Man, GlcNAc, Gal, Sialic acid	11
Deoxyribonuclease C	bovine pancreas	3	Man, GlcNAc	11
Bromelain II	pineapple	3	Man, GlcNAc, Fuc, Xyl	15
Bromelain III	pineapple	2	Man, GlcNAc, Fuc, Xyl	15
Protease b	snake venom	16	Man, GlcNAc, Gal, Sialic acid	43
Protease A	*Saccharomyces carlbergensis*	7	Man, Glc	44
Protease B	*Saccharomyces carlbergensis*	4	Man, Glc	44
Protease C	*Saccharomyces carlbergensis*	20	Man, Glc	44
Protease D	*Saccharomyces carlbergensis*	13	Man, Glc	44
Peroxidase	horseradish	18		17
	Japanese radish	28		18
Glucose oxidase	*Aspergillus niger*	16	Man, GlcNAc, Gal	19,20
Chloroperoxidase	*Caldariomyces fumago*	25	GlcNAc, Ara	21
Monoamine oxidase	bovine plasma	5		23

in their molecular structure include fungal *alpha*-amylase,[4,5,32,33] fungal yeast amylase,[34] glucoamylase,[7,35,36] yeast glucoamylase,[37,38] and fungal glucosaminidase.[39] The glucoamylase from some strains of fungi occurs in isoenzymic forms that appear to differ only in the proportion of carbohydrate component.[40,41] The β-D-galactosidase from the broad bean[16] and the β-D-glucosiduronase from mammalian liver[12,42] are other glycoenzymes that occur in isoenzymic forms having different carbohydrate contents.

Quantitative data on the carbohydrate content of these hydrolases are presented in Table I. This Table also presents information on major types of glycoenzymes other than those just described, together with information on biological sources of glycoenzymes and the types of carbohydrate residues present in them. The data illustrate the large variation in carbohydrate content among glycoenzymes; some contain as little as 2% of carbohydrate, whereas others contain as much as 50%. The monosaccharide residues found most commonly in glycoenzymes are D-mannose and 2-acetamido-2-deoxy-D-glucose. Other monosaccharides often present are D-glucose, D-galactose, D-xylose, L-arabinose, L-fucose, and sialic acid.

2. Nucleic Acid Hydrolases

Many of the nucleic acid hydrolases occur in isoenzymic forms in mammalian tissues. Some of these isoenzymes are glycoenzymes, and information on these is contained in Table I. Ribonuclease B has been studied in the greatest detail. Following the initial observation that the ribonuclease from bovine pancreas occurs in isoenzymic forms,[45] it was found that one form (ribonuclease B) contains an appreciable

(32) V. M. Hanrahan and M. L. Caldwell, *J. Amer. Chem. Soc.*, **75**, 4030 (1953).
(33) M. Anai, T. Ikenaka, and Y. Matsushima, *J. Biochem.* (Tokyo), **59**, 57 (1966).
(34) T. Sawai, *Proc. Symp. Amylase, Osaka, Japan*, 1967, p. 111.
(35) I. D. Fleming and B. A. Stone, *Biochem. J.*, **97**, 13P (1965).
(36) J. H. Pazur and S. Okada, *Carbohyd. Res.*, **4**, 371 (1967).
(37) Y. Hattori, *Staerke*, **17**, 82 (1965).
(38) T. Fukui and Z. Nikuni, *Agr. Biol. Chem.* (Tokyo), **33**, 884 (1969).
(39) T. Mega, T. Ikenaka, and Y. Matsushima, *J. Biochem.* (Tokyo), **68**, 109 (1970).
(40) J. H. Pazur and T. Ando, *J. Biol. Chem.*, **234**, 1966 (1959).
(41) D. R. Lineback, I. J. Russell, and C. Rasmussen, *Arch. Biochem. Biophys.*, **134**, 539 (1969).
(42) B. V. Plapp and R. D. Cole, *Biochemistry*, **6**, 3676 (1967).
(43) G. Oshima, Y. Matsuo, S. Iwanaga, and T. Suzuki, *J. Biochem.* (Tokyo), **64**, 227 (1968).
(44) I. S. Maddox and J. S. Hough, *Biochem. J.*, **117**, 843 (1970).
(45) C. H. W. Hirs, S. Moore, and W. H. Stein, *J. Biol. Chem.*, **200**, 493 (1953).

proportion of carbohydrate,[8] whereas the other form (ribonuclease A) does not. The two ribonucleases possess identical catalytic properties and have essentially the same amino acid composition.[46] The differences between the two ribonucleases in their electrophoretic and chromatographic properties are apparently due to the presence of carbohydrate in one of them. The significance of the carbohydrate moiety in the molecule of ribonuclease B is considered in Section V (see p. 339).

In later work, ribonucleases from other animal species,[9,47] and deoxyribonuclease,[10,11,48] have also been shown to be glycoenzymes. Ribonuclease B from certain sources contains considerably more carbohydrate than the bovine ribonuclease B; for example, porcine ribonuclease B contains nearly four times as much carbohydrate as the bovine variety, and ovine ribonuclease B contains considerably more carbohydrate than does bovine ribonuclease B. Whereas the carbohydrate moiety of ribonuclease B from bovine pancreas is composed of D-mannose and 2-acetamido-2-deoxy-D-glucose residues,[46] the carbohydrate from porcine-pancreatic ribonuclease contains residues of L-fucose, sialic acid (N-glycolylneuraminic acid), and D-galactose, as well as of D-mannose and 2-acetamido-2-deoxy-D-glucose.[9] Several multi-molecular forms of deoxyribonuclease occur in the bovine pancreas.[11] As shown in Table I, the isoenzymic forms of deoxyribonuclease contain characteristic proportions and types of carbohydrates. Not only do the isoenzymes differ in their carbohydrate content, but detailed studies on these isoenzymes show that there are subtle differences in their amino acid compositions.[11] The genetic implications for the control of biosynthesis of glycoenzymes having variations in amino acid composition are manifold.

3. Other Hydrolases

Other glycoenzymes of the hydrolase type listed in Table I include mammalian,[43] yeast,[44] and plant[5,9,49–51] proteases, and mammalian esterases[52] and phosphatases.[53] Several of the lysosomal hydrolases in

(46) T. H. Plummer, Jr., and C. H. W. Hirs, *J. Biol. Chem.*, **239**, 2530 (1964).
(47) R. L. Jackson and C. H. W. Hirs, *J. Biol. Chem.*, **245**, 637 (1970).
(48) P. A. Price, T. Liu, W. H. Stein, and S. Moore, *J. Biol. Chem.*, **244**, 917 (1969).
(49) T. Murachi and H. Neurath, *J. Biol. Chem.*, **235**, 99 (1960).
(50) S. Ota, S. Moore, and W. H. Stein, *Biochemistry*, **3**, 180 (1964).
(51) G. Feinstein and J. R. Whitaker, *Biochemistry*, **3**, 1050 (1964).
(52) G. R. J. Law, *Science*, **156**, 1106 (1967).
(53) F. A. Dugan, R. Radhakrishnamurthy, and G. S. Berenson, *Enzymologia*, **33**, 215 (1967).

mammalian tissues also appear to be glycoenzymes,[13,54] but the exact nature of these enzymes is still under investigation. Except for the proteases, the nature of the carbohydrate components in these glycoenzymes has not been extensively investigated. Of interest is the finding that a plant protease, namely, bromelain from pineapple, contains a deoxy sugar (L-fucose) as well as other monosaccharides more typical of glycoenzymes. The bromelains occur in isoenzymic forms, and differences in the carbohydrate content of isoenzymes of bromelain have been noted.[15]

4. Oxidoreductases

Several enzymes of the oxidoreductase group are glycoenzymes. D-Glucose oxidase[6,19,20] and chloroperoxidase,[21,22] both of fungal origin, have been studied in the greatest detail. Table I presents, for oxidoreductases, information on their biological sources, their carbohydrate contents, and the nature of the carbohydrates. It may be noted that glycoenzymes of the oxidase type have been found in plant[17,18] and mammalian tissues,[23] as well as in microbial extracts. The total carbohydrate content of the oxidoreductases is relatively high. The types of monosaccharide residues in the carbohydrate moiety of these glycoenzymes are very similar to those observed in the hydrolase group of glycoenzymes.

III. CHEMICAL AND PHYSICAL STRUCTURE

1. Criteria of Purity

As polymeric carbohydrates can form complexes with proteins through secondary and ionic forces, it is necessary to employ several types of criteria for establishing that, in enzyme preparations, a covalent linkage does indeed exist between the carbohydrate and the protein. The principal criteria that have been employed for this purpose are: (*a*) a carbohydrate content that is unchanged following repeated precipitation or crystallization; (*b*) a congruent distribution of the carbohydrate and protein in eluates from chromatographic columns; (*c*) identical electrophoretic mobilities of the carbohydrate and protein components; (*d*) identical sedimentation behavior of the carbohydrate and the protein; and (*e*) isolation and structural characterization of glycopeptide fragments.

Examples that illustrate the use of these methods have been taken

(54) A. Goldstone and H. Koenig, *Life Sci.*, **9**, 1341 (1970).

randomly from the literature. In the purification of pineapple bromelain, the two isoenzymes obtained from an ion-exchange column contained 2.5 and 2.1% of carbohydrate, respectively, prior to purification by precipitation with ammonium sulfate.[15] After the precipitation step, the carbohydrate content of the two isoenzymes remained at these values. This constancy in carbohydrate content is indicative of a covalent carbohydrate–protein linkage, especially in view of the fact that polymeric carbohydrates are not generally precipitated by ammonium sulfate. Another type of glycoenzyme, namely, chloroperoxidase, has been prepared in crystalline form;[22] it was recrystallized, and the carbohydrate content was found to remain constant. Results such as those just described for bromelain and chloroperoxidase have been obtained with other glycoenzymes, and are indicative of a carbohydrate–protein linkage in these enzymes.

The purification of many enzymes has been achieved by use of gel-filtration chromatography, ion-exchange chromatography, or other chromatographic procedures. Eluates from such columns may be analyzed for carbohydrate and protein content, as well as for enzymic activity. Such analyses have revealed that, for glycoenzymes, the enzymic activity, the carbohydrate content, and the protein content are distributed identically in the eluates from the columns. Results of this type of experiment for ribonuclease B are shown in Fig. 1. In these experiments,[8] ribonuclease B and a hydrolyzate of the enzyme were subjected to gel filtration on Sephadex, and the eluates were analyzed for protein and for carbohydrate by appropriate colorimetric methods. A control experiment containing ribonuclease A and D-mannose was treated similarly. The data in Fig. 1 show that the carbohydrate component in the unhydrolyzed sample of ribonuclease B

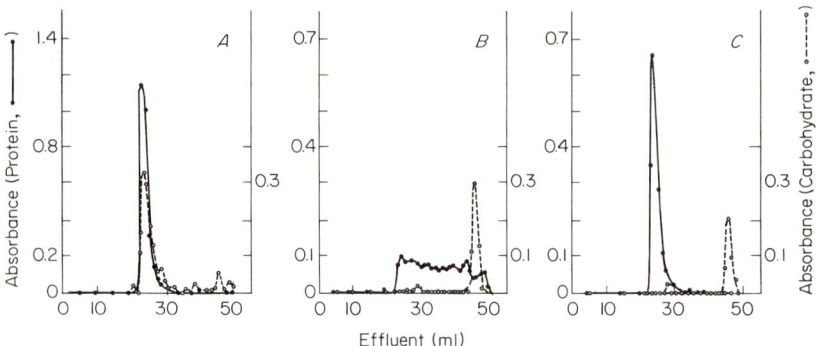

FIG. 1.—Distribution of Carbohydrate and Protein Components in (A) Ribonuclease B, (B) a Hydrolyzate of Ribonuclease B, and (C) a Mixture of Ribonuclease A and D-Mannose, in Effluents from Gel-filtration Columns.[8]

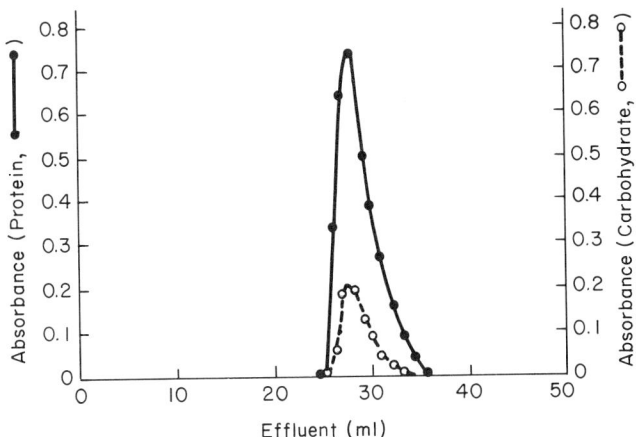

FIG. 2.—Distribution of Carbohydrate and Protein Components in a Preparation of Fungal *alpha*-Amylase in Effluents from an Alumina Column.[4]

migrated through the Sephadex column at the same rate as the ribonuclease itself; however, after hydrolysis of the glycoenzyme, the carbohydrate component migrated at a rate characteristic of D-mannose. Data for another type of filtration experiment[4] are shown in Fig. 2. It may be noted that the protein and carbohydrate components in a preparation of *alpha*-amylase migrated through a column of alumina at the same rates, and appeared in the same fractions of the eluates. Such data indicate protein–carbohydrate linkages in these glycoenzymes.

Electrophoretic techniques have been most valuable for characterizing glycoenzymes. Initially, paper-electrophoretic methods were employed; after electrophoresis of the enzyme for a suitable period, analyses of eluates from the paper strips were performed. With enzymes that contain carbohydrates, it was found that the carbohydrate and the protein components were located in the eluates from the same zone of the paper.[4] Subsequently, other types of electrophoretic procedures have been employed, and an application of curtain electrophoresis is illustrated in Fig. 3. This Figure shows the distribution of carbohydrate and protein components in D-glucose oxidase following curtain electrophoresis of the enzyme,[6] and it is seen quite clearly that the two components of the enzyme migrated at identical rates. The gel-electrophoretic technique[55] for separating proteins, including glycoproteins, is an extremely sensitive procedure for resolving molecules having small differences in net charge; this method,

(55) O. Gabriel, *Methods Enzymol.*, **22**, 565 (1971).

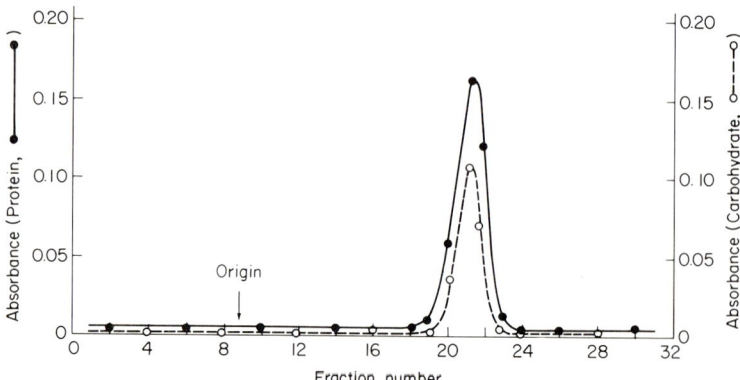

FIG. 3.—Distribution of Carbohydrate and Protein Components in a Preparation of Fungal D-Glucose Oxidase in Fractions from Curtain Electrophoresis.[6]

coupled with a double-staining technique, has proved extremely useful for characterizing glycoenzymes. In such experiments, the developed strips of gel are divided into two longitudinal sections; one section is stained with suitable dyes[55] to reveal the protein, and the other with periodic acid–Schiff reagent to reveal the carbohydrate.[56] Electrophoretic methods can thus be used in a variety of ways for characterizing glycoenzymes.

Several types of sedimentation procedure can be used for characterizing glycoenzymes. Sedimentation of the protein and carbohydrate components at identical rates *may* be indicative of a chemical bond between the two components. However, identical rates of sedimentation would be observed for two components that are not chemically linked but which possess the same molecular weights. Density-gradient centrifugation procedures are especially valuable for separating enzymes of different molecular weights[57,58] and for characterizing glycoenzymes. After centrifugation of the glycoenzyme in the density-gradient tube, the contents of the tube can be fractionated and the fractions analyzed for carbohydrate and protein. Data for the sedimentation of D-glucose oxidase in a glycerol gradient[19] are shown in Fig. 4. It is clear from the Figure that the carbohydrate and protein components of D-glucose oxidase sediment at the same rate. It should be noted that density-gradient centrifugation readily separates mixtures of carbohydrate and protein that are distinct molecular

(56) R. M. Zacharius, T. E. Zell, J. H. Morrison, and J. J. Woodlock, *Anal. Biochem.*, **30**, 148 (1969).
(57) R. G. Martin and B. N. Ames, *J. Biol. Chem.*, **236**, 1372 (1961).
(58) J. H. Pazur, K. Kleppe, and J. S. Anderson, *Biochim. Biophys. Acta*, **65**, 369 (1962).

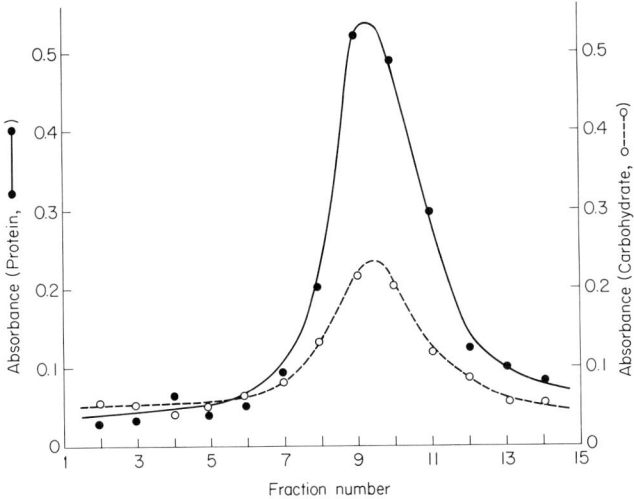

Fig. 4.—Distribution of Carbohydrate and Protein Components in a Preparation of Fungal D-Glucose Oxidase in Fractions from a Density-gradient Tube.[19]

species of different molecular size. Data illustrating this point are shown in Fig. 5 for a bacterial amylase.[59] The enzyme had been purified extensively by chromatography and by fractional precipitation with ammonium sulfate. The purified sample was found to contain a carbohydrate component. However, the data in Fig. 5 show that the carbohydrate and protein components in this enzyme preparation were separable by centrifugation, indicating the absence of a covalent linkage between the carbohydrate and the protein; the bacterial amylase is not, therefore, a glycoenzyme.

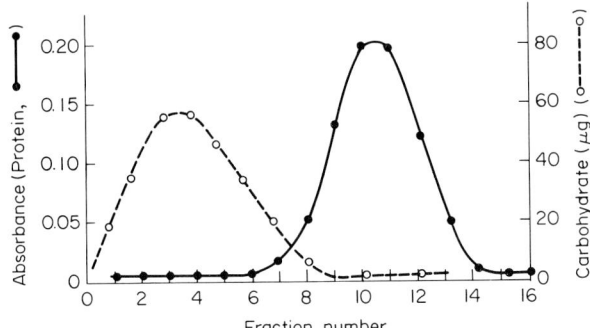

Fig. 5.—Distribution of Carbohydrate and Protein Components in a Preparation of Bacterial *alpha*-Amylase in Fractions from a Density-gradient Tube.[59]

(59) J. H. Pazur and S. Okada, *J. Biol. Chem.*, **241**, 4146 (1966).

Centrifugation data from measurements with an analytical ultracentrifuge are useful for characterizing glycoenzymes. Pure glycoenzymes yield symmetrical patterns when examined by this procedure. Sedimentation measurements can be made at several concentrations of the enzyme, and a number of physical constants characteristic of the molecule can be calculated from these data. On the basis of such measurements and calculations, it was determined that chloroperoxidase, which contains carbohydrate, is homogeneous with respect to molecular size,[21] and it was concluded that the carbohydrate component in the preparation of chloroperoxidase studied is chemically linked to the protein.

The most convincing evidence for the existence of a linkage between carbohydrate and protein in a given enzyme is provided by the isolation and characterization of glycopeptide fragments from the preparation. This approach has been utilized with many glycoenzymes. Application of the technique is dependent on the availability of agents for cleaving the glycoenzyme into suitable fragments that can be isolated and characterized. Such proteolytic enzymes as trypsin, chymotrypsin, and pronase have been used extensively for cleaving glycoenzymes, and conditions for using these degradative enzymes have been described.[60] Glycopeptides produced in digests of glycoenzymes with specific proteases have been isolated by ion-exchange chromatography or by electrophoretic procedures.[60] These glycopeptides can be further degraded by the Edman procedure, and the sequence of amino acids of the polypeptide close to the carbohydrate–protein linkage can be elucidated. Data from experiments of this type also reveal the identity of the amino acid residues that are linked to the carbohydrate. In many such experiments, mixtures of glycopeptides having different amino acid compositions have been obtained, and resolution of the glycopeptides into homogeneous components becomes a major problem. Chemical reactions based on the use of cyanogen bromide and N-bromosuccinimide have been employed for cleaving proteins, and are advantageous for use with glycoenzymes under certain conditions.[60] Similar problems of isolation and purification of the glycopeptides are also encountered when these chemical degradative procedures are used.

2. Protein Component

Proteins may be composed of (a) single polypeptide chains, (b) two or more polypeptide chains joined by cross-linkages, or (c) two or more subunits having single- or multi-chain structures held to-

(60) C. H. W. Hirs, *Methods Enzymol.*, **11**, 411 (1967).

gether by secondary or tertiary forces. In general, molecules of the third type can readily be dissociated into subunits by use of appropriate dissociation-agents. Many glycoenzymes are of relatively low molecular weight, and fall in the structural types of the first two categories. The protein component of glycoenzymes has the single polypeptide-chain folded into a unique tridimensional structure possessing catalytic activity. In some cases, the folded molecule is held together by disulfide cross-linkages. Glycoenzymes of the latter type, which occur in isoenzymic forms, differ markedly from the conventional isoenzymes, which are generally composed of subunits.[61,62]

The amino acid residues that comprise the protein component of glycoenzymes are those found in typical proteins. Table II records the amino acid compositions for some glycoenzymes: deoxyribonuclease,[48] β-D-glucosaminidase,[39] D-glucose oxidase,[19] and monoamine oxidase.[23] Variations among these enzymes in their amino acid compositions do occur, but there does not appear to be any unusual feature in the amino acid composition of glycoenzymes. For those glycoenzymes that occur in isoenzymic forms, it is informative to compare the amino acid compositions of the pairs of isoenzymes. Data for ribonuclease,[8] glucoamylase,[7] and invertase[3] are recorded in Table III. It should be mentioned that the values for ribonuclease are not corrected for losses during hydrolysis, whereas those for the other two enzymes have been so corrected and are rounded to the nearest integer. Within the limits of experimental error, the amino acid compositions of the two forms of ribonuclease are identical. Furthermore, other studies[46] have shown that the two forms of ribonuclease also have essentially the same sequence of amino acids and the same catalytic properties. In view of the foregoing, it is logical to expect that the tridimensional structures of the protein portion of the two isoenzymes of ribonuclease are identical. The carbohydrate moiety in ribonuclease B, adds, of course, an additional conformational aspect to this molecule, but this moiety apparently does not disturb the tertiary structure of the protein moiety of the enzyme.

The data in Table III giving the amino acid compositions of the isoenzymes of glucoamylase indicate a similarity, if not an identity, in the polypeptide chains in the two forms. Information is not available on the amino acid sequences in the glucoamylase isoenzymes. However, the two isoenzymes contain alanine at the N-terminus, and exhibit similar immunological properties.[7,41] All of these observations point to the presence of identical polypeptide chains in the two iso-

(61) C. Markert and F. Moller, *Proc. Nat. Acad. Sci. U. S.*, **45**, 753 (1959).
(62) I. M. Klotz, *Science*, **155**, 697 (1967).

TABLE II

Amino Acid Compositions of Some Glycoenzymes

Component	Deoxyribonuclease[48] M.w. = 31,000	β-D-Glucosaminidase[39] M.w. = 140,000	Glucose oxidase[19] M.w. = 150,000	Monoamine oxidase[23] M.w. = 260,000
Aspartic acid	33	170	130	175
Threonine	15	58	82	133
Serine	30	79	85	142
Glutamic acid	20	120	99	243
Proline	9	69	53	176
Glycine	10	75	111	177
Alanine	22	92	108	155
Cystine/2	4	12	4	19
Valine	27	87	79	160
Methionine	4	10	21	37
Isoleucine	12	73	46	59
Leucine	23	82	96	174
Tyrosine	16	57	52	69
Phenylalanine	12	35	35	118
Lysine	9	30	30	56
Histidine	6	32	37	63
Arginine	12	44	45	97
Tryptophan	4	54	22	36

TABLE III
Amino Acid Composition of Some Isoenzymes

Amino acid	Ribonuclease A[8]	Ribonuclease B[8]	Glucoamylase I[7]	Glucoamylase II[7]	External invertase[63]	Internal invertase[63]
Aspartic acid	15.0[a]	14.6[a]	83[b]	85[b]	178[b]	165[b]
Threonine	8.6	8.4	104	103	84	80
Serine	10.5	10.3	119	121	114	151
Glutamic acid	11.9	11.7	56	54	115	124
Proline	4.3	4.3	29	29	65	63
Glycine	3.1	3.3	59	63	71	115
Alanine	12.1	11.8	78	82	68	84
Cystine/2	6.8	5.3	8	8	5	0
Valine	8.9	8.8	46	46	69	73
Methionine	3.7	3.0	4	4	21	14
Isoleucine	2.9	2.9	27	25	40	38
Leucine	2.0	2.1	53	56	83	77
Tyrosine	5.1	4.5	30	29	65	31
Phenylalanine	2.9	2.9	27	28	80	77
Lysine	10.2	9.8	16	16	60	85
Histidine	4.0	3.8	6	6	16	29
Arginine	3.9	3.8	23	22	27	32
Tryptophan			30	32	33	20

[a] Analyses were performed on a 70-hr hydrolyzate and are not corrected for degradative losses. [b] Values have been corrected for degradative losses, and have been rounded to the nearest integer.

enzymes of glucoamylase, and, consequently, the two forms of glucoamylase are isoglycoenzymes that differ only in their carbohydrate components. Similar structural relationships exist in the β-D-glucosaminidases from mammalian-spleen tissue.[64]

By contrast, the amino acid compositions of the two forms of invertase (internal and external) differ significantly in the number of residues of certain amino acids. These differences occur notably in the content of serine, alanine, cystine, methionine, tyrosine, lysine, and histidine. The significance of these differences is not yet apparent, but these two invertases are obviously not isoglycoenzymes. The synthesis of proteins having such diverse amino acid compositions as have the two invertases would probably necessitate two separate genetic-control mechanisms.

3. Protein–Carbohydrate Linkages

The principal linkages found between the carbohydrate and protein components of glycoenzymes are of the N-glycosyl and the O-glycosyl types. Such linkages are also common in glycoproteins,[65] and many of the experimental techniques developed for glycoproteins have been applied to glycoenzymes. The N-glycosyl bond is formed by elimination of water between the side-chain amide group of asparagine and the hemiacetalic hydroxyl group of a 2-acetamido-2-deoxy-D-glucose residue located at the reducing end of the carbohydrate chain. Glycopeptides having an N-glycosyl linkage between the carbohydrate and an amino acid have been isolated from hydrolyzates of many glycoenzymes, as well as from hydrolyzates of glycoproteins. The simplest such fragment is 2-acetamido-N-L-aspart-4-oyl-2-deoxy-D-glucosylamine. This fragment has been isolated[66] from hydrolyzates of ribonuclease B, but not from hydrolyzates of other glycoenzymes. The structure of the fragment from ribonuclease B has been verified by comparison of its properties with those of aspartamido–2-amino-2-deoxy-D-glucose synthesized by established chemical reactions.[67] The isolation of this fragment from hydrolyzates of glycoenzymes is the best evidence advanced for the occurrence of N-glycosyl linkages in glycoenzymes. The possibility of N-glycosyl linkages to glutamine residues of glycoenzymes exists, particularly as glutamine and aspara-

---

(63) S. Gascon, N. P. Neumann, and J. O. Lampen, *J. Biol. Chem.*, **243**, 1573 (1968).
(64) D. Robinson and J. L. Stirling, *Biochem. J.*, **107**, 321 (1968).
(65) R. D. Marshall and A. Neuberger, *Advan. Carbohyd. Chem. Biochem.*, **25**, 407 (1970).
(66) T. H. Plummer, Jr., A. Tarentino, and F. Maley, *J. Biol. Chem.*, **243**, 5158 (1968).
(67) R. D. Marshall and A. Neuberger, *Biochemistry*, **3**, 1596 (1964).

gine are often present in the glycopeptides isolated from hydrolyzates of glycoenzymes. However, direct evidence for a linkage from a sugar to glutamine has not been presented.

The O-glycosyl linkage in glycoenzymes is formed by abstracting the elements of water from between the hydroxyl group of an L-serine or L-threonine residue of the protein and the hemiacetal hydroxyl group of the carbohydrate residue at the reducing end of the carbohydrate chain. The simplest "glycopeptide" component of a glycoenzyme would thus be an O-glycosylserine or O-glycosylthreonine, although neither of these has been isolated from hydrolyzates of glycoenzymes. Evidence for the existence of O-glycosyl linkages is mainly indirect, and has been obtained by use of an alkali-catalyzed, β-elimination reaction of the carbohydrate residue under reducing conditions.[65] Under such conditions, both the amino acid residue thus modified and the carbohydrate residue released from the linkage-point undergo reduction. After completion of this reaction, the mixture is subjected to hydrolysis, and the hydrolytic products are analyzed. Such analyses performed on glucoamylase show that the serine or threonine content decreases, and a corresponding increase in the content of alanine, glycine, and 2-aminobutyric acid occurs.[68] The alditol in the hydrolyzate was identified as a mannitol.[69] Accordingly, it can be presumed that D-mannose is the bridge residue joining the carbohydrate chain to serine or threonine residues of the protein. These results obtained with glucoamylase are comparable with those obtained with many other glycoproteins.[70–72]

$$\text{ycosyl—Man—O}\overset{\text{Protein—(NHCHCO)—protein}}{\underset{\text{CH}_3}{\text{CH}}} \xrightarrow{\underset{\text{BH}_4^\oplus}{\text{OH}^\ominus}} \overset{\text{Protein—(NHCHCO)—protein}}{\underset{\text{CH}_3}{\text{CH}_2}} + O\text{-glycosylmannitol}$$

$$\downarrow \text{H}^\oplus$$

Mannitol + 2-aminobutyric acid + other amino acids + sugars

Information on the sequence of the amino acids in the region around the point of attachment of the carbohydrate to the protein has

(68) J. H. Pazur, H. R. Knull, and D. L. Simpson, *Biochem. Biophys. Res. Commun.*, **40**, 110 (1970).
(69) D. R. Lineback, *Carbohyd. Res.*, **7**, 106 (1968).
(70) B. Anderson, P. Hoffman, and K. Meyer, *Biochim. Biophys. Acta*, **74**, 309 (1963).
(71) V. Bhavanandan, E. Buddecke, R. Carubelli, and A. Gottschalk, *Biochem. Biophys. Res. Commun.*, **16**, 353 (1964).
(72) N. Payza, S. Rizvi, and W. Pigman, *Arch. Biochem. Biophys.*, **129**, 68 (1969).

been obtained for several glycoenzymes, and such sequences are recorded in Table IV. In general, this information has been derived from studies on the enzymic cleavage of the glycoenzymes into glycopeptide fragments that have been isolated and structurally characterized. Several different types of glycopeptide can be isolated from some glycoenzymes, indicating a heterogeneity for the protein–carbohydrate linkages in a single enzyme.

The sequence of amino acids around the N-glycosyl linkages in glycoenzymes and glycoproteins is generally of the following type:

$$\text{A—Asn—B—Ser(Thr)} \quad \overset{\text{GlcNAc}}{\underset{|}{}}$$

In this depiction, A and B are unspecified amino acid residues; the other abbreviations are the conventional ones. It has been suggested that the L-serine and L-threonine residues of the polypeptide chain are important in that they may function as markers for the transferases responsible for the attachment of a carbohydrate moiety to the polypeptide.[77] It is conceivable that the hydroxyl groups of the L-serine and L-threonine residues, together with the carboxyl and amide groups of L-asparagine residues, may be the groups of the acceptor substrate that participate in formation of the enzyme–substrate complex. If these hydroxy amino acids are not present, the complex cannot form and glycosylation cannot occur. The nature of residue B in the peptide chain also affects the type of carbohydrate that becomes attached to the L-asparagine residues. Thus, if the B residue is a polar one, a carbohydrate moiety of greater complexity is found attached to asparagine.[9] As the polarity of this group might well prolong the life of the enzyme–substrate complex, it is probable that, over a longer period of time, a more complicated carbohydrate chain can be elaborated at this point of the glycoenzyme. The control of biosynthesis of glycoenzymes is considered further in Section IV,2 (see p. 333).

For glycoenzymes containing O-glycosyl bonds, little information is available on the sequence of amino acids in the region of the carbohydrate–protein linkage. An early report[75] indicated that the carbo-

(73) M. Fukuda, T. Muramatsu, F. Egami, N. Takahashi, and Y. Yasuda, *Biochim. Biophys. Acta*, **159**, 215 (1968).
(74) Y. Yasuda, N. Takahashi, and T. Murachi, *Biochemistry*, **9**, 25 (1970).
(75) A. Tsugita and S. Akabori, *J. Biochem.* (Tokyo), **46**, 695 (1959).
(76) H. Yamaguchi, T. Ikenaka, and Y. Matsushima, *J. Biochem.* (Tokyo), **65**, 793 (1969).
(76a) H. Yamaguchi, T. Ikenaka, and Y. Matsushima, *J. Biochem.* (Tokyo), **70**, 587 (1970).
(77) E. H. Eylar, *J. Theor. Biol.*, **10**, 89 (1965).

hydrate in *alpha*-amylase is linked to the L-serine residues of the protein, but later reports have claimed[5,76] that the amylase contains N-glycosyl bonds. Possibly, this enzyme may be one containing both N-glycosyl and O-glycosyl bonds; glycoproteins are known in which both types of linkage occur.[78] The presence of O-glycosyl bonds in fungal glucoamylase has been established,[68,69] and both L-serine and L-threonine residues are known to be the points of attachment of the carbohydrate to the protein. However, the amino acid sequence in the region of the carbohydrate–protein linkage has not been determined. When such information becomes available for glycoenzymes having O-glycosyl bonds, it will be interesting to note whether or not a periodicity also occurs for this class of glycoenzymes in the structures of the glycopeptide at the carbohydrate–protein linkage. As already indicated, periodicity in the amino acid sequence at the carbohydrate–protein linkage may be related to the control mechanisms for the biosynthesis of glycoenzymes.

4. Carbohydrate Component

The carbohydrate compositions and the nature of the monosaccharide residues in typical glycoenzymes have been presented in Table I (see p. 306). It will be noted from this Table that the content of carbohydrate in these glycoenzymes ranges from 2 to 50%. The commonest monosaccharide residues in glycoenzymes are D-mannose and 2-acetamido-2-deoxy-D-glucose. D-Glucose and D-galactose residues are next in abundance, and L-fucose, D-xylose, L-arabinose, and sialic acid occur as structural elements in a number of glycoenzymes. There does not seem to be a direct correlation between the content of carbohydrate and the types of carbohydrate moieties present in the glycoenzymes, nor does there appear to be any correlation between the types of carbohydrate present and the biological origin of the enzyme.

As the molecular weight and carbohydrate content of many glycoenzymes are known, the number of different carbohydrate residues in the enzyme can be calculated. Data from such calculations for a few representative glycoenzymes are presented in Table V. It may be noted that there is a wide variation in the number of sugar residues present in glycoenzymes, varying from 8 in ribonuclease B to 800 in invertase. Other glycoenzymes having many carbohydrate residues include D-glucose oxidase and one of the isoenzymes of glucoamylase.

Although complete details of the molecular structure of the carbo-

(78) G. Dawson and J. R. Clamp, *Biochem. Biophys. Res. Commun.*, **26**, 349 (1967).

TABLE IV
Amino Acid Sequences of Glycopeptides from Some Glycoenzymes

Source	Amino acid sequence	References
Bovine ribonuclease B	—Lys—Ser—Arg—Asn—Leu—Thr—Lys— GlcNAc	44
Porcine ribonuclease B	—Ser—Arg—Arg—Asn—Met—Thr—Gln— GlcNAc	9
	—Ser—Ser—Ser—Asn—Ser—Ser—Asn— GlcNAc	9
	—Tyr—Gln—Ser—Asn—Ser—Thr—Met— GlcNAc	9
Bovine deoxyribonuclease A	—Ser—Asn—Ala—Thr—	10
Pineapple bromelain	Xyl | —Ser—	73
	—Ala—Arg—Val—Pro—Arg—Asn—Asn—Glu—Ser—Ser—Met— GlcNAc	74

Fungal *alpha*-amylase	Xyl | —Ser—Glu—Asp—Gly—	75
	GlcNAc | —Ser—Asn—	76
Fungal glucoamylase	Man | —Ser—	68,69
	Man | —Thr—	68,69
Fungal D-glucose oxidase	GlcNAc | (Asx,Asx,Glx,Ser,Thr)[a]	19

[a] Asx = aspartic acid or its amide; Glx = glutamic acid or its amide.

TABLE V

Number and Types of Carbohydrate Residues for Some Glycoenzymes

Enzyme	GlcNAc	Man	Glc	Gal	Other	References
Glucoamylase I	–	69	16	2	–	7
Glucoamylase II	–	128	20	3	–	7
Glucoamylase R	20	67	–	–	–	7
Invertase	40	760	–	–	–	63
β-D-Glucosaminidase	8	38	–	–	–	39
Ribonuclease B (bovine)	2	6	–	–	–	46,79
Deoxyribonuclease A	2	6	–	–	–	11
Deoxyribonuclease B	3	5	–	1	1^a	11
Deoxyribonuclease C	2	5	–	–	–	11
Protease b (snake venom)	34	10	–	30	9^a	43
Glucose oxidase	19	128	–	3	–	19
Chloroperoxidase	5	–	–	–	68^b	21

^{a} Sialic acid. ^{b} Ara.

hydrate moieties of glycoenzymes have not yet been provided, some information is available on the general types of structures present. These structural types are shown diagrammatically in Fig. 6. It should be noted that 2-acetamido-2-deoxy-D-glucose or D-mannose residues occur at the reducing end of those carbohydrate moieties that have been characterized. The possible occurrence of D-xylose at the reducing end of the chain has been questioned.[76] For purposes of iden-

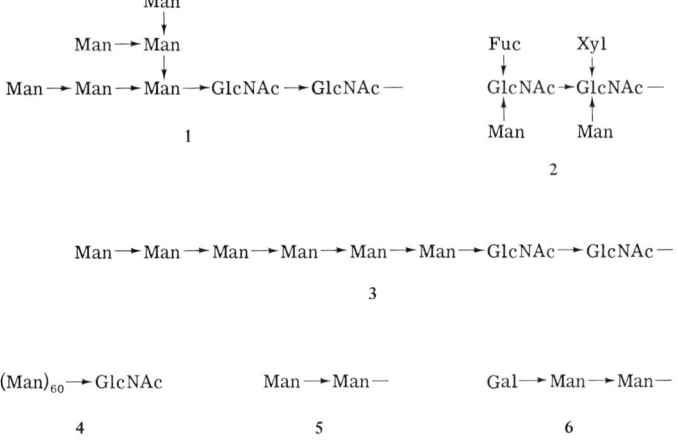

FIG. 6.—Diagrammatic Representation of the Structure of the Carbohydrate Moieties in Glycoenzymes.

tification, these reducing-terminal residues will be referred to as the bridge-carbohydrate residues. It should also be noted that oligosaccharide fragments of low molecular weight are more common than those of high molecular weight. Fragments of the latter type have been found only in invertase, but also may be present in other glycoenzymes that are, as yet, not fully characterized.

Formulas 1 and 2 depict highly branched structures and are very commonly found as components in mammalian glycoproteins, as well as being present in fungal amylase[76a] and pineapple bromelain.[5] The structures of the carbohydrate moieties from these glycoenzymes have been deduced from periodate-oxidation data, from rates of appearance of hydrolytic fragments on controlled hydrolysis by acid, and from the types of O-methylated fragments obtainable from the fully methylated glycoenzymes. The nature of the linkages between monosaccharide residues has not been elucidated, but the linkages are most probably those typical of glycans in general. It is probable that $(1 \rightarrow 3)$, $(1 \rightarrow 4)$, and $(1 \rightarrow 6)$ linkages preponderate, but $(1 \rightarrow 2)$ linkages may also be present. In a few glycoenzymes, the terminal D-mannosyl residues can be removed by an α-D-mannosidase, indicating that these terminal D-mannosyl residues are α-D-linked. The configurations of the other linkages have not yet been elucidated.

Most glycoenzymes contain two or more separate oligosaccharide fragments in their structure. The suggestion of such a multiplicity of carbohydrate moieties is substantiated by the isolation of different types of glycopeptides from hydrolyzates of glycoenzymes.[5,9,11] Variations also exist in the structures of the carbohydrate portion of the molecules.[11] Accordingly, the elucidation of the complete structure of a glycoenzyme will require the use of special techniques.

In some glycoenzymes, such as bovine ribonuclease B, the carbohydrate residues occur as a single chain attached to one amino acid residue of the polypeptide chain; in this particular case, attachment is to the asparagine-34 residue. The carbohydrate moiety of ribonuclease B is an octasaccharide (3, see Fig. 6) composed of six residues of D-mannose and two of 2-acetamido-2-deoxy-D-glucose.[79] The linkage between 2-amino-2-deoxy-D-glucose residues is shown to be β-D-$(1 \rightarrow 4)$ by the isolation of di-N-acetylchitobiose–asparagine as a degradation product of the glycoenzyme. The identity of this degradation product was established by direct comparison with the synthetic compound.[80] The linkage between D-mannose residues of the

(79) A. Tarentino, T. H. Plummer, Jr., and F. Maley, *J. Biol. Chem.*, **245**, 4150 (1970).
(80) M. Spinola and R. W. Jeanloz, *J. Biol. Chem.*, **245**, 4158 (1970).

octasaccharide is probably (1 → 3) on the basis of stereochemical considerations. In view of the observation that five of the D-mannose residues can be removed from the octasaccharide by α-D-mannosidase, the linkages between the D-mannose residues are presumed to have the α-D configuration;[79] the configuration of the linkage between the 2-acetamido-2-deoxy-D-glucose residues is β-D.[79,80] Evidence has been educed for a β-D linkage between D-mannose residues and 2-acetamido-2-deoxy-D-glucose residues.[80a]

Ribonuclease from other mammalian sources may have several carbohydrate fragments in their structures. Thus, ribonuclease B from porcine pancreas contains at least three carbohydrate moieties, and these are attached[9] at asparagine residues 21, 34, and 76. The carbohydrate moieties present at residues 21 and 76 are considerably more complex than that at residue 34, and their structures are under investigation in several laboratories. The carbohydrate portions of the isoenzymes of deoxyribonuclease are similar in structure to those from the ribonucleases.[11] As ribonuclease and deoxyribonuclease originate in the same organ, it is possible that the same pathways and enzymes are utilized for the biosynthesis of the carbohydrate moieties of both enzymes.

Such glycoenzymes as invertase, D-glucose oxidase, and chloroperoxidase contain many carbohydrate fragments. The degree of polymerization of these carbohydrate fragments has not been completely elucidated, but it appears that invertase contains carbohydrate of high molecular weight, whose structure is depicted diagrammatically by formula 4 in Fig. 6. However, the nature of the linkages between the monosaccharide residues, and the molecular architecture of the polymer, are not known.

Two other structural types proposed for the carbohydrate moieties of glycoenzymes are illustrated by formulas 5 and 6 in Fig. 6; these types are di- and tri-saccharides of neutral monosaccharides. The carbohydrate component of glucoamylase contains oligosaccharide fragments of this type. In view of the high carbohydrate content of the glucoamylase, many such carbohydrate fragments must be present in its molecule. Approximately 50% of the carbohydrate residues of glucoamylase resist oxidation by periodate,[1] indicating a (1 → 3) linkage between the monosaccharide residues of the oligosaccharide chains. The D-mannosyl residues in glucoamylase can be partially removed by α-D-mannosidase, indicating that the configuration of the linkages joining the nonreducing, terminal, D-mannose residues

(80a) T. Sukeno, A. L. Tarentino, T. H. Plummer, Jr., and F. Maley, *Biochem. Biophys. Res. Commun.*, **45**, 219 (1971).

to the rest of the molecule is α-D. Terminal D-galactose residues have been detected in the glucoamylase by use of the D-galactose oxidase reaction.[81] A molecular architecture such as that found in glucoamylase may be of fundamental importance in maintaining the tridimensional structure of glycoenzymes, and this suggestion is considered in the next Section.

5. Conformational Structure

The total structure of one isoenzyme of a glycoenzyme, namely, ribonuclease A, has been deduced on the basis of X-ray crystallographic[82-84] and amino acid sequence-data.[85,86] As discussed in an earlier Section (see p. 307), ribonuclease A and ribonuclease B possess the same amino acid composition, the same amino acid sequence, and the same catalytic activity. The two isoenzymes differ in that ribonuclease B contains a carbohydrate moiety at asparagine residue number 34. In view of the fact that the peptide segment in the vicinity of residue 34 can be digested by trypsin and chymotrypsin more readily if the ribonuclease B is heated to a temperature that permits initial unfolding of the molecule,[87,88] it is presumed that this region of the polypeptide chain is situated near the surface of the enzyme molecule. Attachment of a carbohydrate fragment, with its many polar groups, in this region should not disturb the tertiary structure of the protein molecule. Accordingly, it can be assumed that the conformational structure of the protein portion of ribonuclease B is identical with that of ribonuclease A, and the influence of the carbohydrate fragment in ribnuclease B on the conformational structure of the protein part of the molecule is presumed to be minimal.

The molecular architecture of glucoamylase permits some interesting hypotheses to be made in relation to the tridimensional structure of enzymes. As indicated in the preceding Section, this enzyme possesses many short, oligosaccharide moieties attached along its polypeptide chain. It is possible that the spacing of the fragments

(81) W. Karakawa, J. E. Wagner, and J. H. Pazur, *J. Immunol.*, **107**, 554 (1971).
(82) G. Kartha, J. Bello, and D. Harker, *Nature*, **213**, 862 (1967).
(83) H. W. Wyckoff, K. Hardman, N. Allewell, T. Inagami, L. N. Johnson, and F. Richards, *J. Biol. Chem.*, **242**, 3984 (1967).
(84) H. W. Wyckoff, D. Tsernoglou, A. W. Hanson, J. R. Knox, B. Lee, and F. M. Richards, *J. Biol. Chem.*, **245**, 305 (1970).
(85) C. B. Anfinsen, R. D. Redfield, W. L. Choate, J. Page, and W. R. Carroll, *J. Biol. Chem.*, **207**, 201 (1954).
(86) D. G. Smyth, W. H. Stein, and S. Moore, *J. Biol. Chem.*, **238**, 227 (1963).
(87) J. A. Rupley and H. A. Scheraga, *Biochemistry*, **2**, 421 (1963).
(88) T. Ooi, J. A. Rupley, and H. A. Scheraga, *Biochemistry*, **2**, 432 (1963).

along the chain helps to hold the polypeptide folds of the protein in rigid positions. Such carbohydrate fragments may function, therefore, as stabilizers of the tridimensional structure of the enzyme. In support of this suggestion is the observation that glycoenzymes having this type of structure are extremely stable to denaturation on storage,[68] and that isoenzymes of invertase become denatured at rates that decrease as their carbohydrate contents increase.[89] Isoenzymes having a high content of carbohydrate are quite resistant to denaturation, whereas isoenzymes low in carbohydrate are rapidly denatured.

IV. Biosynthetic Pathways

1. Cellular Locale, Reactions, and Mechanisms

Structurally, glycoenzymes are similar to other glycoproteins, and therefore the cellular locale, reactions, and mechanisms of synthesis of the two types of macromolecule are undoubtedly very similar. In mammalian systems, studies with the perfused liver[90] and with isotopically labeled carbohydrates[91] have shown that the liver is the major site where glycoproteins are synthesized. Other mammalian organs and tissues, for example, the pancreas,[92] submaxillary gland,[93] thyroid,[94] retina,[95] kidney,[96] lymph cell,[97] and mammary gland,[98] also effect glycoprotein synthesis. The site of synthesis of glycoproteins in plants and micro-organisms is not yet known, but it is probable that there are involved the same general types of reactions and enzymes as are utilized in animal tissues.[68,99]

For purposes of discussion, the biosynthetic pathway for a glycoprotein will be divided into three phases: *phase 1*, the assembly of the polypeptide chain; *phase 2*, the attachment of the bridge carbohydrate to specific amino acid residues of the polypeptide chain; and *phase 3*, the addition of monosaccharides to the bridge sugar-residues. Studies in the past several years have led to a generalized concept on

(89) W. N. Arnold, *Biochim. Biophys. Acta*, **178**, 347 (1969).
(90) E. Sarcione, *Arch. Biochem. Biophys.*, **100**, 516 (1963).
(91) G. B. Robinson, J. Molnar, and R. J. Winzler, *J. Biol. Chem.*, **239**, 1134 (1964).
(92) J. D. Jamieson and G. E. Palade, *J. Cell Biol.*, **34**, 577 (1967).
(93) E. J. McGuire and S. Roseman, *J. Biol. Chem.*, **242**, 3745 (1967).
(94) M. J. Spiro and R. G. Spiro, *J. Biol. Chem.*, **243**, 6520 (1968).
(95) P. J. O'Brien and C. G. Muellenberg, *Biochim. Biophys. Acta*, **158**, 189 (1968).
(96) G. C. Priestly, M. L. Pruyn, and R. A. Malt, *Biochim. Biophys. Acta*, **190**, 154 (1969).
(97) R. M. Swenson and M. Kern, *Proc. Nat. Acad. Sci. U.S.*, **59**, 546 (1968).
(98) E. J. McGuire, G. W. Jordian, D. M. Carlson, and S. Roseman, *J. Biol. Chem.*, **240**, 4112 (1965).
(99) D. T. A. Lamport, *Ann. Rev. Plant Physiol.*, **21**, 235 (1970).

the interrelations of the three phases, as well as identification of the cellular locale of synthesis of these macromolecules.

The reactions of *phase 1* are the well established processes of template-directed biosynthesis of protein involving the various types of deoxyribonucleic and ribonucleic acids, the chain-initiating and the chain-terminating factors, and the appropriate enzymes for activating the amino acids and assembling them into a polypeptide chain. These processes are multi-type reactions proceeding in an integrated and coordinated manner. The reactions of this phase are excellently presented in a review.[100]

The reactions of *phase 2* relate to the attachment of the bridge-carbohydrate residues to the polypeptide chain. There is evidence showing that this addition occurs while the polypeptide chain is still attached to, or perhaps still being synthesized on, the ribosomes.[101-103] Thus, ^{14}C-labeled 2-amino-2-deoxy-D-glucose, injected into the circulatory system of the rat, was incorporated into protein in the ribosomes of the rough endoplasmic-reticulum of the liver. Administration of puromycin caused release of the ^{14}C-labeled glycoprotein, which could be isolated by acid-precipitation methods. Examination of the radioactivity data revealed that the subcellular structures most actively involved in glycoprotein synthesis were the ribosomes bound to the membrane, and not free polysomes.

As 2-acetamido-2-deoxy-D-glucose is one of the bridge-carbohydrate residues of glycoenzymes, there must be present, in the tissues, enzymes capable of activating and attaching these residues to the polypeptide chain. Nucleotidyl transferases are responsible for the activation of the hexosamine by way of uridine 5′-(2-acetamido-2-deoxy-D-glucopyranosyl pyrophosphate),[104] and glycosyl transferases[105,106] are responsible for attachment of the residues to the polypeptide chain. The latter transferases are particle-bound enzymes and have not yet been obtained in highly purified form. For glycoenzymes containing D-mannose and D-xylose as bridge carbohydrate, the appropriate enzymes for formation of guanosine 5′-(D-mannopyranosyl

(100) J. Lucas-Lenard and F. Lipmann, *Ann. Rev. Biochem.*, **40**, 409 (1971).
(101) G. R. Lawford and H. Schachter, *J. Biol. Chem.*, **241**, 5408 (1966).
(102) T. Hallinan, C. N. Murty, and J. H. Grant, *Arch. Biochem. Biophys.*, **125**, 715 (1968).
(103) J. Molnar, G. B. Robinson, and R. J. Winzler, *J. Biol. Chem.*, **240**, 1882 (1965).
(104) J. L. Strominger and M. S. Smith, *J. Biol. Chem.*, **234**, 1822 (1959).
(105) S. Bouchilloux, O. Chabaud, M. Michel-Bechet, M. Ferrand, and A. M. Athovel-Haon, *Biochem. Biophys. Res. Commun.*, **40**, 314 (1970).
(106) D. Morre, L. M. Merlin, and T. W. Keenan, *Biochem. Biophys. Res. Commun.*, **37**, 813 (1969).

pyrophosphate) and uridine 5′-(D-xylopyranosyl pyrophosphate), and for the transfer of the carbohydrate residue, are known.[1,105,107]

In *phase 3* of the process, the biosynthesis of the glycoprotein is completed by the stepwise addition of carbohydrate residues from nucleoside 5′-(glycosyl pyrophosphates) to the partially glycosylated polypeptide by the mediation of appropriate transferases.[107a] The addition occurs to some extent in the region of the smooth endoplasmic-reticulum,[105,106] but takes place for the most part in the Golgi particles.[108,109] The completed glycoprotein or glycoenzyme accumulates in the Golgi particles prior to secretion into the circulatory system.[110,111] As already indicated, the addition of the carbohydrate residues proceeds in a stepwise manner,[112] with each carbohydrate residue arising from an appropriate nucleoside 5′-(glycosyl pyrophosphate) and being transferred by a specific transferase to the growing chain. A Golgi-particle-rich fraction isolated from rat liver contained high levels of 2-acetamido-2-deoxy-D-glucosyl-, D-galactosyl-, and sialyl-transferase activities.[108,113,114] It should be mentioned that the 2-acetamido-2-deoxy-D-glucosyl transferase present in these particles differs from that in the rough endoplasmic-reticulum.[105] The transferases in the Golgi-particle-rich fraction were capable of sequentially transferring 2-acetamido-2-deoxy-D-glucose, D-galactose, and sialic acid from the appropriate nucleoside 5′-(glycosyl pyrophosphates) to orosomucoid devoid of these three sugars but containing carbohydrate moieties that functioned as acceptors. A diagram[108] that adequately summarizes the general events in the synthesis of a glycoprotein is shown in Fig. 7; the scheme is equally applicable for illustrating the synthesis of glycoenzymes.

An important aspect of the metabolism and synthesis of glycoproteins and glycoenzymes is the secretion of these molecules through the cellular membrane. It is possible that the final sugar residue becomes attached to the molecule at the plasma membrane, during passage of the molecule through the membrane. For example, in the

(107) E. E. Grebner, C. W. Hall, and E. F. Neufeld, *Arch. Biochem. Biophys.*, **116**, 391 (1966).
(107a) H. Nikaido and W. Z. Hassid, *Advan. Carbohyd. Chem. Biochem.*, **26**, 351 (1971).
(108) H. Schachter, I. Jabbal, R. L. Hudgin, L. Pinteric, E. J. McGuire, and S. Roseman, *J. Biol. Chem.*, **245**, 1090 (1970).
(109) R. R. Wagner and M. A. Cynkin, *J. Biol. Chem.*, **246**, 143 (1971).
(110) J. D. Jamieson and G. E. Palade, *J. Cell Biol.*, **39**, 580 (1968).
(111) J. D. Jamieson and G. E. Palade, *J. Cell Biol.*, **39**, 589 (1968).
(112) R. G. Spiro and M. J. Spiro, *J. Biol. Chem.*, **241**, 1271 (1966).
(113) S. Roseman, *Chem. Phys. Lipids*, **5**, 270 (1970).
(114) W. E. Pricer and G. Ashwell, *J. Biol. Chem.*, **246**, 4825 (1971).

GLYCOENZYMES: ENZYMES OF GLYCOPROTEIN STRUCTURE 331

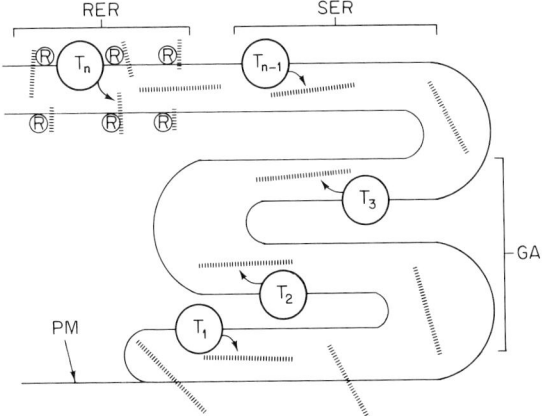

FIG. 7.—Diagrammatic Representation for the Mechanism of Biosynthesis of Glycoproteins and Glycoenzymes. (RER, rough endoplasmic-reticulum; SER, smooth endoplasmic-reticulum; GA, Golgi apparatus; R, ribosome; T_1, T_2,—T_n, transferases concerned with the incorporation of glycose residues into the molecule; PM, plasma membrane; and ||||||, glycoprotein or glycoenzyme at various stages of biosynthesis.[108])

synthesis of γ-globulin by lymph-node cells, it has been shown that the terminal N-acetylneuraminic acid residue is added to the molecules, both inside and outside the cell, at essentially the same rate.[97] This observation can be interpreted as indicating that the sialic acid is added to the glycoprotein at the plasma membrane during passage of the macromolecule through the membrane. Subsequent information indicates that a sialyl transferase is indeed attached to rat-liver plasma-membranes,[114] although, as discussed later, this enzyme might be involved in cell adhesion.[113,115] Similarly, the D-glucosyl transferase involved in attaching terminal D-glucosyl residues in the biosynthesis of collagen was found exclusively on the plasma membrane.[116]

There may be two different mechanisms whereby completed glycoprotein molecules are secreted from cells. In the first, these molecules accumulate in the Golgi membranes and are secreted as packets of molecules by way of Golgi vesicles, a mechanism that is essentially reverse pinocytosis, the process by which extracellular molecules are engulfed by the cell surface in order that these molecules shall end up within a membrane-bound vesicle in the cell cytoplasm. Pancreatic ribonuclease B is an example of a glycoenzyme undergoing this process during the secretion of zymogen granules.[92] In the second

(115) H. B. Bosmann, *Biochem. Biophys. Res. Commun.*, **45**, 1118 (1971).
(116) A. Hagopian, H. B. Bosmann, and E. H. Eylar, *Arch. Biochem. Biophys.*, **128**, 387 (1968).

mechanism, the macromolecules are transported through channels in a continuum of membranes, passing sequentially from rough endoplasmic-reticulum, to smooth endoplasmic-reticulum, to Golgi bodies, and to the plasma membrane. There would occur a successive addition of carbohydrate residues to the molecule during this movement, with terminal sialic acid being added at the last membrane before the glycoprotein is released into the circulatory system.[108] It is noteworthy that glycoproteins from which the terminal residues of sialic acid had been removed enzymically left the circulatory system and entered the hepatocytes immediately after having been injected into a rat, again indicating that the sialic acid moiety acts as a signal for the molecules to remain in the extracellular environment.[117]

Some observations on the synthesis of glycoproteins are not in accord with the depiction in Fig. 7. First of all, mitochondria from liver and brain tissue have been reported capable of synthesizing glycoproteins.[118,119] Secondly, glycopeptides that may be precursors of glycoproteins have been isolated from tissues where glycoproteins are being synthesized,[120] and it has been suggested that glycoproteins may be synthesized in the Golgi particles from these glycopeptides.[121] Such a pathway would require the existence of a "joining" type of enzyme that builds up the glycoprotein from glycopeptides. Finally, some reports have appeared that implicate lipid–carbohydrate intermediates,[122,123] of the types involved in synthesis of cell-wall glycans,[124,125,125a] in the biosynthetic pathways for glycoproteins. In the latter connection, the effect of phospholipids on the glycosyl transferases may also be important.[126,127] The significance of these observations needs to be evaluated further.

(117) A. G. Morell, G. Gregoriadis, I. H. Scheinberg, S. Hickman, and G. Ashwell, *J. Biol. Chem.*, **246**, 1461 (1971).
(118) H. B. Bosmann and S. S. Martin, *Science*, **164**, 190 (1969).
(119) H. B. Bosmann and B. A. Hemsworth, *J. Biol. Chem.*, **245**, 363 (1970).
(120) J. Moschera, E. Mound, N. Payza, W. Pigman, and M. Weiss, *FEBS Lett.*, **6**, 326 (1970).
(121) A. Herp, M. Liska, N. Payza, W. Pigman, and J. Vittek, *FEBS Lett.*, **6**, 321 (1970).
(122) J. F. Caccam, J. J. Jackson, and E. H. Eylar, *Biochem. Biophys. Res. Commun.*, **35**, 505 (1969).
(123) M. Tetas, H. Chao, and J. Molnar, *Arch. Biochem. Biophys.*, **138**, 135 (1970).
(124) W. J. Lennarz and B. Talamo, *J. Biol. Chem.*, **241**, 2707 (1966).
(125) Y. Higashi, J. L. Strominger, and C. C. Sweeley, *Proc. Nat. Acad. Sci. U.S.*, **57**, 1878 (1967).
(125a) F. Shafizadeh and G. D. McGinnis, *Advan. Carbohyd. Chem. Biochem.*, **26**, 297 (1971).
(126) D. A. Vessey and D. Zakim, *J. Biol. Chem.*, **246**, 4649 (1971).
(127) A. Hagopian and E. H. Eylar, *Arch. Biochem. Biophys.*, **129**, 447 (1969).

2. Regulation of Synthesis

Regulation of the synthesis of glycoproteins and glycoenzymes occurs at all three phases of the reaction pathway. Biosynthesis of the polypeptide chain (*phase 1*) is under the primary control of the transcription and translation of the genetic code, as effected by the appropriate nucleic acids. As reviews[128,129] are available on the subject, only one aspect, pertaining to biosynthesis of glycoproteins, will be mentioned. Based on some of the known codons for L-amino acids, it has been speculated that there exists an evolutionary relationship between the various types of glycopeptide linkage.[130] Examination of the data in Table VI indicates that codons for the amino acids involved in the carbohydrate–protein linkage in glycoenzymes could have arisen by single-point mutations from the codon for arginine. Thus, serine, threonine, and lysine (the precursor of hydroxylysine) could be introduced into the growing chain in the region of the sequence where carbohydrate residues ultimately become attached. Glutamine and hydroxyproline are not common as a bridge for carbohydrate residues. It may be noted from Table VI that two-point mutations would be required in order that the codons for these amino acids could be obtained from the codon for asparagine.

The control for *phase 2* of the biosynthesis of glycoenzymes involves regulation by (*a*) the amino acid sequence in the vicinity of the particular peptide fragment that forms the bridge to the oligosaccharide chains, (*b*) the activities of the enzymes involved in the synthesis of the linkages, and (*c*) the concentrations of the precursors. As indicated in an earlier Section (see p. 322), the amino acid sequence at the site of attachment of the polysaccharide in those glycoenzymes having a glycosylamine bridge is of the general type:

$$\begin{array}{c} \text{GlcNAc} \\ | \\ \text{—A—Asn—B—Ser(or Thr)—C.} \end{array}$$

The following aspects of this structural component are important in relation to control and recognition mechanisms: first, a hydroxy-amino acid is located on the carboxyl side of the asparagine residue that forms the site of attachment; second, there is a "spacer" residue (B) in the polypeptide chain; and third, the polarity of residue B appears to control the type of oligosaccharide side-chain that becomes attached to asparagine. If residue B is polar, a complex type of poly-

(128) W. Epstein and J. R. Beckwith, *Ann. Rev. Biochem.*, **37**, 411 (1968).
(129) S. R. Gross, *Ann. Rev. Genetics*, **3**, 395 (1969).
(130) M. Jett and G. A. Jamieson, *Carbohyd. Res.*, **18**, 466 (1971).

TABLE VI

Codons for L-Amino Acids in the Glycosidic Linkages of Glycoproteins

L-Amino acid	Codon
Asn	AAU(C)
Ser	AGU(C)
Thr	ACU(C,A,G)
Lys	AAA(G)
Gln	CAA(G)
Pro	CCU(C,A,G)

saccharide side-chain is added to the protein; whereas, if residue B is apolar, a carbohydrate chain of lesser complexity is elaborated at this site.[9] The functional groups of the amino acids in this region of the polypeptide chain are important, therefore, in establishment of the enzyme–substrate complex leading to formation of the protein–carbohydrate linkage.

For those glycoproteins having the O-glycosyl attachment of carbohydrate, two interesting observations are worthy of mention. First, in IgG immunoglobin, the 2-acetamido-2-deoxy-D-galactose bridge-residue is attached to L-threonine, and the glycopeptide in this section of the molecule possesses the following sequence[131]:

GalNAc
|
—Ser—Lys—Pro—Thr—Cys—Pro—Pro—Pro—.

Second, a myelin protein (the A-1 protein) accepts 2-acetamido-2-deoxy-D-galactose from uridine 5′-(2-acetamido-2-deoxy-D-galactopyranosyl pyrophosphate) when treated with a 2-acetamido-2-deoxy-D-galactose transferase from bovine submaxillary gland, even though the protein normally occurs devoid of carbohydrate.[132] The L-threonine residue number 98 is the site of attachment for the sugar, and the sequence at this region of the peptide is:

GalNAc
|
—Thr—Pro—Arg—Thr—Pro—Pro—Pro—.

It is possible that, in this type of glycoprotein, the amino acid sequence of the peptide (specifically, the unusual tripeptide of proline) is involved in signaling the formation of the O-glycosyl type of bridge linkage between the carbohydrate and the protein. Further examples

(131) D. S. Smyth and S. Utsumi, *Nature*, **216**, 332 (1967).
(132) A. Hagopian, F. C. Westall, J. S. Whitehead, and E. H. Eylar, *J. Biol. Chem.*, **246**, 2519 (1971).

of this structure will be required before attributing a general role for a tripeptide of proline in the control of this phase of glycoprotein synthesis.

Direct control of biosynthesis at the genetic level could exist for phase 2 if glycosylated amino acids indeed provide the means of introducing the bridge-carbohydrate residues into the peptide chain. If such reactions do occur, there would need to exist, for the glycosylated amino acids, appropriate transfer-ribonucleic acids, activating enzymes, and other components of protein synthesis. A triplet code would need to exist in the ribonucleic acid that would carry the message for addition of the glycosylated amino acid residue to the growing polypeptide chain of glycoenzymes. Although investigators have searched for involvement of such transfer-ribonucleic acids, evidence for their existence has not been forthcoming.[133]

An interesting correlation between tripeptide sequences and genetic mutations has been made by compilation of the number of occurrences of the tripeptide sequence Asn—B—Ser and Asn—B—Thr in protein sequence-data at present available from the literature.[134] A total of 18,251 tripeptide sequences from 264 proteins was grouped by computer into the 400 possible tripeptide combinations of the 20 amino acids in which the second position was ignored. The total number of Asn—B—(Ser/Thr) sequences actually present in these proteins was 61, whereas the total number theoretically expected, on the basis of a random distribution within the sequence, would have been 102. The observed data were thus lower than expected by 4 standard deviations. On the other hand, such chemically similar tripeptides as Gln—B—(Ser/Thr), Asp—B—(Ser/Thr), and (Ser/Thr)—B—Asn occurred the number of times expected for random distribution. It was suggested that this low frequency of incidence of the Asn—B—(Ser/Thr) sequence results from a restriction of its occurrence in proteins by a process of natural selection. Any protein that acquired this tripeptide as a result of a mutation would be soon rejected, because carbohydrate would be bound to the asparagine residue and its presence would interfere with the normal metabolic function of the protein.

Control of *phase 3* of the synthesis is influenced by the specificities of the transferases, by the concentrations of substrates and acceptor molecules, and by allosteric effects and feedback-control mechanisms. In order that synthesis of the oligosaccharide chains shall be reproducible, each glycosyl transferase must be present at specific sites and at concentration levels sufficient to effect the addition of the

(133) H. Sinohara and H. H. Sky-Peck, *Biochim. Biophys. Acta*, **101**, 90 (1965).
(134) L. T. Hunt and M. O. Dayhoff, *Biochem. Biophys. Res. Commun.*, **39**, 757 (1970).

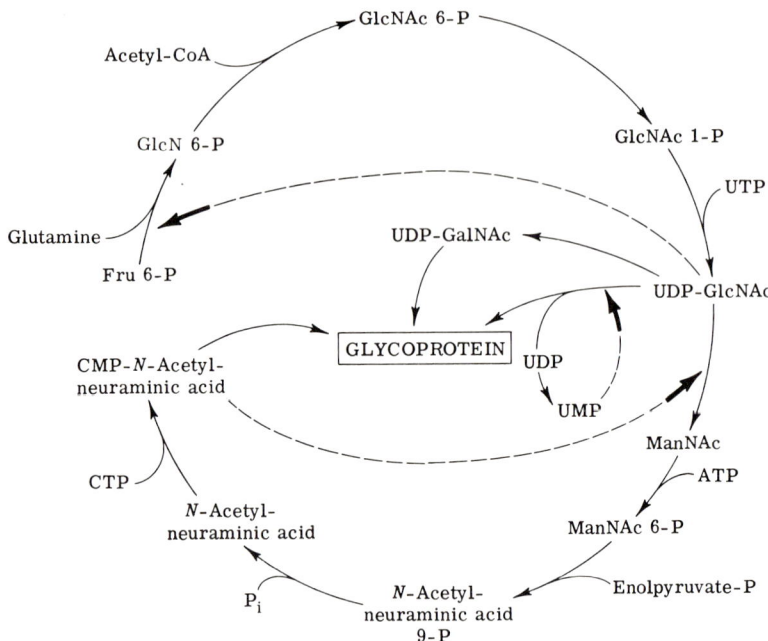

Fig. 8.—Pathway of Biosynthesis of Some "Sugar Nucleotides" That Function as Glycosyl Donors for Biosynthesis of Glycoproteins and Glycoenzymes. (The heavy arrows indicate the sites of feedback control of the reactions.[135])

proper number of carbohydrate residues to the growing chain. The precursors for the synthesis of the chain must be present at appropriate concentration levels. Control of the latter is effected by allosteric mechanisms that have been observed primarily with the amino sugar-containing precursors.

The metabolic interrelations of the amino sugars of glycoproteins are shown[135] diagrammatically in Fig. 8. The pathway leading to the synthesis of these sugars is controlled at the initial reaction-step, in which 2-amino-2-deoxy-D-glucose 6-phosphate is formed from D-fructose 6-phosphate and L-glutamine. This reaction is catalyzed by the enzyme L-glutamine:D-fructose 6-phosphate aminotransferase, which appears to be subject to allosteric control by several different cellular metabolites, including uridine 5'-(2-acetamido-2-deoxy-D-glucopyranosyl pyrophosphate).[135,136] An increase in the level of this nucleoside 5'-(glycosyl pyrophosphate) would consequently decrease the activity

(135) S. Kornfeld, R. Kornfeld, E. F. Neufeld, and P. J. O'Brien, *Proc. Nat. Acad. Sci. U.S.*, **52**, 371 (1964).
(136) P. J. Winterburn and C. F. Phelps, *Biochem. J.*, **121**, 711 (1971).

of the initial enzyme of the sequence, thus decreasing the concentration level of the amino sugar donors. Accordingly, the synthesis of the oligosaccharide side-chains would be completely blocked by this feedback control of a single enzyme. Two other feedback-control mechanisms exist at later stages of the metabolic cycle. Cytidine 5'-(N-acetylneuraminyl phosphate) inhibits the epimerase responsible for the synthesis of 2-acetamido-2-deoxy-D-mannose,[135,137] and uridine 5'-phosphate inhibits the 2-acetamido-2-deoxy-D-glucosyl transferase.[109]

The mechanism whereby the chain-terminating step for oligosaccharide formation is regulated is not well understood. It has been proposed that certain structural features, such as the presence of sialyl or fucosyl residues, and an uncommon anomeric configuration,[138,139] act as signals to halt the synthesis of the oligosaccharide chain. Additional studies are needed in order to clarify the role of these structural features in the chain-terminating process.

The control of phase 3 of the biosynthetic process is not so precise as the genetic system controlling the biosynthesis of the protein part of the molecule. This factor is undoubtedly one of the prime reasons for the heterogeneity observed in the carbohydrate component of glycoenzymes and glycoproteins.[139,140] As the biosynthesis of the peptide portion of glycoenzymes and glycoproteins is under the normal genetic control that regulates protein synthesis, the synthesis of the entire molecule must ultimately be controlled by this system. On the basis of the finding that cytidine 5'-(N-acetylneuraminyl phosphate) synthetase[141] occurs in the cell nucleus, it is conceivable that the "nucleotide sugar" could be involved in effecting control at the gene level, by acting as a co-repressor substance in the nucleus.

V. BIOLOGICAL AND STRUCTURAL SIGNIFICANCE

Much attention has been given to the question of functions for the carbohydrate moieties in glycoproteins and glycoenzymes. The functions that have been proposed fall into two categories, biological and structural. A biological function that has been widely publicized is a role for the carbohydrate residues in the transport of glycoenzymes and glycoproteins through cellular membranes. This suggestion has been based largely on the fact that many cellular proteins that are

(137) W. L. Salo and H. G. Fletcher, Jr., *Biochemistry*, **9**, 882 (1970).
(138) K. O. Lloyd, E. A. Kabat, and E. Licerio, *Biochemistry*, **7**, 2976 (1968).
(139) A. Gottschalk, *Nature*, **222**, 452 (1969).
(140) R. Spiro, *Ann. Rev. Biochem.*, **39**, 599 (1970).
(141) E. L. Kean, *J. Biol. Chem.*, **245**, 2301 (1970).

secreted from the cell are glycoproteins, whereas those cellular proteins that remain in the cell are not.[77] Most of the glycoenzymes listed in Table I are extracellular enzymes, and this generalization appears to hold. The carbohydrate moieties of these molecules can conceivably participate in the transport process by complexing with receptor and carrier substances in the cellular membranes. In the formation of the complex, a conformational change could occur, allowing for passage of the macromolecule through the cellular membrane.

Such a mechanistic function is not necessarily operative in the transport of all glycoproteins. For example, in the secretion of glycoproteins and glycoenzymes by way of Golgi vesicles undergoing a reversal of the pinocytic process, the oligosaccharide chains may serve merely as a signal identifying those molecules to be excreted. Not only may carbohydrate residues function as signals for molecules to be secreted from cellular particles, but they may also function as signals for molecules to enter a particle. In the latter process, neuraminic acid[113] and D-galactose residues[142,143] appear to be important; removal of sialic acid residues, with neuraminidase, from the carbohydrate chains of ceruloplasmin, yielded a molecule that was immediately taken up by the hepatocytes and degraded by the lysosomal enzymes. Other glycoproteins behave similarly upon treatment with neuraminidase.[113] The "signal" residue is D-galactose, and not sialic acid; removal of the exposed D-galactose residues from modified ceruloplasmin by β-D-galactosidase, or oxidation of the D-galactose with D-galactose oxidase, yielded a molecule that had the same survival time in serum as untreated ceruloplasmin.[144] Further experimentation is necessary on the mechanism whereby the carbohydrate moiety influences the transport of proteins across membranes; it may lead to definitive understanding of this important biological process.

A second biological function of the carbohydrates relates to immunological responses to these moieties.[145] Glycans on bacterial cell-surfaces are important antigens, and valuable information on the structures of these glycans can be obtained by immunological techniques.[146] Glycoproteins are often present on the surface of cell mem-

(142) C. J. A. Vanden Hamer, A. G. Morell, I. H. Scheinberg, J. Hickman, and G. Ashwell, *J. Biol. Chem.*, **245**, 4397 (1970).
(143) G. Gregoriadis, A. G. Morell, I. Sternlieb, and I. H. Scheinberg, *J. Biol. Chem.*, **245**, 5833 (1970).
(144) A. G. Morell, R. A. Irvine, I. Sternlieb, I. H. Scheinberg, and G. Ashwell, *J. Biol. Chem.*, **243**, 155 (1968).
(145) M. I. Horowitz, in "The Carbohydrates: Chemistry and Biochemistry," W. Pigman and D. Horton (Eds.), Vol. II B, Academic Press, New York, 1970, p. 685.
(146) J. H. Pazur, J. S. Anderson, and W. W. Karakawa, *J. Biol. Chem.*, **246**, 1793 (1971).

branes, and such molecules function as antigenic substances. Glycoproteins secreted from the cell also function in this way.[147] Although some immunological studies have been reported for glycoenzymes,[7] more investigations need to be made in this area in order to ascertain whether the carbohydrate plays a significant role in the immunological process. Considerable work has been done with some glycoproteins to determine the relative importance of the peptide and oligosaccharide moieties in the immunological process. A report[148] on the relative significance of the two moieties in ovine submaxillary mucin is informative. In this glycoprotein, approximately 50% of the weight of the molecule is composed of disaccharide residues, which can readily be removed enzymically. Removal of the carbohydrate had only a minor effect on the immunological reactions. Evidently, the principal antibody reacting against the mucin is directed toward the protein component, and not toward the carbohydrate component. Similar observations have been made with a fungal glycoenzyme in which the carbohydrate residues had been oxidized by periodate.[19] Thus, in some glycoproteins and glycoenzymes, the carbohydrate components appear to play only minor roles in the immunological process.

The presence of glycoproteins and certain enzymes on cell surfaces has promoted the suggestion of a third biological function for carbohydrates of glycoproteins, namely, that these residues play a role in the adhesion of the cells to one another.[113] The nonreducing termini of sugar residues on the surface of one cell may be bound by specific, glycosyl transferases present on the surface of the adjacent cell. Thus, the specificity of the enzyme–substrate complex would account for compatibility of tissues. Reports are available on the presence of glycosyl transferases in plasma membrane,[113,115] and such enzymes may well be involved in the process of cellular adhesion.

A role for the carbohydrate moieties in the catalytic process of an enzyme was suggested many years ago.[24] However, studies with several different enzymes[19,149,150] have shown that enzymic activity is affected only slightly if the carbohydrate residues are oxidized with periodate. Furthermore, removal of carbohydrate residues with glycosidases,[1,15] followed by measurements of activity, has shown that the resulting, modified enzymes had specific activities that were essentially unchanged. With ribonuclease B, it has also been con-

(147) W. M. Watkins, *Science*, **152**, 172 (1966).
(148) A. Gottschalk, H. Schauer, and G. Uhlenbruck, Z. *Physiol. Chem.*, **352**, 117 (1971).
(149) T. Lee and L. P. Hager, *Fed. Proc.*, **29**, 599 (1970).
(150) Y. Yasuda, N. Takahashi, and T. Murachi, *Biochemistry*, **10**, 2624 (1971).

cluded that the carbohydrate moiety does not function at the active site of the enzyme, from considerations of the known, tridimensional structure of ribonuclease A and location of the active site, and from the similarity in structures of the two isoenzymes. All available evidence argues against the involvement of carbohydrate residues of glycoenzymes in the catalytic process.

With regard to structural roles for the carbohydrate, it should first be mentioned that the carbohydrate residues exert a significant effect on several of the physical properties of the glycoenzymes. The intrinsic viscosity, frictional ratio, diffusion coefficient, and solubility of the protein may be affected. Changes in these properties have been noted for porcine ribonucleases especially.[9] Such changes in certain physical properties will undoubtedly hold true for other pairs of glycoenzymes exhibiting significant differences in their carbohydrate content.

Glycoproteins and glycoenzymes are resistant to hydrolysis by proteolytic enzymes,[151,152] and they are remarkably stable on storage[68] and at elevated temperatures.[89] In the case of susceptibility to proteolysis, the carbohydrate residues apparently interfere with the formation of the enzyme–substrate complex, and hydrolysis of the molecule cannot occur. Removal of some of the carbohydrate residues from the glycoenzymes often makes the molecule much more susceptible to enzymic hydrolysis.[152] However, the glycoproteins that are degraded in the lysosomes appear to be an exception; the peptide components of certain glycoproteins were degraded at the same rate by lysosomal hydrolases, regardless of whether or not there had been prior removal of the oligosaccharide chains from the molecule.[153]

With regard to the stability of glycoenzymes, the suggestion that the carbohydrate residues function as stabilizers of the tridimensional structure has already been noted (see p. 327). This suggestion would hold true especially for those glycoenzymes having a molecular architecture similar to that of glucoamylases. In these glycoenzymes, many carbohydrate side-chains are present on the surface of the molecule, and such chains can be situated along the polypeptide chain in such a way as to minimize molecular transformations. Accordingly, an unfolding of the polypeptide chain is hindered because of the presence of the carbohydrate residues, and the integrity of the glycoenzyme is preserved. Additional support for this suggestion comes from obser-

(151) A. Gottschalk and S. F. deSt. Growth, *Biochim. Biophys. Acta*, **43**, 513 (1960).
(152) J. W. Coffey and C. DeDuve, *J. Biol. Chem.*, **243**, 3255 (1968).
(153) N. N. Aronson and C. DeDuve, *J. Biol. Chem.*, **243**, 4563 (1968).

vations on pepsinogen, which contains covalently linked sugar moieties.[154] These moieties are released upon conversion of the molecule into active pepsin. As the active enzyme is denatured more easily than the zymogen, it appears that the sugar residues stabilize the zymogen in a conformation in which the active site is protected. Additional studies on the role of the carbohydrate moieties in glycoenzymes should lead to further clarification of the functioning of glycoenzymes in biological systems.

(154) H. Neumann, U. Zehavi, and T. D. Tanksley, *Biochem. Biophys. Res. Commun.*, **36**, 151 (1969).

AUTHOR INDEX

Numbers in parentheses are reference numbers and indicate that an author's work is referred to, although his name is not cited in the text.

A

Abadie-Maumert, F. A., 109
Abeles, R. H., 143, 162, 166
Abelsnes, G., 114
Abragam, A., 63
Abraham, R. J., 70, 71(215)
Abraham, S., 173, 179
Abrams, R., 146
Acree, T. E., 120
Adams, A., 94
Adams, M., 302, 305(2)
Adams, W. R., 110
Adkins, G. K., 108, 109
Af Ekenstam, A., 94
Agarwal, R. P., 36, 81(126)
Agawa, M., 123
Akabori, S., 302, 306(4), 307(4), 311(4), 320, 323(75)
Akeson, A., 302, 306(17), 309(17)
Akhtar, M., 161
Albon, N., 94
Albrecht, H. P., 260
Aleksandrov, I. V., 9
Alexeev, Yu. E., 229(9e, f), 230, 232(17), 233(9b, e, f, 17), 236(9b, e, f), 244(17, 23), 245(17, 18, 50), 249(57), 250(18, 50), 251(57), 252(57), 257(18, 23), 263(18)
Alexeeva, V. G., 229(9d, e, f, g), 230, 233(9a, b, c, d, f), 236(9a, b, c, d, e, f, 30), 237(9d), 238(29c, 30), 239(9c), 240(9c, 29c), 241(9d, g), 242(9g), 245(29c), 246(9g), 248(29c), 252(9g, 29c), 271(28a, b), 276(28a, b), 278(28)
Ali, Y., 33,
Allan, G. G., 114, 116(238)
Allentof, N., 193, 194(3)
Allerhand, A., 57(169), 58, 60
Allewell, N., 327
Alvarez, L. W., 128
Amagasa, M., 92
Ames, B. N., 312
Ames, G. R., 85
Anai, M., 306(33), 307

Anderson, A. W., 112
Anderson, B., 319
Anderson, D. R., 54, 55(160), 56(160)
Anderson, J. S., 312, 338
Anderson, L., 28, 67(80), 83(80)
Anderson, W. A., 14, 43,
Ando, T., 307
Anet, E. F. L. J., 119
Anet, F. A. L., 19, 20(51a, d), 70
Anfinsen, C. B., 327
Angerer, J., 229, 233(8), 273(8), 280(8)
Angyal, S. J., 95, 107(69), 137, 138(56), 145, 186(56, 59)
Ankel, H., 168
Arcamone, F., 32
Ariens, E. J., 205, 209, 210(32), 213(46)
Aries, R. S., 106
Arigoni, D., 143
Arnett, E. M., 108
Arni, P. C., 92, 93(30), 122(30)
Arnold, W. W., 328, 340(89)
Aronson, N. N., 340
Aseeva, N. N., 270, 276, 280(86)
Ashida, A., 99, 100, 106(106)
Ashwell, G., 129, 130(12), 140(12), 180(15), 184(12), 185(12), 188(15,223), 330, 331(114), 332, 338
Athovel-Haon, A. M., 329, 330(105)
Attree, R. W., 154
Avigad, G., 129, 170, 183(239), 184(13), 188(13)
Axelrod, M., 225

B

Bachinovsky, A. B., 267, 268(81)
Backinovsky, L. V., 233, 267(27), 268(27a, b, c), 269(27a, b), 270(27)
Badatus, B., 29
Baehler, B., 234, 240(33), 241(33), 271(33), 273(84), 275(84), 276(84), 277(84), 280(83a), 281(84)
Baer, H. H., 33
Baggett, N., 209, 293, 297(92)

343

Baker, B. R., 216
Baker, D. A., 260, 261(76), 262(76)
Baker, D. C., 130(19), 131, 150
Baker, E. B., 19, 20(51)
Baker, E. M., 134, 183(32)
Baker, R. H., Jr., 36, 37(110)
Ball, D. H., 107, 120
Ball, E. M., 302, 306(6), 309(6), 311(6), 312(6)
Bangerter, B. W., 36
Banwell, C. N., 83
Barbieri, G., 32
Barclay, K. S., 91
Barker, R., 30(88), 31, 147, 188(100)
Barker, S. A., 112, 114(249), 149, 188(106), 189(106)
Barlow, C. B., 70, 78(210)
Barlow, G. H., 138, 188(262)
Barlow, R. B., 213
Barnett, J. E. G., 128, 129, 130(11, 19), 131, 137, 140(11, 14, 19), 167(19), 168, 173(19), 174(19), 181, 184(14), 185(11, 19), 186(51), 188(14)
Barnett, W. L., 103
Barrete, J. P., 120
Barrow, G. M., 149
Bartley, J. C., 173
Bartsch, W., 296, 297(97)
Baschang, G., 230, 233(16), 275(16), 279(16)
Basus, V. J., 19, 20(51a),
Batterham, T. J., 72, 81, 145, 161(91)
Bayley, R. J., 139
Bayramova, N. E., 254, 255, 257(68) 261 (64, 70), 263(64, 68), 264(68, 70), 265(68b), 268(81)
Bebbington, A., 208
Becconsall, J. K., 19, 21(48)
Beck, W. S., 161, 162, 186(151)
Becker, E. D., 9, 46, 47(149), 49(149), 54(149), 57, 60(149), 61(149), 71(17)
Beckett, A. H., 205, 210
Beckwith, J. R., 333
Behr, A., 96
Bell, E. M., 134, 183(32)
Bell, L. W., 64
Belleau, B., 208, 209
Bello, J., 327
BeMiller, J. N., 109
Bennett, G. C., 181
Benson, A. A., 153, 184(125)
Bentley, R., 115, 120, 151, 156, 189(265)

Bernstein, H. J., 8, 67(1), 71, 144
Bernstein, R. A., 94, 116(64)
Berthelot, M., 102
Bertland, A. U., 169
Bertsch, H., 115
Bestmann, H. J., 229, 233(8), 273(8), 280(8), 294
Bevill, R. D., 140, 166, 175(69), 184(69), 185(69)
Beynon, P. J., 130(17), 131
Bhacca, N. S., 21, 22, 23, 26, 31, 67(56), 68(56)
Bhate, D. S., 189(265)
Bhattacharjee, S. S., 41
Bhavanandan, V., 319
Bible, R. H., Jr., 9, 12, 15(37), 16(37), 71(5)
Bickart, P., 225
Bieber, M., 150
Binkley, W. W., 31
Binsch, G., 83
Birch, G., 33
Bishop, C. T., 97, 101(86), 109
Bishop, E. O., 70, 71(212), 78(210)
Black, W. A. P., 92, 93(30), 112(33), 122(30)
Blackburn, B. J., 34, 35(107), 81(107), 109
Blake, M. I., 189(267)
Blakley, R. L., 81, 145, 161(91)
Bloch, F. L., 102, 117(128)
Blomberg, C., 193
Bloom, B., 135, 140, 178, 180, 185(72)
Blout, E. R., 251(60)
Blume, R. C., 106, 116(178)
Bobalek, E. G., 115(266)
Bochkov, A. F., 233, 253(26)
Bölcs, J., 113
Böttger, S., 93
Bogert, B. P., 53
Bonner, T. G., 101, 116(121)
Bonner, W. A., 218
Boppel, H., 91
Borggård, M., 121
Bosmann, H. B., 331, 332, 339(115)
Bothner-By, A. A., 19, 20(51), 66, 73, 74(225), 75, 83(229), 195, 329, 330(105)
Botlock, N., 115
Bouchilloux, S., 329, 330(105)
Bourgeois, J. M., 233, 271(29b), 272(29b), 273(29b, 84), 275(29b, 84), 276(29b, 84), 277(29b, 84), 278(29b), 281(84), 282(29b, 85), 283(29b)
Bourne, E. J., 91, 95, 101, 103, 107, 115,

116(121), 149, 188(106), 189(106)
Bovet, D., 208
Bovey, F. A., 9, 71(16)
Boyer, P. D., 185(250), 189(250)
Bradley, C. H., 19, 20(51a, d)
Bradley, R. M., 178
Bradshaw, J. S., 136
Brady, R. F., Jr., 120
Brady, R. O., 178
Bragg, P. D., 105, 120, 122(167)
Brand, J. C. D., 240
Brewster, J. H., 146, 252
Brice, R. E., 137, 186(51)
Bright, H. J., 171, 172(179), 179(179)
Brimacombe, E., 92
Brimacombe, J. S., 10, 65, 66(193), 112, 114(249), 204
Brimblecombe, R. W., 205, 208, 210, 219, 220(33, 51)
Brissaud, L., 104
Broom, A. D., 34, 36(103)
Brossmer, R., 227, 229(3), 253(3)
Brown, D. M., 135(47), 189(47), 230
Brown, F. S., 302, 309(22), 310(22)
Brown, G. B., 187(256)
Brown, J. F., 105, 120(162)
Brown, M. A., 19, 20(51d),
Brown, M. L., 95, 115(73), 123
Browne, F., 98
Brownson, C., 81, 145, 161(91)
Buchanan, J. G., 30(86), 81(86)
Buck, K. W., 209, 222, 223(65, 67), 224(65, 68), 225(67)
Buddecke, E., 319
Bugge, B., 169
Buller, M., 94
Bunge, W., 99
Buras, E. M., 103, 106
Burkhardt, F., 26
Burrows, E. P., 193
Burton, J. S., 204, 214(22), 216(22), 217 (22)
Burton, R., 12, 13(38)
Buss, D. H., 216
Butterworth, R. F., 130(18), 131, 164(18)

C

Caccam, J. F., 332
Cadenas, R. A., 23
Cadenbach, G., 93
Cahn, R. S., 141

Caldwell, M. L., 307
Campbell, J. R., 147
Campbell, T. W., 230
Carbon, J. A., 114, 123(242, 243)
Cardinal, E. V., 188(262)
Carey, P. R., 70, 78(210)
Carlson, D. M., 328
Carlsson, O., 120
Carr, R. H., 102
Carrol, M., 163, 164(157a)
Carroll, W. R., 327
Carubelli, R., 319
Caspary, W. F., 181
Cassinelli, G., 32
Castellano, S., 74, 75, 80, 83(229)
Casu, B., 40, 41(136), 42(136), 43(136), 149
Casu, M., 124
Casy, A. F., 205, 213(29)
Caterall, W. A., 36, 38(121)
Catley, B. J., 302, 306(10), 308(10), 322(10)
Catsoulacos, P., 260, 261(75, 76), 262(75, 76)
Causa, A. G., 115(266)
Cepure, A., 302, 306(7, 19), 307(7), 309(19), 312(9), 313(19), 315(7, 19), 316(19), 317(7), 323(19), 324(7, 19), 339(7, 19)
Černý, M., 112
Červinka, O., 199
Chabaud, O., 329, 330(105)
Chaikoff, I. L., 179
Chalet, J. M., 233, 271(29a), 272(29a), 276(29a), 277(29a), 278(29a), 281(29a), 282(29a), 283(29a)
Chan, S. I., 36, 37(122)
Chaney, A., 104
Chao, H., 332
Chapman, D., 9
Chapman, O. L., 124
Charalampous, F. C., 173
Chen, I. W., 173
Cheng, A., 19, 20(51a)
Chenoweth, M., 179
Cherniak, Ya. A., 293
Chia-Chen, C., 94
Chiu, T. H., 183(238)
Chizhov, O. S., 150, 233, 241, 253(26)
Choate, W. L., 327
Christie, P. H., 238
Christman, D. R., 137, 138(55), 188(55)
Cifonelli, J. A., 180, 184(245), 185(245)
Ciotti, M. M., 178

Clamp, J. R., 321
Clark, G. L., 94
Clarke, H. T., 103
Clarke, J. G., 109
Cleveland, E. A., 100
Cleveland, J. H., 100
Clitherow, J. W., 210
Clode, D. M., 150
Cochran, D. W., 57
Codington, J. F., 80
Coffey, J. W., 340
Cohen, A., 64
Cohen, E., 119
Cohen, H. L., 193
Cohen, S. S., 123
Cole, R. D., 302, 306(12, 42), 307(12)
Collings, W. J., 115(266)
Collins, P. M., 29, 130(17), 131
Commelin, J. W., 119
Cone, C., 67
Conner, A. H., 28, 67(80), 83(80)
Conrad, C. M., 115(326), 121
Cooley, J. W., 52, 53(156, 157)
Cooper, A. S., 106
Cooper, F. P., 97, 101(86)
Coops, J., 193
Corey, E. J., 28, 68(77)
Corina, D. L., 129, 130(11, 19), 131, 140 (11, 19), 167(19), 168, 173(19), 174 (19), 180, 185(11, 19)
Corio, P. L., 9, 71(9)
Cornforth, J. W., 170
Cornforth, R. M., 170
Cornog, R., 128
Couch, D. H., 100
Covarribas, C. B., 239
Cox, G. J., 103
Cox, H. C., 208
Coxon, B., 10, 17, 18, 23(43), 30(85), 33, 48(45), 51(45), 57(45), 58(45), 63(43), 66, 67(34, 43, 94), 68(43, 198), 69(198), 77(43, 85, 196, 197, 200, 201), 78(197, 200, 201), 79(85, 196), 80(94), 147
Cram, D. J., 136, 193
Crane, R. K., 181, 183(236), 184(236), 185(236), 186(236)
Crespi, H. L., 189(267)
Crofford, O. B., 181
Crowell, T. I., 95
Cunningham, J., 237, 241(40)
Curran, P. F., 181

Cushley, R. J., 54, 55(160), 56(160), 76, 80
Cushny, A. R., 205
Cynkin, M. A., 330, 337(109)

D

Dacons, J. C., 103
Dadok, J., 19, 20(51)
Dahlquist, F. W., 176
Dalton, J. G., 35
Dalton, L. K., 93, 106, 107(54), 108(54), 114(54), 123(54)
David, S., 131, 189(22)
Davidson, E. A., 160
Davis, H. E., 103
Davis, J., 168
Davis, L., 130(20), 131, 160(20), 169(20), 183(20), 185(20), 188(20), 189(20)
Davis, T. C. M., 115(267)
Davydova, L. P., 230, 233(15a), 274(15a), 276(15a), 279(15a)
Dawoud, A. F., 103
Dawson, G., 150, 321
Dayhoff, M. O., 335
Deane, C. C., 209
DeDuve, C., 340
Defaye, J., 63
Deferrari, J. O., 23
de Garilhe, M. P., 123
Degering, E. F., 94, 116(65)
de Grandchamp-Chaudun, A., 120
de Groote, M., 106
Dehn, W. M., 117, 118(273)
DeJongh, D. C., 150
DeLeo, A. B., 163
Demarco, P. V., 33
De Mendoza, A. P., 115(266)
Dénes, G., 164(157b)
Derevitskaya, V. A., 122
de Robichon-Szulmajster, H., 159
de Rudder, J., 123
Deslauriers, R., 35, 81(109)
de St. Growth, S. F., 340
Deupree, J. D., 161
Dewar, E. T., 92, 93(30), 112(33), 122(30)
de Whalley, H. C. S., 94
Dey, P. M., 302, 306(16), 307(16)
Diehl, P., 75
Dietrich, C. P., 41, 42(142)
Dinovo, E. C., 185(250), 189(250)
Dirksen, H. W., 113

Ditchfield, R., 78
Dmitriev, B. A., 228, 229(5), 233, 253(5, 26, 63), 254(5), 255(63, 65), 257(68), 261(63, 64, 66, 67, 70), 263(64, 68), 264(68, 70), 265(63, 67, 68b, 70), 266(66, 67, 79), 267(27), 268(27a, b, c, 81), 269(27a, b), 270(27), 293
Doddrell, D., 57(169), 58, 60
Dodds, M. L., 103
Doebereiner, U., 167
Doelker, E., 234, 240(33), 241(33), 271(33), 276(33)
Doerr, E. L., 115(269a)
Doganges, P. T., 130(17), 131
Domke, J., 105
Domous, K. B., 95
Donninger, C., 170
Dorfman, A., 180
Dorofeenko, G. N., 228, 229(4), 230, 232(17), 233(17, 24), 237(25), 244(17, 23, 24, 25), 245(17, 24, 25, 39, 50, 53), 246(24, 53), 247(24, 25, 54, 55), 248(24, 54), 249(24, 57), 250(50), 251(57), 252(57), 253(4, 25, 39), 254(53), 256(25), 257(23)
Dorrer, H. D., 140, 151, 179, 183(74, 213, 241, 242), 184(74), 185(242), 186(74, 242), 187(74), 188(74)
Douty, F., 108
Driesen, H. E., 108, 124(191), 125(191)
Drydale, R., 183(236), 184(236), 185(236), 186(236)
Dubose, A., 95
Dugan, F. A., 308
Dugas, H., 35, 81(109)
Dunn, A., 179, 185(211)
Durette, P. L., 10, 22, 28, 67(75, 76), 144, 145(87a)
Durham, L. J., 24, 30(87), 145, 161(92)
Duxbury, J. M., 222, 224(70)
Dwek, R. A., 29, 63(82), 67(82), 73(82)
Dziengel, K., 103

E

Ebata, M., 302, 306(23), 309(23), 315(23), 316(23)
Ebert, K. H., 140, 183(74), 184(74), 185(247), 186(74), 187(74), 188(74, 247)
Eberwein, H., 302, 309(22), 310(22)

Egami, F., 320, 322(73)
Eglinton, G., 240
Elhafez, F. A. A., 193
Eliel, E. L., 196, 206, 215(34)
Elizarova, A. N., 289
Ellenbroek, B. W. J., 205, 210(32)
Elliot, W. H., 146, 151, 166(96), 171(96), 175(96)
Ellis, G. P., 115
Ellzey, S. E., 110, 114(218), 123(218)
Emoto, S., 200, 201, 202, 212
Emsley, J. W., 9, 46(4), 64(4), 71(4)
Endres, H., 108
Englar, J. R., 40
Englard, S., 132, 170, 183(239)
Ennor, K. S., 104
Epstein, W., 333
Epstein, W. W., 109
Ernst, R. R., 16, 18(42), 21(42), 43(42), 45(145), 46(145), 47(145), 48(145), 49(145), 52(42, 145), 53(42), 55, 56(42, 145, 163), 57(163), 62(145)
Ershov, B. A., 9
Eustache, J., 131, 189(22)
Evans, E. A., 139
Evans, M. E., 115
Eveleigh, D. E., 41
Everett, W. W., 109
Eylar, E. H., 320, 331, 332, 334, 338(7)

F

Fabryova, A., 199
Fahim, F. A., 222, 223(67), 225(67)
Falk, M., 40
Fang, K. N., 36
Farrar, T. C., 46, 47(149), 49(149), 54(149), 57, 60(149), 61(149)
Fatiadi, A. J., 64, 73(187), 129, 183(6), 184(6, 244), 185(6, 244), 186(6, 244), 187(6)
Faulkner, I. J., 104, 120(161)
Feast, A. A. J., 204, 215(25)
Fedoroňko, M., 120
Fedtke, C., 179, 183(213)
Feeney, J., 9, 46(4), 64(4), 71(4)
Feingold, D. S., 140, 165(76), 174(76), 183(238), 184(76), 187(76)
Feinstein, G., 308
Feldman, I., 36, 81(126)
Felling, J., 305

Ferguson, J. H., 103
Ferguson, R. C., 74
Fernandez, C. M., 137, 138(56), 140(56), 186(56, 59)
Fernez, A., 121
Ferrand, M., 329, 330(105)
Ferrante, N. D., 137, 138(55), 188(55, 261)
Ferreti, J. A., 57, 75
Ferrier, R. J., 10, 101, 120, 215
Feuge, R. O., 95, 115(73), 123
Fey, M. W., 98
Fiedler, F., 175
Fields, M., 251(60)
Finchler, A., 115(262)
Fischer, E., 207
Fischer, E. H., 305
Fitzgerald, F. F., 91, 112(18)
Flaherty, B., 204
Fleming, I. D., 307
Fletcher, H. G., Jr., 69, 158, 186(141b), 337
Fletcher, R., 30(86), 81(86)
Fleury, G., 104
Floss, H. G., 163, 164(157a)
Follman, H., 161
Fondy, T. P., 143
Ford, Y. A., 259
Foster, A. B., 65, 66(193), 209, 222, 223(65, 66, 67), 224(65, 66, 68, 69, 70), 225(67)
Foster, D. W., 140, 178, 185(72)
Foster, J. F., 98, 109
Fourneau, E., 208
Fox, J. J., 80
Franklin, E. C., 91
Fraser-Reid, B., 28, 29, 131, 132, 146(23), 189(23)
Frech, M. E., 178
Fredenhagen, K., 93
Freeman, R., 14, 45, 49(148), 50(148), 52, 56(148), 57, 60, 61(175), 62(175)
Freudenberg, K., 91
Freund, E. H., 95
Frey, P. A., 143, 168
Friebolin, H., 39, 40(127, 128, 129),
Friedberg, F., 129, 153(5), 185(5)
Fries, I., 151
Fritz, H., 230, 233(16), 275(16), 279(16)
Frush, H. L., 129, 140(7), 153, 154(126), 176, 183(6, 7, 188), 184(6, 7, 244)
Frye, G. H., 187(259)

185(6, 9, 244), 186(6, 7, 9, 188, 244), 187(6, 7, 9, 188), 188(7)
Fuertes, M., 82(255), 83
Fukuda, M., 320, 322(73)
Fukui, T., 307

G

Gabe, E. J., 142
Gabriel, O., 129, 130(12), 140(12), 141(71), 166, 168, 170(169), 184(12, 169), 185(12, 246), 188(12), 311, 312(55)
Gaertner, V. R., 93, 115(269a)
Gagnaire, D., 63
Gallo, G. G., 40
Garbisch, E. W., Jr., 71
García González, F., 106
Garcia-Muñoz, G., 82(255), 83
Garnett, J. L., 137, 138(56), 140(56), 186(56)
Garwin, R. L., 53
Gascon, S., 317(63), 318, 324(63)
Gasselseder, H., 113
Gates, M., 110, 114(206, 207)
Gavrilenko, O. A., 244, 245(51), 247(51), 249(51), 255(51), 261(51), 294(51)
Gawron, O., 143
Gazzard, V. J., 75
Gee, P., 302, 306(23), 309(23), 315(23), 316(23)
Geissler, G., 227
Genich, A. P., 239
Gestblom, B., 52
Ghambeer, R. K., 81, 145, 161(91)
Gibson, Q. H., 171, 172(179), 179(179)
Gigg, J., 237, 241(40), 258, 261(71), 262(71), 271(71)
Gigg, R., 237, 238(40d), 241(40), 258, 261(71), 262(71), 271(71)
Gillis, J., 97
Glaser, L., 130(20), 131, 146, 159(95), 160(20), 166(96), 169(20), 171(95,96), 175(96), 183(20), 185(20), 186(254), 188(20), 189(20)
Glass, C. A., 40
Glen, W. L., 107
Goldfrank, M., 105
Goldner, A. M., 181
Goldsack, R. J., 99, 101(101), 107(101)
Goldsmith, H. A., 121

Goldstein, I. J., 94
Goldstone, A., 308
Golovkina, L. S., 241
Goncalves, R. P., 181
Gordon, M., 187(256)
Gorin, P. A. J., 41
Gottesman, M. M., 161, 186(151)
Gottschalk, A., 319, 337, 339, 340
Govil, G., 75
Graf, R., 273, 282(85)
Grant, G. A., 107
Grant, J. H., 329
Grassner, H., 114
Graves, D. J., 176
Gray, G. R., 30(88), 147, 188(100)
Grebner, E. E., 330
Green, D. M., 205, 210, 220(33, 51)
Greenwood, C. T., 108, 109
Gregoriadis, G., 332, 338
Greiling, H., 185(249), 188(249), 190, 305
Greiling, P., 305
Gresser, W., 110
Grey, A. A., 34, 35(104, 105, 107), 81(104, 105, 107)
Griffin, C. E., 146
Grimaldi, J., 59
Grindley, T. B., 65, 66, 73(192)
Grisebach, H., 167
Grötsch, H., 114, 123(244, 245, 246)
Gross, D., 94
Gross, J. I., 184(245), 185(245)
Gross, S. R., 333
Grossman, H., 102, 117(128)
Grundschober, F., 92
Grunler, S., 305, 339(24)
Grunwald, E., 18
Grunwald, E. A., 94
Günther, H., 74, 75
Guenther, T., 185(249), 188(249)
Guisley, K., 122
Gunning, J. W., 97
Gupta, S. K., 230, 233(21), 245(21), 246(21), 253(21), 254(21), 256(21), 261(69)
Guthrie, R. D., 10, 70, 78(210), 103

H

Haas, H. B., 93, 115(263, 264)
Hager, L. P., 302, 306(21), 309(21, 22), 310(22), 314(21), 324(21), 339

Hagge, W., 115(268)
Hagopian, A., 331, 332, 334
Haigh, C. W., 23, 67(63), 73(63), 75, 83(237)
Hall, C. W., 330
Hall, L. D., 9, 12, 13(38), 21, 64, 66(2), 69, 70(2), 71(2), 80, 82, 143, 147
Hallinan, T., 329
Hamaleinen, C., 103, 106
Hamilton, F. D., 146
Hamor, T. A., 222, 224(71)
Hanafusa, H., 302, 306(4), 307(4), 311(4)
Hanessian, S., 27, 130(18), 131, 150, 164 (18), 210, 215, 216(56)
Hanrahan, V. M., 307
Hanson, A. W., 327
Hanson, K. R., 141
Hardegger, E., 207, 208
Harding, T. S., 102
Hardman, K., 327
Harker, D., 327
Harmon, R. E., 231, 233(21), 245(21), 246(21), 253(21), 254(21), 256(21), 261(69)
Harnden, M. R., 112
Harper, N. J., 210
Harris, M. J., 122
Harris, R. K., 9, 75
Harteck, P., 128
Harting, H., 105
Hasan, Q. H., 222, 223(64, 66), 224(64, 66)
Haskell, T. H., 210
Hassid, W. Z., 330
Hatton, L. R., 101
Hattori, Y., 307
Haubenstock, H., 196
Haug, A., 160(148a), 161
Hauska, V. G., 135(43), 183(43), 185(43)
Hawkins, D. R., 222, 223(66), 224(66)
Hawkins, H., 183(236), 184(236), 185(236), 186(236)
Haworth, W. N., 121, 207
Hays, J. T., 298
Hecht, H. G., 9
Hedgley, E. J., 120
Helferich, B., 93, 100, 107, 120, 305, 339(24)
Heller, A., 132, 186(27), 189(27)
Hemsworth, B. A., 332
Heneka, H., 114

Henglein, F. A., 114
Hengst, G., 99
Hermann, K., 166
Herp, A., 332
Hers, H. G., 179
Herscovics, A., 180
Herstein, K. M., 85, 86(1), 89(1), 96(1), 107(1), 108(1), 114(1), 117(1), 118(1), 119(1)
Herz, W., 107
Hess, K., 103, 121
Hessel, F. A., 106
Heubach, G., 119
Heublein, G., 113
Heuser, E., 94
Heyns, K., 119, 150, 188(264)
Hickman, J., 338
Hickman, S., 332
Higashi, Y., 332
Hilbert, G. E., 91
Hill, A. S., 120
Hillery, P., 95
Hinton, J., 10
Hirano, S., 42, 204
Hirs, C. H. W., 302, 306(8, 9), 307, 308(8, 9), 310(8), 314, 315(8, 46), 317(8), 320(9), 322(9), 324(46), 325(9), 326(9), 334(9), 340(9)
Hirst, E. L., 103
Hixon, R. M., 98, 100
Hobart, S. R., 106
Hochstetter, A., 105, 106(164), 110(164)
Hockett, R. C., 122
Hodge, J. E., 91, 119
Hoffman, H., 259
Hoffman, P., 319
Hoffman, R. A., 70
Hoffmann, R. C., 175
Hoffmann-Ostenhof, O., 135(43), 183(43), 185(43)
Hogenkamp, H. P. C., 137, 156, 161, 183(52), 187(52)
Hogness, D. S., 159
Hokin, L. E., 186(251)
Hokin, M. R., 186(251)
Holland, C. V., 22, 67(56), 68(56)
Hollis, D. P., 34, 36(103), 37(113), 38(113, 121)
Holt, N. B., 129, 140(7), 183(7), 184(7), 186(7), 187(7), 188(7)
Holty, J. G., 117, 118(274)

Honda, S., 41, 115, 122(256)
Honeyman, J., 104, 105
Hooper, I. R., 112
Hopper, S. P., 146
Horecker, B. L., 156
Horner, L., 259
Horner, W. H., 177
Horowitz, M. I., 338
Horton, D., 10, 22, 23, 28, 31, 32, 63, 67(56, 75, 76), 68(56), 130(19), 131, 133, 147(31), 150, 189(30, 31, 115a), 204, 215, 230, 234
Hoskinson, R. M., 138
Hossain, S., 109
Hough, J. S., 306(44), 307, 308(44), 322(44)
Hough, L., 33, 63, 67, 110, 121
Hourston, D. J., 109
How, M. J., 10
Howard, J., 143, 144(80), 189(80), 220
Howarth, G. B., 230, 235(11), 241(11)
Howgate, P., 258, 260(72), 261(72)
Hribar, J. D., 150
Hruska, F. E., 34, 35(104, 105, 107), 81(104, 105, 107)
Huber, L. M., 19, 20(51)
Huber, W. F., 110
Hudgin, R. L., 330, 331(108), 332(108)
Hudson, C. S., 96, 97, 98(81), 102(78), 106(78), 302, 305(2)
Hughes, J. B., 147, 234
Hullar, T. L., 94
Hunt, L. T., 335
Hyman, H. H., 93

I

Ikenaka, T., 302, 306(4, 33, 39), 307(4), 311(4), 315(39), 316(39), 320, 321(76), 323(76), 324(39, 76), 325(76a)
Inagami, T., 327
Inch, T. D., 10, 124, 194, 201, 204, 205, 209, 210, 211, 214, 215(23, 28), 217(28), 218, 219(23, 60), 219, 220(33, 51), 222, 224(68, 69, 70), 234
Ingold, C. K., 141
Inouye, S., 27
Inouye, Y., 202
Intrieri, O. M., 187(256)
Ionin, B. I., 9
Irvine, J. C., 93, 100(43), 103, 110(45)
Irvine, R. A., 338

AUTHOR INDEX, VOLUME 27 351

Isbell, H. S., 96, 129, 140(7), 141, 150, 153, 154(126, 127), 155(128, 129), 176, 183(6, 7, 188), 184(6, 7, 244), 185(6, 9, 244), 186(6, 7, 9, 188, 244), 187(6, 7, 9, 188), 188(7), 231, 234
Iwanaga, S., 306(43), 307, 308(43), 324(43)

J

Jabbal, I., 330, 331(108), 332(108)
Jaccard-Thorndal, S., 271, 276(83a), 277(83a), 280(83a)
Jacin, H., 114
Jackman, L. M., 9, 71(19)
Jackson, J. J., 332
Jackson, R. L., 302, 306(9), 308(9), 320(9), 322(9), 325(9), 326(9), 334(9), 340(9)
Jacobson, B., 160, 176
Jacobus, J., 225
James, K., 204, 216(27)
Jamieson, G. A., 333
Jamieson, J. D., 328, 330, 331(92)
Jaques, L. B., 41
Jardetzky, O., 36, 37(111), 39(111)
Jarvis, J. A., 112, 114(247, 248, 249)
Jeanloz, R. W., 325, 326(80)
Jefferis, R., 209
Jennings, H. J., 33, 67(94), 80(94), 105
Jensen, M. A., 70, 78(210)
Jett, M., 333
Jewell, J. S., 133, 147(31), 189(30, 31), 234, 235(32), 240(32), 240, 243(32, 47)
Jochims, J. C., 108, 124(191), 125(191)
Jochmann, C., 135(44), 183(44)
Johannesen, R. B., 75
Johnson, A. W., 298
Johnson, C. K., 142
Johnson, L. F., 19, 20(50), 24(50), 30(85), 40, 41(136), 42(136), 43(136), 77(85), 79(85)
Johnson, L. N., 327
Johnson, P. G., 114, 116(238)
Jones, A. S., 258, 260(72), 261(72)
Jones, B. D., 65, 66(193), 92, 93(39), 122(39)
Jones, D. M., 116
Jones, G. H., 135(47), 189(47), 230, 260
Jones, J. K. N., 105, 120, 122(167), 230, 234(10), 234, 235(10, 11, 32), 240(32), 241(11), 242(10), 243(32)
Jones, W. M., 138

Jordian, G. W., 328
Jumper, C. F., 18
Just, E. K., 133, 147(31), 150, 189(31, 115a), 204, 230

K

Kabat, E. A., 337
Kaiser, R., 55
Kaiser, S., 209
Kalckar, H. M., 169
Kale, J. K., 102
Kameda, K., 302, 306(18), 309(18)
Kamprath-Scholtz, U., 297, 298(99b), 299(99b, c)
Kaplan, L., 129, 153(5), 178, 185(5)
Kaplan, N. O., 36, 37(112, 116, 118), 38(112, 116, 118)
Kapo, G., 115(266)
Karabatsos, G. C., 143
Karakawa, W., 327, 338
Karay, S., 150
Karcz, M., 98
Karjala, S. J., 91
Karplus, M., 144
Kartha, G., 327
Kashimura, N., 204
Kato, H., 100
Katz, J., 135, 178, 179, 185(42, 211)
Katz, J. J., 93, 189(267)
Kavanagh, L. W., 41
Kawana, M., 200, 201, 202
Kaysen, G., 170
Kean, E. L., 337
Keenan, T. W., 329, 330(106)
Keil, K. D., 141, 152, 185(123), 186(123)
Keilich, G., 39, 40(127, 128, 129)
Kelleher, W., 167
Keller, H., 138, 140(62), 184(62), 185(62)
Kellerhals, H. P., 75
Kellermeyer, R., 179, 189(214)
Kelly, S., 177
Kemp, R. G., 135, 178, 179, 185(42, 199), 186(199)
Kennedy, J. F., 10
Kent, P. W., 29, 63(82), 67(82), 73(82)
Kern, M., 180, 328, 331(97)
Khan, A. A., 70
Khetrapal, C. L., 75
Kindl, H., 135(43), 177, 183(43), 185(43)
King, A. H., 111

King, R. W., 124
Kintner, W. B., 181
Kirkwood, S., 140, 168, 170(168), 175(69), 184(69), 185(69, 240), 186(240)
Kiss, J., 26
Kisters, R., 305
Klebe, J. F., 152, 183(124), 185(124)
Kleppe, K., 302, 306(6, 19), 309(6, 19), 311(6), 312(6, 19), 313(19), 315(19), 316(19) 323(19), 324(19), 339(19)
Klesse, P., 121
Klopfenstein, C. E., 75
Klotz, I. M., 315
Knock, M., 107
Knox, J. R., 327
Knull, H. R., 301, 302, 306(7), 307(7), 315(7), 317(7), 319, 321(68), 323(68), 324(7), 326(1), 328(68), 330(1), 339(7)
Koch, H. J., 28, 130(21), 131, 149(21), 189(21)
Kochetkov, N. K., 122, 150, 228, 229(5), 233, 253(5, 26, 63), 254(5), 255(63, 65), 257, 261(63, 64, 66, 70), 261, 263(64), 264(70), 265(63, 67, 70), 266(66, 67, 79), 267(27), 268(27a,b,c, 81), 269(27a, b), 270(27), 293
Kocourek, J., 112
Kögl, F., 208
Koenig, H., 309
Kohlenberg, L., 119
Kohn, B. D., 175, 184(243), 185(243)
Kohn, P., 175, 184(243), 185(243)
Kohtes, L., 305
Kollonitsch, V., 85, 93(5), 115(5), 120(5)
Komori, S., 117, 118(272), 119(272), 121(272), 122(272), 123
Komoroski, R., 57(169), 58, 60
Kondo, N. S., 36,
Kononenko, O. K., 85, 86(1), 89(1), 96(1), 107(1), 108(1), 114(1), 117(1), 118(1), 119(1)
Kopecky, K. R., 193
Korn, H. F., 115, 121(258)
Kornfeld, R., 336, 337(135)
Kornfeld, S., 336, 337(135)
Kornilov, V. I., 233, 278(28)
Korolchenko, G. A., 244, 245(53), 246(53), 254(53)
Korpium, O., 225
Kost, A. A., 254, 261(64), 263(64)
Kostelnik, R., 80

Kraus, C. A., 91
Kravchenko, G. I., 244, 247(55)
Krzeminsky, Z. S., 101, 116(122)
Kuhn, R., 114, 121, 227, 229(3), 253(3)
Kullnig, R. K., 8, 67(1), 144
Kuzuhara, H., 212

L

Lai, Y.-Z., 114, 116(238)
Lake, W. H. G., 207
Lamdin, E., 179
Lammers, J. N. J. J., 114
Lampen, J. O., 302, 305, 306(29), 317(63), 318, 324(63)
Lamport, D. T. A., 328
Lamprecht, H., 100
Lamprecht, W., 136(44), 183(44)
Lance, D. G., 230, 234, 235(11, 31), 240(31), 241(11)
Landor, S. R., 196, 197
Lang, 99
Lark, C., 189(255)
Larsen, B., 160(148a), 161
Larsson, A., 145, 161(92), 175(150)
László, E. D., 121
Lauterbach, J. H., 32, 147
Lavoie, J. L., 209
Law, G. R. J., 308
Lawford, G. R., 329
Leatherwood, J. M., 158
Leblond, C. P., 181
Lee, B., 327
Lee, C. K., 33
Lee, J. B., 109
Lee, T., 339
Lee, Y. C., 302, 305, 306(5, 15), 307(5), 308(5), 309(15), 310(15), 321(5), 325(5)
Legler, 183(237)
Lehmann, J., 136, 153, 166, 184(49, 125), 185(49, 50), 186(49), 293
Le-Hong, N., 271, 272(83b), 275(83b), 276(83b), 277(83b), 281(83b)
Leloir, L. F., 157(138)
Lemieux, R. U., 8, 21, 22, 28, 31, 36, 37(114), 67(1, 59, 80), 68(59), 70(59), 83(80), 115(265), 120, 132, 133, 143, 144(80), 145, 189(28, 80), 207, 220
Lennard, J., 94
Lennarz, W. J., 332
Leskina, L. P., 244, 245(51), 247(51), 249(51, 52), 255(51), 261(51), 294(51)

AUTHOR INDEX, VOLUME 27

Levene, P. A., 97
Levine, M., 98
Levine, S., 132, 189(28), 278
Levisalles, J., 227
Levy, H. R., 145, 146(88), 170
Lewis, G. J., 194, 204, 211, 215(28), 217(28)
Lewis, J. E., 70, 78(210)
Lewis, J. J., 213
Lewis, P. A. W., 52, 53(157)
Lewis, R. N., 298
Ley, R. V., 201, 204, 215(23), 218, 219(23, 60)
Libby, R. A., 109
Licerio, E., 337
Liechti, P., 208
Lienhard, G. E., 135(45), 143(45), 165(45), 184(45), 187(45)
Lienhard, K., 114
Likhosherstov, L. M., 122
Lindet, M. L., 97
Lindquist, L. C., 166
Lineback, D. R., 145, 307, 315(41), 319, 321(69), 323(69)
Linek, K., 120
Link, T., 19, 20(51)
Linn, C. J., 94
Lipmann, F., 329
Lipsky, S. R., 54, 55(160), 56(160)
Liska, M., 332
Liu, T., 308, 315(48), 316(48)
Lloyd, K. O., 337
Lobry de Bruyn, C. A., 97, 100
Löw, I., 114
Loewus, F. A., 140, 145, 146(88), 170, 177(73)
Lohmar, R. L., 121
Lohse, F., 207, 208
Long, J. S., 121
Long, L., Jr., 115, 120
Lorand, E. J., 104
Lowe, I. J., 49
Lowenstein, J. M., 134, 178, 179(196), 185(196)
Lown, J. W., 36, 37(114)
Lowry, T. M., 104, 120(161)
Lucas, F. A., 111
Lucas-Lenard, J., 329
Ludowieg, J. J., 26
Ludwig, E., 92
Lüttringhaus, A., 113

Lung, K. H., 135, 188(39)
Lusebrink, T. R., 52
Lutz, P., 108, 124(191), 125(191)
Lynden-Bell, R. M., 9
Lyu Dyc Shoung, 233, 278(28)

M

McCarthy, J. F., 103
McCasland, G. E., 24, 30(87)
McCausland, J. H., 19, 20(51d)
McDonald, R. N., 230
McDonough, M. W., 135(46), 143(46), 158(46), 161(46), 187(46)
McGinnis, G. D., 332
McGuire, E. J., 328, 330, 331(108), 332(108)
McInnes, A. G., 40, 115(265)
McIvor, M. C., 19, 21(48)
Mack, C. H., 110, 114(218), 123(218)
McKelvy, J. F., 302, 306(5), 307(5), 308(5), 321(5), 325(5)
McKensie, A., 201
Mackie, D. M., 41, 42(142)
Mackie, W., 28, 124, 129, 149(10), 188(10), 189(10, 266)
McLachlan, J., 40
McLauchlan, K. A., 33, 67(94), 80(94)
McPherson, J. A., 109
Maddox, I. S., 306(44), 307, 308(44), 322(44)
Madroñero, R., 82(255), 83
Maercker, A., 227, 252(2f), 259(2f)
Magnin, A. A., 67, 83(207)
Magnus, P. D., 9
Magus, M. J., 41
Mahler, H. P., 36, 37(110)
Mainas, F., 115
Maitra, U. S., 168
Makita, M., 120
Maley, F., 318, 324(79), 325, 326(79)
Malm, C. J., 103
Malt, R. A., 328
Manville, J. F., 9, 21, 82
Marawou, A., 103
Margolis, R. U., 132, 186(27), 189(27)
Markert, C., 315
Markovitz, A., 180, 184(245), 185(245)
Marquardt, D. W., 74
Marshall, R. D., 318, 319(65)

Martin, J. B., 114, 122
Martin, J. C., 21, 74
Martin, R. G., 312
Martin, S. S., 332
Marzluf, G. A., 188(260)
Masamune, H., 107
Massey, V., 302, 306(20), 309(20)
Mast, W. E., 147
Matsuo, Y., 306(43), 307, 308(43), 324 (43)
Matsushima, Y., 306(33, 39), 307, 315(39), 316(39), 320, 321(76), 323(76), 324(39, 76), 325(76a)
Matthaeus, G., 115(268)
Matthes, K. J., 179
Mattich, L. R., 120
Mavel, G., 9
Maxwell, E. S., 159
Mazurek, M., 41
Medina, R.. 158, 179, 186(253)
Mega, T., 306(39), 307, 315(39), 316(39), 324(39)
Mehler, A. H., 164(158)
Mehltretter, C. L., 101
Meiboom, S., 18
Melchers, F., 180
Melo, A., 146, 166(96), 171(96), 175(96)
Meloche, H. B., 165, 184(159), 186(159), 189(159)
Memory, J. D., 9
Merlin, L. M., 329, 330(106)
Merten, R., 99
Meshi, T., 186(252), 188(263)
Messmer, E., 121
Metzenberg, R. L., 188(260)
Meyer, K., 319
Meyer, W. L., 36, 37(110)
Meyer zu Reckendorf, W., 297, 298(99b), 299(99b, c)
Micheel, F., 94, 110
Michel-Bechet, M., 329, 330(105)
Miller, C. O., 91
Miller, M. J., 22, 67(56), 68(56)
Miller, P. S., 36
Mirzayanova, M. N., 230, 233(15a, b, c), 274(15a, b), 276(15a), 279(15a, b)
Mislow, K., 63, 64(179), 192, 225
Mitchell, B. J., 196, 197
Mitts, E., 100
Mock, W. L., 19, 20(51d)
Moffatt, J. G., 130(16), 131, 260

Mohan, V., 189(267)
Moller, F., 315
Molnar, J., 328, 329, 332
Montezin, G., 208
Moore, M. C., 36
Moore, R. H., 149, 188(106), 189(106)
Moore, S., 302, 306(10, 11), 307, 308(10, 11), 315(48), 316(48), 322(10), 324(11), 325(11), 326(11), 327
Morell, A. G., 129, 180(15), 188(15), 332, 338
Morgan, J. W. W., 105
Morita, Y., 302, 306(18), 309(18)
Morre, D., 329, 330(106)
Morris, D. R., 302, 306(21), 309(21, 22), 310(22), 314(21), 324(21)
Morrison, J. H., 312
Morse, D. E., 156
Mortimer, F. S., 64
Mortlock, R. P., 158, 164(140), 184(140), 186(140), 187(140)
Moschera, J., 332
Moser, H. C., 185(248), 188(248)
Mosher, H. S., 193
Moshy, R. J., 114
Moss, G., 129, 185(8)
Mound, E., 332
Mowery, D. F., 100
Moye, C. J., 87, 96(6, 7, 8, 9, 10, 11, 12), 98(6, 8, 9), 99(6, 7, 8, 9, 10, 11, 12), 100(7, 9, 11), 101(100, 101), 106(7, 8, 9, 10, 11), 107(6, 7, 8, 9, 10, 11, 101), 113, 116(122), 117(12)
Moyer, J. D., 129, 140(7), 150, 183(7), 184(7), 185(9), 186(7, 9), 187(7, 9), 188(7)
Muellenberg, C. G., 328
Müller, D., 119, 150, 188(264)
Muesser, M., 63
Mullhofer, G., 158, 175, 183(242), 185(242), 186(242)
Munday, K. A., 180, 181
Munroe, P. A., 33
Murachi, T., 302, 308, 320, 322(74), 339
Muramatsu, T., 180, 320, 322(73)
Murray, L., 110
Murty, C. N., 329
Musher, J. I., 28, 68(77)
Muskat, I. E., 92, 97
Mustafa, A., 103

Myers, G. S., 107
Mynott, R. J., 57, 58(168a)

N

Naar-Colin, C., 73, 74(225)
Nachtergaele, H., 97
Naegele, W., 43, 45(145), 46(145), 47(145), 48(145), 49(145), 52(145), 56(145), 62(145)
Nagarajan, R., 31
Nakanishi, Y., 181
Nakhre, P., 114
Nathenson, S. G., 180
Naumann, M. O., 24, 30(87)
Nebbia, G., 121
Nelsestuen, G. L., 168, 170(168), 175, 185(240), 186(240)
Nelson, D. C., 110
Nelson, F. H., 18, 19(47), 21(47)
Nelson, J. H., 36, 37(122)
Neuberger, A., 318, 319(65)
Neufeld, E. F., 177, 330, 336, 337(135)
Neumann, H., 341
Neumann, N. P., 302, 305, 306(29), 317 (63), 318, 324(63)
Neurath, H., 308
Nguyen, L., 230, 259(13, 14), 261(13, 14), 262(13, 14)
Niemann, C., 298
Niida, T., 27
Nikaido, H., 330
Nikuni, Z., 307
Nitsch, E., 96
Nivard, R. J. F., 205, 210(32)
Nonaka, M., 251
Norberg, R. E., 49
Nordin, J. M., 140, 175(69), 184(69), 185(69)
Nordin, P., 185(248), 188(248)

O

O'Brien, P. J., 328, 336, 337(135)
O'Connell, E. L., 134, 143(35), 157(34), 158(34, 35), 164(34, 140), 183(35), 184(40), 185(34), 186(140), 187(35, 140), 188(34), 189(34)
Ohlenbusch, H., 305
Ohrui, H., 212
Oji, N., 179

Okada, S., 307, 313
Okahara, M., 117, 118(272), 119(272), 121(272), 122(272)
Okamoto, K., 121
Olds, D. W., 91
Olin, S. M., 207
Oliphant, M. L. E., 128
Onderka, D. K., 163, 164(157a)
Onikura, N., 92
Onodera, K., 204
Ooi, T., 327
Oppelt, M., 108
Orth, P., 89
Oshima, G., 306(43), 307, 308(43), 324(43)
Osipow, L., 115(262, 263), 115
Osolin, A. E., 244, 245(53), 246(53), 254(53)
Ota, S., 308
Otake, T., 124
Otey, F. H., 101
Overend, W. G., 94, 101, 120, 130(17), 131, 204, 214(22), 215(25), 216(22), 217(22)

P

Pacák, J., 112
Pachler, K. G. R., 67, 83(207)
Page, J., 327
Palade, G. E., 328, 330, 331(92)
Pardoe, W. D., 222, 224(68)
Parker, A. J., 89
Parrish, F. W., 115, 120
Parry, K., 30(86), 81(86)
Pastuska, G., 123
Patel, D. J., 36, 38(116)
Paterson, J. C., 92, 93(30), 122(30)
Patt, S. L., 59
Paulsen, H., 23, 210, 296, 297(97)
Payne, J. H., 96
Payza, N., 319, 332
Pazur, J. H., 102, 103, 301, 302, 306(6, 7, 19), 307(7), 309(6, 19), 312(6, 19), 313(19), 315(7, 19), 316(19), 317(7), 319, 321(68), 323(19, 68), 324(7, 19), 326(1), 327, 328(68), 330(1), 338, 339(7, 19)
Peat, S., 207
Pedersen, C., 82
Pellet, H., 98
Pellet, L., 98
Perlin, A. S., 28, 40, 41(136), 42(136, 142),

43(136), 124, 129, 130(21), 131, 149 (10, 21), 188(10), 189(10, 21, 266)
Pernet, A. G., 27
Perret, F., 271, 272(83b), 275(83b), 276(83b), 277(83b), 281(83b)
Perry, A. R., 222, 223(65, 67), 224(65), 225(67)
Peters, D., 251(59)
Peters, O., 93
Pfitzinger, W., 102
Pfitzner, K. E., 130(16), 131
Pflughaupt, K. W., 119
Phelps, C. F., 336
Phillips, G. O., 139
Philips, K. D., 147
Pietsch, K., 92
Pigman, W., 100, 305, 319, 332
Pinteric, L., 330, 331(108), 332(108)
Pippen, E. L., 251
Plapp, B. V., 302, 306(12, 42), 307(12)
Plato, F., 105
Plummer, T. H., Jr., 302, 306(8), 308(8), 310(8), 315(8, 46), 317(8), 318, 324(46, 79), 325, 326(79)
Pol, C., 32
Polenov, V. A., 229(9f), 230, 233(9f), 236(9f), 244, 248(56), 284(7a, b, c, d, e, f, g, h, i), 285(7a, c, e, f, g, i), 286(7a, c, e, f, g, i), 287(7c, f), 288(7a, c, e, f, g, i), 289(7a, f, g), 290(7a, f, i)
Pollman, W., 114, 123(244, 245, 246)
Polyakov, A. I., 276, 280(86)
Popenoe, E. A., 137, 138(55), 188(55, 261)
Popjack, G., 170
Porter, C. R., 121
Portoghese, P. S., 195
Portsmouth, D., 166
Posternak, T., 189(255, 268)
Prelog, V., 141, 193, 199(4)
Press, S. H., 150
Prey, V., 85, 92, 112, 120(2)
Price, C. C., 110
Price, P. A., 308, 315(48), 316(48)
Pricer, W. E., 330, 331(114)
Priddle, J. E., 110, 121
Pridham, J. B., 302, 306(16), 307(16)
Priestly, G. C., 328
Prilezhaeva, E. P., 239
Primas, H., 14, 83
Pruyn, M. L., 328
Puranen, J., 208
Purdie, T., 93, 100(43)

Q

Qadir, M. H., 222, 223(67), 224(68, 69), 225(67)
Quadulieg, M., 115(268)

R

Raban, M., 63, 64(179), 192
Rabin, B., 121
Radatus, B., 28, 131, 132, 146(23), 189(23)
Rader, C. M., 53
Radford, T., 150
Radhakrishnamurthy, R., 308
Rafferty, G. A., 215
Raftery, M. A., 176
Rajabalee, F., 33
Ralph, A., 181
Ramer, R. M., 187(257)
Randall, M. H., 209
Randle, D. G., 98
Rand-Meir, T., 176
Rao, C. N. R., 239
Rasheed, A., 168
Raske, K., 207
Rasmussen, C., 307, 315(41)
Rauschenbach, P., 135(44), 183(44)
Reber, L. A., 98
Redfern, G. L., 102
Redfield, R. D., 327
Reeder, J. A., 92, 93(38), 122(38)
Reeds, R. D., 204, 216(27)
Rees, B. H., 209
Rees, D. A., 40
Reeves, W. A., 110
Reggiani, M., 40, 124
Rehpenning, W., 120
Reichard, P., 145, 161(92)
Reilly, C. A., 74, 83(230)
Reinefeld, E., 115, 121(258)
Reist, E. J., 238
Rendleman, J. A., 100
Renold, A. E., 181
Reymond, D., 189(255, 268)
Rheineck, A. E., 121
Rich, P., 204, 215(23), 218, 219(23, 60)
Richards, F., 327
Richards, J. H., 156, 171(137)
Richardson, A. C., 33

Richter, J., 185(247), 188(247)
Richter, W., 305, 339(24)
Richtmyer, N. K., 100
Rieder, S. V., 135, 175, 179(185), 183(37, 185), 185(185), 188(185)
Rist, C. E., 101, 119, 121
Rizvi, S., 319
Roberts, H. J., 120
Roberts, J. D., 44, 80, 149
Robins, R. K., 35, 81(109)
Robinson, D., 318
Robinson, G. B., 328, 329
Rodman, S., 70
Rognstad, R., 135, 178, 179, 185(42)
Rogovin, Z. A., 121
Romy, P. R., 105, 120(163)
Ropp, A., 154
Rose, I. A., 134, 135(45), 141(40), 142, 143(35, 45), 156, 157(34), 158(34, 35), 164(34, 40, 136, 140, 168), 165(45), 172, 175, 178, 179(185), 183(35, 37, 41, 185), 184(45, 140), 185(34, 185, 199), 186(140, 199), 187(35, 45, 140), 188(34, 185), 189(34, 214)
Roseman, S., 328, 330, 331(108, 113), 332(108), 338(113), 339(113)
Rosenblum, C., 186(258)
Rosenthal, A., 150, 230, 234, 235(34, 35), 240(34, 35), 241(34), 242(35), 243(35, 49), 259(13, 14), 260, 261(13, 14, 75, 76), 262(13, 14, 75, 76), 292
Rosenthal, L., 121
Ross, V., 36, 37(112), 38(112)
Routledge, D., 93, 110(45)
Rouzeau, C., 131, 189(22)
Rowe, J. J. M., 10
Rowe, K. L., 10
Rowland, F. S., 138, 140(61, 62), 184(61, 62), 185(61, 62)
Rudenko, N. Z., 97
Ruffini, G., 111
Ruoff, P. M., 122
Rupley, J. A., 327
Ruskiewicz, M., 101, 116(121)
Russ, G. A., 177
Russell, I. J., 307, 315(41)
Rutherford, D., 92, 93(30), 122(30), 145
Rutherford, E., 128
Rutter, W. J., 135, 188(39)
Ryback, G., 170
Ryhage, R., 151

S

Saeki, H., 110
Saha, N. C., 151
Sainsbury, G. L., 194
Salas, M., 158
Salemink, C. A., 208
Salnikow, J., 302, 306(11), 308(11), 324(11), 325(11), 326(11)
Salo, W. L., 158, 186(141b), 337
Sammon, P. J., 137, 138(55), 188(55)
Samokhvalov, G. I., 230, 233(15a, b, c), 274(15a, b, c, d), 276(15a), 279(15a, b)
Sanders, R. D., 180, 181(227), 184(227)
Sanderson, G. R., 40, 41(136), 42(136), 43(136)
Sangster, A. W., 24
Sarcione, E., 328
Sarkanen, K. V., 114, 116(238)
Sarma, R. H., 36, 37(112, 116, 118), 38(112, 116, 118, 120), 39(120), 57, 58(168a)
Sasson, S., 111
Sato, Y., 188(263)
Sattler, L., 94
Sawai, T., 307
Sawyer, H. L., 102
Schachter, H., 329, 330, 331(108), 332(108)
Schäfer, H., 293
Schaeffer, L. D., 179
Schafer, W., 100
Schaffer, R., 18, 231, 234, 296
Scharmann, H., 150
Schauer, H., 339
Scheibler, C., 98
Scheinberg, I. H., 129, 180(15), 188(15), 332, 338
Scheinost, K., 114
Schenkein, I., 180
Scheraga, H. A., 327
Schiff, N., 102
Schiweck, H., 85
Schmid, L., 92
Schneebeli, P., 110, 114(216, 217), 123(216, 217)
Schneider, J. J., 26
Schneider, P., 92, 93, 108
Schneider, W. G., 8, 67(1), 144
Schöllkopf, U., 227
Schoepfer, G. L., Jr., 170
Scholda, R., 177
Schopfer, W. H., 189(255, 268)
Schramm, G., 114, 123(244, 245, 246)

Schrefeld, O., 98
Schubert, D., 180
Schuching, S., 187(259)
Schultz, H., 294
Schultz, S. G., 181
Schupp, O. E., 259
Schutzbach, J. S., 140, 165(76), 174(76), 184(76), 187(76)
Schwartz, J., 305
Schwartz, J. H., 121
Schweiger, R. G., 114, 116(239)
Schweizer, M. P., 34, 36(103)
Scocca, J., 302, 306(15), 309(15), 310(15)
Seeliger, A., 108, 124(191), 125(191)
Segal, L., 112
Segur, J. B., 98
Seibl, J., 143
Sellers, D. J., 194
Senne, J. K., 185(248), 188(248)
Sephton, H. H., 120
Seus, E. Y., 259
Seyama, Y., 169
Shafizadeh, F., 100, 332
Shallenberger, R. S., 120
Sherman, W. R., 171
Sherry, R. H., 91, 94(17), 112(17)
Shibaev, V. N., 233, 253(26)
Shimizu, S., 181
Shinsugi, E., 117, 118(272), 119(272), 121(272), 122(272)
Shkurina, T. N., 239
Shockley, W., 94
Shorygin, P. P., 239
Shostakovsky, M. F., 239, 276, 280(86)
Shreeve, W. W., 179
Siddall, T. H., 64
Siefert, E., 39, 40(127, 128, 129)
Siehrs, A. E., 91
Simon, H., 119, 140, 151, 152, 158, 179, 183(74, 124, 241, 242), 184(74), 185(123, 124, 242), 186(74, 123, 253), 187(74), 188(74)
Simpson, D. L., 301, 319, 321(68), 323(68), 326(1), 328(68), 330(1)
Singleton, M. F., 120
Sinohara, H., 335
Sircar, A. K., 115(326), 121
Skell, P. S., 93, 99
Sky-Peck, H. H., 335
Slansky, J. M., 114
Slavinski, R., 179

Slessor, K. N., 81
Sloan, J., 305
Sloan, J. W., 121
Smirnyagin, V., 101
Smith, A. A., 35
Smith, B., 120
Smith, D. G., 40
Smith, F., 140, 175(69), 184(69), 185(69)
Smith, I. C. P., 34, 35(104, 105, 107), 81(104, 105, 107, 109)
Smith, M. S., 329
Smyth, D. G., 327
Smyth, D. S., 334
Smythe, B. M., 87, 96(6, 7, 8, 9, 10, 11), 98(6, 8, 9), 99(6, 7, 8, 9, 10, 11), 100(7, 9, 11), 106(7, 8, 9, 10, 11), 107(6, 7, 8, 9, 10, 11)
Snell, F. D., 115(262, 263)
Sniegoski, L. T., 140, 141, 153, 154(126, 127), 155(128, 129), 176, 183(188), 186(188), 187(188)
Snyder, W. H., 110
Sols, A., 158
Somers, P. J., 10
Spedding, H., 149
Spencer, J. F. T., 41
Spencer, W. W., 116
Spinola, M., 325, 326(80)
Spiro, M. J., 328, 330
Spiro, R. G., 328, 330, 337
Sprecher, R. 19, 20(51)
Sprinson, D. B., 163
Sprinzl, M., 234, 235(34, 35), 240(34, 35), 241(34), 242(35), 243(35, 49), 292
Srinivasan, P. R., 163
Stacey, M., 91, 103, 149, 188(106), 189(106)
Stärker, A., 93
Stafford, H. A., 170
Stahl, P. D., 302, 309(13)
Staněk, J., 112
Stanford, F. G., 139
Stanonis, D. J., 115(326), 121
Staub, M., 164(157b)
Stein, W. H., 302, 306(10, 11), 307, 308(10, 11), 315(48), 316(48), 322(10), 324(11), 325(11), 326(11), 327
Steinbach, L. H., 209
Steiner, P. R., 82, 147
Stephen, A. M., 67, 83(207)
Stern, B. K., 170, 180(175), 189(175)

Sternhell, S., 9, 71(19)
Sternlieb, I., 338
Stetten, D., 132, 133(26), 137, 140(26), 141(26), 188(26), 189(26)
Stetten, M. R., 137, 189(57)
Stevens, J. D., 22, 67(59), 68(59), 69, 70(59), 133
Stevens, W. H., 154
Stewart, M. A., 171
Stewart, W. E., 64
Stift, A., 98
Stirling, C. E., 181
Stirling, J. L., 318
Stjernholm, R., 179, 189(214)
Stockham, T. G., Jr., 53
Stoddart, J. F., 10, 64(33), 65, 66, 73(192), 81(192)
Stoffyn, P. J., 121
Stolarova, L. G., 239
Stone, B. A., 307
Stoolmiller, A. C., 166
Strahs, S., 179
Strobach, D. R., 28
Stroh, H. H., 100
Strohmer, F., 98
Strominger, J. L., 329, 332
Stud, M., 82(255), 83
Suhadolnik, R. J., 187(257)
Sukeno, T., 326
Sun, C., 80
Sutcliffe, L. H., 9, 46(4), 64(4), 71(4)
Suzuki, B., 100, 106(106)
Suzuki, S., 181
Suzuki, T., 306(43), 307, 308(43), 324(43)
Swain, C. G., 105, 120(162)
Swalen, J. D., 74, 75, 83(230, 241)
Swan, J., 123
Swartz, M. N., 178
Sweat, F. W., 109
Sweeley, C. C., 120, 150, 151, 332
Swenson, R. M., 180, 328, 331(97)
Swezey, F. H., 106, 116(178)
Swoboda, B. E. P., 302, 306(20), 309(20)
Sykes, B. D., 59
Szarek, W. A., 65, 66, 73(192), 81(192), 230, 234(10), 235(10, 11, 31, 32), 240(31, 32), 241(11), 242(10), 243(32, 47)
Szczerek, J., 234, 235(32), 240(32), 243 (32)
Szmant, H. H., 104, 122

T

Taigel, G., 108, 124(191), 125(191)
Takahashi, N., 320, 322(73, 74), 339
Takahashi, T., 186(252)
Takuira, K., 115, 122(256)
Talamo, B., 332
Talley, E. A., 121
Tanksley, T. D., 341
Tarelli, E., 33
Tarentino, A. L., 318, 324(79), 325, 326(79)
Tatchell, A. R., 196, 197, 204, 216(27)
Tatlow, J. C., 103
Taylor, M. R., 142
Tedder, J. M., 103
Terajima, K., 108
Tetas, M., 332
Thakkar, A. L., 33
Theobald, R. S., 105, 106(164), 110(164)
Theorell, H., 302, 306(17), 309(17)
Thomas, S., 24
Thomas, W. A., 30(86), 81(86)
Thompson, A., 112
Thompson, P. B., 205, 220(33)
Thomson, J. K., 215
Tindall, C. G., Jr., 63, 130(19), 131
Tipson, R. S., 64, 92, 107, 119(184), 120
Tittensor, J. R., 258, 260(72), 261(72)
Tolbert, B. M., 134, 183(32)
Tomita, S., 108
Tomoskozi, I., 298
Topper, Y. J., 132, 133(26), 134, 135, 140(26), 141(26), 157(33), 188(26), 189(26, 33)
Touster, O., 302, 309(13)
Tovey, G. P., 103
Tracey, A. S., 81
Trams, E. C., 178
Treadway, R. H., 121
Trebst, A., 151, 175, 179, 183(213, 241)
Trey, H., 97, 98(80)
Triggle, D. J., 209
Trippett, S., 244
Trischmann, H., 114
Tronchet, J., 273, 282(85)
Tronchet, J. M. J., 233, 234, 240(33), 241(33), 271(29a, b, 33), 272(29a, b, 83b), 273(29b, 84), 275(29b, 83b, 84), 276(29a, b, 83, 84), 277(29a, b, 83a, b, 84), 278(29a, b), 280(83a), 281(29a, 83b, 84), 282(29a, b, 85), 283(29a, b)

Tsernoglou, D., 327
Ts'O, P. O. P., 34, 36(103)
Tsugita, A., 320, 323(75)
Tsuji, M., 181
Tsuruoka, T., 27
Tu, J. I., 176
Tucker, N. B., 110, 114, 121
Tukey, J. W., 52, 53(156)
Turner, J. C., 105, 122(167)
Turner, W. N., 147
Turton, C. N., 138, 140(61), 184(61), 185(61)
Turvey, J. R., 122
Tyler, T. R., 158

U

Uematsu, T., 187(257)
Uhlenbruck, G., 339
Uhr, J. W., 180
Ulsperger, E., 115
Unger, W., 85, 112, 120(2)
Urey, H. C., 128
Uslova, L. A., 228, 229(4, 9f), 230, 232, 233(9f, 24), 236(9f), 237(25), 244(24, 25), 245(24, 25, 39, 51, 52, 53), 246(24, 53), 247(24, 25, 51, 54, 55), 248(24, 54), 249(24, 51, 52), 253(25, 39), 254(53), 255(51), 256(25), 261(51), 267, 268(80), 278(28), 294(51), 295(96)
Usov, A. I., 233
Utsumi, S., 334

V

Van den Hamer, C. J. A., 129, 180(15), 188(15), 338
van Gorkom, M., 64
Van Lenten, L., 180, 188(223)
van Rossum, J. M., 205, 210(32)
Vennesland, B., 145, 146(88), 170, 180(175), 189(175)
Verhaar, G., 107
Vessey, D. A., 332
Vigevani, A., 40
Vinuela, E., 158
Vittek, J., 332
Vogel, H., 117
Vogele, P., 305

W

Wade, C. W. R., 184(244), 185(244), 186(244)
Wade, R. H., 103, 110
Wade-Jardetzky, N. G., 36, 37(111), 39(111)
Wadsworth, W. S., 259
Wagner, A., 94
Wagner, J. J., 19, 20(51d)
Wagner, J. E., 327
Wagner, R. R., 330, 337(109)
Wahlgren, S., 98
Walborsky, H. M., 202
Walkers, D. M., 244
Wall, H. M., 215
Walter, C. F., 36, 38(121)
Walton, D. J., 81
Wander, J. D., 67, 133, 147(31), 150, 189(31, 115a)
Wang, P. Y., 41
Wang, S. F., 168, 170(169), 184(169)
Ward, K., 110
Ward, L., 146, 159(95), 171(95)
Warren, C. D., 237, 238(40d), 241(40), 258, 261(71), 262(71), 271(71)
Waschkau, A., 92
Waser, P. G., 207
Washburn, E. W., 128
Watkin, D. J., 222, 224(71)
Watkins, W. M., 339
Waugh, J. S., 57
Weaver, H. E., 18, 19(47), 21(47)
Webb, M., 91
Webber, J. M., 209, 222, 223(65, 66, 67), 224(65, 66, 68, 69, 70), 225(67), 293, 297(92)
Weber, H., 143
Wee, T. G., 168
Weigel, H., 149, 188(106), 189(106)
Weigert, F. J., 149
Weil, C. M., 98
Weinhold, P. A., 180, 181(227), 184(227)
Weiss, M., 332
Weiss, T. J., 95, 115(73), 123
Welch, F. J., 193
Welch, P. D., 52, 53(157)
Wellman, G., 231, 233(21), 245(21), 246(21), 253(21), 254(21), 256(21), 261(69)
Wells, W. W., 120
Wenzel, M., 185(249), 188(249)
Werner, J., 106
Wernicke, A., 102
West, E. S., 106

Westall, F. C., 334
Wetherell, J., 219
Weygand, F., 119, 141, 152, 183(124), 185(123, 124), 186(123)
Wheeler, O. H., 239
Whiffen, D. H., 75, 149, 188(106), 189 (106)
Whistler, R. L., 109, 111, 116
Whitaker, J. R., 308
Whitehead, J. S., 334
Whitemore, N. K., 293, 297(92)
Whyte, J. N. C., 40
Wicker, R. J., 110, 114(206, 207)
Wiedenhof, N., 114
Wight, W., 40
Wilcox, G. N., 112, 119
Wilkins, C. L., 75
Willard, J. J., 65, 66(193)
Williams, E. H., 230, 234(10), 235(10), 242(10)
Williams, J. M., 23, 67(62, 63), 71(62), 73(63), 75(62), 114(249)
Williams, N., 219
Williams, N. R., 204, 214(22), 215(25), 216(22), 217(22)
Williams, R. H., 204, 216(27)
Willstaedt, H., 121
Wilson, D. B., 159
Wilson, T. H., 181
Wilton, D. C., 159
Wilzbach, K. E., 137, 188(54)
Winterburn, P. J., 336
Winzler, R. J., 328, 329
Wippel, H. G., 259
Wittig, G., 227, 234(2b)
Wolff, I. A., 91, 121
Wolfgang, R., 138, 140(61), 184(61), 185(61)
Wolfrom, M. L., 41, 96, 103, 104, 112, 207, 215, 216(56)
Wood, H. B., Jr., 96
Wood, H. G., 179, 189(214)
Wood, W. A., 135(46), 143(46), 158(46), 161(46), 161, 165, 184(159), 186(159), 189(159)
Woodlock, J. J., 312
Woodman, C. M., 75
Woodward, R. B., 239
Wright, G. F., 193, 194(3)
Wulf, H. D., 179
Wulfson, N. S., 150, 241
Wyckoff, H. W., 327

Y

Yamada, H., 302, 306(23), 309(23), 315(23), 316(23)
Yamaguchi, H., 320, 321(76), 323(76), 324 (76), 325(76a)
Yamane, T., 34
Yanovskaya, L. A., 227
Yanovsky, E., 97, 98(81), 121
Yasuda, Y., 302, 320, 322(73, 74), 339
Yasui, M., 302
Yasunobu, K., 302, 306(23), 309(23), 315(23), 316(23)
York, C., 115(262, 263)
Young, H. H., 100
Young, R. C., 29, 63(82), 67(82), 73(82)
Yu, R. J., 109
Yunker, M., 28, 131, 146(23), 189(23)

Z

Zacharius, R. M., 312
Zagalack, B., 143
Zagoren, B. L., 101
Zakim, D., 332
Zanlungo, A. B., 23
Zehavi, U., 341
Zell, T. E., 312
Zemplén, G., 121
Zerban, F. W., 94
Zeringue, H. J., 95, 115(73), 123
Zhdanov, Yu. A., 228, 229(4, 9d, f), 230, 232(17), 233(9a, c, d, f, 17, 24), 236(9a, c, d, f), 237(9d, 25), 238(29c), 239(9c), 240(9c, 29c), 241(9d), 244(17, 23, 24, 25), 245(17, 24, 25, 29c, 39, 50, 51, 53), 246(24, 53), 247(24, 25, 51, 54, 55), 248(24, 29c, 54, 56), 249(24, 51, 57), 250(50), 251(57), 252(29c, 57), 253(4, 25, 39), 254(53), 255(51), 256(25), 257(23), 261(51), 267, 268(80), 271(28a, b), 276(28a, b), 278(28), 284(7a, c, e, f, g, h), 285(7a, c, e, f, g), 286(7a, c, e, f, g), 287(7c, f), 288(7a, c, e, f, g), 289(7a, f, g), 290(7a, f), 294(51), 295(96)
Zief, M., 121, 122
Zilkha, A., 111
Zimmer, H. J., 115(269b), 116
Zinbo, M., 171
Zinner, H., 120
Zitko, V., 109
Zolotarev, B. M., 150

SUBJECT INDEX

A

Acetic acid
  as solvent for sugars, 102
  —, chloro-, as solvent for sucrose, 104
Acetone
  solubility of sucrose in, 107
  solubility of sugars in, effect of zinc chloride on, 95
Acetonitrile, trichloro-, as solvent for sugars, 113
Acids
  hydroxy, of high optical purity, preparation of, 199–201
  organic, as solvents and reaction media, 102
  α,β-unsaturated, chemical and physical properties, 260–267
    synthesis by Wittig reaction, 253–260
Acrolein, reaction with sucrose, effect of zinc chloride on, 95
Adenine, dinucleotide with nicotinamide, proton magnetic resonance spectroscopy of, 37, 57
Adenosine
  5'-phosphate, proton magnetic resonance spectroscopy of, 81
  spin-lattice relaxation of ^{13}C nuclei in, 59
—, adenylyl-(3'→5')-, proton magnetic resonance spectroscopy of, 36
—, 2'-deoxy-, proton magnetic resonance spectroscopy of, 81
Alanine, L-, configurational relationship with 2-amino-2-deoxy-D-glucose, 206
Alcohols
  alkoxy, solvation of sugars by, 88
  as solvents for sucrose, 87
  as solvents for sugars, 96, 99
  ω-aminoalkyl, as solvents for sugars, 99
  configuration determination of, 201
  optically active secondary, preparation of, 196
  polyhydric, as solvents for sugars, 98
    as reaction media, 99
    as solvents for sugars, 96–102
Ald-3-enulosonic acids
  alkyl esters, chemical properties of, 268
  synthesis by Wittig reaction, 267

Alditols, unsaturated, synthesis by Wittig reaction, 237
Alditols-1-t, preparation of, 129
Alditols-2-t, preparation of, 129
Aldohexopyranosides, 6-deoxy-, proton magnetic resonance spectroscopy of, 70
Aldolases, and hydrogen-isotope action on sugars, 156
Aldonic acids, α,β-unsaturated, synthesis of, 261
Aldoses
  oxidation of tritiated, mechanism of, 153–155
—, anhydro-C-(p-methoxyphenyl)-, synthesis and properties of, 284–292
Aldoses-1-t, preparation of, 129
Alkanolamines, as solvents for sugars, 99
Allofuranose, 3-C-(cyanomethyl)-3-deoxy-1,2:5,6-di-O-isopropylidene-α-D-, preparation of, 262
—, 3-deoxy-1,2:5,6-di-O-isopropylidene-3-C-(methoxycarbonylmethyl)-α-D-, preparation of, 262
—, 3-deoxy-1, 2:5, 6-di-O-isopropylidene-3-C-methyl-α-D-, preparation of, 241
—, 3-deoxy-1,2:5,6-di-O-isopropylidene-3-C-(nitromethyl)-α-D-, preparation of, 243
—, 3-deoxy-3-C-(hydroxymethyl)-1,2:5,6-di-O-isopropylidene-α-D-, preparation of, 243
Allofuranose-3-d, 1,2:5,6-di-O-isopropylidene-α-D-, nuclear magnetic resonance spectroscopy of, 147
Allose, 2-deoxy-D-, synthesis of, 4
Amides, as solvents for sugars, 112
Amines, aliphatic, as solvents for sugars, 112
Amino acids
  in protein components of glycoenzymes, 315–318
  sequences of glycopeptides in glycoenzymes, 319–323, 333
Aminolysis, of sucrose, 99
Ammonia
  liquid, metallation of sugars in, 92
  as solvent for carbohydrates, 91

reaction with α,β-unsaturated acids, 262–264
Ammonolysis, of sucrose in liquid ammonia, 93
Amylase, fungal yeast, carbohydrate content of, 307
α-Amylase
  carbohydrate content of, 306, 307
  chromatography and sedimentation of, 311, 313
  protein-carbohydrate linkages in, 323
Amylopectin, solvents for, 108
Amylose
  solvents for, 108
  structure and proton magnetic resonance spectroscopy of, 39
—, 2,3,6-tri-O-acetyl-, proton magnetic resonance spectroscopy of, 39
Anticholinergic drugs, preparation of, 209, 219
Arabinitol, 1-deoxy-1-C-heptyl-L-, preparation of, 241
—, 2,3:4,5-di-O-cyclohexylidene-1-deoxy-1-C-heptyl-L-, preparation of, 241
Arabinopyranose, 1-thio-α-L-, tetraacetate, proton magnetic resonance spectroscopy of, 21, 22
Arabinopyranosyl fluoride, tri-O-acetyl-β-D-, proton magnetic resonance spectroscopy of, 82
Arabinose,
  D-, solubility in methanol, 97
  L-, glycoenzyme constituent, 302, 307, 321
  solubility in liquid ammonia, 91
—, 2-deoxy-2-deuterio-D-, synthesis of, 131
Arabinose-2-d, preparation of, 133
Arabinose-5,5-d_2, L-, nuclear magnetic resonance spectroscopy of, 148
Ascorbic acid-4-t
  L-$arabino$-, preparation of, 134
  L-$xylo$-, preparation of, 134
Atrolactic acid, asymmetric synthesis of, 193, 200
Atropine-like drugs, preparation of, 213, 219
Autoradiography, histochemical detection and localization of sugars in tissue by, 181
Aziridines, sugar, preparation of, 257, 264

B

Benzyl alcohol, α-methyl-, R and S isomers, preparation of, 197
Betaines, formation in Wittig reaction, 290–292
Biochemistry
  hydrogen movement within the cell and incorporation into cell components, 177
  incorporation of labeled monosaccharides into polysaccharides, 179
  mechanistic, hydrogen-labeled sugars in study of, 155–176
  pathway of enzymic transformations in tritiated compounds, 176
Biologically active compounds, configuration and purity of, 205–221
Biosynthesis
  of glycoenzymes and glycoproteins, 303, 308, 328–337
  of nucleotides, 336
  of L-streptidine, 177
Biotin
  absolute configuration of, 212
—, dethio-, synthesis and absolute configuration of, 212
Boltzmann distribution, in nuclear magnetic resonance spectroscopy, 44
Bromelains
  glycoenzymes, 309
  pineapple, carbohydrate components of, 325
  purification of pineapple, 310
Bromination, of unsaturated sugars, 241, 249
Butanol, 4-methoxy-, solubility of sucrose in, 87, 89
2-Butanol, 2-phenyl-, asymmetric synthesis of, 193, 194
Butyl alcohol, solubility of sucrose in, 89, 98

C

Calcium chloride, effect on sugar solubility in methanol, 95
Carbodiimide, N,N'-dicyclohexyl-, effect on solubility of sugars in methyl sulfoxide, 109
Carbohydrate chemistry, Wittig reaction in, 227–299

Carbohydrates
  in glycoenzymes, 302, 306, 321–327
  nuclear relaxation times, measurement of, 59
  proton magnetic resonance spectroscopy of, 7–83
  solubility in liquid ammonia, 91
  solvents (non-aqueous) for, 85–125
  as sources of asymmetric carbon atoms, 205
  in synthesis and configurational assignments of optically active non-carbohydrate compounds, 191–225
Carbon, isotope ^{13}C, identification in sugars by use of deuterium, 149
5β-Card-20(22)-enolide, 3β-(2-deoxy-β-D-lyxo-hexopyranosyl)oxy-14β-hydroxy-, synthesis of, 4
Cardenolides, synthesis of, 3
Cellulose
  methylation of, in liquid ammonia, 91
  solubility in anhydrous hydrogen fluoride, 93
  solubilization by thiocyanates, 95
  solvents for, 113
  structure and proton magnetic resonance spectroscopy of, 39
—, 2,3,6-tri-O-acetyl-, proton magnetic resonance spectroscopy of, 39
Centrifugation, in glycoenzyme characterization, 312, 314
Chloral, as solvent for sugars, 107
Chloroform, as solvent for sugars, 105
Chloroperoxidase
  carbohydrate components of, 326
  glycoenzyme, 309
  purification of, 310
Cholesterol-4-^{14}C, thesis, 2
Choline, acetyl-α- and -β-methyl-, enantiomeric, preparation and effect on nervous systems, 210
Cholinergic drugs, structure–activity relationships, 208, 209
Chondroitin, 4- and 6-sulfate, proton magnetic resonance spectroscopy of, 41, 43
Chromatography
  gas–liquid, of deuterated sugars, 151
  glycoenzyme purification by, 310
  of sugar solutions, 123
Collagen, biosynthesis of, 331

Configuration
  of biologically important non-carbohydrate compounds, proof of, 205–221
  carbohydrate use in asymmetric synthesis and proof of, 191–204
Conformation, of glycoenzymes, 327
Convallatoxin, 6-hydroxy-, preparation and cardiotonic activity of, 4
Cram rules, in asymmetric synthesis, 193, 263
Cresol, as solvent for mutarotation of D-glucose, 104
Cyclobutene, sugar derivative, synthesis of, 299
Cycloheptaamylose, complex with p-hydroxybenzoic acid, proton magnetic resonance spectroscopy of, 32
Cyclohexaneglycolic acid, α-phenyl-, preparation of, 201
Cyclohexanol, as solvent in reduction of D-glucose, 99
Cytidine, proton magnetic resonance spectroscopy of, 33, 146
Cytidine-2'-d, 2'-deoxy-, proton magnetic resonance spectroscopy of, 145
Cytosine, proton magnetic resonance spectroscopy of derivatives, 28, 29, 146

D

Degradation, localization of hydrogen isotopes by chemical, 140
Deoxyribonuclease
  amino acids of, 315, 316
  as glycoenzyme, 308
Dermatan sulfate, proton magnetic resonance spectroscopy of, 41, 43
Dethiobiotin, synthesis and absolute configuration of, 212
Deuterium, determination of location of, by nuclear magnetic resonance spectroscopy, 143
Dextran
  proton magnetic resonance spectroscopy of derivatives of, 40
  solubility in methanol, 95
Dextrin, solubility in methanol, 95
Dialdoses, α,β-unsaturated, Wittig reaction with, 251

Diastereotopism, proton magnetic resonance spectra, 63
Dielectric constant
  effect on solubility of sucrose, 90
  of solvents, effect on sugar solubility, 87
Diethanolamine, from aminolysis of sucrose, 99
Diethyl ether, as solvent for sucrose, 87
Di-D-fructopyranose 2',1:2,1'-dianhydride, from D-fructose in anhydrous hydrogen fluoride, 94
Digitoxigenin
  α-L-rhamnopyranoside, preparation of, 4
  α- and β-D-rhamnosides, synthesis of, 4
Digitoxose, D-, 1,3,4-tris(p-nitrobenzoate), preparation of, 3
Digitoxoside, mono-, preparation and properties of, 3
Dinitrogen tetraoxide, as solvent for sugars, 94
Dinucleosides, monophosphates, proton magnetic resonance spectroscopy of, 36
Dinucleotides, monophosphates, proton magnetic resonance spectroscopy of, 36
p-Dioxane, as solvent for sugars, 107
1,3-Dioxan-5-ol, as solvent for sugars, 88, 89
1,3-Dioxolane, 2-(α-cyclohexyl-α-hydroxybenzyl)-4-(dimethylaminomethyl)-, isomers, preparation of anticholinergic drug, 219
—, cis-2,4-dimethyl-, nuclear magnetic resonance spectroscopy reference compound, 209
—, 4-[(dimethylamino)methyl]-2-methyl-, methiodide, isomers, cholinergic potency of, 208
—, 4-(hydroxymethyl)-2-methyl-, relative configurations of, 208
1,3-Dioxolane-4-methanol, as solvent for sugars, 88, 89
Dipole moment, of solvents, effect on sugar solubility, 87
Dipropyl sulfoxide, as solvent for sugars, 108
Disaccharides
  solubility in liquid ammonia, 91
  tritiated, preparation and position of tritium, 188

E

Electron availability, of solvent, effect on sugar solubility, 87
Electronic absorption spectra
  of anhydro-C-(p-methoxyphenyl)-aldoses and anhydro-1,3-dideoxyketoses, 286
  of C-arylated unsaturated hexitols, 239
  of dienic carbonyl sugars, 251
  of furan derivatives from sugars, 289
  of thioenol sugars, 275
  of α,β-unsaturated acids, 260
  of α,β-unsaturated carbonyl sugars, 248
Electrophoresis, of glycoenzymes, 311
Enzymes
  of glycoprotein structure, 301-341
  isotope-effects in reactions, 171-176
  mechanistic studies, hydrogen isotope labeled sugars in, 155-162
  in solvent-exchange labeling of sugars with hydrogen isotopes, 134
  stereochemistry determination at a chiral methylene group by, 141
Enzymolysis, of polysaccharides and proton magnetic resonance spectroscopy, 41
Epimerases, and hydrogen-isotope action on sugars, 156
Erythran, as solvent for sugars, 88
Erythritol, 1,4-anhydro-, as solvent for sugars, 88
Esterases, mammalian, glycoenzyme, 308
Esters
  α-oxo, of carbohydrates, Grignard reaction with, 199
  of carbohydrates, reduction reactions of, 201
  as solvents for sugars, 106
α,β-unsaturated carbohydrate, conjugate-addition reactions of Grignard reagents with, 202-204
Ethanol
  solubility of sucrose in, 87, 89
  as solvent for sugars, 96, 97
—, 2-acetoxy-, solubility of sucrose in, 89
—, 2-amino-, as solvent for sucrose, 99
—, 2-butoxy-, solubility of sucrose in, 89
—, (R)-(+)-1-cyclohexyl-1-phenyl-, prep-

aration of, 194
—, (S)-(−)-1-cyclohexyl-1-phenyl-, preparation of, 194
—, 2,2',2'',2'''-(ethylenedinitrilo)tetra-, from aminolysis of sucrose, 99
—, 2,2'-iminodi-, from aminolysis of sucrose, 99
—, 2-methoxy-, solubility of sucrose in, 87, 89
as solvent for sugars, 96
—, 2-(2-methoxyethoxy)-, solubility of sucrose in, 89
—, 1-phenyl-, R and S isomers, preparation of, 197
—, 2,2',2'',2'''-(propylenedinitrilo)tetra-, from aminolysis of sucrose, 99
Ethanol-1-d, (+)-, absolute configuration of, 220
Ethanol-2-d
configuration of, 144, 146
in determination of hydrogen configuration in sugars, 143
Ethylene glycol, as solvent for sugars, 98
Ethyl lactate, as solvent for sugars, 106
Evomonoside, preparation of, 4

F

Formamide, N,N-dimethyl-, as solvent for sugars, 89, 109, 113–116
sucrate production in, 93
Formic acid, as solvent for sugars, 102
D-Fructopyranose D-fructofuranose 2',1:2,1'-dianhydride, from D-fructose in anhydrous hydrogen fluoride, 94
Fructose
D-, anomeric equilibria in aqueous solution and Fourier-transform spectra, 60
solubility in alkoxy alcohols, 99
in anhydrous hydrogen fluoride, 94
in liquid ammonia, 91
in methanol, effect of calcium chloride on, 95
β-D-, solubility in methanol, 97
Fructose $1(R)$-t
D-
mechanism of osazone formation, 151
6-phosphate, preparation of, 134, 158
Fructose-$1(S)$-t,
D-
mechanism of osazone formation, 151

6-phosphate, preparation of, 134
Fructose-5-t, D-, preparation of, 132, 135
Fructose-3,4,5-t_3, D-, preparation of, 135
Fucose, L-, glycoenzyme constituent, 302, 307, 321
2-Furaldehyde, 5-(hydroxymethyl)-, preparation of, and derivatives, 101, 104
Furan
C-glycosylated, synthesis by Wittig reaction, 288
—, 2,5-bis(hydroxymethyl)tetrahydro-, half-ethers, as solvents for sugars, 88
—, 2-(p-bromophenyl)-, synthesis by Wittig reaction, 288
—, 3-(ethylthio)-2-[(ethylthio)methyl]-, from D-xylose, 101
—, 2-(hydroxymethyl)-5-(p-methoxyphenyl)-, synthesis by Wittig reaction, 288
—, 2-(p-methoxyphenyl)-, synthesis by Wittig reaction, 288
—, tetrahydro-, solubility of sucrose in, 107
2,5-Furandimethanol, tetrahydro-, alkoxyalkyl monoethers, as solvents for sugars, 107
Furan-2(3)-one, dihydro-, as solvent for sugars, 106
Furfuryl alcohol
as solvent for sugars, 88
—, tetrahydro-, solubility of sucrose in, 89
as solvent for sugars, 88, 96, 107
Furoic acid, methyl ester, Fourier proton magnetic resonance spectroscopy of, 62

G

Galactofuranose, 3-deoxy-1,2:5,6-di-O-isopropylidene-3-C-methyl-D-*galacto*-, synthesis of, 282
Galactofuranosylamine, N-acetyl-β-D-, proton magnetic resonance spectroscopy of, 23
Galactopyranose
β-D-, pentaacetate, proton magnetic resonance spectroscopy of, 70
—, 2-acetamido-2-deoxy-3-O-(4-deoxy-α-L-*threo*-hex-4-enopyranosyluronic acid)-α-D-, proton magnetic resonance spectroscopy of, 42
—, 6-deoxy-1,2:3,4-di-O-isopropylidene-

6-phthalimido-α-D-, proton free-induction decay signal, 48, 49, 51, 57, 58
—, 1-thio-β-D-, pentaacetate, proton magnetic resonance spectroscopy of, 68
Galactopyranoside, methyl 4,6-O-benzylidene-2,3-O-methylene-D-, proton magnetic resonance spectroscopy of, 66
Galactose, D-
  in glycoenzymes, 302, 307, 321
  solubility in liquid ammonia, 91
  in methanol, 97
  in methanol, effect of calcium chloride on, 95
  in trichloroacetonitrile, 113
Galactose-4-t, D-, preparation of, 136
γ-Globulin, synthesis of, 331
Glucal-3(S)-d, 4,6-O-benzylidene-3-deoxy-D-, synthesis of, 131
Glucitol, 5-amino-1,5-anhydro-5-deoxy-D-, proton magnetic resonance spectroscopy of, 27
—, 1-deoxy-1-C-methyl-L-, preparation of, 242
Glucoamylase
  amino acids of, 315, 317
  carbohydrate content of, 306, 307
  structure of, 327
Glucofuranose, 3-O-benzyl-1,2-O-cyclohexylidene-α-D-, complex with lithium aluminum hydride, stereoselectivity of reductions with, 196–199
—, 3-O-crotonyl-1,2:5,6-di-O-cyclohexylidene-α-D-, Grignard reaction with, 202
—, 1,2-O-cyclohexylidene-α-D-, and 3-methyl ether, complexes with lithium aluminum hydride, reductions with, 196
—, 3-deoxy-3-deuterio-1,2:5,6-di-O-isopropylidene-α-D-, preparation of, 135
—, 6-deoxy-1,2:3,5-di-O-isopropylidene-6-phthalimido-α-D-, proton magnetic resonance spectroscopy of, 29, 79
—, 6-deoxy-1,2:3,5-di-O-isopropylidene-6-phthalimido-¹⁵N-α-D-, proton magnetic resonance spectroscopy of, 29, 76, 79
—, 6-deoxy-1,2-O-isopropylidene-5-C-phenyl-α-D-, preparation of, 214–216
—, 1,2:5,6-di-O-cyclohexylidene-3-O-(phenylglyoxylyl)-α-D-, reduction of, 201
—, 1,2:5,6-di-O-isopropylidene-α-D-, stereoselectivity of Grignard reaction in presence of, and its derivatives, 194, 195
—, 1,2-O-isopropylidene-α-D-, and 3-ethers, complexes with lithium aluminum hydride, reduction with, 196
—, 1,2-O-isopropylidene-3,5-O-[(endo-methoxy)methylidene]-6-O-p-tolylsulfonyl-α-D-, proton magnetic resonance spectra, iterative analysis of, 78
Glucofuranose-d_{12}, 6-deoxy-1,2:3,5-di-O-isopropylidene-α-D-, proton magnetic resonance spectra, iterative analysis of, 79
Glucofuranoside, 3,6-anhydro-α-D-glucofuranosyl 3,6-anhydro-α-D-, tetrabenzoate, proton magnetic resonance spectroscopy of, 33
Glucofuranosylamine, N-acetyl-α-D-, proton magnetic resonance spectroscopy of, 23
Glucopyranose
  D-, proton magnetic resonance spectroscopy of derivatives of, 26
—, 1,2:4,6-di-O-benzylidene-3-O-(methylsulfonyl)-α-D-, proton magnetic resonance spectroscopy of, 67, 78
—, tetra-O-methyl-α-D-, mutarotation of, in cresol and pyridine, 104
Glucopyranose 6-sulfate, 2-deoxy-4-O-(4-deoxy-α-L-threo-hex-4-enopyranosyluronic acid 2-sulfate)-2-sulfoamino-D-, from heparin, 41
Glucopyranoside, methyl α-D-
  preparation of, 101
  tetraacetate, proton magnetic resonance spectroscopy of, 32
—, methyl 2-O-acetyl-4,6-O-benzylidene-3-deoxy-3-phenylazo-α-D-, proton magnetic resonance spectroscopy of, 69, 78
—, methyl 2,3- and 3,4-anhydro-, proton magnetic resonance spectra of, 81
—, methyl 4,6-O-benzylidene-α-D-, complex with lithium aluminum hydride, reduction with, 196

—, methyl 4,6-O-benzylidene-2,3-O-methylene-α-D-, proton magnetic resonance spectroscopy of, 65
—, methyl 2,3-di-O-acetyl-4,6-O-benzylidene-α-D-, proton magnetic resonance spectroscopy of, 68
Glucopyranoside-2-d, methyl 2-deoxy-α-D- and -β-D-, preparation of, 132
β-D-Glucopyranosiduronate, methyl cholest-5-en-3β-yl 2,3,4-tri-O-acetyl-, proton magnetic resonance spectroscopy of, 26
Glucopyranosiduronates, D-, proton magnetic resonance spectroscopy of, 26
β-D-Glucopyranosylmalonate, diethyl 2,3,4,6-tetra-O-acetyl-, proton magnetic resonance spectroscopy of, 27
Glucosaminidase
 amino acids of, 315, 316
 carbohydrate content of, 306, 307
Glucosan, from cellulose in anhydrous hydrogen fluoride, 93
Glucose,
 D-
  glycoenzyme constituent, 302, 307, 321
  reduction in alcohols, 99, 100
  solubility in alkoxy alcohols, 99
  in anhydrous hydrogen fluoride, 94
  in ethanol, 97
  in isobutyl alcohol, 98
  in liquid ammonia, 91
  in methanol, 97
  in methanol, effect of calcium chloride on, 95
  in thiocyanates, 95
  Wittig five-membered anhydro products from, and its derivatives, 284
 α-D-
  mutarotation and proton magnetic resonance spectroscopy, 59
  mutarotation in cresol and pyridine or pyridinone, 104
 α-D- and β-D-
  infrared spectroscopy of, 149
  proton magnetic resonance spectroscopy of, 28
—, 2-acetamido-2-deoxy-D-, glycoenzyme constituent, 302, 307, 321, 329
—, 2-amino-2-deoxy-D-, configurational relationship to L-alanine, 206

 conversion into D-(−)-muscarine, 208
—, 2-amino-2-deoxy-L-, conversion into L-(+)-muscarine, 207
—, 2,3,4,5,6-penta-O-acetyl-*aldehydo*-D-, Wittig reaction with, 231
Glucose-5-d
 D-
  nuclear magnetic resonance spectroscopy of, 149
  preparation of, 129
Glucose-5-d, α,β-D-, proton magnetic resonance spectroscopy of, 28
Glucose-5,6,6-d_3,
 D-, nuclear magnetic resonance spectroscopy of, 148
 α-D- and α,β-D-, proton magnetic resonance spectroscopy of, 28
Glucose-d_7, β-D-, mass-spectrometric analysis of, 151
Glucose-t, 6-deoxy-D-, detection and localization in tissue by autoradiography, 181
Glucose-1-t, D-
 osazone formation, mechanism of, 152
 preparation of, 129
Glucose-2-t, D-
 6-phosphate, preparation of, 158
 preparation of, 134
Glucose-3-t, D-, preparation of, 129, 135
Glucose-4-t, D-, preparation of, 135
Glucose-5-t, D-
 6-phosphate, preparation of, 135
 preparation of, 129
Glucose-6-t, D-, preparation of, 129
D-Glucose oxidase
 amino acid content of, 315, 316
 carbohydrate components of, 326
 electrophoresis and sedimentation of, 312, 313
 glycoenzyme, 309
Glucosidase, β-D-, purification and carbohydrates of, 305
Glucosiduronase, carbohydrate content of, 306, 307
Glucosyl arsenate, fast Fourier-transform proton magnetic resonance spectroscopy in preparation of, 59
D-Glucosyl-3-t phosphate, synthesis of, 130
Glutaraldehydic-2,4-d_2 acid, 2-oxo-, preparation of, 165

Glyc-3-enuloses, 1-bromo-1,3,4-trideoxy-, synthesis of, 295
Glyceraldehyde
  D-, configurational correlation with L-serine, 207
  Wittig reaction with, 245, 288, 289
Glycerol
  as solvent for sucrose, 98
  —, 1,3-O-methylidene-, as solvent for sugars, 88, 89
  —, 2,3-O-methylidene-, as solvent for sugars, 88, 89
Glycoenzymes
  amino acid sequences in glycopeptides of, 319–323, 333
  amino acids of, 315–318
  biological and structural significance of, 337–341
  biosynthesis of, 303, 308
    mechanism of, 328–337
  carbohydrate components, 306, 321–327
  cellular locale, reactions and mechanisms, 328–332
  conformational structures, 327
  of glycoprotein structure, 301–341
  protein–carbohydrate linkages in, 318–323
  purity criteria for, 309–314
  stability of, 340
  types of, 305–307
Glycofuranosides, methyl, synthesis of, 4
Glycogen
  solubility in liquid ammonia, 91
  solubilization by methyl sulfoxide, 109
Glycolaldehyde
  Wittig reaction with, 288
  —, 2-cyclohexyl-2-phenyl-(R)-(−)- and (S)-(+)-, preparation of, 219
Glycolic acid, aminoalkyl and 3-piperidinyl esters, preparation and anticholinergic potency of, 209
Glycolic-2-d acid, preparation and configuration of, 142
Glycolic-2-t acid, preparation and stereochemistry of, 141
Glycopeptides
  amino acid sequences in, 333
  in glycoenzyme characterization, 314, 319
Glycoproteins
  biological and structural significance

of, 337–341
  biosynthesis of, 303, 328–337
  carbohydrate components of, 325
  enzymes with structure of, 301–341
  stability of, 340
Glycosaminoglycans, proton magnetic resonance spectroscopy of, 41
Glycosidation, of sugars in methanol, 97
Glycosides, methyl 2(or 3)-amino-4,6-O-benzylidene-2(or 3)-deoxy-α-D-, proton magnetic resonance spectra, iterative analyses of, 78
Glycosuloses
  reactions in which carbonyl groups are converted into asymmetric centers, 193, 204
  —, anhydro-1,3-dideoxy-, synthesis by Wittig reaction and properties, 284–288
Glycosyl halides, acylated, preparation of, 3
Glyculoses, Wittig reaction with, 231
Glyculosonic acids, 3-deoxy-, synthesis by Wittig reaction, 267, 274
Glyoxylic acid, phenyl-, esters with sugars, Grignard reaction with, 200
Golgi particles, in glycoenzyme biosynthesis, 303, 330
Grignard reaction
  with acyclic sugar derivatives for synthesis of compounds containing asymmetric benzylic carbon atoms, 217
  asymmetric benzylic carbon atoms created by, 213
  conjugate-addition with α,β-unsaturated carbohydrate esters, 202–204
  with glycosuloses, new asymmetric centers by, 204
  with α-oxo esters of optically active alcohols, 199–201
  in solvents containing carbohydrate derivatives, 193–195
Gulofuranose, 3-deoxy-1,2:5,6-di-O-isopropylidene-3-C-methyl-α-D-, synthesis of, 282

H

Heparin
  labeling with hydrogen isotopes, 138

proton magnetic resonance spectroscopy of, 41
Heparitin, proton magnetic resonance spectroscopy of, 41, 42
Hept-5-enodialdo-1,4-furanose, 7-C-aryl-1,2-O-cyclohexylidene-5,6-dideoxy-α-D-*xylo*-, syntheses by Wittig reaction, 247
—, 3-O-benzyl-1,2-O-cyclohexylidene-5,6-dideoxy-α-D-*xylo*-, synthesis by Wittig reaction, 244
Wittig reaction with, 251
Hept-5-eno-1,4-furanuronate, butyl 3-O-benzoyl-1,2-O-cyclohexylidene-5,6-dideoxy-α-D-*xylo*-, synthesis by Wittig reaction, 257
—, methyl 3-O-benzyl-1,2-O-cyclohexylidene-5,6-dideoxy-β-D-*xylo*-, synthesis by Wittig reaction, 255
Hept-2-enonamide, 4,6:5,7-di-O-benzylidene-2,3-dideoxy-D-*ribo*-, synthesis by Wittig reaction, 256
Hept-2-enonate, ethyl 2-bromo-2,3-dideoxy-L-*arabino*-, 4,5,6,7-tetraacetate, synthesis by Wittig reaction, 257
—, ethyl 2-bromo-2,3-dideoxy-4,5:6,7-di-O-isopropylidene-L-*arabino*-, synthesis by Wittig reaction, 257
—, ethyl 2-bromo-2,3-dideoxy-4,6:5,7-di-O-isopropylidene-D-*ribo*-, synthesis by Wittig reaction, 257
—, ethyl 3-deoxy-2-O-ethyl-4,5:6,7-di-O-isopropylidene-D-*arabino*-, synthesis by Wittig reaction, 274
—, ethyl 4,6:5,7-di-O-benzylidene-2,3-dideoxy-D-*ribo*-, synthesis by Wittig reaction, 254
—, ethyl 2,3-dideoxy-D- and -L-*arabino*-, synthesis by Wittig reaction, 254
—, ethyl 2,3-dideoxy-D-*ribo*-, synthesis by Wittig reaction, 254
—, ethyl 2,3-dideoxy-4,5:6,7-di-O-isopropylidene-L-*arabino*-, synthesis by Wittig reaction, 254
—, methyl 2,3-dideoxy-3-C-(hydroxymethyl)-, 3¹,4,5,6,7-pentaacetate, synthesis by Wittig reaction, 255, 256
Hept-2-enononitrile, 4,6:5,7-di-O-benzylidene-2,3-dideoxy-D-*ribo*-, synthesis by Wittig reaction, 256
Hept-6-enopyranose, 6,7-dideoxy-1,2:3,4-di-O-isopropylidene-α-D-*galacto*-, synthesis by Wittig reaction, 293
Hept-2-enose, 4,6:5,7-di-O-benzylidene-2,3-dideoxy-1-C-phenyl-D-*ribo*-, synthesis by Wittig reaction, 246
—, 4,5:6,7-di-O-cyclohexylidene-2,3-dideoxy-*aldehydo*-L-*arabino*-, synthesis by Wittig reaction, 245
—, 4,5:6,7-di-O-cyclohexylidene-2,3-dideoxy-*aldehydo*-D-*xylo*-, synthesis by Wittig reaction, 245
—, 4,5:6,7-di-O-cyclohexylidene-2,3-dideoxy-1-C-phenyl-L-*arabino*-, synthesis by Wittig reaction, 246
Wittig reaction with, 252
Heptonamide, 3-acetamido-2,3-dideoxy-4,5:6,7 di-O-isopropylidene-L-*arabino*-, preparation of epimers, 263
Heptonate, ethyl 2,3-(N-benzylepimino)-2,3-dideoxy-L-*glycero*-L-*galacto*-, 4,5,6,7-tetraacetate, preparation of, 264
Heptopyranose, 6,7-dideoxy-1,2:3,4-di-O-isopropylidene-7-(diethoxymethylsilyl)-α-D-*galacto*-, synthesis of, 293
Heptose, D-*glycero*-D-*galacto*-, synthesis of, 294
—, D-*glycero*-L-*ido*-, synthesis of, 294
—, 3,6-anhydro-2-deoxy-1-C-(p-methoxyphenyl)-D-*xylo*-, synthesis by Wittig reaction, 285
Heptulosonate, ethyl 3-deoxy-4,5-O-isopropylidene-D-*arabino*-, synthesis of, 279
Heterocyclic compounds, as solvents for sugars, 107, 111, 117
1-Hexene-L-*arabino*-1,3,4,5,6-pentol, 3,4:5,6-di-O-cyclohexylidene-1-O-methyl-, synthesis by Wittig reaction, 271
—, 3,4:5,6-di-O-isopropylidene-1-O-methyl-, synthesis by Wittig reaction, 271
—, 3,4:5,6-di-O-isopropylidene-1-S-methyl-1-thio-, synthesis by Wittig reaction, 271
—, 3,4:5,6-di-O-isopropylidene-1-O-p-tolyl-, synthesis by Wittig reaction, 271
1-Hexene-D-*ribo*-1,3,4,5,6-pentol, 3,6-anhydro-4,5-O-isopropylidene-1-S-methyl-1-thio-, *cis*- and *trans*-, syn-

thesis by Wittig reaction, 271
1-Hexene-D-*xylo*-1,3,4,5,6-pentol, 3,4:5,6-di-O-cyclohexylidene-1-O-methyl-, synthesis by Wittig reaction, 271
1-Hexene-3,4,5,6-tetrol, 3,4:5,6-di-O-isopropylidene-L-*arabino*-, synthesis by Wittig reaction, 234
Hex-1-enitol, 1,5-anhydro-4,6-O-benzylidene-2,3-dideoxy-3-(iodomethyl)-D-*arabino*-, proton magnetic resonance spectroscopy of, 29
—, 1,5-anhydro-4,6-O-benzylidene-2,3-dideoxy-3-(iodomethyl)-D-*ribo*-, proton magnetic resonance spectroscopy of, 29
—, 1-C-*p*-anisyl-3,4:5,6-di-O-cyclohexylidene-1,2-dideoxy-L-*arabino*-, synthesis by Wittig reaction, 236
—, 1-C-*p*-anisyl-3,4:5,6-di-O-cyclohexylidene-1,2-dideoxy-D-*xylo*-, synthesis by Wittig reaction, 236
—, 1-C-butyl-3,4:5,6-di-O-cyclohexylidene-1,2-dideoxy-D-*xylo*-, synthesis by Wittig reaction, 236
—, 3,4:5,6-di-O-cyclohexylidene-1,2-dideoxy-1-C-hexyl-L-*arabino*-, synthesis by Wittig reaction, 236
—, 3,4:5,6-di-O-cyclohexylidene-1,2-dideoxy-1-C-hexyl-D-*xylo*-, synthesis by Wittig reaction, 236
—, 3,4:5,6-di-O-cyclohexylidene-1,2-dideoxy-1-C-methyl-L-*arabino*-, synthesis by Wittig reaction, 236
—, 3,4:5,6-di-O-cyclohexylidene-1,2-dideoxy-1-C-methyl-D-*xylo*-, synthesis by Wittig reaction, 236
—, 3,4:5,6-di-O-cyclohexylidene-1,2-dideoxy-1-C-1-naphthyl-L-*arabino*-, synthesis by Wittig reaction, 236
—, 3,4:5,6-di-O-cyclohexylidene-1,2-dideoxy-1-C-(*p*-nitrophenyl)-L-*arabino*-, synthesis by Wittig reaction, 236
—, 3,4:5,6-di-O-cyclohexylidene-1,2-dideoxy-1-C-(*p*-nitrophenyl)-D-*xylo*-, synthesis by Wittig reaction, 236
—, 3,4:5,6-di-O-cyclohexylidene-1,2-dideoxy-1-C-pentyl-L-*arabino*-, synthesis by Wittig reaction, 236
—, 3,4:5,6-di-O-cyclohexylidene-1,2-dideoxy-1-C-pentyl-D-*xylo*-, synthesis by Wittig reaction, 236
—, 3,4:5,6-di-O-cyclohexylidene-1,2-dideoxy-1-C-phenyl-L-*arabino*-, synthesis by Wittig reaction, 236
—, 3,4:5,6-di-O-cyclohexylidene-1,2-dideoxy-1-C-phenyl-D-*xylo*-, synthesis by Wittig reaction, 236
—, 3,4:5,6-di-O-cyclohexylidene-1,2-dideoxy-1-C-propyl-L-*arabino*-, synthesis by Wittig reaction, 236
—, 1,2-dideoxy-1-C-hexyl-L-*arabino*-, synthesis by Wittig reaction, 237
—, 1,2-dideoxy-1-C-pentyl-L-*arabino*-, synthesis by Wittig reaction, 237
—, 3,4,5,6-tetra-O-acetyl-1-deoxy-1-nitro-D-*ribo*-, proton magnetic resonance spectroscopy of, 23
—, 3,4,5,6-tetra-O-acetyl-1-deoxy-1-nitro-D-*xylo*-, proton magnetic resonance spectroscopy of, 23
—, 3,4,5,6-tetra-O-acetyl-1,2-dideoxy-1-nitro-D-*arabino*-, proton magnetic resonance spectroscopy of, 71, 72
—, 3,4,5,6-tetra-O-acetyl-1,2-dideoxy-1-nitro-D-*xylo*-, proton magnetic resonance spectroscopy of, 72, 77
Hex-5-enofuranose, 3-O-benzyl-5,6-dideoxy-1,2-O-isopropylidene-α-D-*xylo*-, synthesis by Wittig reaction, 293
—, 5,6-dideoxy-1,2-O-isopropylidene-α-D-*xylo*-, and acetate, synthesis by Wittig reaction, 234, 293
Hex-3-enulose, 1,3,4-trideoxy-D-*glycero*-, synthesis by Wittig reaction, 245, 290
Hexitol, 3,6-anhydro-1,2-dideoxy-4,5-O-isopropylidene-D-*ribo*-, synthesis of, 281
—, 1,2-dideoxy-1-(diethylphosphono)-3,4:5,6-di-O-isopropylidene-D-*arabino*-, synthesis of, 297
—, 2,3,4,5,6-penta-O-acetyl-1-O-10-anthranylidene-1-deoxy-D-*galacto*-, synthesis by Wittig reaction, 237
—, 2,3,4,5,6-penta-O-acetyl-1-deoxy-1-O-9-fluorenylidene-D-*galacto*-, synthesis by Wittig reaction, 237
—, 2,3,4,5,6-penta-O-acetyl-1-deoxy-1-C-furfurylidene-D-*gluco*-, synthesis by Wittig reaction, 237

Hexitols, C-alkylated and C-arylated unsaturated, physical properties of, 239–241

Hexofuranose, 3-deoxy-1,2:5,6-di-O-isopropylidene-3-C-methylene-α-D-*ribo*-, preparation by Wittig reaction, 235

Hexofuranos-3-uloses, 1,2:5,6-di-O-isopropylidene-D-, proton magnetic resonance spectra, analyses of, 81

Hexopyranoside, methyl 2,3-C-(aminomethylene)-4,6-O-benzylidene-2,3-dideoxy-α-D-*manno*-, synthesis (attempted) of, 299

—, methyl 2,3-C-[(aminomethyl)methylene]-4,6-O-benzylidene-2,3-dideoxy-α-D-*manno*-, synthesis of, 298

—, methyl 4,6-O-benzylidene-2,3-C-(cyanomethylene)-2,3-dideoxy-α-D-*manno*-, synthesis of, 299

—, methyl 4,6-O-benzylidene-2-deoxy-3-C-methyl-α-L-*arabino*-, preparation of, 242

—, methyl 4,6-O-benzylidene-2,3-dideoxy-2,3-C-(formylmethylene)-α-D-, synthesis of, 298

—, methyl 4,6-O-benzylidene-2,3-dideoxy-2,3-C-[(hydroxymethyl)methylene]-α-D-*manno*-, synthesis of, 299

—, methyl 4,6-O-benzylidene-2,3-dideoxy-2,3-C-(methoxycarbonylmethylene)-α-D-*manno*-, synthesis of, 298

—, methyl 4,6-O-benzylidene-2,3-dideoxy-3-C-methylene-α-D-*threo*-, synthesis by Wittig reaction, 234, 235

—, methyl 4,6-O-benzylidene-2,3-dideoxy-2,3-C-vinylene-α-D-*manno*-, synthesis of, 299

—, methyl 2,3-dideoxy-3-C-[(methoxycarbonyl)methyl]-D-*ribo*-, preparation of, 262

—, methyl 2,3-dideoxy-3-C-methylene-α-L-*erythro*-, synthesis by Wittig reaction, 235

Hexopyranos-4-ulose-3-*d*, 1,6-anhydro-2,3-O-isopropylidene-β-D-*lyxo*-, preparation of, 133

Hexopyranosyl fluoride, poly-O-acyl-D-, proton magnetic resonance spectroscopy of, 21

Hexose, 2-deoxy-L-*arabino*-, synthesis of, 280

—, 2-deoxy-D-*ribo*-, methyl glycosides, synthesis of, 4

—, 3,4:5,6-di-O-cyclohexylidene-2-deoxy-*aldehydo*-L-*arabino*-, synthesis of, 278

—, 3,4:5,6-di-O-cyclohexylidene-2-deoxy-*aldehydo*-D-*xylo*-, synthesis of, 278

—, 2,6-dideoxy-D-*arabino*-, synthesis of, 4

—, 2,6-dideoxy-D-*ribo*-, 1,3,4-tris(*p*-nitrobenzoate), preparation of, 3

—, 2,6-dideoxy-3-C-methyl-L-*arabino*-, synthesis of, 242

Hexose-2-*t*, 2-deoxy-D-*arabino*-, preparation of, 137

Hexosyl bromide, 2-deoxy-3,4,6-tri-O-(*p*-nitrobenzoyl)-α-D-*lyxo*-, synthesis of, 4

—, 2-deoxy-3,5,6-tri-O-(*p*-nitrobenzoyl)-D-*ribo*-, synthesis of, 4

Hexulose, 1,3-dideoxy-D-*erythro*-, synthesis of, 4

Hexulosonic acid, 3-deoxy-D-*erythro*-, and ethyl ester, synthesis of, 279

Hyaluronic acid
  proton magnetic resonance spectroscopy of, 41, 43
  synthesis and assay of, 181

Hydration, of C-glycosylated alkenes, 241

Hydrazine, phenyl-, reactions with α,β-unsaturated sugars, 249

Hydrazino group, oxidation in alkaline deuterium oxide, labeling of sugars with hydrogen isotopes by, 135

Hydroboration, of alkene sugars, 242

Hydrocarbons
  aliphatic, as solvents for sugars, 105
  aromatic, as solvents for sugars, 105

Hydrocinnamic acid, β-methyl-, (+), (−), (R)-(−), and (S)-(+), preparation of, 202, 203

Hydrogen
  isotopes, hydride and proton exchange, 156–162
    intermolecular transfer, 166
    localization of proton movement, 162–166

sugars labeled with, 127–190
transfer to cofactor, 169
Hydrogenation
  of α,β-unsaturated acids, 261–263
  of unsaturated sugars, 241
Hydrogen fluoride, anhydrous, as solvent for carbohydrates, 93
Hydrogen sulfide, as solvent for sugars, 94
Hydrolases
  carbohydrate, 305–307
  nucleic acid, 307
Hydroxylation
  of α,β-unsaturated aldonic esters, 266
  of unsaturated sugars, 241

I

Iditol, 2,3:4,5-dianhydro-D-, proton magnetic resonance spectroscopy of, and esters, 64
Idofuranose, 3,6-anhydro-5-deoxy-5-fluoro-1,2-O-isopropylidene-α-L-, proton magnetic resonance spectroscopy of, 82
—, 6-deoxy-1,2-O-isopropylidene-5-C-phenyl-β-D-, preparation of, 214–216
Idopyranose, α-D-, pentaacetate, proton magnetic resonance spectroscopy of, 23
Indoline, 1-(2,3,4,6-tetra-O-acetyl-β-D-glucopyranosyl)-, proton magnetic resonance spectrum, analysis of, 83
Infrared absorption spectra
  of alkyl ald-3-enulosonates, 268
  of C-alkylated unsaturated sugars, 240
  of anhydro-C-(p-methoxyphenyl)aldoses and anhydro-1,3-dideoxyketoses, 286
  of deuterated sugars, 149
  of furan derivatives from sugars, 289
  of thioenol sugars, 276
  of α,β-unsaturated acids, 261
  of α,β-unsaturated sugars, 248
Inositol, myo-, 1-phosphate, from D-glucose 1-phosphate, 167
Inositol-2-d, myo-, preparation of, 132
Inositol-2-t, myo-, degradation in higher plants, 177
Inositols
  labeling with hydrogen isotopes, 137, 138

proton magnetic resonance spectroscopy of, 64
myo-Inosose-2, phenylosotriazole derivative, proton magnetic resonance spectroscopy of, 64, 73
Invertase
  amino acids of, 315, 317
  carbohydrate components of, 326
  purification and carbohydrates of, 305
Isobutyl alcohol, solubility of D-glucose in, 98
Isoenzymes, amino acid components of, 315, 317
Isoglycoenzymes, structure of, 301
Isomerases, and hydrogen-isotope action on sugars, 156
Isopropyl alcohol, as solvent for sugars, 96
Isosucrose, permethylation of, in liquid ammonia, 93
Isotopes
  effect in enzymic reactions, 171–176
  hydrogen, radiochemical and chemical stability of labeled sugars, 138
  sugars labeled with, 127–190

K

Karplus equation
  in localization of hydrogen isotopes, 144, 145
  and proton magnetic resonance spectroscopy, 35
Keratan sulfate, proton magnetic resonance spectroscopy of, 41, 43
1-Kestose, hendecaacetate, proton magnetic resonance spectroscopy of, 30, 31
Ketoses, anhydro-1,3-dideoxy-, synthesis and properties of, 284–292

L

Lactic acid, ethyl ester, as solvent for sugars, 106
Lactones, as solvents for sugars, 106
Lactose
  solubility in liquid ammonia, 91
  in methanol, 95, 97
Lithium aluminum deuteride, for reduction and deuteration of sugar derivatives, 131

Lithium aluminum hydride, reductions with complexes of, in presence of optically active sugars, 195–199, 216
Lithium borohydride-*t*, tritiation of sugars by, 129
Lyxopyranose, 1-thio-α-D-, tetraacetate, proton magnetic resonance spectroscopy of, 28
Lyxopyranoside, methyl tri-*O*-acetyl-α-D-, proton magnetic resonance spectroscopy of, 28
—, methyl tri-*O*-benzoyl-α-D-, proton magnetic resonance spectroscopy of, 28

M

Malic-3-*d* acid, in determination of hydrogen configuration in sugars, 143
Maltose
  solubility in liquid ammonia, 91
  in methanol, effect of calcium chloride on, 95
Mandelic acid, (*R*)-(−)- and (*S*)-(+)-, preparation of, 201, 202
Mannan, proton magnetic resonance spectroscopy of derivatives of, 40
Mannitol, 3,4-*O*-benzylidene-2,5-*O*-methylene-D-, proton magnetic resonance spectroscopy of, and derivatives, 65
—, 1-deoxy-1-*C*-methyl-L-, preparation of, 242
—, 1,6-dideoxy-2,5-*O*-methylene-D-, proton magnetic resonance spectroscopy of 65, 73
—, 2,5-*O*-ethylidene-1,3:4,6-di-*O*-methylene-D-, proton magnetic resonance spectroscopy of, 65
—, 2,5-*O*-isopropylidene-1,3:4,6-di-*O*-methylene-D-, proton magnetic resonance spectroscopy of, 65
—, 1,3,4,6-tetra-*O*-acetyl-2,5-*O*-methylene-*1,1,6,6-d₄*-D-, proton magnetic resonance spectroscopy of, 73, 81
—, 1,3:2,5:4,6-tri-*O*-ethylidene-D-, proton magnetic resonance spectroscopy of, 65
—, 1,3:2,5:4,6-tri-*O*-methylene-D-, proton magnetic resonance spectroscopy of, 65
Mannitol-*1,1,6,6-d₄*, 1,3,4,6-tetra-*O*-acetyl-2,5-*O*-methylene-D-, proton magnetic resonance spectroscopy of, 73, 81
Mannopyranoside, methyl 4-deoxy-2,3-*O*-isopropylidene-4-*C*-methyl-6-*O*-methyl-α-D-, synthesis of, 281
Mannose, D-
  glycoenzyme constituent, 302, 307, 321
  solubility in methanol, 97
  in trichloroacetonitrile, 113
Mass spectrometry, of deuterated sugars, 146, 149
Melezitose, solubility in methanol, effect of calcium chloride on, 95
Mercuration–demercuration, of alkene sugars, 242, 280
Metallation, of sugars in liquid ammonia, 92
Methane, bis(ethylsulfonyl)-(2,3,4-tri-*O*-acetyl-β-D-ribopyranosyl)-, proton magnetic resonance spectra, nonequivalence of nuclei in, 63
—, dichloro-, as solvent for sugars, 105
Methanol
  solubility of sugars in, effect of salts on, 95
  as solvent for sugars, 96, 97
Methylation, of sugars in liquid ammonia, 93
C-Methyl groups, proton magnetic resonance spectroscopy of, 29
Methylsulfonyl group, deshielding effect in sugars, 283
Methyl sulfoxide
  as solvent for sugars, 89, 108
  sucrate production in, 93
Monoamine oxidase, amino acids of, 315, 316
Monosaccharides
  branched-chain, synthesis by Wittig reaction, 294
  deuterated, preparation and position of deuterium, 188, 189
  glycoenzyme constituents, 302, 307
  labeled, incorporation into polysaccharides, 179
  proton magnetic resonance spectroscopy of, 21–30
  tritiated, preparation and localization of tritium, 183–187
Morpholine
  as solvent for sucrose, 117, 118
  as solvent for sugars, 107
—, *N*-acetyl-, solubility of sucrose in, 119

—, N-butyryl-, solubility of sucrose in, 119
—, N-formyl-, solubility of sucrose in, 119
—, N-methyl-, as solvent for sucrose, 117, 118
—, N-propionyl-, solubility of sucrose in, 119
Muscarine
  D-(−)-, from 2-amino-2-deoxy-D-glucose, 208
  L-(+)-, from 2-amino-2-deoxy-L-glucose, 207
Mutarotation
  of α-D-glucose in cresol and pyridine, 104
  of sugars, solvents for, 120
Mytilitol, hexaacetate, proton magnetic resonance spectroscopy of, 24

N

Nicotinamide adenine dinucleotide
  for enzymic reduction and labeling with hydrogen isotope, 132
  proton magnetic resonance spectroscopy of reduced, 37, 57
Nicotinamide mononucleotide, monodeuteriated, proton magnetic resonance spectroscopy of, 38
Nitriles, as solvents for sugars, 112
Nitrogen compounds, as solvents for sugars, 112–123
Nojirimycin, proton magnetic resonance spectroscopy of reduction product of, 27
Non-2-enonate, ethyl 5-acetamido-3,5-dideoxy-D-*glycero*-D-*galacto*-, synthesis by Wittig reaction, 274
—, ethyl 2,3-dideoxy-D-*glycero*-D-*gluco*-, synthesis by Wittig reaction, 255
Non-3-enulose, 1,3,4-trideoxy-D-*galacto*-, 5,6,7,8-tetraacetate, synthesis by Wittig reaction, 245
—, 1,3,4-trideoxy-D-*gluco*-, 5,6,7,8-tetraacetate, preparation of, 245
Non-3-enulosonic acid, 3,4-dideoxy-D-*galacto*-, and methyl and isobutyl esters, pentaacetates, syntheses by Wittig reaction, 267
—, 3,4-dideoxy-D-*gluco*-, methyl ester pentaacetate, synthesis by Wittig reaction, 267
Nonulose, 3,7-anhydro-6,8-O-benzylidene-2-deoxy-1-C-(p-methoxyphenyl)-D-*gluco*-, synthesis by Wittig reaction, 288
—, 4,7-anhydro-1,3-dideoxy-D-*glycero*-L-*gluco*-, synthesis by Wittig reaction, 286, 287
—, 4,7-anhydro-1,3-dideoxy-D-*glycero*-L-*manno*-, synthesis by Wittig reaction, 286, 287
—, 4,7-anhydro-1,3-dideoxy-5,6:8,9-di-O-isopropylidene-D-*glycero*-D-*galacto*-, synthesis by Wittig reaction, 285
—, 4,7-anhydro-1,3-dideoxy-5,6:8,9-di-O-isopropylidene-D-*glycero*-D-*talo*-, synthesis by Wittig reaction, 285
—, 4,8-anhydro-1,3,9-trideoxy-L-*manno*-, synthesis by Wittig reaction, 288
Nuclear magnetic resonance spectra
  of alkyl ald-3-enulosonates, 268
  of deuterated compounds, 147
  deuterium location, determination by, 143
  cis-2,4-dimethyl-1,3-dioxolane as reference compound for, 209
  Fourier-transform, acquisition of the free-induction decay signal, 48, 49
  applications of, 56–62
  digital processing of the free-induction decay signal, 50
  distinction of continuous-wave and pulsed techniques, 44
  fast Fourier transformation, 52–54
  noise-stimulated resonance, 55
  phase correction, 54
  pulse methods, 45–49
  techniques, 43
  solvents for sugars for, 108, 124
  of α,β-unsaturated acids, 261
  of unsaturated furanoses, 240
  of unsaturated sugar phosphonates, 297
  of vinylthio ethers of sugars, 276
Nuclear relaxation time, measurement of, 59
Nucleic acid hydrolases, 307
Nucleoside, 2-deoxyhexofuranosyl-, synthesis of, 4
—, 1′-methyl 2′-deoxy-, synthesis of, 4
Nucleosides
  book on chemistry of, 5
  proton magnetic resonance spectroscopy of, 33–39, 81

synthesis of branched-chain, by Wittig reaction, 234, 235, 243
Nucleotides
biosynthesis of sugar, 336
book on chemistry of, 5
proton magnetic resonance spectroscopy of, 33–39
synthesis of, 260
tritiated, stability and storage of, 139
Nystose, tetradecaacetate, proton magnetic resonance spectroscopy of, 31

O

Obituary, William Werner Zorbach, 1–6
Octasaccharide, of ribonuclease B, 325
Oct-5-eno-1,4-furanose, 3-O-benzoyl-1,2-O-cyclohexylidene-5,6,8-trideoxy-α-D-*xylo*-, 250
Oct-5-eno-1,4-furanos-7-ulose, 3-O-benzoyl-1,2-O-cyclohexylidene-5,6,8-trideoxy-α-D-*xylo*-, reactions of, 250
synthesis by Wittig reaction, 245
—, 3-O-benzyl-1,2-O-cyclohexylidene-5,6,8-trideoxy-α-D-*xylo*-, synthesis by Wittig reaction, 245
Oct-2-enonamide, 2,3-dideoxy-D-*gluco*-, 4,5,6,7,8-pentaacetate, synthesis by Wittig reaction, 256
Oct-2-enonate, ethyl 4-acetamido-3,4-dideoxy-2-O-ethyl-5,6:7,8-di-O-isopropylidene-D-*gluco*-, synthesis by Wittig reaction, 274
—, ethyl 4-acetamido-2,3,4-trideoxy-D-*gluco*-, synthesis by Wittig reaction, 255
—, ethyl 2-bromo-2,3-dideoxy-D-*galacto*-, 4,5,6,7,8-pentaacetate, synthesis by Wittig reaction, 257
—, ethyl 2-bromo-2,3-dideoxy-D-*gluco*-, 4,5,6,7,8-pentaacetate, synthesis by Wittig reaction, 257
—, ethyl 2,3-dideoxy-D-*galacto*-, pentaacetate, synthesis by Wittig reaction, 253, 255
synthesis by Wittig reaction, 255
4,5,6,8-tetraacetate, synthesis by Wittig reaction, 255
—, ethyl 2,3-dideoxy-D-*gluco*-, pentaacetate, synthesis by Wittig reaction, 253

synthesis by Wittig reaction, 254, 255
—, ethyl 2,3-dideoxy-2-C-phenyl-D-*gluco*-, 4,5,6,7,8-pentaacetate, synthesis by Wittig reaction, 256
—, methyl 4-acetamido-6,8-O-benzylidene-2,3,4-trideoxy-D-*gluco*-, synthesis by Wittig reaction, 255
—, methyl 4-acetamido-2,3,4-trideoxy-D-*gluco*-, synthesis by Wittig reaction, 255
Oct-3-enonate, dimethyl 3-(carboxymethyl)-2,3,4-trideoxy-5,6:7,8-di-O-isopropylidene-, synthesis by Wittig reaction, 255, 256
Oct-2-enononitrile, 2,3-dideoxy-D-*gluco* , 4,5,6,7,8-pentaacetate, synthesis by Wittig reaction, 256
Oct-2-enose, 2,3-dideoxy-1-C-phenyl-D-*galacto*-, synthesis by Wittig reaction, 246
Oct-6-enose, 6,7,8-trideoxy-1,2:3,4-di-O-isopropylidene-α-D-*galacto*-, synthesis by Wittig reaction, 235
Octenoses, 1-C-aryl *aldehydo*-, pentaacetates, syntheses by Wittig reactions, 246, 247
Oct-3-enulose, 5,7:6,8-di-O-benzylidene-1,3,4-trideoxy-D-*ribo*-, synthesis by Wittig reaction, 245
Oct-3-enulosonic acid, 3,4-dideoxy-D-*arabino*-, and *tert*-butyl ester pentaacetate, synthesis by Wittig reaction, 267, 268
—, 3,4-dideoxy-L-*arabino*-, and methyl and isobutyl esters, pentaacetates, synthesis by Wittig reaction, 267, 268
Octonate, methyl 4-acetamido-2,3,4-trideoxy-D-*gluco*-, preparation of, 261
Octonic acid, D-*threo*-L-*galacto*-, preparation of, and 1,4-lactone, 266
—, D-*threo*-L-*ido*-, preparation of, 266
—, 4-acetamido-6,8-O-benzylidene-4-deoxy-D-*threo*-L-*galacto*-, methyl ester, preparation of, 266
Octose, D-*threo*-L-*galacto*-, preparation of, 266
—, 3,6-anhydro-2-deoxy-4,5:7,8-di-O-isopropylidene-1-C-(*p*-methoxyphenyl)-D-*glycero*-D-*galacto*-, synthesis by Wittig reaction, 285
—, 3,6-anhydro-2-deoxy-4,5:7,8-di-O-

isopropylidene-1-C-(p-methoxyphenyl)-D-*glycero*-D-*talo*-, synthesis by Wittig reaction, 285
—, 3,6-anhydro-2-deoxy-1-C-(p-methoxyphenyl)-D-*galacto*-, synthesis by Wittig reaction, 285
—, 3,7-anhydro-2,8-dideoxy-1-C-(p-methoxyphenyl)-L-*manno*-, synthesis by Wittig reaction, 288
Octulose, 4,8-anhydro-7,9-O-benzylidene-1,3-dideoxy-D-*gluco*-, synthesis by Wittig reaction, 288
—, 4,7-anhydro-1,3-dideoxy-D-*ribo*-, synthesis by Wittig reaction, 285
—, 4,7-anhydro-1,3-dideoxy-D-*xylo*-, synthesis by Wittig reaction, 285
Oleic acid, lithium, potassium, and sodium salts, as catalysts and solubilizers for sugars, 96
Oligosaccharides
  in glycoenzymes, 325
  proton magnetic resonance spectroscopy of, 30
Olivomycose, synthesis of, 235, 242
Optical rotation, and configuration of deuterated compounds, 146
Osazones, formation mechanism of, hydrogen isotope effect on, 151
1,4-Oxathiane, 6-(hydroxymethyl)-2-methoxy-, S-oxide, isomers, preparation of, 222
Oxidation
  mechanism of, hydrogen isotope effect on aldose, 153–155
  of vinylthio sugar derivatives, 283
Oxidoreductases, glycoenzymes, 309

P

Palmitic acid, lithium, potassium, and sodium salts, as catalysts and solubilizers for sugars, 96
Pent-1-enitol, 4-O-acetyl-1,5-anhydro-3-(6-chloro-9-purinyl)-1,2,3-trideoxy-D-*threo*-, proton magnetic resonance spectra, analysis of, 82
Pent-1-enitol-3-ulose, 1,2-dideoxy-5,6-O-isopropylidene-D-*glycero*-, proton magnetic resonance spectra, analysis of, 80
Pent-2-enonate, ethyl 2,3-dideoxy-D-*glycero*-, synthesis by Wittig reaction, 254
—, ethyl 2,3-dideoxy-4,5-O-isopropylidene-D-*glycero*-, synthesis by Wittig reaction, 253
Pent-2-enopyranoside, methyl 3,4-dichloro-4-deoxy-α-D- and -β-D-*glycero*-, proton magnetic resonance spectra of, analysis of, 80
Pent-2-enose, 1-C-(p-bromophenyl)-2,3-dideoxy-D-*glycero*-, synthesis by Wittig reaction, 246
—, 2,3-dideoxy-1-C-(p-methoxyphenyl)-D-*glycero*-, synthesis by Wittig reaction, 246, 290
—, 2,3-dideoxy-1-C-(p-nitrophenyl)-D-*glycero*-, synthesis by Wittig reaction, 246
—, 2,3-dideoxy-1-C-phenyl-D-*glycero*-, synthesis by Wittig reaction, 246
Pentitol, 1-C-cyclohexyl-2,3:4,5-di-O-isopropylidene-, preparation of isomers, 218
—, 2,3:4,5-di-O-isopropylidene-1-C-methyl-, preparation of isomers, 218
—, 2,3:4,5-di-O-isopropylidene-1-C-phenyl-D-*gluco*-, preparation and configuration of, 218
Pentodialdo-1,4-furanose, 3-O-benzyl-1,2-O-isopropylidene-α-D-*xylo*-, reaction with phenylmagnesium bromide, 214
Pentofuranose, 5-O-benzyl-3-deoxy-1,2-O-isopropylidene-3-C-methylene-α-D-*erythro*-, synthesis by Wittig reaction, 235
Pentofuranoside, methyl 2-deoxy-β-D-*erythro*-, proton magnetic resonance spectroscopy of, 28, 83
Pentono-1,4-lactone, 5-S-acetyl-2,2'-anhydro-3-deoxy-2-C-(hydroxymethyl)-5-thio-D-*erythro*-, proton magnetic resonance spectroscopy of, 28
Pentopyranoside, methyl 2-deoxy-3,4-O-isopropylidene-2-C-methyl-D-*ribo*-, synthesis by Wittig reaction, 293
—, methyl 2-deoxy-3,4-O-isopropylidene-2-C-methylene-β-D-*erythro*-, synthesis by Wittig reaction, 292
Pentopyranosyl fluoride, poly-O-acyl-D-,

proton magnetic resonance spectroscopy of, 21
Pentose, 2-deoxy-2(S)- and 2(R)-deuterio-D-*erythro*-, cytosine derivatives, proton magnetic resonance spectroscopy of, 28
Pentose-2-*d*, 2-deoxy-D-*erythro*-, proton magnetic resonance spectroscopy of, 146
Pentose-2(S)-*d*, 4,6-O-benzylidene-2-deoxy-D-*erythro*-, synthesis of, 131
—, 2-deoxy-D-*erythro*-, synthesis of, 131
Pentulose-*1*(R)-*t*, D-*erythro*-, 5-phosphate, preparation of, 135
Pentulose-*1*(S)-*t*, D-*erythro*-, 5-phosphate, preparation of, 135
Phenols, as solvents for sugars, 104
Phosphatases, glycoenzymes, 308
Phosphonate, diphenyltriphenylphosphoranylidenemethyl, Wittig reactions with, 260, 296
Phosphonates, of unsaturated sugars, synthesis by Wittig reaction, 296
Phosphonium ylides, in Wittig reaction with carbohydrates, 228–239
Phosphorane
 sugar, synthesis of, 295
—, (bromoacetylmethylene)triphenyl-, synthesis and reactions of, 294
—, formylmethylenetriphenyl-, preparation of, 244
—, methylene-, derivatives, in Wittig reaction with carbohydrates, 228–239
Phosphoric acid, as solvent for sugars, 94
Phosphoric triamide, hexamethyl-, as solvent for sugars, 123
Phosphorus compounds, as solvents for sugars, 123
Phytosphingosines, synthesis of, 237, 258, 262
Piperazine, 2-methyl-, solubility of sucrose in, 118
Piperidine
 as solvent for sugars, 117
—, N-acetyl-, solubility of sucrose in, 118
—, N-formyl-, solubility of sucrose in, 118
3-Piperidinol, (S)-(−)-, absolute configuration of, 209
2-Piperidinone
 as solvent for sugars, 119

—, 1-methyl-, as solvent for sugars, 118, 119
Polyethylene glycol, as solvent for sugars, 98
Polysaccharides
 enzymolyses and proton magnetic resonance spectroscopy, 41
 incorporation of labeled monosaccharides into, 179
 liquid ammonia effect on, 91
 proton magnetic resonance spectroscopy of, 39–43
 tritiated, preparation and position of tritium, 188
Prelog's rule, for configurations, 193, 199–201
Propanol, 3-methoxy-, solubility of sucrose in, 87, 89
1-Propanol, 2-(dimethylamino)-, enantiomers and benzilic and glycolic acid esters, effect on nervous systems, 210
2-Propanol, 1-(dimethylamino)-, enantiomers and benzilic and glycolic esters, effect on nervous system, 210
2-Propanone-3-*t*, 1,3-dihydroxy-, 1-phosphate, configuration of, 141
Propionic-2-*d* acid, in determination of hydrogen configuration in sugars, 143
Propyl alcohol, solubility of sucrose in, 89, 98
Propylene glycol, as solvent for sucrose, 98
Propyl sulfoxide, as solvent for sugars, 108
Proteases, glycoenzymes, 308
Protein–carbohydrate linkages, in glycoenzymes, 318–323
Proteins, in glycoenzymes, 314–318
Proton magnetic resonance spectra
 analysis, assignment of transitions in, 77
 computerized iterative, 73–77
 criteria for good, 80
 experimental errors and variation of parameters in, 77–80
 limitations of first-order, 66–71
 manual, 71–73
 non-equivalence of nuclei in, 62–66
 of carbohydrates, 7–83
 and diastereotopism, 63
 enantiotopic nuclei, 64

instrumentation, automatic control and data acquisition, 17
field-frequency stabilization, 14–16
magnets and probes, 11–13
signal-averaging techniques, 16
spectrometer consoles, 13
superconducting solenoids, 18–21
magnetic nonequivalence, 64
of monosaccharides, 21–30
virtual coupling in, 68–71
Pseudouridine, α- and β-, proton magnetic resonance spectroscopy of, 34
Pullulan, tri-O-benzoyl-, proton magnetic resonance spectroscopy of, 40
Purine, 9-(4-O-acetyl-2,3-dideoxy-α-D- and -β-D-*glycero*-pent-2-enopyranosyl)-6-chloro-, proton magnetic resonance spectra, analysis of, 82
Pyran-2-one, tetrahydro-, as solvent for acylation of cellulose, 106
Pyrazine
  solubility of sucrose in, 118
  as solvent for sugars, 96
—, 2-methyl-, solubility of sucrose in, 118
Pyridine
  as solvent for mutarotation of D-glucose, 104
  as solvent for sugars, 117, 118
2-Pyridinone, as solvent for mutarotation of D-glucose, 105
Pyrrolidine, N-acetyl-, solubility of sucrose in, 118
—, N-formyl-, solubility of sucrose in, 118

Q

Quercitols, proton magnetic resonance spectroscopy of, 24, 25

R

Radiochemical stability, of hydrogen-isotope labeled compounds, 138
Raffinose
  solubility in liquid ammonia, 91
  in methanol, 97
  in methanol, effect of calcium chloride on, 95
  spin-lattice relaxation of ^{13}C nuclei in, 60
Reaction media
  acids (organic) as, 102

alcohols as, 99
Reactions, mechanism of chemical, hydrogen isotope effect on, 151
Reduction
  asymmetric, with complexes of lithium aluminum hydride, 195–199, 216
  enzymic, and labeling with hydrogen isotopes, 132
  by hydride reagents and labeling of sugars with hydrogen isotopes, 129–132
  by hydrogen in deuterium oxide, 132
  of α-oxo esters of carbohydrate derivatives, 201
Rhamnose
  L-, solubility in methanol, 97
    synthesis of, 4
  α-L-, hydrate, solubility in ethanol, 98
Ribitol, 2,4-O-benzylidene-1-deoxy-1-(2,5-dioxo-3-pyrrolylidene)-D-, synthesis by Wittig reaction, 256
—, 2,4:3,5-di-O-benzylidene-1-deoxy-1-(2,5-dioxo-N-phenyl-3-pyrrolylidene)-D-, synthesis by Wittig reaction, 256
—, 2,4:3,5-di-O-benzylidene-1-deoxy-1-(2,5-dioxo-3-pyrrolylidene)-D-, synthesis by Wittig reaction, 256
Ribofuranose, 5-O-benzyl-3-deoxy-3-C-(hydroxymethyl)-1,2-O-isopropylidene-α-D-, preparation of, 243
—, 5-O-benzyl-1,2-O-isopropylidene-3-C-methyl-α-D-, preparation of, 243
—, 3-deoxy-1,2:5,6-di-O-isopropylidene-3-C-methyl-D-, synthesis of, 282
—, 1,3,5-tri-O-benzoyl-α-D-, proton magnetic resonance spectra, virtual coupling in, 69
N-β-D-Ribofuranosylcyanuric acid, proton magnetic resonance spectroscopy of, 35
Ribofuranosyl fluoride, tri-O-benzoyl-α-D- and -β-D-, proton magnetic resonance spectroscopy of, 82
Ribonucleases
  amino acids of, 315, 317
  carbohydrates of, 307, 325, 326, 340
  conformational structure of, 327
  deoxy, amino acids of, 315, 316
  as glycoenzyme, 308
  purification by chromatography, 310

Ribonucleotide reductase, and hydrogen-isotope action on sugars, 156
Ribopyranose
  β-D-, tetraacetate, proton magnetic resonance spectroscopy of, 22
  —, 3-O-benzoyl-1,2,4-O-benzylidyne-α-D-, proton magnetic resonance spectra, iterative analysis of, 78
  —, 1-thio-β-D-, tetraacetate, proton magnetic spectroscopy of, 68
Ribopyranosyl cyanide, tri-O-benzoyl-, proton magnetic resonance spectroscopy of conformers of, 23
  —, 2,3,4-tri-O-benzoyl-β-D-, proton magnetic resonance spectroscopy of, 68
Ribose, 1,3,4-tri-O-benzoyl-2-deoxy-2-fluoro-D-, proton magnetic resonance spectra, analysis of, 80
  —, 2,4:3,5-di-O-benzylidene-*aldehydo*-D-, Wittig reaction with, 231
Ribose-*t*, D-, detection and distribution in rat tissue, autoradiography in, 181
Ribose-3-*t*, D-, preparation of, 137

S

Salts, effect on sugar solubility, 95
Sedimentation, of glycoenzymes, 312
Serine, L-, configurational correlation with D-glyceraldehyde, 207
Showdomycin, synthesis of, 256
Sialic acid
  glycoenzyme constituent, 302, 307, 321
  synthesis of, 275
Sodium amide, sucrate production by, in N,N-dimethylformamide or methyl sulfoxide, 93
Sodium borohydride-*d*, deuteration of sugars with, 129
Sodium borohydride-*t*, tritiation of sugars by, 129
Sodium hydride, sucrate production by, in N,N-dimethylformamide or methyl sulfoxide, 93
Sodium methoxide, sucrate production by, in N,N-dimethylformamide or methyl sulfoxide, 93
Solubility
  determination of, of sugars, 90
  solvation and, of sugars, 86–90

Solvation, solubility and, of sugars, 86–90
Solvent-exchange
  base-catalyzed, in labeling sugars with hydrogen isotopes, 133
  enzyme-catalyzed, in labeling sugars with hydrogen isotopes, 134
  hydride and proton exchange related to enediol or oxido-reduction mechanisms, 156–162
  of hydrogen isotopes, mechanism of, 153
  localization of proton movement, 162–171
  on oxidation in deuterium oxide of hydrazino group, labeling of sugars with hydrogen isotopes by, 135
  platinum-catalyzed in deuterium oxide, 137
Solvents
  acids (organic) as, for sugars, 102
  aliphatic amines as, for sugars, 112
  aprotic, for sugars, 105–123
  heterocyclic, for sugars, 117
  inorganic, for sugars, 90–96
  non-aqueous, for carbohydrates, 85–125
  organic, for proton magnetic resonance spectroscopy of polysaccharides, 40
  for sugars, 96–123
  protic, for sugars, 96–105
Sorbose,
  L-
    pentaacetate, Wittig reaction with, 247
    solubility in methanol, effect of calcium chloride on, 95
Sphingosine
  synthesis and structure of, 238, 241
  —, dihydro-, synthesis by Wittig reaction, 238
Stachyose, spin-lattice relaxation of ^{13}C nuclei in, 60
Starch
  hydrogen fluoride as solvent for, 93
  solubility in methyl sulfoxide, 108, 109
Stereochemistry
  enzymes in determination of, 141
  of hydride exchange, 159
  of hydrogen transfer, 170
  of Wittig normal and abnormal products, 290–292

Stereoselective synthesis, of asymmetric sulfoxides, 222
Stereoselectivity
of ammonia addition to α,β-unsaturated acids, 264
carbohydrates in asymmetric syntheses, 192–204, 214
in hydrogenation, hydroboration, and mercuration–demercuration of unsaturated sugars, 241–244
Streptidine, L-, biosynthesis of, 177
Strophanthidin
α- and β-D-rhamnosides, synthesis of, 4
—, 3-O-(α-L-mannosyl)-, preparation and cardiotonic activity of, 4
Sucrates, preparation of, 92, 110
Sucrose
aminolysis of, 99
ammonolysis of, in liquid ammonia, 93
metallation of, in liquid ammonia, 93
octaacetate, proton magnetic resonance spectra of, 30, 31
reaction with acrolein, effect of zinc chloride on, 95
with fatty acid esters, salts of palmitic and oleic acids as catalysts and solubilizers in, 96
solubility in acetone, 107
in acetone, effect of zinc chloride on, 95
in alcohols, 89
in alkoxy alcohols, 99
in anhydrous hydrogen fluoride, 93
in ethanol, 98
in ethyl and methyl carbamates, 113
in heterocylic solvents, 117–119
in liquid ammonia, 91
in methanol, 97
in methanol, effect of calcium chloride on, 95
in methyl sulfoxide and in propyl sulfoxide, 108
in polyhydric alcohols, 98
in sulfolanes, 111
in thiocyanates, 95
solvents for, 85, 87
spin-lattice relaxation of ^{13}C nuclei in, 59
Sugars
acyclic, Grignard reaction with, 217

aldehydo and *keto*, Wittig reactions with, 229–239
C-alkylated and C-arylated unsaturated, physical properties of, 239–241
3-amino-2,3-dideoxy, synthesis of, 262, 263
configuration of branched-chain, 215
cyclobutene derivative, 299
deoxy, synthesis of, 274, 281
enol, thioenol, and enamino groupings as precursors of carbonyl sugars, 270–283
C-glycosylated alkenes, preparation and properties of, 233–244
labeled with isotopes of hydrogen, 127–190
metallation of, in liquid ammonia, 92
solubility determination of, 90
tritiated, for assay of enzyme activity, 180
stability of, 139
unsaturated, chemical properties of, 249
physical properties of, 248
Wittig reaction in synthesis of, 227–239, 244–248
Wittig reaction with free and partially protected, 284–292
Sulfolanes, solubility of sucrose in, 87, 111
Sulfoxides, stereoselective synthesis of asymmetric, 222
Sulfur dioxide, liquid, as solvent for sugars, 94
Surface activity, of C-alkylated hexitols, 241
Synthesis
carbohydrate use in asymmetric, and configuration proof, 191–225
stereoselective, of asymmetric sulfoxides, 222

T

Talopyranoside, methyl 4-deoxy-2,3-O-isopropylidene-4-C-methyl-6-O-methyl-α-D-, synthesis of, 281
Temperature, effect on solubility of sucrose, 89
Terpenes, C-glycosylated, syntheses by Wittig reaction, 238
Tetrahydrofuran-2-one, as solvent for

sugars, 106
Tetralin, as reaction medium for reduction of D-glucose, 106
Tetrofuranose, 4-C-isopropyl-1,2-O-isopropylidene-α-D-*xylo*-, proton magnetic resonance spectra, non-equivalence of nuclei in, 63
Thiazoles, amino-, C-glycosylated, synthesis of, 295
Thiocyanates, as solvents for sugars, 95
Thiophene, tetrahydro-, 1,1-dioxide, solubility of sucrose in, 87
S-oxide, as solvent for sugars, 111
—, tetrahydro-3,4-dimethyl-, S-oxide, as solvent for sucrose, 111
—, tetrahydro-3-methyl-, S-oxide, as solvent for sucrose, 111
Thiophene-2-aldehyde, 3-bromo-, Fourier proton magnetic resonance spectroscopy of, 62
Thymine, 2-deoxyglucosyl-, as inhibitor of pyrimidine nucleoside phosphorylase, 4
Toluene, as dispersing medium for reaction of sucrose with phosgene, 106
s-Triazine, 2,4,6-tris(methoxy)-, as solvent for sugars, 117, 119
Tritiation
  by addition to double bond of unsaturated sugar, 136
  of heparin by tritium gas, 138
  recoil labeling, 138
Tritium, localization and chemical degradation, 140
Tritium compounds, stability of, 139
Turanose, D-, anomeric equilibria in aqueous solution and Fourier-transform spectra, 60

U

Ultraviolet absorption spectra, of arylated unsaturated hexitols, 239
Urea
  and derivatives as solvents for sugars, 112
  effect on solubility of sugars in methanol, 97
—, tetramethyl-, as solvent for sugars, 113
Urethans, as solvents for sugars, 112

Uridine
  proton magnetic resonance spectroscopy of, 33
—, 5'-(carboxymethyl)-5'-deoxy-, synthesis of, 258
—, 5'-deoxy-5'-(diphenylphosphinylmethyl)-, synthesis of, 260

V

Vitamin B$_{12}$, spin-lattice relaxation of ^{13}C nuclei in, 60

W

Wilzbach reaction, for labeling sugars with hydrogen isotopes, 137, 138
Wittig reaction
  in carbohydrate chemistry, 227–299
  in dethiobiotin synthesis, 212

X

Xylene, as dispersion medium in preparation of sucrose esters, 106
Xylofuranose, 3-O-benzyl-5-deoxy-5-fluoro-1,2-O-isopropylidene-α-D-, proton magnetic resonance spectroscopy of, 29
—, 3-O-benzyl-5-deoxy-5-iodo-1,2-O-isopropylidene-α-D-, proton magnetic resonance spectroscopy of, 29, 72
—, 5-O-benzyl-1,2-O-isopropylidene-3-C-methyl-α-D-, preparation of, 243
—, 3-O-crotonyl-5-deoxy-1,2-O-isopropylidene-α-D-, Grignard reaction with, 203
—, 5-deoxy-5-fluoro-1,2-O-isopropylidene-α-D-, proton magnetic resonance spectroscopy of, 29
—, 5-deoxy-5-iodo-3-O-*p*-nitrobenzyl-1,2-O-isopropylidene-α-D-, proton magnetic resonance spectroscopy of, 29, 72
Xylopyranose-5-*d*, β-D-, tetraacetate, preparation of, 220
Xylopyranose-(R)- and -(S)-5-*d*, tetra-O-acetyl-, preparation of, 144

Xylose, D-, glycoenzyme constituent, 302, 307, 321

Y

Yeast, spin-lattice relaxation of ^{13}C nuclei in thermally denatured, 60

Z

Zinc chloride, effect on sugar solubility in acetone, 95

Zorbach, William Werner, obituary, 1–6

Zorbach Memorial Prize, 6

CUMULATIVE AUTHOR INDEX FOR VOLS. 1-27

A

ADAMS, MILDRED. See Caldwell, Mary L.
ALEXEEV, YU. E. See Zhdanov, Yu. A.
ALEXEEVA, V. G. See Zhdanov, Yu. A.
ANDERSON, ERNEST, and SANDS, LILA, A Discussion of Methods of Value in Research on Plant Polyuronides, **1**, 329-344
ANDERSON, LAURENS. See Angyal, S. J.
ANET, E. F. L. J., 3-Deoxyglycosuloses (3-Deoxyglycosones) and the Degradation of Carbohydrates, **19**, 181-218
ANGYAL, S. J., and ANDERSON, LAURENS, The Cyclitols, **14**, 135-212
ARCHIBALD, A. R., and BADDILEY, J., The Teichoic Acids, **21**, 323-375
ARONSON, N. N., JR. See Pazur, John H.
ASPINALL, G. O., Gums and Mucilages, **24**, 333-379
ASPINALL, G. O., The Methyl Ethers of Hexuronic Acids, **9**, 131-148
ASPINALL, G. O., The Methyl Ethers of D-Mannose, **8**, 217-230
ASPINALL, G. O., Structural Chemistry of the Hemicelluloses, **14**, 429-468

B

BADDILEY, J. See Archibald, A. R.
BAER, HANS H., The Nitro Sugars, **24**, 67-138
BAER, HANS H., [Obituary of] Richard Kuhn, **24**, 1-12
BAILEY, R. W., and PRIDHAM, J. B., Oligosaccharides, **17**, 121-167
BALL, D. H., and PARRISH, F. W., Sulfonic Esters of Carbohydrates:
Part I, **23**, 233-280
Part II, **24**, 139-197
BALLOU, CLINTON E., Alkali-sensitive Glycosides, **9**, 59-95
BANKS, W., and GREENWOOD, C. T., Physical Properties of Solutions of Polysaccharides, **18**, 357-398

BARKER, G. R., Nucleic Acids, **11**, 285-333
BARKER, S. A., and BOURNE, E. J., Acetals and Ketals of the Tetritols, Pentitols and Hexitols, **7**, 137-207
BARNETT, J. E. G., Halogenated Carbohydrates, **22**, 177-227
BARNETT, J. E. G., and CORINA, D. L., Sugars Specifically Labeled with Isotopes of Hydrogen, **27**, 127-190
BARRETT, ELLIOTT, P., Trends in the Development of Granular Adsorbents for Sugar Refining, **6**, 205-230
BARRY, C. P., and HONEYMAN, JOHN, Fructose and Its Derivatives, **7**, 53-98
BAYNE, S., and FEWSTER, J. A., The Osones, **11**, 43-96
BEÉLIK, ANDREW, Kojic Acid, **11**, 145-183
BELL, D. J., The Methyl Ethers of D-Galactose, **6**, 11-25
BEMILLER, J. N., Acid-catalyzed Hydrolysis of Glycosides, **22**, 25-108
BEMILLER, J. N. See also, Whistler, Roy L.
BHAT, K. VENKATRAMANA. See Zorbach, W. Werner.
BINKLEY, W. W., Column Chromatography of Sugars and Their Derivatives, **10**, 55-94
BINKLEY, W. W., and WOLFROM, M. L., Composition of Cane Juice and Cane Final Molasses, **8**, 291-314
BIRCH, GORDON G., Trehaloses, **18**, 201-225
BISHOP, C. T., Gas-liquid Chromatography of Carbohydrate Derivatives, **19**, 95-147
BLAIR, MARY GRACE, The 2-Hydroxyglycals, **9**, 97-129
BOBBITT, J. M., Periodate Oxidation of Carbohydrates, **11**, 1-41
BÖESEKEN, J., The Use of Boric Acid for the Determination of the Configuration of Carbohydrates, **4**, 189-210
BONNER, T. G., Applications of Trifluoro-

acetic Anhydride in Carbohydrate Chemistry, **16,** 59–84

BONNER, WILLIAM A., Friedel–Crafts and Grignard Processes in the Carbohydrate Series, **6,** 251–289

BOURNE, E. J., and PEAT, STANLEY, The Methyl Ethers of D-Glucose, **5,** 145–190

BOURNE, E. J. *See also,* Barker, S. A.

BOUVENG, H. O., and LINDBERG, B., Methods in Structural Polysaccharide Chemistry, **15,** 53–89

BRADY, ROBERT F., JR., Cyclic Acetals of Ketoses, **26,** 197–278

BRAY, H. G., D-Glucuronic Acid in Metabolism, **8,** 251–275

BRAY, H. G., and STACEY, M., Blood Group Polysaccharides, **4,** 37–55

BRIMACOMBE, J. S. *See* How, M. J.

BUTTERWORTH, ROGER F., and HANESSIAN, STEPHEN, Tables of the Properties of Deoxy Sugars and Their Simple Derivatives, **26,** 279–296

C

CAESAR, GEORGE V., Starch Nitrate, **13,** 331–345

CALDWELL, MARY L., and ADAMS, MILDRED, Action of Certain Alpha Amylases, **5,** 229–268

CANTOR, SIDNEY M., [Obituary of] John C. Sowden, **20,** 1–10

CANTOR, SIDNEY M. *See also,* Miller, Robert Ellsworth.

CAPON, B., and OVEREND, W. G., Constitution and Physicochemical Properties of Carbohydrates, **15,** 11–51

CARR, C. JELLEFF, and KRANTZ, JOHN C., JR., Metabolism of the Sugar Alcohols and Their Derivatives, **1,** 175–192

CHIZHOV, O. S. *See* Kochetkov, N. K.

CHURMS, SHIRLEY C., Gel Chromatography of Carbohydrates, **25,** 13–51

CLAMP, JOHN R., HOUGH, L., HICKSON, JOHN L., and WHISTLER, ROY L., Lactose, **16,** 159–206

COMPTON, JACK, The Molecular Constitution of Cellulose, **3,** 185–228

CONCHIE, J., LEVVY, G. A., and MARSH, C. A., Methyl and Phenyl Glycosides of the Common Sugars, **12,** 157–187

CORINA, D. L. *See* Barnett, J. E. G.

COURTOIS, JEAN EMILE, [Obituary of] Emile Bourquelot, **18,** 1–8

COXON, BRUCE, Proton Magnetic Resonance Spectroscopy of Carbohydrates, **27,** 7–83

CRUM, JAMES D., The Four-carbon Saccharinic Acids, **13,** 169–188

D

DAVIES, D. A. L., Polysaccharides of Gram-negative Bacteria, **15,** 271–340

DEAN, G. R., and GOTTFRIED, J. B., The Commercial Production of Crystalline Dextrose, **5,** 127–143

DE BELDER, A. N., Cyclic Acetals of the Aldoses and Aldosides, **20,** 219–302

DEFAYE, J., 2,5-Anhydrides of Sugars and Related Compounds, **25,** 181–228

DEITZ, VICTOR R. *See* Liggett, R. W.

DEUEL, H. *See* Mehta, N. C.

DEUEL, HARRY J., JR., and MOREHOUSE, MARGARET G., The Interrelation of Carbohydrate and Fat Metabolism, **2,** 119–160

DEULOFEU, VENANCIO, The Acylated Nitriles of Aldonic Acids and Their Degradation, **4,** 119–151

DIMLER, R. J., 1,6-Anhydrohexofuranoses, A New Class of Hexosans, **7,** 37–52

DOUDOROFF, M. *See* Hassid, W. Z.

DUBACH, P. *See* Mehta, N. C.

DURETTE, PHILIPPE L., and HORTON, D., Conformational Analysis of Sugars and Their Derivatives, **26,** 49–125

DUTCHER, JAMES D., Chemistry of the Amino Sugars Derived from Antibiotic Substances, **18,** 259–308

E

ELDERFIELD, ROBERT C., The Carbohydrate Components of the Cardiac Glycosides, **1,** 147–173

EL KHADEM, HASSAN, Chemistry of Osazones, **20,** 139–181

EL KHADEM, HASSAN, Chemistry of Osotriazoles, **18,** 99–121

EL KHADEM, HASSAN, Synthesis of Nitrogen Heterocycles from Saccharide Derivatives, **25**, 351–405
ELLIS, G. P., The Maillard Reaction, **14**, 63–134
ELLIS, G. P., and HONEYMAN, JOHN, Glycosylamines, **10**, 95–168
EVANS, TAYLOR H., and HIBBERT, HAROLD, Bacterial Polysaccharides, **2**, 203–233
EVANS, W. L., REYNOLDS, D. D., and TALLEY, E. A., The Synthesis of Oligosaccharides, **6**, 27–81

F

FERRIER, R. J., Unsaturated Sugars, **20**, 67–137; **24**, 199–266
FEWSTER, J. A. See Bayne, S.
FLETCHER, HEWITT G., JR., The Chemistry and Configuration of the Cyclitols, **3**, 45–77
FLETCHER, HEWITT G., JR., and RICHTMYER, NELSON K., Applications in the Carbohydrate Field of Reductive Desulfurization by Raney Nickel, **5**, 1–28
FLETCHER, HEWITT G., JR. See also, Jeanloz, Roger W.
FORDYCE, CHARLES R., Cellulose Esters of Organic Acids, **1**, 309–327
FOSTER, A. B., Zone Electrophoresis of Carbohydrates, **12**, 81–115
FOSTER, A. B., and HORTON, D., Aspects of the Chemistry of the Amino Sugars, **14**, 213–281
FOSTER, A. B., and HUGGARD, A. J., The Chemistry of Heparin, **10**, 335–368
FOSTER, A. B., and STACEY, M., The Chemistry of the 2-Amino Sugars (2-Amino-2-deoxy-sugars), **7**, 247–288
FOSTER, A. B., and WEBBER, J. M., Chitin, **15**, 371–393
FOX, J. J., and WEMPEN, I., Pyrimidine Nucleosides, **14**, 283–380
FOX, JACK J. See also, Ueda, Tohru.
FRENCH, DEXTER, The Raffinose Family of Oligosaccharides, **9**, 149–184
FRENCH, DEXTER, The Schardinger Dextrins, **12**, 189–260
FREUDENBERG, KARL, Emil Fischer and his Contribution to Carbohydrate Chemistry, **21**, 1–38

G

GARCÍA GONZÁLEZ, F., Reactions of Monosaccharides with *beta*-Ketonic Esters and Related Substances, **11**, 97–143
GARCÍA GONZÁLEZ, F., and GÓMEZ SÁNCHEZ, A., Reactions of Amino Sugars with *beta*-Dicarbonyl Compounds, **20**, 303–355
GOEPP, RUDOLPH MAXIMILIAN, JR. See Lohmar, Rolland
GOLDSTEIN, I. J., and HULLAR, T. L., Chemical Synthesis of Polysaccharides, **21**, 431–512
GÓMEZ SÁNCHEZ, A. See García González, F.
GOODMAN, IRVING, Glycosyl Ureides, **13**, 215–236
GOODMAN, LEON, Neighboring-group Participation in Sugars, **22**, 109–175
GORIN, P. A. J., and SPENCER, J. F. T., Structural Chemistry of Fungal Polysaccharides, **23**, 367–417
GOTTFRIED, J. B. See Dean, G. R.
GOTTSCHALK, ALFRED, Principles Underlying Enzyme Specificity in the Domain of Carbohydrates, **5**, 49–78
GREEN, JOHN W., The Glycofuranosides, **21**, 95–142
GREEN, JOHN W., The Halogen Oxidation of Simple Carbohydrates, Excluding the Action of Periodic Acid, **3**, 129–184
GREENWOOD, C. T., Aspects of the Physical Chemistry of Starch, **11**, 335–385
GREENWOOD, C. T., The Size and Shape of Some Polysaccharide Molecules, **7**, 289–332; **11**, 385–393
GREENWOOD, C. T., The Thermal Degradation of Starch, **22**, 483–515
GREENWOOD, C. T., and MILNE, E. A., Starch Degrading and Synthesizing Enzymes: A Discussion of Their Properties and Action Pattern, **23**, 281–366
GREENWOOD, C. T. See also, Banks, W.
GURIN, SAMUEL, Isotopic Tracers in the

Study of Carbohydrate Metabolism, **3**, 229-250

GUTHRIE, R. D., The "Dialdehydes" from the Periodate Oxidation of Carbohydrates, **16**, 105-158

GUTHRIE, R. D., and MCCARTHY, J. F., Acetolysis, **22**, 11-23

H

HALL, L. D., Nuclear Magnetic Resonance, **19**, 51-93

HANESSIAN, STEPHEN, Deoxy Sugars, **21**, 143-207

HANESSIAN, STEPHEN. See also, Butterworth, Roger F.

HARRIS, ELWIN E., Wood Saccharification, **4**, 153-188

HASKINS, JOSEPH F., Cellulose Ethers of Industrial Significance, **2**, 279-294

HASSID, W. Z., and DOUDOROFF, M., Enzymatic Synthesis of Sucrose and Other Disaccharides, **5**, 29-48

HASSID, W. Z. See also, Neufeld, Elizabeth F.

HASSID, W. Z. See also, Nikaido, H.

HAYNES, L. J., Naturally Occurring C-Glycosyl Compounds, **18**, 227-258; **20**, 357-369

HAYNES, L. J., and NEWTH, F. H., The Glycosyl Halides and Their Derivatives, **10**, 207-256

HEHRE, EDWARD J., The Substituted-sucrose Structure of Melezitose, **8**, 277-290

HELFERICH, BURCKHARDT, The Glycals, **7**, 209-245

HELFERICH, BURCKHARDT, Trityl Ethers of Carbohydrates, **3**, 79-111

HEYNS, K., and PAULSEN H., Selective Catalytic Oxidation of Carbohydrates, Employing Platinum Catalysts, **17**, 169-221

HIBBERT, HAROLD. See Evans, Taylor H.

HICKSON, JOHN L. See Clamp, John R.

HILTON, H. W., The Effects of Plant-growth Substances on Carbohydrate Systems, **21**, 377-430

HINDERT, MARJORIE. See Karabinos, J. V.

HIRST, E. L., [Obituary of] James Colquhoun Irvine, **8**, xi-xvii

HIRST, E. L., [Obituary of] Walter Norman Haworth, **6**, 1-9

HIRST, E. L., and JONES, J. K. N., The Chemistry of Pectic Materials, **2**, 235-251

HIRST, E. L., and ROSS, A. G., [Obituary of] Edmund George Vincent Percival, **10**, xiii-xx

HODGE, JOHN E., The Amadori Rearrangement, **10**, 169-205

HONEYMAN, JOHN, and MORGAN, J. W. W., Sugar Nitrates, **12**, 117-135

HONEYMAN, JOHN. See also, Barry, C. P.

HONEYMAN, JOHN. See also, Ellis, G. P.

HORTON, D., [Obituary of] Alva Thompson, **19**, 1-6

HORTON, D., [Obituary of] Melville Lawrence Wolfrom, **26**, 1-47

HORTON, D., Tables of Properties of 2-Amino-2-deoxy Sugars and Their Derivatives, **15**, 159-200

HORTON D., and HUTSON, D. H., Developments in the Chemistry of Thio Sugars, **18**, 123-199

HORTON, D. See also, Durette, Philippe L.

HORTON, D. See also, Foster, A. B.

HOUGH, L., and JONES, J. K. N., The Biosynthesis of the Monosaccharides, **11**, 185-262

HOUGH, L., PRIDDLE, J. E., and THEOBALD, R. S., The Carbonates and Thiocarbonates of Carbohydrates, **15**, 91-158

HOUGH, L. See also, Clamp, John R.

HOW, M. J., BRIMACOMBE, J. S., and STACEY, M., The Pneumococcal Polysaccharides, **19**, 303-357

HUDSON, C. S., Apiose and the Glycosides of the Parsley Plant, **4**, 57-74

HUDSON, C. S., The Fischer Cyanohydrin Synthesis and the Configurations of Higher-carbon Sugars and Alcohols, **1**, 1-36

HUDSON, C. S., Historical Aspects of Emil Fischer's Fundamental Conventions for Writing Stereo-formulas in a Plane, **3**, 1-22

HUDSON, C. S., Melezitose and Turanose, **2**, 1-36

HUGGARD, A. J. See Foster, A. B.
HULLAR, T. L. See Goldstein, I. J.
HUTSON, D. H. See Horton, D.

I

INCH, T. D., The Use of Carbohydrates in the Synthesis and Configurational Assignments of Optically Active, Non-carbohydrate Compounds, **27**, 191–225
ISBELL, HORACE S., and PIGMAN, WARD, Mutarotation of Sugars in Solution: Part II, Catalytic Processes, Isotope Effects, Reaction Mechanisms, and Biochemical Aspects, **24**, 13–65
ISBELL, HORACE S. See also, Pigman, Ward.

J

JAMIESON, G. A., [Obituary of] William Werner Zorbach, **27**, 1–6
JEANLOZ, ROGER W., [Obituary of] Kurt Heinrich Meyer, **11**, xiii–xviii
JEANLOZ, ROGER W., The Methyl Ethers of 2-Amino-2-deoxy Sugars, **13**, 189–214
JEANLOZ, ROGER W., and FLETCHER, HEWITT G., JR., The Chemistry of Ribose, **6**, 135–174
JEFFREY, G. A., and ROSENSTEIN, R. D., Crystal-structure Analysis in Carbohydrate Chemistry, **19**, 7–22
JONES, DAVID M., Structure and Some Reactions of Cellulose, **19**, 219–246
JONES, J. K. N., and SMITH, F., Plant Gums and Mucilages, **4**, 243–291
JONES, J. K. N. See also, Hirst, E. L.
JONES, J. K. N. See also, Hough, L.
JONSEN, J., and LALAND, S., Bacterial Nucleosides and Nucleotides, **15**, 201–234

K

KARABINOS, J. V., Psicose, Sorbose and Tagatose, **7**, 99–136
KARABINOS, J. V., and HINDERT, MARJORIE, Carboxymethylcellulose, **9**, 285–302

KENT, P. W. See Stacey, M.
KERTESZ, Z. I., and McCOLLOCH, R. J., Enzymes Acting on Pectic Substances, **5**, 79–102
KISS, J., Glycosphingolipids (Sugar–Sphingosine Conjugates), **24**, 381–433
KLEMER, ALMUTH. See Micheel, Fritz.
KOCHETKOV, N. K., and CHIZHOV, O. S., Mass Spectrometry of Carbohydrate Derivatives, **21**, 39–93
KORT, M. J., Reactions of Free Sugars with Aqueous Ammonia, **25**, 311–349
KOWKABANY, GEORGE N., Paper Chromatography of Carbohydrates and Related Compounds, **9**, 303–353
KRANTZ, JOHN C., JR. See Carr, C. Jelleff.

L

LAIDLAW, R. A., and PERCIVAL, E. G. V., The Methyl Ethers of the Aldopentoses and of Rhamnose and Fucose, **7**, 1–36
LALAND, S. See Jonsen, J.
LEDERER, E., Glycolipids of Acid-fast Bacteria, **16**, 207–238
LEMIEUX, R. U., Some Implications in Carbohydrate Chemistry of Theories Relating to the Mechanisms of Replacement Reactions, **9**, 1–57
LEMIEUX, R. U., and WOLFROM, M. L., The Chemistry of Streptomycin, **3**, 337–384
LESPIEAU, R., Synthesis of Hexitols and Pentitols from Unsaturated Polyhydric Alcohols, **2**, 107–118
LEVI, IRVING, and PURVES, CLIFFORD B., The Structure and Configuration of Sucrose (alpha-D-Glucopyranosyl beta-D-Fructofuranoside), **4**, 1–35
LEVVY, G. A., and MARSH, C. A., Preparation and Properties of β-Glucuronidase, **14**, 381–428
LEVVY, G. A. See also, Conchie, J.
LIGGETT, R. W., and DEITZ, VICTOR R., Color and Turbidity of Sugar Products, **9**, 247–284
LINDBERG, B. See Bouveng, H. O.
LOHMAR, ROLLAND, and GOEPP, RUDOLPH MAXIMILIAN, JR., The

Hexitols and Some of Their Derivatives, **4**, 211–241

M

MAHER, GEORGE G., The Methyl Ethers of the Aldopentoses and of Rhamnose and Fucose, **10**, 257–272
MAHER, GEORGE G., The Methyl Ethers of D-Galactose, **10**, 273–282
MALHOTRA, OM PRAKASH. *See* Wallenfels, Kurt.
MANNERS, D. J., Enzymic Synthesis and Degradation of Starch and Glycogen, **17**, 371–430
MANNERS, D. J., The Molecular Structure of Glycogens, **12**, 261–298
MARCHESSAULT, R. H., and SARKO, A., X-Ray Structure of Polysaccharides, **22**, 421–482
MARSH, C. A. *See* Conchie, J.
MARSH, C. A. *See* Levvy, G. A.
MARSHALL, R. D., and NEUBERGER, A., Aspects of the Structure and Metabolism of Glycoproteins, **25**, 407–478
MCCARTHY, J. F. *See* Guthrie, R. D.
MCCASLAND, G. E., Chemical and Physical Studies of Cyclitols Containing Four or Five Hydroxyl Groups, **20**, 11–65
MCCLOSKEY, CHESTER M., Benzyl Ethers of Sugars, **12**, 137–156
MCCOLLOCH, R. J. *See* Kertesz, Z. I.
MCDONALD, EMMA J., The Polyfructosans and Difructose Anhydrides, **2**, 253–277
MCGALE, E. H. F., Protein–Carbohydrate Compounds in Human Urine, **24**, 435–452
MCGINNIS, G. D. *See* Shafizadeh, F.
MEHLTRETTER, C. L., The Chemical Synthesis of D-Glucuronic Acid, **8**, 231–249
MEHTA, N. C., DUBACH, P., and DEUEL, H., Carbohydrates in the Soil, **16**, 335–355
MESTER, L., The Formazan Reaction in Carbohydrate Research, **13**, 105–167
MESTER, L., [Obituary of] Géza Zemplén, **14**, 1–8
MICHEEL, FRITZ, and KLEMER, ALMUTH, Glycosyl Fluorides and Azides, **16**, 85–103
MILLER, ROBERT ELLSWORTH, and CANTOR, SIDNEY M., Aconitic Acid, a By-product in the Manufacture of Sugar, **6**, 231–249
MILLS, J. A., The Stereochemistry of Cyclic Derivatives of Carbohydrates, **10**, 1–53
MILNE, E. A. *See* Greenwood, C. T.
MONTGOMERY, JOHN A., and THOMAS, H. JEANETTE, Purine Nucleosides, **17**, 301–369
MONTGOMERY, REX, [Obituary of] Fred Smith, **22**, 1–10
MOODY, G. J., The Action of Hydrogen Peroxide on Carbohydrates and Related Compounds, **19**, 149–179
MOREHOUSE, MARGARET G. *See* Deuel, Harry J., Jr.
MORGAN, J. W. W. *See* Honeyman, John.
MORI, T., Seaweed Polysaccharides, **8**, 315–350
MOYE, C. J., Non-aqueous Solvents for Carbohydrates, **27**, 85–125
MUETGEERT, J., The Fractionation of Starch, **16**, 299–333
MYRBÄCK, KARL, Products of the Enzymic Degradation of Starch and Glycogen, **3**, 251–310

N

NEELY, W. BROCK, Dextran: Structure and Synthesis, **15**, 341–369
NEELY, W. BROCK, Infrared Spectra of Carbohydrates, **12**, 13–33
NEUBERG, CARL, Biochemical Reductions at the Expense of Sugars, **4**, 75–117
NEUBERGER, A. *See* Marshall, R. D.
NEUFELD, ELIZABETH F., and HASSID, W. Z., Biosynthesis of Saccharides from Glycopyranosyl Esters of Nucleotides ("Sugar Nucleotides"), **18**, 309–356
NEWTH, F. H., The Formation of Furan Compounds from Hexoses, **6**, 83–106
NEWTH, F. H. *See also*, Haynes, L. J.
NICKERSON, R. F., The Relative Crystallinity of Celluloses, **5**, 103–126

NIKAIDO, H., and HASSID, W. Z., Biosynthesis of Saccharides from Glycopyranosyl Esters of Nucleoside Pyrophosphates ("Sugar Nucleotides"), **26**, 351–483

NORD, F. F., [Obituary of] Carl Neuberg, **13**, 1–7

O

OLSON, E. J. See Whistler, Roy L.

OVEREND, W. G., and STACEY, M., The Chemistry of the 2-Desoxy-sugars, **8**, 45–105

OVEREND, W. G. See also, Capon, B.

P

PACSU, EUGENE, Carbohydrate Orthoesters, **1**, 77–127

PARRISH, F. W. See Ball, D. H.

PAULSEN, H., Cyclic Acyloxonium Ions in Carbohydrate Chemistry, **26**, 127–195

PAULSEN, H., and TODT, K., Cyclic Monosaccharides Having Nitrogen or Sulfur in the Ring, **23**, 115–232

PAULSEN, H. See also, Heyns, K.

PAZUR, JOHN H., and ARONSON, N. N., JR., Glycoenzymes: Enzymes of Glycoprotein Structure, **27**, 301–341

PEAT, STANLEY, The Chemistry of Anhydro Sugars, **2**, 37–77

PEAT, STANLEY. See also, Bourne, E. J.

PERCIVAL, E. G. V., The Structure and Reactivity of the Hydrazone and Osazone Derivatives of the Sugars, **3**, 23–44

PERCIVAL, E. G. V. See also, Laidlaw, R. A.

PERLIN, A. S., Action of Lead Tetraacetate on the Sugars, **14**, 9–61

PERLIN, A. S., [Obituary of] Clifford Burrough Purves, **23**, 1–10

PHILLIPS, G. O., Photochemistry of Carbohydrates, **18**, 9–59

PHILLIPS, G. O., Radiation Chemistry of Carbohydrates, **16**, 13–58

PIGMAN, WARD, and ISBELL, HORACE S., Mutarotation of Sugars in Solution: Part I. History, Basic Kinetics, and Composition of Sugar Solutions, **23**, 11–57

PIGMAN, WARD. See also, Isbell, Horace S.

POLGLASE, W. J., Polysaccharides Associated with Wood Cellulose, **10**, 283–333

PRIDDLE, J. E. See Hough, L.

PRIDHAM, J. B., Phenol–Carbohydrate Derivatives in Higher Plants, **20**, 371–408

PRIDHAM, J. B. See also, Bailey, R. W.

PURVES, CLIFFORD B. See Levi, Irving.

R

RAYMOND, ALBERT L., Thio- and Selenosugars, **1**, 129–145

REES, D. A., Structure, Conformation, and Mechanism in the Formation of Polysaccharide Gels and Networks, **24**, 267–332

REEVES, RICHARD E., Cuprammonium–Glycoside Complexes, **6**, 107–134

REICHSTEIN, T., and WEISS, EKKEHARD, The Sugars of the Cardiac Glycosides, **17**, 65–120

RENDLEMAN, J. A., JR., Complexes of Alkali Metals and Alkaline-earth Metals with Carbohydrates, **21**, 209–271

REYNOLDS, D. D. See Evans, W. L.

RICHTMYER, NELSON K., The Altrose Group of Substances, **1**, 37–76

RICHTMYER, NELSON K., The 2-(aldo-Polyhydroxyalkyl)benzimidazoles, **6**, 175–203

RICHTMYER, NELSON K. See also, Fletcher, Hewitt G., Jr.

ROSENSTEIN, R. D. See Jeffrey, G. A.

ROSENTHAL, ALEX, Application of the Oxo Reaction to Some Carbohydrate Derivatives, **23**, 59–114

ROSS, A. G. See Hirst, E. L.

S

SANDS, LILA. See Anderson, Ernest.

SARKO, A. See Marchessault, R. H.

SATTLER, LOUIS, Glutose and the Unfermentable Reducing Substances in Cane Molasses, **3**, 113–128

SCHOCH, THOMAS JOHN, The Fractionation of Starch, **1**, 247–277

SHAFIZADEH, F., Branched-chain Sugars of Natural Occurrence, **11**, 263–283

SHAFIZADEH, F., Formation and Cleavage of the Oxygen Ring in Sugars, **13**, 9–61

SHAFIZADEH, F., Pyrolysis and Combustion of Cellulosic Materials, **23**, 419–474

SHAFIZADEH, F., and MCGINNIS, G. D., Morphology and Biogenesis of Cellulose and Plant Cell-walls, **26**, 297–349

SIDDIQUI, I. R., The Sugars of Honey, **25**, 285–309

SMITH, F., Analogs of Ascorbic Acid, **2**, 79–106

SMITH, F. *See also*, Jones, J. K. N.

SOLTZBERG, SOL, Alditol Anhydrides, **25**, 229–283

SOWDEN, JOHN C., The Nitromethane and 2-Nitroethanol Syntheses, **6**, 291–318

SOWDEN, JOHN C., [Obituary of] Hermann Otto Laurenz Fischer, **17**, 1–14

SOWDEN, JOHN C., The Saccharinic Acids, **12**, 35–79

SPECK, JOHN C., JR., The Lobry de Bruyn–Alberda van Ekenstein Transformation, **13**, 63–103

SPEDDING, H., Infrared Spectroscopy and Carbohydrate Chemistry, **19**, 23–49

SPENCER, J. F. T. *See* Gorin, P. A. J.

SPRINSON, D. B., The Biosynthesis of Aromatic Compounds from D-Glucose, **15**, 235–270

STACEY, M., The Chemistry of Mucopolysaccharides and Mucoproteins, **2**, 161–201

STACEY, M., and KENT, P. W., The Polysaccharides of *Mycobacterium tuberculosis*, **3**, 311–336

STACEY, M. *See also*, Bray, H. G.

STACEY, M. *See also*, Foster, A. B.

STACEY, M. *See also*, How, M. J.

STACEY, M. *See also*, Overend, W. G.

STOLOFF, LEONARD, Polysaccharide Hydrocolloids of Commerce, **13**, 265–287

STRAHS, GERALD, Crystal-structure Data for Simple Carbohydrates and Their Derivatives, **25**, 53–107

SUGIHARA, JAMES M., Relative Reactivities of Hydroxyl Groups of Carbohydrates, **8**, 1–44

T

TALLEY, E. A. *See* Evans, W. L.

TEAGUE, ROBERT S., The Conjugates of D-Glucuronic Acid of Animal Origin, **9**, 185–246

THEANDER, OLOF, Dicarbonyl Carbohydrates, **17**, 223–299

THEOBALD, R. S. *See* Hough, L.

THOMAS, H. JEANNETTE. *See* Montgomery, John A.

TIMELL, T. E., Wood Hemicelluloses:
Part I, **19**, 247–302
Part II, **20**, 409–483

TIPSON, R. STUART, The Chemistry of the Nucleic Acids, **1**, 193–245

TIPSON, R. STUART, [Obituary of] Harold Hibbert, **16**, 1–11

TIPSON, R. STUART, [Obituary of] Phoebus Aaron Theodor Levene, **12**, 1–12

TIPSON, R. STUART, Sulfonic Esters of Carbohydrates, **8**, 107–215

TODT, K. *See* Paulsen, H.

TURVEY, J. R., [Obituary of] Stanley Peat, **25**, 1–12

TURVEY, J. R., Sulfates of the Simple Sugars, **20**, 183–218

U

UEDA, TOHRU, and FOX, JACK J., The Mononucleotides, **22**, 307–419

V

VERSTRAETEN, L. M. J., D-Fructose and Its Derivatives, **22**, 229–305

W

WALLENFELS, KURT, and MALHOTRA, OM PRAKASH, Galactosidases, **16**, 239–298

WEBBER, J. M., Higher-carbon Sugars, **17,** 15–63
WEBBER, J. M. *See also,* Foster, A. B.
WEIGEL, H., Paper Electrophoresis of Carbohydrates, **18,** 61–97
WEISS, EKKEHARD. *See* Reichstein, T.
WEMPEN, I. *See* Fox, J. J.
WHISTLER, ROY L., Preparation and Properties of Starch Esters, **1,** 279–307
WHISTLER, ROY L., Xylan, **5,** 269–290
WHISTLER, ROY L., and BEMILLER, J. N., Alkaline Degradation of Polysaccharides, **13,** 289–329
WHISTLER, ROY L., and OLSON, E. J., The Biosynthesis of Hyaluronic Acid, **12,** 299–319
WHISTLER, ROY L. *See also,* Clamp, John R.
WHITEHOUSE, M. W. *See* Zilliken, F.
WIGGINS, L. F., Anhydrides of the Pentitols and Hexitols, **5,** 191–228
WIGGINS, L. F., The Utilization of Sucrose, **4,** 293–336

WILLIAMS, NEIL R., Oxirane Derivatives of Aldoses, **25,** 109–179
WISE, LOUIS E., [Obituary of] Emil Heuser, **15,** 1–9
WOLFROM, M. L., [Obituary of] Claude Silbert Hudson, **9,** xiii–xviii
WOLFROM, M. L., [Obituary of] Rudolph Maximilian Goepp, Jr., **3,** xv–xxiii
WOLFROM, M. L. *See also,* Binkley, W. W.
WOLFROM, M. L. *See also,* Lemieux, R. U.

Z

ZHDANOV, YU. A., ALEXEEV, YU. E., and ALEXEEVA, V. G., The Wittig Reaction in Carbohydrate Chemistry, **27,** 227–299
ZILLIKEN, F., and WHITEHOUSE, M. W., The Nonulosaminic Acids—Neuraminic Acids and Related Compounds (Sialic Acids), **13,** 237–263
ZORBACH, W. WERNER, and BHAT, K. VENKATRAMANA, Synthetic Cardenolides, **21,** 273–321

CUMULATIVE SUBJECT INDEX FOR VOLS. 1–27

A

Acetals,
  cyclic, of the aldoses and aldosides, **20**, 219–302
  of ketoses, **26**, 197–277
  of hexitols, pentitols, and tetritols, **7**, 137–207
Acetic acid, trifluoro-, anhydride, applications of, in carbohydrate chemistry, **16**, 59–84
Acetolysis, **22**, 11–23
Aconitic acid, **6**, 231–249
Action pattern,
  of starch degrading and synthesizing enzymes, **23**, 281–366
Acyloxonium ions,
  cyclic, in carbohydrate chemistry, **26**, 127–195
Adsorbents,
  granular, for sugar refining, **6**, 205–230
Alcohols,
  higher-carbon sugar, configurations of, **1**, 1–36
  unsaturated polyhydric, **2**, 107–118
Alditols, anhydrides of, **25**, 229–283
Aldonic acids,
  acylated nitriles of, **4**, 119–151
Aldopentoses,
  methyl ethers of, **7**, 1–36; **10**, 257–272
Aldoses, oxirane derivatives of, **25**, 109–179
Aldoses and aldosides,
  cyclic acetals of, **20**, 219–302
Alkaline degradation,
  of polysaccharides, **13**, 289–329
Altrose,
  group of compounds related to, **1**, 37–76
Amadori rearrangement, **10**, 169–205
Amino sugars. See Sugars, 2-amino-2-deoxy.
Ammonia, aqueous, reactions with free sugars, **25**, 311–349
Amylases,
  certain alpha, **5**, 229–268

Analysis,
  conformational, of sugars and their derivatives, **26**, 49–125
  of crystal structure, in carbohydrate chemistry, **19**, 7–22
Anhydrides,
  2,5-, of sugars, **25**, 181–228
  difructose, **2**, 253–277
  of alditols, **25**, 229–283
  of aldoses, **25**, 109–179
  of hexitols, **5**, 191–228
  of pentitols, **5**, 191–228
Anhydro sugars. See Sugars, anhydro.
Animals,
  conjugates of D-glucuronic acid originating in, **9**, 185–246
Antibiotic substances,
  chemistry of the amino sugars derived from, **18**, 259–308
Apiose, **4**, 57–74
Ascorbic acid,
  analogs of, **2**, 79–106
Aromatic compounds,
  biosynthesis of, from D-glucose, **15**, 235–270

B

Bacteria,
  glycolipids of acid-fast, **16**, 207–238
  nucleosides and nucleotides of, **15**, 201–234
  polysaccharides from, **2**, 203–233; **3**, 311–336
  polysaccharides of Gram-negative, **15**, 271–340
Benzimidazoles,
  2-(*aldo*-polyhydroxyalkyl)-, **6**, 175–203
Benzyl ethers,
  of sugars, **12**, 137–156
Biochemical aspects,
  of mutarotation of sugars in solution, **24**, 13–65
Biochemical reductions,
  at the expense of sugars, **4**, 75–117

393

Biogenesis,
  and morphology, of cellulose and plant cell-walls, **26**, 297–349
Biosynthesis,
  of aromatic compounds from D-glucose, **15**, 235–270
  of hyaluronic acid, **12**, 299–319
  of the monosaccharides, **11**, 185–262
  of saccharides, from glycopyranosyl esters of nucleoside pyrophosphates ("sugar nucleotides"), **18**, 309–356; **26**, 351–483
Blood groups,
  polysaccharides of, **4**, 37–55
Boric acid,
  for determining configuration of carbohydrates, **4**, 189–210
Bourquelot, Emile,
  obituary of, **18**, 1–8
Branched-chain sugars. *See* Sugars, branched-chain.

C

Cane juice,
  composition of, **8**, 291–314
Cane molasses. *See* Molasses, cane.
Carbohydrates,
  action of hydrogen peroxide on, **19**, 149–179
  application of reductive desulfurization by Raney nickel, in the field of, **5**, 1–28
  application of trifluoroacetic anhydride in chemistry of, **16**, 59–84
  application of the Oxo reaction to some derivatives of, **23**, 59–114
  as components of cardiac glycosides, **1**, 147–173
  carbonates of, **15**, 91–158
  chemistry of, Emil Fischer and his contribution to, **21**, 1–38
    the Wittig reaction in, **27**, 227–299
  complexes of, with alkali metals and alkaline-earth metals, **21**, 209–271
  compounds with proteins, in human urine, **24**, 435–452
  constitution of, **15**, 11–51
  crystal-structure analysis of, **19**, 7–22
  crystal-structure data for, **25**, 53–107
  cyclic acyloxonium ions, in the chemistry of, **26**, 127–195
  degradation of, **19**, 181–218
  determination of configuration of, with boric acid, **4**, 189–210
  dicarbonyl, **17**, 223–299
  enzyme specificity in the domain of, **5**, 49–78
  formazan reaction, in research on, **13**, 105–167
  Friedel–Crafts and Grignard processes applied to, **6**, 251–289
  gas–liquid chromatography of derivatives of, **19**, 95–147
  gel chromatography of, **25**, 13–51
  halogen oxidation of simple, **3**, 129–184
  halogenated, **22**, 177–227
  infrared spectra of, **12**, 13–33
  infrared spectroscopy of, **19**, 23–49
  mass spectrometry of derivatives of, **21**, 39–93
  mechanisms of replacement reactions in chemistry of, **9**, 1–57
  metabolism of, **2**, 119–160; **3**, 229–250
  non-aqueous solvents for, **27**, 85–125
  orthoesters of, **1**, 77–127
  paper electrophoresis of, **18**, 61–97
  periodate oxidation of, **11**, 1–41
    the "dialdehydes" from, **16**, 105–158
  phenol derivatives, in higher plants, **20**, 371–408
  photochemistry of, **18**, 9–59
  physicochemical properties of, **15**, 11–51
  proton magnetic resonance spectroscopy of, **27**, 7–83
  radiation chemistry of, **16**, 13–58
  and related compounds, action of hydrogen peroxide on, **19**, 149–179
  paper chromatography of, **9**, 303–353
  relative reactivities of hydroxyl groups of, **8**, 1–44
  selective catalytic oxidation of, employing platinum catalysts, **17**, 169–221
  in the soil, **16**, 335–355
  stereochemistry of cyclic derivatives of, **10**, 1–53
  sulfonic esters of, **8**, 107–215; **23**, 233–280; **24**, 139–197
  systems, effects of plant-growth substances on, **21**, 377–430

CUMULATIVE SUBJECT INDEX FOR VOLS. 1–27 395

thiocarbonates of, **15**, 91–158
trityl ethers of, **3**, 79–111
use of, in the synthesis and configurational assignments of optically active, non-carbohydrate compounds, **27**, 191–225
zone electrophoresis of, **12**, 81–115
Carbonates,
of carbohydrates, **15**, 91–158
Carboxymethyl ether,
of cellulose, **9**, 285–302
Cardenolides. *See also*, Glycosides, cardiac.
synthetic, **21**, 273–321
Catalysts,
effects of, in mutarotation of sugars in solution, **24**, 13–65
platinum, in selective catalytic oxidation of carbohydrates, **17**, 169–221
Cellulose,
carboxymethyl-, **9**, 285–302
esters of, with organic acids, **1**, 309–327
ethers of, **2**, 279–294
molecular constitution of, **3**, 185–228
and plant cell-walls, morphology and biogenesis of, **26**, 297–349
of wood, polysaccharides associated with, **10**, 283–333
Celluloses,
relative crystallinity of, **5**, 103–126
some reactions of, **19**, 219–246
structure of, **19**, 219–246
Cellulosic materials,
combustion and pyrolysis of, **23**, 419–474
Chemistry,
of the amino sugars, **14**, 213–281
of the 2-amino sugars, **7**, 247–288
of anhydro sugars, **2**, 37–77
of carbohydrates, applications of trifluoroacetic anhydride in, **16**, 59–84
Emil Fischer and his contribution to, **21**, 1–38
crystal-structure analysis in, **19**, 7–22
infrared spectroscopy and, **19**, 23–49
some implications of theories relating to the mechanisms of replacement reactions in, **9**, 1–57
the Wittig reaction in, **27**, 227–299
of the cyclitols, **3**, 45–77

of cyclitols containing four or five hydroxyl groups, **20**, 11–65
of the 2-deoxy sugars, **8**, 45–105
of heparin, **10**, 335–368
of mucopolysaccharides and mucoproteins, **2**, 161–201
of the nucleic acids, **1**, 193–245
of osazones, **20**, 139–181
of osotriazoles, **18**, 99–121
of pectic materials, **2**, 235–251
of ribose, **6**, 135–174
of streptomycin, **3**, 337–384
of thio sugars, **18**, 123–199
physical, of carbohydrates, **15**, 11–51
of starch, **11**, 335–385
radiation, of carbohydrates, **16**, 13–58
stereo-, of cyclic derivatives of carbohydrates, **10**, 1–53
structural, of fungal polysaccharides, **23**, 367–417
of the hemicelluloses, **14**, 429–468
of polysaccharides, **15**, 53–89
Chitin, **15**, 371–393
Chromatography,
column. *See* Column chromatography.
gas–liquid. *See* Gas–liquid chromatography.
gel. *See* Gel chromatography.
paper. *See* Paper chromatography.
Color,
of sugar products, **9**, 247–284
Column chromatography,
of sugars and their derivatives, **10**, 55–94
Combustion, of cellulosic materials, **23**, 419–474
Complexes,
of carbohydrates, with alkali metals and alkaline-earth metals, **21**, 209–271
cuprammonium–glycoside, **6**, 107–134
Composition,
of sugar solutions, **23**, 11–57
Configuration,
of carbohydrates, determination of, **4**, 189–210
of cyclitols, **3**, 45–77
of higher-carbon sugar alcohols, **1**, 1–36
of sucrose, **4**, 1–35
Configurational assignments, of optically active, non-carbohydrate compounds,

use of carbohydrates in, **27**, 191–225
Conformation,
in formation of polysaccharide gels and networks, **24**, 267–332
Conformational analysis,
of sugars and their derivatives, **26**, 49–125
Conjugates,
of D-glucuronic acid, **9**, 185–246
of sugars with sphingosines, **24**, 381–433
Constitution,
of carbohydrates, **15**, 11–51
Crystallinity,
relative, of celluloses, **5**, 103–126
Crystal-structure,
analysis, in carbohydrate chemistry, **19**, 7–22
data, for simple carbohydrates and their derivatives, **25**, 53–107
Cuprammonium–glycoside complexes, **6**, 107–134
Cyanohydrin synthesis,
Fischer, **1**, 1–36
Cyclic acetals,
of the aldoses and aldosides, **20**, 219–302
of hexitols, pentitols, and tetritols, **7**, 137–207
of ketoses, **26**, 197–277
Cyclic acetoxonium ions,
in carbohydrate chemistry, **26**, 127–195
Cyclic derivatives,
of carbohydrates, stereochemistry of, **10**, 1–53
Cyclic monosaccharides,
having nitrogen or sulfur in the ring, **23**, 115–232
Cyclitols, **14**, 135–212
chemistry and configurations of, **3**, 45–77
containing four or five hydroxyl groups, chemical and physical studies of, **20**, 11–65

D

Degradation,
of acylated nitriles of aldonic acids, **4**, 119–151
of carbohydrates, **19**, 181–218
enzymic, of glycogen and starch, **3**, 251–310; **17**, 407–430
thermal, of starch, **22**, 483–515
3-Deoxyglycosones. *See* Glycosuloses, 3-deoxy-.
3-Deoxyglycosuloses. *See* Glycosuloses, 3-deoxy-.
Deoxy sugars. *See* Sugars, deoxy.
Desulfurization,
reductive, by Raney nickel, **5**, 1–28
Dextran,
structure and synthesis of, **15**, 341–369
Dextrins,
the Schardinger, **12**, 189–260
Dextrose,
commercial production of crystalline, **5**, 127–143
"Dialdehydes,"
from the periodate oxidation of carbohydrates, **16**, 105–158
Dicarbonyl derivatives,
of carbohydrates, **17**, 223–299
Difructose,
anhydrides, **2**, 253–277
Disaccharides,
enzymic synthesis of, **5**, 29–48
trehalose, **18**, 201–225

E

Electrophoresis, of carbohydrates,
paper, **18**, 61–97
zone, **12**, 81–115
Enzymes. *See also*, Amylases, Galactosidases, β-Glucuronidase, Glycoenzymes.
acting on pectic substances, **5**, 79–102
degradation by, of starch and glycogen, **3**, 251–310; **17**, 407–430
of glycoprotein structure, **27**, 301–341
specificity of, in the domain of carbohydrates, **5**, 49–78
starch degrading and synthesizing, **23**, 281–366
synthesis by, of glycogen and starch, **17**, 371–407
of sucrose and other disaccharides, **5**, 29–48
Esters,
of cellulose, with organic acids, **1**, 309–327
glycopyranosyl, of nucleoside pyro-

phosphates, **18**, 309-356; **26**, 351-483
beta-ketonic (and related substances),
  reactions with monosaccharides, **11**, 97-143
nitric, of starch, **13**, 331-345
of starch, preparation and properties of, **1**, 279-307
sulfonic, of carbohydrates, **8**, 107-215; **23**, 233-280; **24**, 139-197
Ethanol, 2-nitro-,
  syntheses with, **6**, 291-318
Ethers,
  benzyl, of sugars, **12**, 137-156
  carboxymethyl, of cellulose, **9**, 285-302
  of cellulose, **2**, 279-294
  methyl,
    of the aldopentoses, **7**, 1-36; **10**, 257-272
    of 2-amino-2-deoxy sugars, **13**, 189-214
    of fucose, **7**, 1-36; **10**, 257-272
    of D-galactose, **6**, 11-25; **10**, 273-282
    of D-glucose, **5**, 145-190
    of hexuronic acids, **9**, 131-148
    of D-mannose, **8**, 217-230
    of rhamnose, **7**, 1-36; **10**, 257-272
  trityl, of carbohydrates, **3**, 79-111

F

Fat,
  metabolism of, **2**, 119-160
Fischer, Emil, and his contribution to carbohydrate chemistry, **21**, 1-38
Fischer, Hermann Otto Laurenz,
  obituary of, **17**, 1-14
Formazan reaction,
  in carbohydrate research, **13**, 105-167
Formulas,
  stereo-, writing of, in a plane, **3**, 1-22
Fractionation,
  of starch, **1**, 247-277; **16**, 299-333
Friedel-Crafts process,
  in the carbohydrate series, **6**, 251-289
Fructans, **2**, 253-277
Fructofuranoside,
  α-D-glucopyranosyl β-D-, **4**, 1-35
Fructosans, poly-. *See* Fructans.
Fructose,
  and its derivatives, **7**, 53-98; **22**, 229-305

di-, anhydrides, **2**, 253-277
Fucose,
  methyl ethers of, **7**, 1-36; **10**, 257-272
Fungal polysaccharides,
  structural chemistry of, **23**, 367-417
Furan compounds,
  formation from hexoses, **6**, 83-106

G

Galactose,
  methyl ethers of D-, **6**, 11-25; **10**, 273-282
Galactosidases, **16**, 239-298
Gas-liquid chromatography,
  of carbohydrate derivatives, **19**, 95-147
Gel chromatography,
  of carbohydrates, **25**, 13-51
Gels,
  polysaccharide, **24**, 267-332
Glucose. *See also*, Dextrose.
  biosynthesis of aromatic compounds from D-, **15**, 235-270
  methyl ethers of D-, **5**, 145-190
Glucuronic acid, D-,
  chemical synthesis of, **8**, 231-249
  conjugates of, of animal origin, **9**, 185-246
  in metabolism, **8**, 251-275
β-Glucuronidase,
  preparation and properties of, **14**, 381-428
Glutose, **3**, 113-128
Glycals, **7**, 209-245
—, 2-hydroxy-, **9**, 97-129
Glycoenzymes: enzymes of glycoprotein structure, **27**, 301-341
Glycofuranosides, **21**, 95-142
Glycogens,
  enzymic degradation of, **3**, 251-310; **17**, 407-430
  enzymic synthesis of, **17**, 371-407
  molecular structure of, **12**, 261-298
Glycolipids,
  of acid-fast bacteria, **16**, 207-238
Glycoproteins. *See* Proteins, glyco-.
Glycopyranosyl esters,
  of nucleoside pyrophosphates ("sugar nucleotides"), biosynthesis of saccharides from, **18**, 309-356; **26**, 351-483

Glycoside–cuprammonium complexes, **6**, 107–134
Glycosides,
  acid-catalyzed hydrolysis of, **22**, 25–108
  alkali-sensitive, **9**, 59–95
  cardiac, **1**, 147–173
    the sugars of, **17**, 65–120
  methyl, of the common sugars, **12**, 157–187
  of the parsley plant, **4**, 57–74
  phenyl, of the common sugars, **12**, 157–187
C-Glycosides. See C-Glycosyl compounds.
Glycosiduronic acids,
  of animals, **9**, 185–246
  poly-, of plants, **1**, 329–344
Glycosphingolipids, **24**, 381–433
Glycosones, 3-deoxy-. See Glycosuloses, 3-deoxy-.
Glycosuloses, 3-deoxy-, and the degradation of carbohydrates, **19**, 181–218
Glycosylamines, **10**, 95–168
Glycosyl azides, **16**, 85–103
C-Glycosyl compounds,
  naturally occurring, **18**, 227–258; **20**, 357–369
Glycosyl fluorides, **16**, 85–103
Glycosyl halides,
  and their derivatives, **10**, 207–256
Goepp, Rudolph Maximilian, Jr.,
  obituary of, **3**, xv–xxiii
Grignard process,
  in the carbohydrate series, **6**, 251–289
Gums (see also, Hydrocolloids), **24**, 333–379
  commercial, **13**, 265–287
  of plants, **4**, 243–291

H

Halogen oxidation. See Oxidation, halogen.
Halogenated carbohydrates, **22**, 177–227
Haworth, Walter Norman,
  obituary of, **6**, 1–9
Hemicelluloses,
  structural chemistry of, **14**, 429–468
  of wood, **19**, 247–302; **20**, 409–483
Heparin,
  chemistry of, **10**, 335–368

Heuser, Emil,
  obituary of, **15**, 1–9
Hexitols,
  acetals of, **7**, 137–207
  anhydrides of, **5**, 191–228
  and some of their derivatives, **4**, 211–241
  synthesis of, **2**, 107–114
Hexofuranoses,
  1,6-anhydro-, **7**, 37–52
Hexosans, **7**, 37–52
Hexoses. See also, Hexofuranoses.
  formation of furan compounds from, **6**, 83–106
Hexuronic acids,
  methyl ethers of, **9**, 131–148
Hibbert, Harold,
  obituary of, **16**, 1–11
History,
  of mutarotation, **23**, 11–57
Honey, the sugars of, **25**, 285–309
Hudson, Claude Silbert,
  obituary of, **9**, xiii–xviii
Hyaluronic acid,
  biosynthesis of, **12**, 299–319
Hydrazones,
  of sugars, **3**, 23–44
Hydrocolloids,
  commercial, polysaccharidic, **13**, 265–287
Hydrogen, isotopes of,
  sugars specifically labeled with, **27**, 127–190
Hydrogen peroxide,
  action on carbohydrates and related compounds, **19**, 149–179
Hydrolysis,
  acid-catalyzed, of glycosides, **22**, 25–108
Hydroxyl groups,
  relative reactivities of, **8**, 1–44

I

Infrared spectra,
  of carbohydrates, **12**, 13–33
Infrared spectroscopy,
  and carbohydrate chemistry, **19**, 23–49
Irvine, James Colquhoun,
  obituary of, **8**, xi–xvii
Isotopes,

effects of, in mutarotation of sugars in solution, **24**, 13–65
of hydrogen, sugars specifically labeled with, **27**, 127–190
Isotopic tracers. *See* Tracers, isotopic.

K

Ketals. *See* Acetals.
Ketoses,
  cyclic acetals of, **26**, 197–277
Kinetics, basic,
  of mutarotation, **23**, 11–57
Kojic acid, **11**, 145–183
Kuhn, Richard,
  obituary of, **24**, 1–12

L

Lactose, **16**, 159–206
Lead tetraacetate,
  action of, on the sugars, **14**, 9–61
Levene, Phoebus Aaron Theodor,
  obituary of, **12**, 1–12
Lipids,
  glycosphingo-. *See* Glycosphingolipids.
Lobry de Bruyn–Alberda van Ekenstein transformation, **13**, 63–103

M

Maillard reaction, **14**, 63–134
Mannose,
  methyl ethers of D-, **8**, 217–230
Mass spectrometry,
  of carbohydrate derivatives, **21**, 39–93
Materials,
  cellulosic, combustion and pyrolysis of, **23**, 419–474
Mechanism,
  in the formation of polysaccharide gels and networks, **24**, 267–332
  of replacement reactions in carbohydrate chemistry, **9**, 1–57
Melezitose, **2**, 1–36
  structure of, **8**, 277–290
Metabolism,
  of carbohydrates, **2**, 119–160
    use of isotopic tracers in studying, **3**, 229–250
  of fat, **2**, 119–160

of the sugar alcohols and their derivatives, **1**, 175–192
D-glucuronic acid in, **8**, 251–275
Methane, nitro-,
  syntheses with, **6**, 291–318
Methods,
  in structural polysaccharide chemistry, **15**, 53–89
Methyl ethers. *See* Ethers, methyl.
Meyer, Kurt Heinrich,
  obituary of, **11**, xiii–xviii
Molasses,
  cane, **3**, 113–128
  cane final, composition of, **8**, 291–314
Molecular structure,
  of glycogens, **12**, 261–298
Mononucleotides, **22**, 307–419
Monosaccharides,
  biosynthesis of, **11**, 185–262
  cyclic, having nitrogen or sulfur in the ring, **23**, 115–232
  reactions of, with *beta*-ketonic esters and related substances, **11**, 97–143
Morphology,
  and biogenesis of cellulose and plant cell-walls, **26**, 297–349
Mucilages (*see also*, Hydrocolloids), **24**, 333–379
  commercial, **13**, 265–287
  of plants, **4**, 243–291
Mucopolysaccharides. *See* Polysaccharides, muco-.
Mucoproteins. *See* Proteins, muco-.
Mutarotation,
  of sugars in solution:
    Part I. History, basic kinetics, and composition of sugar solutions, **23**, 11–57
    Part II. Catalytic processes, isotope effects, reaction mechanisms, and biochemical aspects, **24**, 13–65
Mycobacterium tuberculosis,
  polysaccharides of, **3**, 311–336

N

Neighboring-group participation, in sugars, **22**, 109–175
Networks, polysaccharide, **24**, 267–332
Neuberg, Carl,
  obituary of, **13**, 1–7

Neuraminic acids, and related compounds, **13**, 237–263
Nickel, Raney. *See* Raney nickel.
Nitrates,
 of starch, **13**, 331–345
 of sugars, **12**, 117–135
Nitriles,
 acetylated, of aldonic acids, **4**, 119–151
Nitrogen heterocycles, synthesis from saccharide derivatives, **25**, 351–405
Nitro sugars. *See* Sugars, nitro.
Non-aqueous solvents for carbohydrates, **27**, 85–125
Nonulosaminic acids, **13**, 237–263
Nuclear magnetic resonance, **19**, 51–93
Nucleic acids, **1**, 193–245; **11**, 285–333
Nucleosides,
 bacterial, **15**, 201–234
 purine, **17**, 301–369
 pyrimidine, **14**, 283–380
 pyrophosphates, glycopyranosyl esters of, **18**, 309–356; **26**, 351–483
Nucleotides,
 bacterial, **15**, 201–234
 mono-, **22**, 307–419

O

Obituary,
 of Emile Bourquelot, **18**, 1–8
 of Emil Fischer, **21**, 1–38
 of Hermann Otto Laurenz Fischer, **17**, 1–14
 of Rudolph Maximilian Goepp, Jr., **3**, xv–xxiii
 of Walter Norman Haworth, **6**, 1–9
 of Emil Heuser, **15**, 1–9
 of Harold Hibbert, **16**, 1–11
 of Claude Silbert Hudson, **9**, xiii–xviii
 of James Colquhoun Irvine, **8**, xi–xvii
 of Richard Kuhn, **24**, 1–12
 of Phoebus Aaron Theodor Levene, **12**, 1–12
 of Kurt Heinrich Meyer, **11**, xiii–xviii
 of Carl Neuberg, **13**, 1–7
 of Stanley Peat, **25**, 1–12
 of Edmund George Vincent Percival, **10**, xiii–xx
 of Clifford Burrough Purves, **23**, 1–10
 of Fred Smith, **22**, 1–10
 of John Clinton Sowden, **20**, 1–10
 of Alva Thompson, **19**, 1–6
 of Melville Lawrence Wolfrom, **26**, 1–47
 of Géza Zemplén, **14**, 1–8
 of William Werner Zorbach, **27**, 1–6
Oligosaccharides, **17**, 121–167
 the raffinose family of, **9**, 149–184
 synthesis of, **6**, 27–81
Orthoesters,
 of carbohydrates, **1**, 77–127
Osazones,
 chemistry of, **20**, 139–181
 of sugars, **3**, 23–44
Osones, **11**, 43–96
Osotriazoles,
 chemistry of, **18**, 99–121
Oxidation,
 halogen, of simple carbohydrates, **3**, 129–148
 lead tetraacetate, of sugars, **14**, 9–61
 periodate, of carbohydrates, **11**, 1–41
  the "dialdehydes" from, **16**, 105–158
 selective catalytic, of carbohydrates, employing platinum catalysts, **17**, 169–221
Oxirane derivatives, of aldoses, **25**, 109–179
Oxo reaction,
 application to some carbohydrate derivatives, **23**, 59–114
Oxygen ring,
 formation and cleavage of, in sugars, **13**, 9–61

P

Paper chromatography,
 of carbohydrates and related compounds, **9**, 303–353
Paper electrophoresis,
 of carbohydrates, **18**, 61–97
Parsley,
 glycosides of the plant, **4**, 57–74
Participation,
 neighboring-group, in sugars, **22**, 109–175
Peat, Stanley,
 obituary of, **25**, 1–12
Pectic materials,
 chemistry of, **2**, 235–251
 enzymes acting on, **5**, 79–102

Pentitols,
  acetals of, **7,** 137–207
  anhydrides of, **5,** 191–228
  synthesis of, **2,** 107–118
Percival, Edmund George Vincent,
  obituary of, **10,** xiii–xx
Periodate oxidation. *See* Oxidation,
  periodate.
Phenol–carbohydrate derivatives,
  in higher plants, **20,** 371–408
Photochemistry,
  of carbohydrates, **18,** 9–59
Physical chemistry,
  of carbohydrates, **15,** 11–51
  of starch, **11,** 335–385
Physical properties,
  of solutions of polysaccharides, **18,**
    357–398
Physical studies,
  of cyclitols containing four or five
    hydroxyl groups, **20,** 11–65
Plant-growth substances,
  effect on carbohydrate systems, **21,**
    377–430
Plants,
  cell walls of, morphology and bio-
    genesis of, **26,** 297–349
  glycosides of parsley, **4,** 57–74
  gums of, **4,** 243–291
  mucilages of, **4,** 243–291
  polyuronides of, **1,** 329–344
Platinum. *See* Catalysts.
Pneumococcal polysaccharides, **19,** 303–
  357
Polyfructosans. *See* Fructans.
Polyglycosiduronic acids. *See* Glycosiduronic acids, poly-.
Polysaccharides. *See also,* Carbohydrates,
  Cellulose, Dextran, Dextrins, Fructans, Glycogen, Glycosiduronic acids
  (poly-), Pectic materials, Starch, and
  Xylan.
  alkaline degradation of, **13,** 289–329
  associated with wood cellulose, **10,**
    283–333
  bacterial, **2,** 203–233; **15,** 271–340
  blood-group, **4,** 37–55
  chemical synthesis of, **21,** 431–512
  fungal, structural chemistry of, **23,**
    367–417
  gels and networks, role of structure,
    conformation, and mechanism, **24,**
    267–332
  hydrocolloidal, **13,** 265–287
  methods in structural chemistry of, **15,**
    53–89
  muco-, chemistry of, **2,** 161–201
  of Gram-negative bacteria, **15,** 271–340
  of *Mycobacterium tuberculosis,* **3,** 311–
    336
  of seaweeds, **8,** 315–350
  physical properties of solutions of, **18,**
    357–398
  pneumococcal, **19,** 303–357
  shape and size of molecules of, **7,** 289–
    332; **11,** 385–393
  x-ray structure of, **22,** 421–482
Polyuronides,
  of plants, **1,** 329–344
Preparation,
  of esters of starch, **1,** 279–307
  of β-glucuronidase, **14,** 381–428
Properties,
  of 2-amino-2-deoxy sugars and their
    derivatives, **15,** 159–200
  of deoxy sugars and their simple deriv-
    atives, tables of, **26,** 279–296
  of esters of starch, **1,** 279–307
  of β-glucuronidase, **14,** 381–428
  physical,
    of solutions of polysaccharides, **18,**
      357–398
  physicochemical, of carbohydrates, **15,**
    11–51
Proteins,
  compounds with carbohydrates, in
    human urine, **24,** 435–452
  glyco-, aspects of the structure and
    metabolism of, **25,** 407–478
    enzymes (glycoenzymes), **27,** 301–341
  muco-, chemistry of, **2,** 161–201
Proton magnetic resonance spectroscopy,
  of carbohydrates, **27,** 7–83
Psicose, **7,** 99–136
Purines,
  nucleosides of, **17,** 301–369
Purves, Clifford Burrough,
  obituary of, **23,** 1–10
Pyrimidines,
  nucleosides of, **14,** 283–380
Pyrolysis, of cellulosic materials, **23,**
  419–474

R

Radiation,
  chemistry of carbohydrates, **16**, 13–58
Raffinose,
  family of oligosaccharides, **9**, 149–184
Raney nickel,
  reductive desulfurization by, **5**, 1–28
Reaction,
  the formazan, in carbohydrate research, **13**, 105–167
  the Maillard, **14**, 63–134
  the Oxo, application to some carbohydrate derivatives, **23**, 59–114
  the Wittig, in carbohydrate chemistry, **27**, 227–299
Reactions,
  mechanisms of, in mutarotation of sugars in solution, **24**, 13–65
  of amino sugars with *beta*-dicarbonyl compounds, **20**, 303–355
  of cellulose, **19**, 219–246
  of free sugars with aqueous ammonia, **25**, 311–349
  of monosaccharides with *beta*-ketonic esters and related substances, **11**, 97–143
Reactivities,
  relative, of hydroxyl groups of carbohydrates, **8**, 1–44
Rearrangement,
  the Amadori, **10**, 169–205
Reductions,
  biochemical, at the expense of sugars, **4**, 75–117
Replacement reactions,
  mechanisms of, in carbohydrate chemistry, **9**, 1–57
Rhamnose,
  methyl ethers of, **7**, 1–36; **10**, 257–272
Ribose,
  chemistry of, **6**, 135–174

S

Saccharides,
  biosynthesis of, from glycopyranosyl esters of nucleoside pyrophosphates ("sugar nucleotides"), **18**, 309–356; **26**, 351–483
  synthesis of nitrogen heterocycles from, **25**, 351–405
Saccharification,
  of wood, **4**, 153–188
Saccharinic acids, **12**, 35–79
  four-carbon, **13**, 169–188
Schardinger dextrins, **12**, 189–260
Seaweeds,
  polysaccharides of, **8**, 315–350
Seleno sugars. *See* Sugars, seleno.
Shape,
  of some polysaccharide molecules, **7**, 289–332; **11**, 385–393
Sialic acids, **13**, 237–263
Size,
  of some polysaccharide molecules, **7**, 289–332; **11**, 385–393
Smith, Fred,
  obituary of, **22**, 1–10
Soil,
  carbohydrates in, **16**, 335–355
Solutions,
  of polysaccharides, physical properties of, **18**, 357–398
  of sugars, mutarotation of, **23**, 11–57; **24**, 13–65
Solvents, non-aqueous, for carbohydrates, **27**, 85–125
Sorbose, **7**, 99–136
Sowden, John Clinton,
  obituary of, **20**, 1–10
Specificity,
  of enzymes, in the domain of carbohydrates, **5**, 49–78
Spectra, infrared,
  of carbohydrates, **12**, 13–33
Spectrometry, mass,
  of carbohydrate derivatives, **21**, 39–93
Spectroscopy,
  infrared, and carbohydrate chemistry, **19**, 23–49
  nuclear magnetic resonance, **19**, 51–93
  proton magnetic resonance, of carbohydrates, **27**, 7–83
Sphingosines, conjugates with sugars, **24**, 381–433
Starch,
  degrading and synthesizing enzymes, **23**, 281–366

enzymic degradation of, **3**, 251–310; **17**, 407–430
enzymic synthesis of, **17**, 371–407
fractionation of, **1**, 247–277; **16**, 299–333
nitrates of, **13**, 331–345
physical chemistry of, **11**, 335–385
preparation and properties of esters of, **1**, 279–307
thermal degradation of, **22**, 483–515

Stereochemistry,
of cyclic derivatives of carbohydrates, **10**, 1–53
formulas, writing of, in a plane, **3**, 1–22

Streptomycin,
chemistry of, **3**, 337–384

Structural chemistry,
of fungal polysaccharides, **23**, 367–417
of the hemicelluloses, **14**, 429–468

Structure, molecular,
of cellulose, **19**, 219–246
of dextran, **15**, 341–369
of glycogens, **12**, 261–298
of polysaccharide gels and networks, **24**, 267–332
of sucrose, **4**, 1–35
x-ray, of polysaccharides, **22**, 421–482

Sucrose. See also, Sugar.
enzymic synthesis of, **5**, 29–48
structure and configuration of, **4**, 1–35
utilization of, **4**, 293–336

Sugar,
aconitic acid as by-product in manufacture of, **6**, 231–249

Sugar alcohols. See also, Alditols, Hexitols, Pentitols, Tetritols.
higher-carbon, configurations of, **1**, 1–36
and their derivatives, metabolism of, **1**, 175–192

"Sugar nucleotides." See Nucleoside pyrophosphates, glycopyranosyl esters of.

Sugar products,
color and turbidity of, **9**, 247–284

Sugar refining,
granular adsorbents for, **6**, 205–230

Sugars,
action of lead tetraacetate on, **14**, 9–61
amino,
aspects of the chemistry of, **14**, 213–281
derived from antibiotic substances, **18**, 259–308
methyl ethers of, **13**, 189–214
properties of, **15**, 159–200
reactions with beta-dicarbonyl compounds, **20**, 303–355
2-amino. See Sugars, 2-amino-2-deoxy.
2-amino-2-deoxy, **7**, 247–288
2,5-anhydrides of, **25**, 181–228
anhydro,
chemistry of, **2**, 37–77
benzyl ethers of, **12**, 137–156
biochemical reductions at the expense of, **4**, 75–117
branched-chain, of natural occurrence, **11**, 263–283
of the cardiac glycosides, **17**, 65–120
conjugates, with sphingosines, **24**, 381–433
deoxy, **21**, 143–207
tables of the properties of, and their simple derivatives, **26**, 279–296
2-deoxy, **8**, 45–105
free, reactions with aqueous ammonia, **25**, 311–349
higher-carbon, **17**, 15–63
configurations of, **1**, 1–36
of honey, **25**, 285–309
hydrazones of, **3**, 23–44
methyl glycosides of the common, **12**, 157–187
neighboring-group participation in, **22**, 109–175
nitrates of, **12**, 117–135
nitro, **24**, 67–138
osazones of, **3**, 23–44
oxygen ring in, formation and cleavage of, **13**, 9–61
phenyl glycosides of the common, **12**, 157–187
related to altrose, **1**, 37–76
seleno, **1**, 144–145
solutions of, mutarotation of, **23**, 11–57; **24**, 13–65
specifically labeled with isotopes of hydrogen, **27**, 127–190
sulfates of the simple, **20**, 183–218

and their derivatives, column chromatography of, **10**, 55–94
conformational analysis of, **26**, 49–125
thio, **1**, 129–144
developments in the chemistry of, **18**, 123–199
unsaturated, **20**, 67–137; **24**, 199–266
Sulfates,
of the simple sugars, **20**, 183–218
Sulfonic esters,
of carbohydrates, **8**, 107–215; **23**, 233–280; **24**, 139–197
Synthesis,
biochemical, of monosaccharides, **11**, 185–262
of cardenolides, **21**, 273–321
chemical, of D-glucuronic acid, **8**, 231–249
of polysaccharides, **21**, 431–512
of dextran, **15**, 341–369
enzymic,
of glycogen and starch, **17**, 371–407
of sucrose and other disaccharides, **5**, 29–48
of nitrogen heterocycles from saccharide derivatives, **25**, 351–405
and configurational assignments of optically active, non-carbohydrate compounds, use of carbohydrates in, **27**, 191–225

T

Tables,
of the properties of deoxy sugars and their simple derivatives, **26**, 279–296
Tagatose, **7**, 99–136
Teichoic acids, **21**, 323–375
Tetritols,
acetals of, **7**, 137–207
Thiocarbonates,
of carbohydrates, **15**, 91–158
Thio sugars. *See* Sugars, thio.
Thompson, Alva,
obituary of, **19**, 1–6

Tracers,
isotopic, **3**, 229–250
Transformation,
the Lobry de Bruyn–Alberda van Ekenstein, **13**, 63–103
Trehaloses, **18**, 201–225
Trityl ethers,
of carbohydrates, **3**, 79–111
Turanose, **2**, 1–36
Turbidity,
of sugar products, **9**, 247–284

U

Unsaturated sugars. *See* Sugars, unsaturated.
Ureides, glycosyl, **13**, 215–236
Urine, human,
protein–carbohydrate compounds in, **24**, 435–452

W

Wittig reaction,
in carbohydrate chemistry, **27**, 227–299
Wolfrom, Melville Lawrence,
obituary of, **26**, 1–47
Wood,
hemicelluloses of, **19**, 247–302; **20**, 409–483
polysaccharides associated with cellulose of, **10**, 283–333
saccharification of, **4**, 153–188

X

X-Rays, crystal-structure analysis by, **19**, 7–22
Xylan, **5**, 269–290

Z

Zémplen, Géza,
obituary of, **14**, 1–8
Zone electrophoresis,
of carbohydrates, **12**, 81–115
Zorbach, William Werner,
obituary of, **27**, 1–6

ERRATA

Volume 25

Page 216, lines 11–13 should read: Treatment of 2-amino-2-deoxy-L-gluconic acid (**100**) with nitrous acid, followed by esterification with diazomethane, gave the methyl ester (**101**) of 2,5-anhydro-L-gluconic acid;

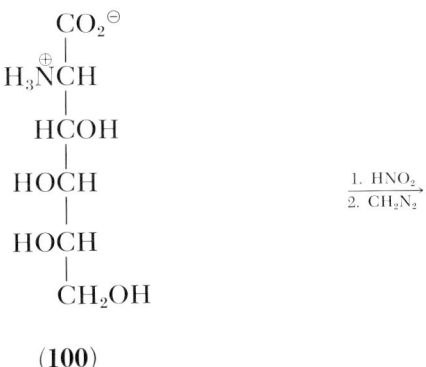

(**100**)

Page 461, Table XI, column 2, first entry. Delete "-α"; change "β" to "α"; insert "-β" after "(1 → 3)."

Page 461, Table XI, column 2, second and third entries. Delete first "-β"; insert "-β" after "(1 → 3)."

Page 461, Table XI, column 2, fourth entry. Delete first "-β"; change second "β" to "α"; insert "-α" after second "(1 → 4)."

QD
321
A2
v.27
1972

FEB 22 1973